THE NEW BOOK OF
POPULAR SCIENCE

THE NEW BOOK OF
POPULAR SCIENCE

VOLUME 4

Plant Life
Animal Life

COPYRIGHT © 1979 BY

INCORPORATED
DANBURY, CONNECTICUT

COPYRIGHT IN CANADA 1979 BY GROLIER LIMITED

COPYRIGHT PHILIPPINES 1979 BY GROLIER INTERNATIONAL, INC.

COPYRIGHT REPUBLIC OF CHINA 1979 BY GROLIER INTERNATIONAL, INC.

Library of Congress Cataloging in Publication Data

Main entry under title:
The New Book of Popular Science.

    SUMMARY: Discusses the major sciences and their application in today's world.
    Bibliography: end of each volume
    1. Science. 2. Technology. 3. Natural History.
[1. Science. 2. Technology. 3. Natural history]
Q162.B68 1976   50   78-21369
ISBN 7172-1209-2

No part of THE NEW BOOK OF POPULAR SCIENCE may be reproduced without special permission in writing from the publishers

PRINTED IN U.S.A.

Volume 4

Contents

Plant Life 1 - 175

Animal Life 176 - 489

PLANT LIFE

Left: Toni Angermayer/PR
Above: Exxon

Plants—from the impressive tree towering over a field to a delicate water lily floating in a pond—are beautiful, peaceful. But also much more. Green plants are essential for all life, providing oxygen and food for all living things.

2-7	What Is A Plant?
8-10	The Plant Kingdom
11-21	Soil
22-31	Bacteria
32-42	Algae
43-53	Fungi
54-60	Mosses and Ferns
61-70	Seed Plants
71-83	Vital Processes of Plants
84-88	Photosynthesis
89-98	Roots and Stems
99-102	Leaves
103-110	Flowers
111-118	Fruits and Seeds
119-125	Seed and Fruit Dispersal
126-132	From Seed To Seedling
133-144	Plant Adaptation
145-148	Vegetables
149-153	Cactus Plants
154-161	Houseplants
162-175	Trees

WHAT IS A PLANT?

The plants of the world are living things and differ in several important respects from things that are not alive. For one thing, they are made up chiefly of *protoplasm,* a substance in which all the processes of life take place. This protoplasm is contained in compartments called *cells.* Plants also display irritability—that is, they respond to outer stimuli. Plants transform food into plant tissues, and they reproduce.

Plants are generally more or less clearly distinguished from animals, except in the case of certain borderline forms. The chief point of difference is that most plants can manufacture their own food, while animals cannot. Again, most plants (not all) are fixed in one spot, while most animals (not all) can move more or less freely from place to place. Finally, plant cells have comparatively rigid walls, while the covering of practically all animal cells consists of a very thin, elastic membrane.

PLANT CELLS

The cells, the basic units of plant life, vary greatly in size and form, but they all have certain features in common. The *cell wall* is made up of several compounds, secreted by the protoplasm. The commonest of these compounds is *cellulose,* which is chemically akin to sugars and starches. Just inside the cell wall is a very thin membrane—the *plasma membrane.* The protoplasm within the cell has a variety of visible structures. In most plants, each cell contains a *nucleus,* or core, separated from the rest of the protoplasm by a thin membrane. Within the nucleus is a material called *chromatin,* which is concerned with the transmission of hereditary traits. The protoplasm outside the nucleus is called *cytoplasm.* In the mature plant cell, much of the cytoplasm is taken up by a nonliving transparent structure—the *vacuole*—which is filled with a watery solution—the cell sap. The cytoplasm also contains a variety of living structures, including the bodies called *plastids.* In the plastids are found various pigments. As we shall see, the pigment known as *chlorophyll* plays a vital part in the manufacture of food.

PLANT TISSUES

In some cases, as in bacteria and various algae, the entire organism consists of a single cell, in which all the activities of life go on. Most plants, however, are made up of a great many cells, specialized to perform different tasks. Thus, for example, there are root cells that absorb water and minerals from the soil; tuber cells that serve for the storage of starch; and leaf cells in which the manufacture of food takes place.

In multicellular, or many-celled, plants, cells are combined to form tissues. A tissue may be described as a group of similar cells, performing the same task or tasks. There are various kinds. *Meristematic tissue* is found in actively growing regions, such as the tips of roots, buds, and certain parts of the stem. As a result of such growth, meristematic tissue is gradually transformed into other kinds, called *permanent tissues.* These include the *epidermis,* or surface layer, of leaves and flower parts; the *phellogen,* or cork tissue, which forms the outer bark of the stems and roots of woody plants; and the *conducting tissues, xylem* and *phloem,* which transport water, minerals, and foods throughout the plant. Tissues in turn are combined to form organs. There are three kinds of organs in plants—roots, stems, and leaves. The reproductive structures—cones and flowers—of higher plants are sometimes referred to as separate organs. Actually, they are specialized stems.

To keep alive, plants must obtain food, breathe, and carry on a variety of activities. Most animals can move from one place to another in search of food. Most plants are unable to do so, since they are rooted in the soil. They generally obtain food by manufacturing it themselves from natural raw materials.

PHOTOSYNTHESIS

The starting point in this natural manufacture of food substance is the process known as *photosynthesis*. The green pigment chlorophyll, which most plants possess, is essential for this process; so is sunlight or light of comparable wavelengths. The raw materials required are water and the gas carbon dioxide. Water is drawn from the soil by the plant roots. Carbon dioxide is obtained from the atmosphere. It passes into the plant through pores, called *stomata,* in the leaves. In the presence of chlorophyll and sunlight, water and carbon dioxide react chemically to form oxygen and a sugar known as glucose — a food element of the carbohydrate group. A part of the oxygen remains within the plant. The rest ultimately makes its way through the stomata of the leaves to the outer air.

Light is as necessary as chlorophyll

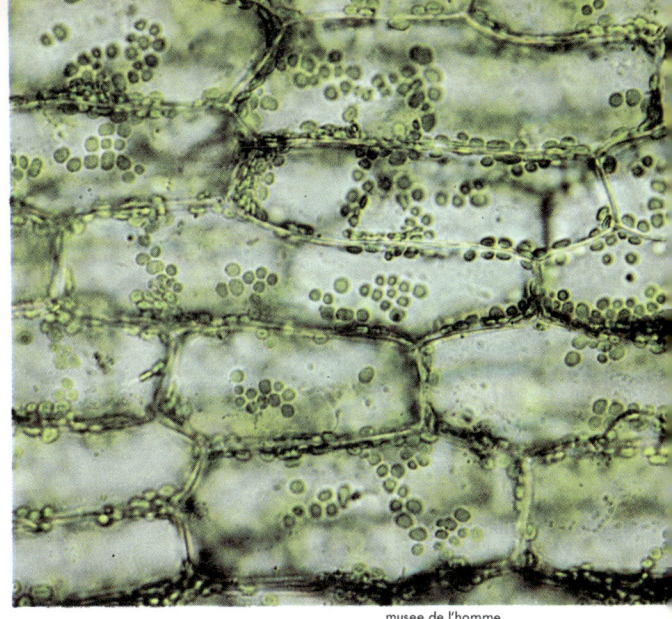

musee de l'homme

Plants are able to manufacture their own food. Chlorophyll is an essential part of the plant's food-making machinery. Above chlorophyll-containing chloroplasts on the leaf of a plant.

in the process of photosynthesis. Hence the name, which means in Greek "putting together in light." Through photosynthesis, carried on in the presence of chlorophyll, the plant captures the energy of sunlight and stores it up for future needs. Bacteria,

Fields such as the one below are not only beautiful but are also essential. Green plants provide food and oxygen for all life.

Nuridsany

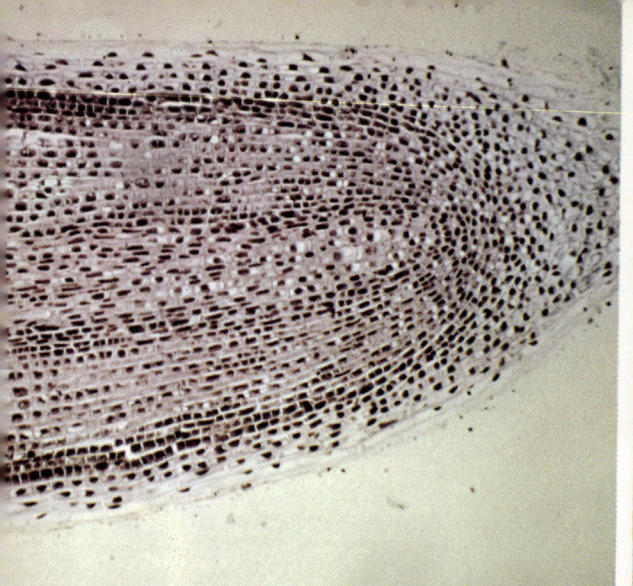 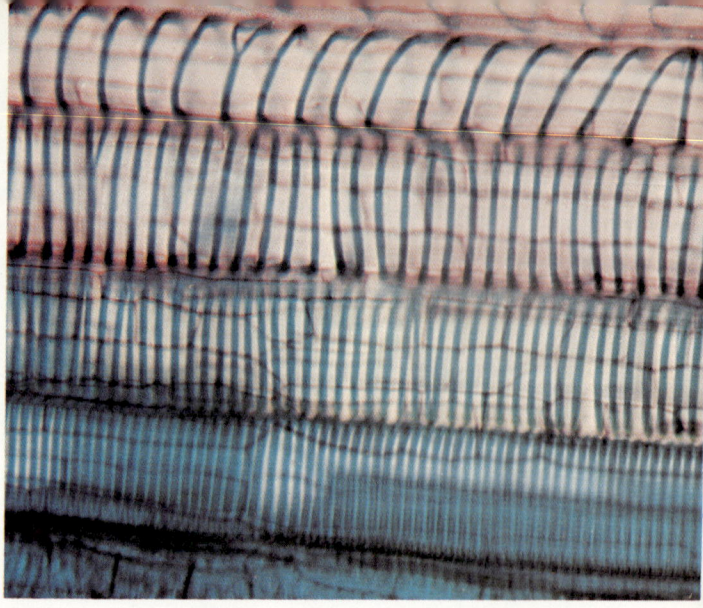

In multicellular plants, cells are combined to form tissues that perform a specific task. Left: meristematic tissue of a young root. Right: conducting tissue that transports water, minerals, and food throughout the plant.

molds, and fungi do not possess chlorophyll and cannot make their own food. They obtain it from decaying plant and animal life.

As we pointed out, photosynthesis is the starting point in the plant's manufacture of food. Some of the glucose produced in the process is utilized directly by the plant. Some of it is used in making other kinds of food, and some of it serves in the synthesis, or building up, of new protoplasm and of chemical compounds essential for growth.

RESPIRATION

Food manufactured by the plant is used in the process of breathing, or respiration—for plants breathe as truly as animals do. The process takes place in the cells. It involves a series of chemical reactions in which oxygen enters into chemical combination with sugars and other foods; the gas carbon dioxide is formed; and energy is released. Respiration is a form of combustion. It may be compared with the burning of coal, which releases the energy that has been stored in this fuel.

There are two kinds of respiration: *aerobic* and *anaerobic*. In aerobic respiration, free oxygen, derived from photosynthesis, takes part in the reaction. No free oxygen is involved in anaerobic respiration. Energy is released by a series of chemical changes in which oxygen is transferred from one type of chemical compound to another.

GROWTH AND MOVEMENT

Of the energy that is released as a result of respiration, some escapes as heat. The rest is utilized in the various activities of the plant. One of these is the synthesis of proteins and other organic compounds from simpler materials. Another essential activity is growth. This may represent the formation of new cells by the process of mitosis, or cell division; or the enlargement of the cells formed in this way; or the development of tissues.

Growth is closely associated with another plant activity—movement—since movement often results from unequal growth in the different parts of a plant organ. There are various types of plant movements. Some of them, such as the twining of morning-glory tendrils, result chiefly from internal stimuli. Others are due to external factors. For example, plants bend toward light; are directed downward in response to the force of gravity; and move toward sources of water, particularly when growing in dry soil. Certain plants move as they react to various contacts. A well-known example is the closing of the two halves of the leaf of a Venus's-flytrap plant when an insect touches a part of the leaf's inner surface.

In sexually reproducing plants, pollen is transferred from one plant to another in a variety of ways. Insects (top) may carry the pollen on their legs or wings, or the wind may simply carry off exposed pollen grains (bottom).

As a result of this particular movement, the insect is trapped; it is then digested by the Venus's-flytrap.

REPRODUCTION

One of the most important activities of the plant is reproduction, which ensures the survival of the species. Bacteria reproduce by simple division, or *fission*. Each bacterium divides into halves, and each half forms a new individual. In the case of yeasts and certain other one-celled plants, a small *bud* appears on the surface and gradually grows until it is as large as the original cell. It becomes a cell in its own right and separates from the original cell. Other plants, including various fungi, develop special cells called *spores,* and these give rise to new plants.

Certain plants practice *sexual reproduction*. These plants possess sex organs, which produce male sex cells (sperm) and female sex cells (eggs, or ova). When a sperm unites with an egg in the process of fertilization, the resulting cell gives rise to a new individual. This type of reproduction is found in the seed plants, which make up the great majority of plant species. In plants such as ferns and mosses, spores and sex cells are produced in *alternating generations*.

Sometimes new plants are formed through the process of *vegetative reproduction*. This involves the growth and development of new plants from various parts of old ones—runners, bulbs, tubers, leaves, and the like.

Whatever the method of reproduction, each plant transmits its traits to the following generation through the mechanism of heredity. Thus the immense variety of plants is preserved.

SIZES AND SHAPES

Plants range in size from such microscopic forms as bacteria, which may be less than 1/10,000 centimeter in length, up to

the giant redwoods of California, which may reach a height of over 100 meters and a diameter of 10 meters or so. As a general thing, there is an average range of size for a given plant species. Under unusual conditions, though, plants may be considerably larger or smaller than the average.

WHAT IS A PLANT?

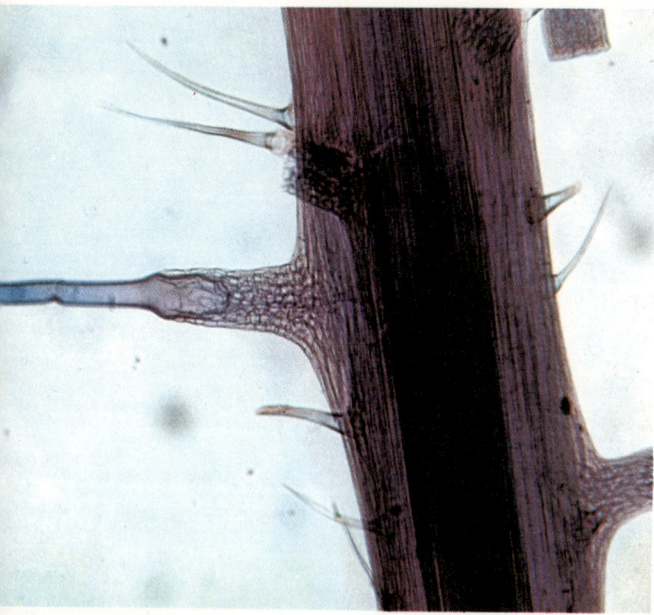

Plants differ widely in their size, shape, and adaptations to climate, habitat, and other plant and animal neighbors. Top: a nettle with one of its prickly, liquid-filled, hairlike projections that is irritating to many animals. Bottom: a vine with tendrils forming cuplike projections to cling to its wall support.

There are startling contrasts, too, in the life span of plants. A bacterium may exist as an individual for only half an hour or so. At the end of that time, it may divide and form two new individuals. Some plants, called annuals, live for one growing season; others—biennials—for two; still others—perennials—for a number of years. Certain perennials may live for hundreds or even thousands of years. The California redwoods may live more than 3,000 years.

The forms of plants show great variety. Some are trees; others, shrubs; still others, vines or herbs. There are many differences in the methods of branching; in the shape of the leaves, which may be smooth-edged or indented, broad or narrow; in the structure and color of flowers. Certain plants consist of a simple body called a *thallus*. They have neither stems, nor leaves nor roots.

Plants differ, too, in their adaptations to various environments. Cacti and yucca plants are found only in arid regions; lianas, only in the rain-drenched forests of the tropics; and so on.

FIELDS OF STUDY

We study the different aspects of plant life in the science called botany (from the Greek *botane,* meaning "plant"). Modern botany is a vast field with many subdivisions. *Plant anatomy* goes into the details of plant structure; *plant histology* is the microscopic study of tissues. The life activities that go on in plants form the subject matter of *physiology. Plant taxonomy,* or *systematic botany,* considers the classification of plants. *Plant genetics* deals with inheritance; *plant cytology,* with plant cells; *plant pathology,* with plant diseases; *plant ecology,* with the relation of plants to their environments; *paleobotany,* with plants of earlier geological periods, as revealed by the fossils left in various rock layers. Some branches of botany take up particular plant groups. Thus *mycology* has to do with fungi; *bacteriology,* with bacteria; *algology,* with algae.

There are also various fields of applied plant science. For example, *horticulture* is concerned with crops grown in orchards, gardens and greenhouses; *forestry,* with forests; *agronomy,* with field-crop production: *floriculture,* with ornamental flowering plants; *olericulture,* with vegetable crops; *pomology,* with fruit crops. *Landscape architecture* deals with the arrangement of vegetation so as to produce a pleasing effect.

ECONOMIC IMPORTANCE

The study of botany is truly a vital one, because plants are essential for our very existence. We pointed out that they can create their own food, while animals cannot. Like all other members of the animal kingdom, humans must obtain food, directly or indirectly, from plants.

Plants are the only major source we possess of the gas oxygen, which we need for respiration. As we saw, free oxygen is produced in photosynthesis, and much of it passes into the atmosphere through the leaves of plants. If the oxygen supply of the atmosphere were not replenished in this way, it would soon be exhausted.

Plants prevent the erosion of the soil. Where a thick plant cover exists, the full force of rain strikes the vegetation and not the soil itself. Besides, the root system of plants binds the soil particles together so that they are not easily carried away by rainwater coursing down slopes. Plants also help conserve the water supply. If rainfall is caught by forest cover, grass cover, or crop cover, a good deal of it will ultimately make its way into our underground water reserves.

As bacteria feed and breathe, they often bring about important changes in foodstuffs and industrial materials. Such changes are involved in the manufacture of vinegar, butter, and certain kinds of cheese, as well as in the tanning of leather and the curing of tobacco. Some of the waste products that bacteria release are valuable industrial materials. They include acetone, butyl alcohol, and citric acid. Certain bacteria make nitrogen available to plants by transforming nitrogen gas into nitrogen compounds that the plant can use.

There are many other ways in which plants serve us. Cotton, flax, hemp, jute, and other plants provide fibers used for clothing, rope, twine, fishing lines, and so on. The wood derived from various trees supplies lumber for houses, as well as paper for books, magazines and newsprint. The plastics called cellulosics are obtained from the cellulose found in the cell walls of plants. Certain plants provide essential medicines. Quinine, useful in the prevention and treatment of malaria, is derived from cinchona bark. The heart stimulant digitalis comes from the dried leaves of the foxglove. The antibiotics, which have revolutionized medical practice, are made from various molds, bacteria, algae, and higher plants. Plants also give us beverages (tea, coffee, coca); perfumes (attar of roses, lavender, rosemary); spices (mace, nutmeg, cloves); and rubber, used for automobile tires, conveyer belts, surgical gloves, paints, and numerous other products.

Certain plants are harmful to man. Various species of bacteria are responsible for such human diseases as tuberculosis, pneumonia, and cholera. Other species cause disease in cultivated plants, including potatoes, squash, tomatoes, and pears. The plants called rusts and smuts infect cereals, apples, potatoes, peaches, and other useful plants. Some mushroom species are wholesome human foods. Others, such as the fly amanita and the destroying angel, are deadly poisons. Various skin ailments are caused by contact with poison ivy, poison oak, poison sumac, and other plants.

Many plants are economically important, providing man with food, fibers, and many other products. Left: rice, one of the most important foods in the world. Right: a field of cotton, an extremely important fiber.

Mallet/Jacana   Proffit

THE PLANT KINGDOM

C. Nuredsary

Living things are generally grouped as plants or animals and classified in the Plant Kingdom or Animal Kingdom. In general, plants share certain characteristics—the ability to make their own food, for example—while animals, in general, share other features—the ability to move from place to place, for example. Some one-celled organisms that have both plant and animal characteristics are placed by some biologists in a third kingdom—that of the Protista. In general, however, a classification into two kingdoms is convenient and is what we have followed in *The New Book of Popular Science*. The subdivisions of the Plant Kingdom are given below. The most important groups of plants are discussed in separate articles.

CYANOPHYTA PHYLUM—"Dark-blue plants"
Minute one-celled algae, blue-green, sometimes red; have chlorophyll. Reproduction asexual. Live in water and soil. OSCILLATORIA, GLOEOCAPSA.

EUGLENOPHYTA PHYLUM—"Plants with true eyes"
Minute one-celled algae, often with chlorophyll; yellow-green in color. Have eyespot and flagella. Resemble animals in that they move and in some cases take in food from the outside. Reproduction mostly asexual. Live in fresh water. EUGLENA.

CHLOROPHYTA PHYLUM—"Green plants"
Green algae, single-celled or multicellular; have chlorophyll; cell wall of cellulose; provided with flagella; some move; often threadlike or in sheetlike colonies. Reproduce by fission, by spores, or sexually. Found in water, snow, ice, soil; some parasitic. CHLAMYDOMONAS, VOLVOX, ULVA.

CHRYSOPHYTA PHYLUM—"Golden plants"
Yellow-green and brown algae; one-celled, multicellular, or colonial. Have chlorophyll and other pigments. Cell walls may be siliceous. Reproduce by fission, by spores, or sexually. Found in fresh and salt water, soil, damp places, and on various objects; may live on other plants. DIATOM, MALLOMONAS, CHLOROMONAS.

PYRROPHYTA PHYLUM—"Fiery plants"
Small one-celled algae, single or colonial; yellow-green to brown pigments; some with shell-like cell walls. Usually have flagella and are capable of motion. Reproduce by fission, by spores, or sexually. Mostly marine. DINOFLAGELLATE.

PHAEOPHYTA PHYLUM—"Dusky plants"
Brown algae; usually multicellular; have brown pigments and chlorophyll; very small to a hundred or more meters long; complex structures; float bladders. Alternation of generations. Mostly marine. LAMINARIA, SARGASSUM, FUCUS.

RHODOPHYTA PHYLUM—"Rose-colored plants"
Red algae; multicellular; have chlorophyll and other pigments, including red pigments. Alternation of generations. Mostly marine. SEA MOSS.

SCHIZOMYCOPHYTA PHYLUM—"Dividing fungal plants"
The bacteria, one-celled single or colonial plants with no chlorophyll or nuclei; often

parasitic and disease-producing; sometimes saprophytic; some make their own food; live with or without air; some are beneficial to man. Reproduce asexually. Found everywhere. BACILLUS, COCCUS, SPIRILLUM.

MYXOMYCOPHYTA PHYLUM — "Slime fungal plants"

Slime fungi or slime molds; without chlorophyll; saprophytic; plasmodium visible to naked eye. Alternation of generations. Generally small, but not always microscopic. Inhabit soil and dead vegetation; some parasitic. Sometimes classed with animals, under the phylum Protozoa. BADHAMIA.

EUMYCOPHYTA PHYLUM — "True fungal plants"

Mushrooms, molds, mildews, smuts, rusts, yeasts. Body made up of filaments (hyphae); saprophytic or parasitic; no chlorophyll; pale to brilliantly colored; microscopic to large. Reproduce by spores, by budding, or sexually. Inhabit damp, often dark, places; found in air, water, soil, or wherever organic matter abounds. May live with algae and other green plants in mutually beneficial relationship.

Phycomycetes Class Algal fungi	Hyphae loose and shapeless; microscopic to large; one-celled to many-celled; many parasitic. BREAD MOLD, MILDEW.
Ascomycetes Class Sac fungi	Spores in sac or on hyphae; one-celled to many-celled; microscopic to large; reproduction sexual and asexual. YEAST, PENICILLIUM, BLUE MOLD, SPHERE FUNGUS, POWDERY MILDEW.
Basidiomycetes Class Club fungi	Spores on basidium, or clublike extension of hypha. Multicellular; often large; may be parasitic. FIELD MUSHROOM, PUFFBALL, SMUT.
Deuteromycetes Class Secondary fungi	Many parasitic and disease-causing. Reproduction asexual. RINGWORM FUNGUS.

BRYOPHYTA PHYLUM — "Moss plants"

The mosses and liverworts; green land plants; small, multicellular, but with few organized structures; alternation of generations. Inhabit damp soil, swamps, and streams.

Musci Class Mosses	The mosses. Gametophyte body is upright, with leaflike extension; sporophyte is stalklike, extending above gametophyte. SPHAGNUM.
Hepaticae Class Liverworts	The liverworts. Gametophyte body is horizontal on ground; sporophyte erect. MARCHANTIA, CONOCEPHALUS.
Antherocerate Class Flower horns	The hornworts. Gametophyte body with complex cylindrical sporophyte. ANTHOCEROS.

PSILOPSIDA PHYLUM — "Naked-appearing plants"

Earliest, most primitive land plants with definite but simple structures; stem with rootlike and leaflike extensions; very simple vascular system; elementary wood

Two types of non-flowering plants typically found in damp, dark areas. Left: mushrooms, a type of fungus. Right: sporophyte of a typical moss.

C. Nuredsary

structure. Alternation of generations. Most extinct. PSILOTUM.

LYCOPSIDA PHYLUM — "Wolf-resembling plants"
The club mosses and quillworts. Sporophyte with roots, stem, and leaves; spore structures on leaves. Alternation of generations. Generally small, but in the Carboniferous were often large, contributing to coal deposits. LYCOPODIUM, SELAGINELLA.

SPHENOPSIDA PHYLUM — "Wedge-resembling plants"
The horsetails, or scouring rushes. Sporophyte with stem, roots, and leaves; spores on cones; leaves in whorls; jointed hollow stem; gametophyte ribbonlike. Generally small, but in the Carboniferous contributed to coal deposits. EQUISETUM.

PTEROPSIDA PHYLUM — "Fern-resembling plants"
Sporophyte with well-differentiated roots, stems, and leaves; gametophyte independent or a microscopic part of the sporophyte. True seeds and flowers in higher members of the phylum. In some classifications, the Pteropsida are grouped together with the Psilopsida, Lycopsida, and Sphenopsida to form the phylum Tracheophyta.

Filicineae Class Ferns	The ferns. Sporophyte with leaves but no seeds or flowers; gametophyte small. Small to treelike; prefer damp soil and shady areas.
Gymnospermae Class Naked-seed plants	True, uncovered seeds, developing after pollen grain in male gametophyte fertilizes egg cell in female gametophyte; many-celled embryo, or sporophyte-to-be, inside seed; gametophytes form part of sporophyte, which has stem or trunk, leaves and roots; cambium layer in trunk. Mostly trees, bearing cones.
Angiospermae Class Vessel-seed plants	True, covered seeds, developing after pollen grains fertilize eggs. Many-celled embryo in protected seed gives rise to sporophyte with stem or trunk, leaves, and roots; gametophytes form microscopic part of sporophyte. Cambium in most woody dicotyledonous plants. Flowers and fruits.
Monocotyledoneae Subclass Plants with one cotyledon	Plants with one coytledon; flowers usually in three or six parts, sometimes in two or four; one plant may have either male or female flowers, or both kinds of flowers or bisexual flowers, or it may combine all types. Leaves with parallel veins; wood fibers not concentrated.
Dicotyledoneae Subclass Plants with two cotyledons	Two cotyledons. Flowers usually in four or five parts, sometimes in two parts or multiples of two and five. Flowers male, female, or bisexual. Leaves veined; wood fibers of stem usually concentrated in ring. Many trees.

Left: female cone of a maritime pine. Right: male cone of a silver pine. Both are gymnosperms, or conebearing plants.

Angiosperms, or flowering plants, are the largest group of plants. Below: flower bud, open flower, and fruit of a corn poppy.

SOIL

The plants that grow upon the earth's land areas could not exist without the thin, loose surface layer that we call the soil. It provides plants with mechanical support, or anchorage. It offers mineral nutrients for plant growth. Water and air, trapped within it, are made available to the roots. Not only plant life, but animal life as well, is dependent on the soil, for animal foods are derived directly or indirectly from plants.

Weathered rock particles make up a large part of the film of soil. The rest is formed of the decayed products of organic matter—that is, living or formerly living substances. The soil is not an inert, unchanging body. It is constantly being transformed by chemical and physical processes and by the activities of living organisms.

It takes but a moment for you to dig up a spadeful of garden soil. But it has taken the processes of nature countless centuries to create the soil. The first step in soil-building is the weathering of rocks that lie at the earth's surface. The result is a vast accumulation of rock debris above the bedrock. This debris is called the *regolith*, from two Greek words meaning "stone blanket". It is not soil itself but rather the forerunner, or parent, of soil.

W. Schwarz

HOW THE REGOLITH IS FORMED

The weathering that results in the formation of the regolith is due to many different factors:

Temperature. Among the most important causes of weathering are variations in temperature. During the day, exposed rock is heated by the sun and expands. At night, it cools and contracts. The rock is composed of different minerals, which expand and contract at different rates. Unequal stresses are built up as a result, and cracks appear. Water gains entrance by way of these cracks and begins its eroding and dissolving action. If freezing occurs, the expansion of the water as it freezes exerts tremendous pressure. This causes the rock mass to break up into smaller pieces. Differences in temperature between the exposed outer part of a rock and the protected inner portion also bring about unequal stresses, causing the surface layers to peel off.

Water. The loose material of the regolith is shifted, sorted, and transported by rainwater beating upon the land and running downhill. Water from melting snow has a similar effect. If the running water is carrying a load of debris, it cuts, or erodes, the bedrock, giving rise to more debris.

The sediment may be carried by rivers many kilometers before it is finally deposited as potential soil in alluvial fans, terraces, and deltas. Sometimes debris is laid down in lakes, gulfs, and oceans. Here it remains submerged until the elevation of the land or the lowering of the water level exposes it as parent material for a new soil.

SOIL   11

Glacial ice. Glaciers are other important agents of soil transportation, erosion, and deposition. As the glacier moves, it picks up rocks and grinds them into fragments against one another. Rock particles imbedded in the lower level and sides of the glacier scour rock powder from the bedrock of the glacier's channel. When the ice melts, the rock fragments and powder are either immediately deposited or carried for many kilometers by the water flowing from the glacial front.

Wind. Wind picks up the finer particles of the regolith. Armed with this grit, it abrades more fragments from rocks through the action of sandblasting. The winds carry sand and dust for many kilometers.

Chemical changes. Rock is also affected by a variety of chemical changes. Water combines with carbon dioxide to form carbonic acid, which is a powerful solvent and attacks the minerals contained in rock. In the process of oxidation, oxygen combines with various minerals. These are decomposed as a result of the reaction. The rock mass in which they are found is weakened, so that it readily crumbles. In hydration, various minerals take up water molecules. These minerals increase in size and become soft. They are then broken down by weathering and erosion. Hydrolysis is another chemical process that affects rock. It occurs as salts are dissolved in water. The salts undergo ionization. That is, they are converted into electrically charged atoms or groups of atoms called ions. The ions then react with hydroxyl and hydronium ions, which are always present in water.

FORMATION OF TRUE SOIL

As we have pointed out, the regolith is only the forerunner of soil. True soil is formed only when living organisms become active in the rock debris. Sooner or later, hardy lichens and mosses gain a foothold on the exposed, weathering surfaces of the regolith. These relatively simple plants catch and hold the finer mineral particles carried by wind and water, and they absorb mineral nutrients. They also give off carbon dioxide which, in the presence of water, helps dissolve minerals in the regolith.

As parts of the plants die, bacteria and molds appear. They feed on and break down the plant remains. Nutrients in the dead plants are liberated and can be used by living plants. The decaying vegetation also yields organic acids, which dissolve many of the minerals in the regolith.

Eventually, a thin film of organic materials combined with mineral matter is built upon the regolith. This represents the beginning of true soil. Larger and bushier mosses now invade the soil. They trap and hold more minerals and contribute their remains to the growing soil mat. After a time the soil is firmly enough established to provide essential nutrients, water, and firm support for grasses, herbaceous plants, shrubs, and, perhaps, trees.

As these higher plants gradually establish themselves, a layer of *humus* develops. Humus represents the very dark material that is usually so noticeable at the surface of most soils. It is the result of complicated chemical reactions carried on by the soil microbes. Part of the humus is a combination of protein and lignin (one of the building blocks of plant cell walls). Part of it, apparently, is composed of the tissues of dead soil microbes.

The presence of organic humus means the difference between a productive soil and one that is sterile. Humus holds most of the soil's nitrogen and much of its sulfur and phosphorus. It attracts and retains water. It also holds the gases and ions that are needed by plants. It causes soil particles to clump, providing a workable soil. When humus itself is decomposed, heat is produced and the soil is warmed. Under such conditions, the seeds of the higher plants can readily germinate.

THE SOIL PROFILE

As soil is formed, particularly if water drainage is good, horizontal layers develop in the regolith. If you dig a trench into soil down to bedrock or examine any fresh road cut, you can see these layers. They are called *horizons* by soil scientists. The horizons taken together make up the *soil profile*.

There are three major horizons in any

soil profile: the A, B, and C horizons. The A horizon, or, roughly, the surface soil, is the zone where organic materials have become incorporated within the mineral matter. Often there is a litter of fresh plant remains at the surface and, below this, a layer of very dark humus. Under the humus is a lighter-colored mineral layer. It is the result of *leaching* — the removal of soluble substances by percolating water. In cultivated soil, the layers of humus and mineral matter are mixed together. A considerable part of the nutrients and water available for plants is to be found in the A horizon. Here we find the most extensive root development.

The B horizon, or subsoil, has little organic material in it, though the mineral matter is noticeably weathered. This horizon contains substances derived from the A horizon above and the C horizon below. Iron, aluminum, and sometimes calcium compounds accumulate here; so does clay that is washed down from the surface layers.

The lowest, or C, horizon is composed of the same loose parent material from which the true soil above has developed. It is more or less weathered, and its upper layers gradually become part of the true soil.

Often the horizons are not distinct. There are zones of transition between the A and B horizons and the B and C horizons. When the soil is mature, its profile does not change much, even though weathering and soil organisms are constantly at work. There is normal erosion at the surface, balanced by the creation of new soil from the C horizon below.

FACTORS INFLUENCING SOIL DEVELOPMENT

The nature of mature soil depends on several factors: the sizes and groupings of the rock fragments of which the regolith is composed, the time it takes for this material to be transformed into soil, the lay of the land, the vegetation, and the climate.

Sizes of soil particles. The regolith consists of particles of various sizes. This variation determines the texture that the

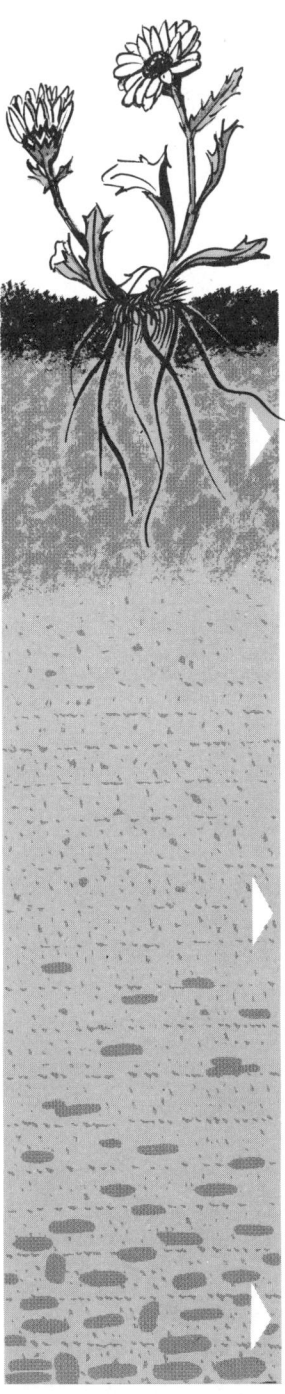

A horizon. This is the surface soil. Organic materials have been incorporated in the mineral matter derived from the original rock.

B horizon. Also called the subsoil, the B horizon has little organic material. Its mineral elements have been heavily weathered.

C horizon. In this, the parent rock material has undergone but little weathering. The C horizon may be absent in certain soils.

The art above shows the horizons, or horizontal layers of soil, that make up a typical soil profile. There are three major horizons — A, B, and C. There are also some subdivisions, but these are not shown in this art.

Soil particles differ in size. Above are particles of gravel. They are irregularly shaped rock fragments, having a diameter greater than two millimeters.

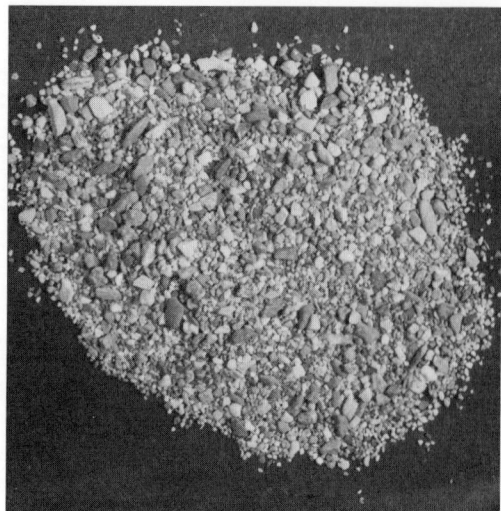

Sand particles consist of grains smaller than gravel. These particles may be rounded or irregularly shaped. They have a diameter of 1/16 millimeter to two millimeters.

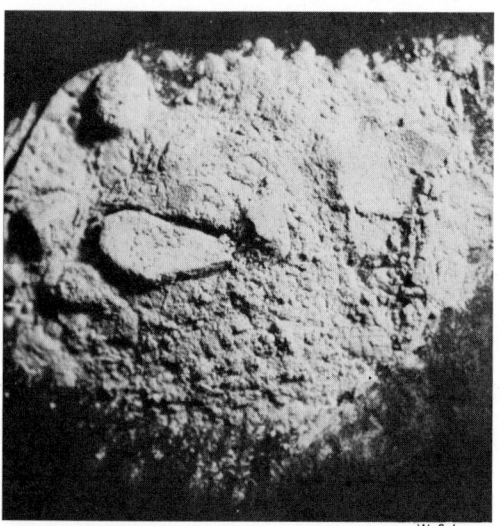

Next in size are silt particles. Silt particles have a variety of shapes. They range in size from 1/256 millimeter to 1/16 millimeter.

Clay particles have the smallest diameter of all. They are made up of complex minerals and have a diameter of less than 1/256 millimeter.

soil ultimately develops. The largest particles are stones and gravel, which are irregularly shaped rock fragments having a diameter of more than two millimeters. Sand particles range from two millimeters to $1/16$ millimeter in diameter. The sand grains are rounded or irregular in shape. They are usually of quartz, though they may also be of feldspar, mica, magnetite, and garnet. Silt particles are from $1/16$ millimeter to $1/256$ millimeter. They are most often quartz and have different shapes. Clay particles have a diameter of less than $1/256$ millimeter. They are plate-shaped. They may be of quartz but are more often of compounds such as kaolinite, illite, and montmorillonite.

If sand makes up 70 per cent or more of the soil by weight, the soil is said to be sandy. If the sand grains are not coated with clay, they do not stick together and are not plastic, or capable of being molded. Sand does not hold water well and has large pore spaces between the grains. Consequently, a sandy soil is exceedingly loose, drains water rapidly, and encourages the movement of air through the soil.

Soils with at least 35 to 40 per cent clay are called clay soils. A great part of the chemical reactions in soil takes place at the surface of clay particles. Because clay is made of such finely divided particles, a tremendous surface area is available for the adsorption of water, gases, and ions. Adsorption is the adhesion, in a thin film, of molecules and ions to the surfaces of solids. The compact clay soils do not allow rapid movement of air and water through them. Clay is sticky and highly plastic when wet, and very hard and cloddy when dry.

The soils known as *loams* are a mixture of sand, silt, and clay. They combine the desirable properties of both sand and clay. There are various other kinds of soils, depending on the size of the predominating particles: sandy loams, silty loams, clay loams, sandy clays, silty clays, and so on.

Groupings of soil particles. Individual particles of sand, silt, and clay are clustered together in a variety of aggregates, or particle groupings. This gives the soil its characteristic structure. There may be only one kind of structural pattern in a soil profile. More often, however, the pattern differs from one horizon to the next.

The structure of the soil is very important. It influences water drainage, the transfer of heat through the soil, aeration, and the degree to which plant roots can penetrate the soil and spread through it.

When thin, flat aggregates, lying horizontally, pile up one on top of another, they form a platelike structure. This structure may be found in any part of the soil profile. In semiarid and arid regions, the subsoil often shows a columnar pattern or a prismatic pattern in which the aggregates are shaped like pillars. A blocklike or nutlike structure is found in subsoils of humid re-

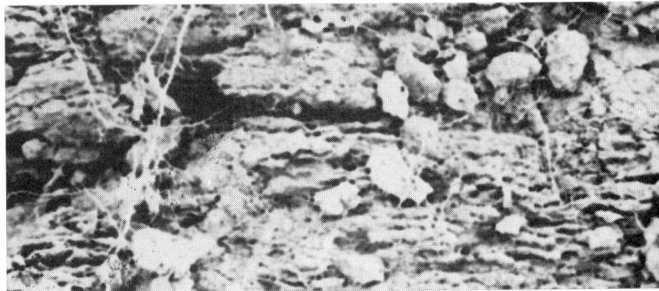

Roy W. Simonson

Soil structures. When thin, flat particle groupings are piled on top of one another, they form platelike structures. In semi-arid and arid regions, the subsoil may show a prismatic pattern, suggesting pillars set side by side. A blocklike structure is found in the subsoils of humid regions. When small rounded aggregates are loosely combined, they make up a granular structure. Examples of each type are shown above. Top to bottom: platy, prismatic, blocklike, granular.

SCHEMATIC MAP OF MAJOR SOIL GROUPS

- **INCEPTISOLS:** Early soils of cold climates, with moss and permafrost.
- **SPODOSOLS AND HISTOSOLS:** Forest soils of wet, cool climates.
- **ALFISOLS AND INCEPTISOLS:** Forest soils of wet, temperate climates.
- **ULTISOLS:** Older forest soils of wet, warm climates.
- **MOLLISOLS:** Fertile grass soils, often in dry to moist plains.
- **VERTISOLS:** Clay soils of warm climates with wet and dry seasons.
- **ARIDISOLS:** Soils of dry to desert climates, with sparse growth.
- **OXISOLS, INCEPTISOLS, ULTISOLS:** Soils of wet, warm to tropical climates.
- **MOUNTAIN SOILS — ENTISOLS AND INCEPTISOLS:** Thin and stony soils.
- **RIVER SOILS — INCEPTISOLS AND ENTISOLS:** Rich soils in floodplains and deltas.

The map shows the distribution of the ten major types of soil. Climate and geography affect soil distribution. Note that Greenland is covered by an ice cap, with little or no exposed soil.

gions; the aggregates are in the shape of cubes or rounded cubes, lying close together. Small rounded aggregates, loosely combined, make up a granular or crumb soil structure. This formation occurs in many surface soils, especially those rich in organic materials. In some soils, sands are so predominant that there is no structure. The soil is formed of single grains. In other soils, such as those high in clay, the units are quite massive, and they likewise have no distinct structure.

The time factor. The nature of soil depends to a certain extent on the length of time that the parent materials have been exposed to weathering and have supported vegetation. In areas where debris has been deposited by wind, water, glaciers, or volcanic action, there may not have been enough time for the formation of a mature soil.

The lay of the land. The topography of the land influences soil formation to a certain extent. If the land is hilly, water will be drained off more rapidly than it would be if the surface were more or less flat. As the water runs downhill, it takes some of the soil with it. In gently rolling country, a deep soil often develops if the climate is mild and humid. There is enough drainage to bring about important chemical changes, but not enough to remove much soil. On the other hand, a mature soil cannot be formed if water collects in a level area and stands for most of the year, as in bogs.

Vegetation. In its development, soil is greatly influenced by the plants that grow upon it. Plants take up mineral elements

from the soil by their roots. When these plants die, their remains decay and the mineral nutrients are released to the surface soil. The minerals are leached from this area. They are taken up by the roots of living plants and again released to the surface soil when the plants die. This cycle varies according to the plants taking part in it, their rooting habits, and their mineral needs.

Plants influence the soil in other ways. Roots hold the soil in place, thwarting the efforts of wind and water to carry it away. Branching roots penetrate and fracture hard soil lumps and layers; they make the soil loose and open for the circulation of air and water.

Climate. Developing soils are influenced most of all, perhaps, by climate. The temperature and amount of moisture control the degree of physical weathering. They also regulate the speed at which chemical decomposition takes place. In an excessively dry climate, for example, temperature changes are extreme. Wind and running-water erosion predominate over chemical weathering, for there is not enough constant moisture to allow vigorous chemical reactions. As a result, the parent soil material tends to be coarse. In humid regions, chemical action is greatly speeded up, and the potential soil material may become exceedingly fine in texture.

The soil organisms that decompose vegetable matter are also affected by moisture conditions and temperature. The climate dictates the kind of vegetation that will live in any given region. As we have seen, this vegetation, in turn, influences the soil in which it grows.

CLASSIFYING SOILS

A number of soil classifications have been devised. These are based largely on the stages of development, or ages, of different soils as reflected in the presence or absence of various horizons. Also important are such factors as climate, topography, minerals, acidity, amount of organic matter, and water content.

The latest classification puts the world's soils into ten orders. The last syllable, "sol," in each name is Latin for "soil." The first elements of the names indicate outstanding characteristics of the soils.

Entisols, or "recent soils". These soils are very young, with few or no horizons. They often develop on the floodplains of rivers and on stationary sand dunes. Entisols contain little organic matter. However, they may be fertile and yield bumper crops. More than 12 per cent of the world's land area is covered by entisols of various types.

Inceptisols, or "beginning soils". Inceptisols are somewhat older than entisols and their horizons are slightly more developed. In cold places, such as northern tundras, inceptisols are acidic and have much organic matter. They occur in other lands, also, and are sometimes used for agriculture or pasturage. They cover nearly 16 per cent of the earth's land surface.

Aridisols, or "dry soils". Such soils, formed in deserts and semideserts, are often reddish. They contain some horizons, with a layer of lime below the surface. There is little water or organic material, and vegetation is usually scant. Aridisols are commonly used for grazing. If irrigated, they may yield fantastic crops. Most of the world's soils are aridisols, which cover more than 19 per cent of the total land area.

Mollisols, or "soft soils". These soils are soft, even when dry. They are brown, red, dark, or black and contain horizons, organic matter, and water. Mollisols arise in regions less arid than deserts, such as plains. They are the most fertile soils, yielding grass or bountiful crops. They cover 9 per cent of the world land area.

Spodosols, or "ashy soils". These mature soils have horizons with an A layer that is ash-gray in color. Spodosols develop in cold, humid climates and underlie much of the world's northern forests. They contain much silica, aluminum, iron, and organic matter. Acidic and with little water-holding ability, spodosols are unfit for most farming. Less than $5\frac{1}{2}$ per cent of the world land surface is spodosol.

Alfisols, or "aluminum-iron soils". Alfisols are rich in aluminum, iron, water, organic matter, and substances that give a basic chemical reaction. Gray to brown in color, their horizons contain layers of clay. Alfisols are second only to mollisols in fertility and underlie many agricultural and forest areas. They cover nearly 15 per cent of the earth's land surface.

Vertisols, or "turned soils". These soils contain certain clays that, due to seasonal expansion and contraction, turn them over completely. Vertisols mature in warmer climates with marked wet and dry seasons. They crack widely during the dry spells and fill with water when it rains. These soils are deep, containing considerable clay, organic matter, and lime. They are fine for agriculture, particularly with fertilizer. Vertisols cover only 2 per cent of the world's land area.

Histosols, or "organic soils". These mature soils are saturated with decaying plant matter, organic muck, and water. They originate in lakes or marshes that are gradually filling with vegetation, mud, and peat. Eventually, the water plants are replaced by land plants, such as pine trees. Histosols, when partly drained, make excellent farmland. They are the rarest soils, taking up somewhat less than 1 per cent of the earth's land surface.

Ultisols, or "last, or ultimate, soils". Ultisols are soils in an advanced stage of development, or old age. They contain clay layers and much aluminum, with little organic matter and few substances giving a basic chemical reaction. Ultisols have also been leached of numerous minerals by water. They develop in moist, mild, and tropical climates. Unless fertilized, they are unsuitable for permanent agriculture.

About 8.5 per cent of the world's land area is covered by ultisols.

Oxisols, or "oxide soils". These very ancient soils have a high content of oxygen compounds—oxides—of aluminum and iron, along with much clay. They are often brick red in color, with hard lumps of iron oxide. The more soluble minerals have long since been dissolved by water, leaving residues that do not have a basic chemical reaction. Organic material is low. Oxisols are unsuitable for farming. They are common in warm to tropical climates with high rainfall and cover over 9 per cent of the earth's land surface.

LOCAL VARIATIONS

Soil variations that are often found within a region are due to local conditions. For example, where drainage is poor, marshes, swamps, and bogs may develop. Mosses, pondweeds, sedges, reeds, shrubs, and even trees live and die here. Their remains form thick mats of partially decayed organic matter, eventually producing peaty soils. Where drainage is poor in arid regions, salts accumulate in the surface soil, giving rise to salt and alkali flats. Only a thin surface soil may be present in areas where there are steep slopes. Immature soils occur on steeply sloping glacial drift, on sand dunes, on wind-carried deposits, and on river-carried deposits along river banks and in valley bottoms.

NUTRIENTS IN SOILS

All soils provide the vegetation that flourishes on them with a variety of nutrients. Water, nitrogen, phosphorus, sulfur, potassium, calcium, and magnesium are generally present in relatively large amounts. Other vital soil elements—absorbed in smaller quantities and therefore called trace elements—include iron, manganese, boron, zinc, and copper. Some plants seem to require still other nutrients, such as sodium, molybdenum, chlorine, fluorine, iodine, silicon, strontium, barium, and cobalt.

The vital nutrients are more abundant in some soils than in others. However, the quantity of nutrients is not as important as the ease with which they are made available to plants. This depends to a large extent on how acid or alkaline the soil is. In highly acid soils, for example, phosphorus may become unavailable because it forms compounds with aluminum and iron, and these compounds do not dissolve in soil water. Plants may not obtain enough calcium from soils that are too acid, because it has been leached from the topsoil. Microbes do not flourish in such soils. As a result, organic matter will not be decomposed and such essential nutrients as nitrogen, sulfur, and phosphorus will not be released. If the soil is too alkaline, iron, manganese, copper, and zinc will become unavailable to plants. However, when soil reaches an extreme degree of alkalinity, these metals become too readily available and plants may be poisoned by them.

Nitrogen is among the most essential of the soil elements required by plants. Most of it is held in plant remains in the form of such compounds as proteins and amino acids. Soil microbes decompose these to ammonium salts. Then certain bacteria oxidize the ammonium to form nitrite and nitrate salts by the process called nitrification. Plants can use both ammonium and nitrate and absorb these substances from the soil solution.

Leguminous plants, such as alfalfa, clover, cowpeas, beans, and lupines, get essential nitrogen compounds from bacteria that live in nodules on the plants' roots. These nitrogen-fixing bacteria obtain free nitrogen from air in the soil and synthesize, or fix, it in complex forms, which are then absorbed by the legumes. When the legumes die, their nitrogen-containing tissues become part of the soil's organic matter. Other bacteria take nitrogen from the air found in the soil and incorporate it into their bodies. When they die, they decompose, releasing nitrogen compounds for use by higher plants. Small quantities of nitrogen, in the form of ammonium and nitrate, are brought to the soil by rain water. These compounds have been produced through electrical activity in the atmosphere.

Phosphorus occurs in the soil in the mineral apatite and in phosphates of calcium, iron, and aluminum. It is also found in various organic compounds. It is released from these materials as various chemical changes occur. Phosphorus is taken up by roots in the form of phosphate ions.

Sulfur is found in the minerals pyrite and gypsum. These minerals are decomposed chemically and the product is acted on by soil bacteria. Sulfates that can be absorbed by plants then become available.

Potassium, calcium, and magnesium are metals. Potassium is contained in feldspar and mica; calcium in feldspar, amphibole, calcite, and dolomite; magnesium in mica, amphibole, dolomite, and serpentine. As these minerals react with water or with other substances in the soil, they liberate potassium, calcium, and magnesium ions, which all have a positive electric charge. Humus and clay soil particles normally carry a negative electrical charge. They attract the positively charged ions, which are absorbed on the surface of the soil particles.

When carbonic acid and other acids in the soil ionize, they free positively charged

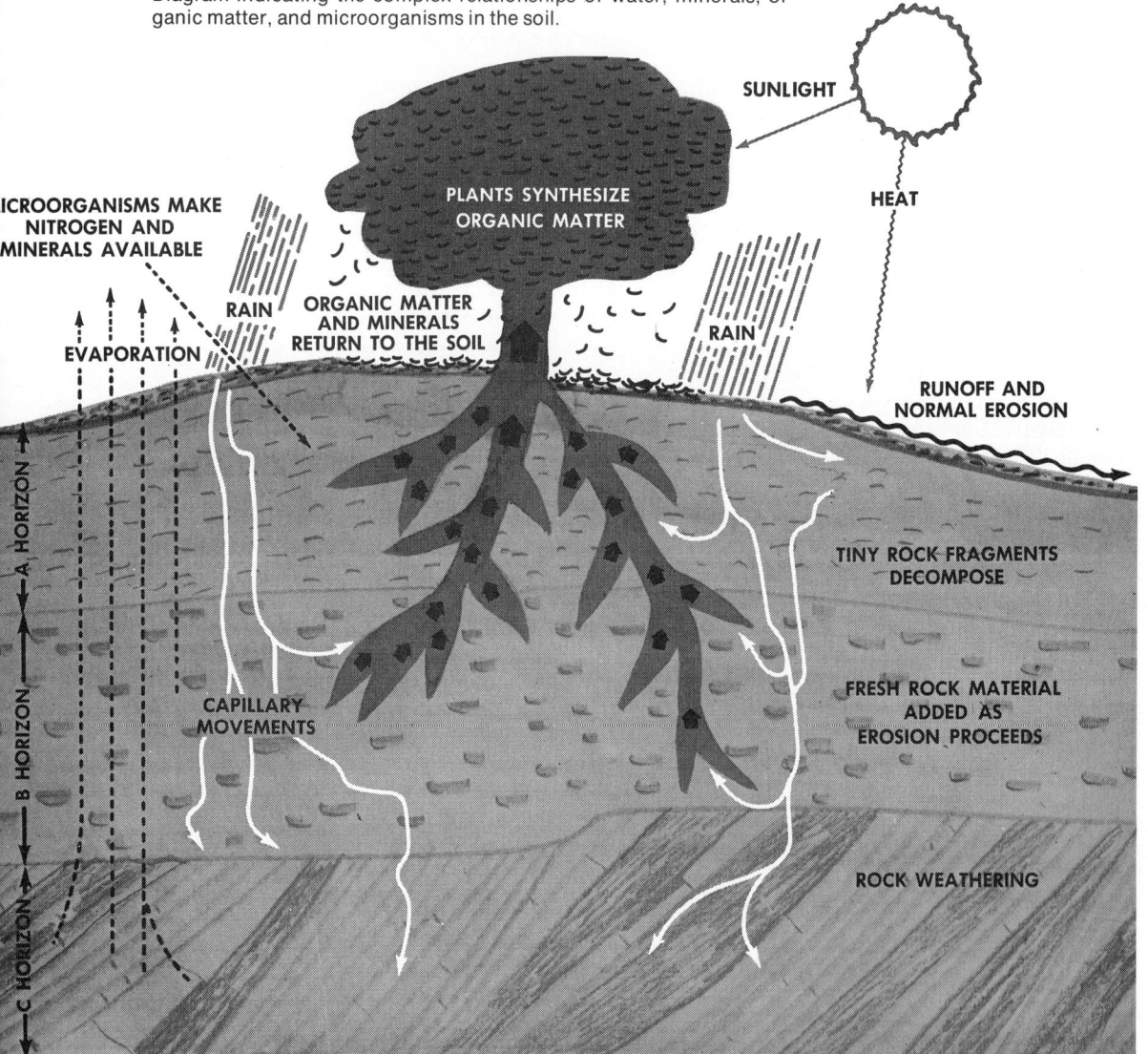

Diagram indicating the complex relationships of water, minerals, organic matter, and microorganisms in the soil.

hydrogen ions. If these are present in quantity, they replace the potassium, calcium, and magnesium ions on the soil particles, since hydrogen ions are more strongly attracted to the particles than are other positively charged ions.

If hydrogen ions are not numerous, the potassium, calcium, and magnesium ions remain on the surface of the soil particles. Root hairs come into contact with the particles, but they cannot at once absorb the metallic ions that have collected there. A striking phenomenon now takes place. A series of chemical reactions occur at the boundary where root tissues and soil particles meet. As a result, positively charged hydrogen ions are formed. These ions are attracted to the negatively charged soil particles, thus driving away the potassium, calcium, and magnesium ions, which are then absorbed by the root hairs.

WATER AND AIR IN SOILS

The water contained in soil is essential for chemical and other activities that take place there, as well as for the functioning of plant life. The abundance or lack of soil water depends on many factors. Among them are the quantity of rain, snow, and other forms of precipitation; the ability of soil particles to hold on to moisture; the amount of drainage; the slope of the soil surface; the amount of water removed by plants; and the amount lost by evaporation.

Water enters soil and circulates through it by way of the innumerable spaces between the soil particles. Earthworms, insects, moles, rodents, and other burrowing animals provide other paths for water in the soil. So do the roots of plants when they die and decay. After a heavy rain or quick thaw of snow or ice, a great deal of water percolates downward through all these channels under the influence of gravity and is generally lost to the surface soil. This water is unavailable to plants except as it passes over their roots in its passage. If the water moves rapidly, it is likely to wash down plant nutrients—especially nitrate, sulfate, calcium, and potassium ions—with it.

Some of the soil water is held in the form of a surface film by the soil particles. This is called *capillary water,* since it adheres to the particles through an attraction known as capillary force. Capillary water forms the true soil solution. It is readily absorbed by roots, and in it nutrient substances, available for plants, are dissolved.

The air in soil is also important. Atmospheric air is a mixture of gases: about 21 per cent oxygen, 78 per cent nitrogen, and 1 per cent of other gases, including a small amount of carbon dioxide. These gases are adsorbed on the soil particles, dispersed through the pore spaces, or dissolved in the soil solution.

Roots and soil organisms require oxygen for respiration. Roots need it, too, in order to absorb nutrients and water adequately. Oxygen is essential for various activities of bacteria in the soil, including the important activities of nitrogen fixation and nitrification. Because of these oxygen requirements, the amount of the gas is usually lower in the soil air than in the outside atmosphere. If the soil is porous, this imbalance may be somewhat lessened, as oxygen diffuses into the soil from the atmosphere.

Carbon dioxide is much more concentrated in the soil and tends to diffuse to the outside. As for the nitrogen that forms the principal part of air, it is important in the soil only as it is fixed by bacteria.

In a soil where plants are to flourish, there must be a suitable balance between water and air. If the soil is waterlogged, there is not enough air in it. As a result, roots and soil organisms suffer. If the water supply is inadequate, the essential chemical processes that take place in the soil are slowed, and food manufacture and growth are impeded.

Any exposed soil is subject to the hazards of erosion if it is not provided with an adequate cover of vegetation. Soils such as ultisols, which occur mainly in forested tropical regions, are especially fragile. When divested of their forest cover, they are subject to the effects of alternate wetting and drying. As a result, the clay-rich soils become soft and doughy and then harden to a slaglike material.

BACTERIA

by Harry J. Fuller

For many years, the harmful part played by bacteria in the transmission of diseases has been thoroughly publicized. Bacterial diseases such as tuberculosis, tetanus, diphtheria, and cholera have taken millions of lives. As a result, many people think of all bacteria as enemies of mankind—lurking in our food and drink, on the ground, and in the air, waiting to pounce upon us, their victims. However, we shall see, the benefits that we derive from bacteria outweigh the harm some of them may do.

Bacteria have a wider distribution than any other living organisms. They occur in large numbers in air, soils, and water. They live in and on the bodies of other living organisms, and in dead and nonliving organic materials such as cadavers, dung, garbage, humus, and milk. Bacteria have been found at depths of many meters in soils and also in the ooze of ocean beds, far below the surface of the sea.

Early investigators of bacteria thought of them as tiny animals and generally grouped them together with the microscopic animals called protozoans. Today, practically all biologists regard bacteria as plants. They are classified in the phylum Schizomycophyta. The name means "fungus plants that divide," and refers to the most common method of reproduction among bacteria.

The exact relationship of bacteria to the fungi and to other plant groups is uncertain. Various bacteria show undoubted similarities with true fungi. Certain members of both groups have a substance called chitin in their cell walls. Both bacteria and fungi can obtain nourishment directly from organic matter. Bacteria are microscopic in size. Some fungi are also extremely small.

Some bacteria, however, seem to be more closely related to the lower algae, particularly the blue-green algae. Both bacteria and blue-green algae reproduce mainly by fission. Also, the cells of both lack definitely organized nuclei. Some blue-green algae have shapes resembling those of bacteria. Some are colorless, which is also a characteristic of bacteria.

GENERAL CHARACTERISTICS

Most bacteria are one-celled plants of extremely small size. They rarely exceed 0.005 millimeter in their greatest dimension. Some average only 0.00015 millimeter. These plants are the smallest known living organisms (excluding viruses, which may or may not be living organisms). They can be observed only under microscopes of very great magnifying power.

There are three common bacterial body forms: spherical or ovoid (coccus forms), rod-shaped or cylindrical (bacillus forms), and spiral or screw (spirillum forms). Some species of bacteria have much-branched, threadlike bodies, but these species are few in comparison with those that take the form of spheres, rods, or spirals.

A bacterium has a very thin cell wall, consisting chiefly of cellulose or chitin. This is often surrounded by a slimy, transparent capsule, secreted by the bacterium itself. Within the cell wall is a mass of the living material called protoplasm. When observed under the optical microscope, which uses ordinary light, bacterial protoplasm appears simple in structure. However, research with the electron microscope, which uses beams of electrons, indicates that this protoplasm is as complex as that found in other plants.

Bacteria that live in liquids often have long, threadlike processes called *flagella*. Rhythmic movements of its flagella propel a bacterium through the liquid, usually in a twisting fashion. The numbers and arrangement of flagella vary greatly in the different species of bacteria and are used as a basis for identification. A few species of mobile bacteria lack flagella. They move about by snakelike, twisting movements of the entire cell.

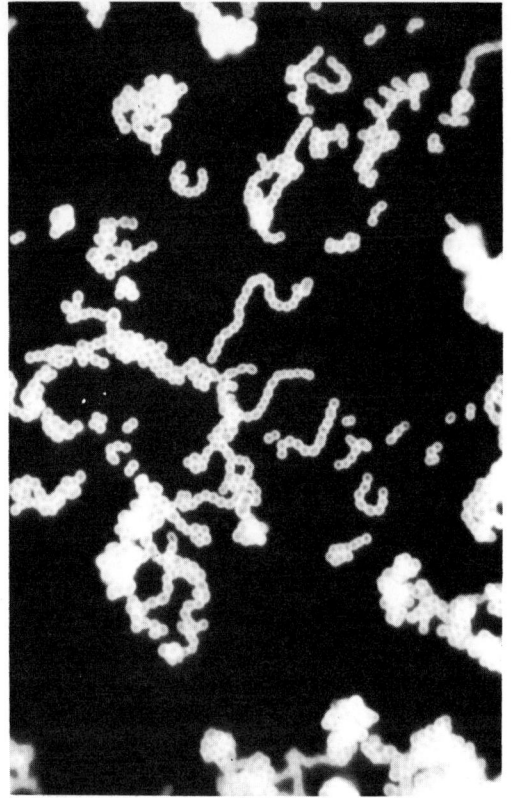

Institut Pasteur, Paris

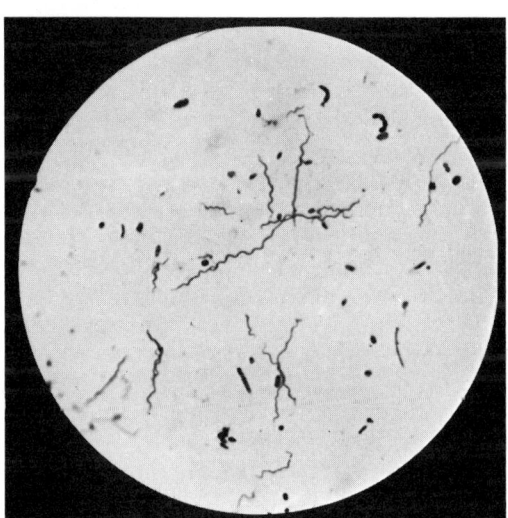

Institut Pasteur, Paris

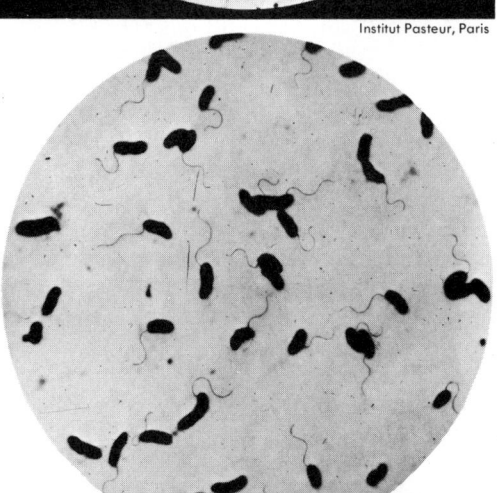

BACTERIAL REPRODUCTION

Reproduction in bacteria is largely asexual. Most bacteria reproduce by *fission*. This is a simple process of cell division, in which one bacterium splits into two new ones. Fission may occur with incredible rapidity—as often as every 15 or 20 minutes under particularly favorable conditions of temperature, moisture, and food supply.

It is estimated that within the space of 24 hours a cholera bacterium, reproducing by fission at its most rapid rate, could produce an enormous number of offspring weighing more than 2,000 metric tons. This extraordinary rate of reproduction is only theoretically possible. For one thing, there would not be nearly enough food or moisture in any one place for such enormous numbers.

There are other asexual methods of bacterial reproduction. For example, some species form very tiny structures called *spores* (one spore per bacterium). This method of reproduction is often, but not always, the consequence of certain environmental conditions, such as temperature extremes or inadequate food supply. Such conditions do not favor the active, vigorous growth and fission of bacteria. A spore is usually formed by the condensation of protoplasm within a bacterial cell into a spherical or egg-shaped body. Spores have a lower water content than active bacterial cells. Because of this, they are more resistant to unfavorable environmental conditions than are active bacteria. Pasturizing milk, for example, kills most of the active bacteria present. But the bacterial spores in the milk survive the heat treatment. Fortunately, most of the bacteria that cause serious diseases in human beings do not form resistant spores.

When bacterial spores encounter favorable conditions of temperature and food

There are three common bacteria body shapes. Spherical or ovoid forms (upper left) are known as cocci. Spiral or screwlike forms (middle left) are known as spirilla. Rod-shaped or cylindrical forms (lower left) are known as bacilli. The streptococci bacteria (upper left) have joined to form a long chain.

BACTERIA 23

supply, they germinate, or sprout. Each spore then grows into an active bacterium. Since only one spore is produced per bacterium, no increase in the total number of bacteria results from the formation and germination of spores.

Some bacteria are also known to exhibit a type of sexual reproduction. Members of certain species contain a viruslike agent known as the *fertility,* or *F, factor.* The cell containing this factor is designated F+ and called "male." The cell lacking it is designated F− and called "female." During a type of mating known as *conjugation,* male and female bacteria attach themselves by means of a bridge. The male then transfers the F factor to the female. The bacteria then separate and each undergoes fission. This unusual type of reproduction in bacteria permits a certain amount of recombination of hereditary material in bacteria. The bacteria that are produced after conjugation contain hereditary material from both "parents."

MAKING AND GETTING FOOD

Most bacteria are unable to manufacture their own foods. Instead, they feed on organic compounds manufactured by other organisms. Such bacteria are given the name *heterotrophic,* which means "other nourishment" in Greek.

Heterotrophic bacteria that derive their food from dead plant and animal bodies or from dung and other waste products of organisms are known as *saprophytes.* Other heterotrophic bacteria, called *parasites,* obtain food directly from the tissues of living plants or living animals. This is really another way of saying that the plants or animals in question have a bacterial disease. Many bacterial species are exclusively saprophytes. Others are exclusively parasites. Still others may live saprophytically or parasitically, depending upon the nature of the environment and the organic materials available to them.

Many heterotrophic bacteria are cosmopolitan in their tastes, feeding upon a variety of organic compounds. Other species are highly selective. Certain bacteria, for example, feed mainly on cellulose. Others ingest only sugars. Some parasitic bacteria live only on blood. Others feed on specific tissues of animal or plant bodies.

Like other organisms, bacteria produce regulatory chemicals called enzymes. These promote digestion—that is, the conversion of complex, water-insoluble foods into simple, water-soluble foods.

A few bacterial species are able to manufacture foods from simple inorganic substances such as carbon dioxide. Such bacteria are called *autotrophic,* or self-nourishing. There are two kinds of autotrophic bacteria: chemosynthetic and photosynthetic. Chemosynthetic bacteria obtain the energy required for food manufacture by oxidizing (combining with oxygen) various chemicals. Photosynthetic bacteria contain purple or greenish pigments that enable them to absorb and use light energy in food manufacture. This process of photosynthesis involves pigments somewhat different from the chlorophyll of higher plants.

The chemosynthetic bacteria are particularly important in the scheme of nature. Sulfur bacteria, for example, convert hydrogen sulfide, a product of protein decay, to sulfur and then to sulfuric acid. The acid undergoes chemical reactions in soils to form sulfates, which are the principal source of the sulfur needed by higher plants for normal growth and reproduction. Iron bacteria oxidize certain types of iron compounds into other iron compounds. The

Electron micrograph of streptococci. The scale at lower left indicates 1 micron, or approximately 1/10,000 of a centimeter.

Dr. Stuart Mudd, Society of American Bacteriologists

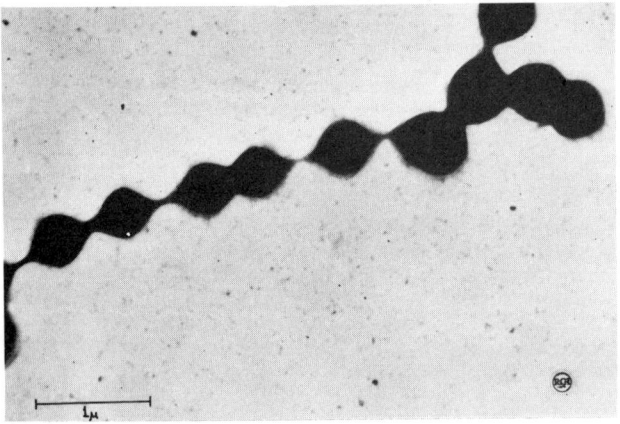

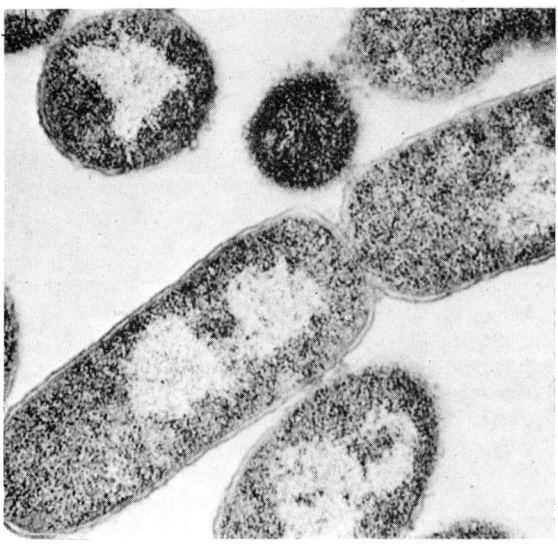
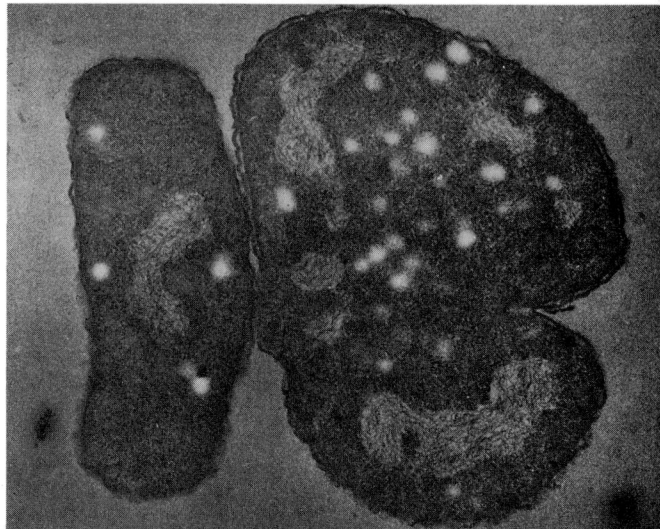

Institut Pasteur, Paris

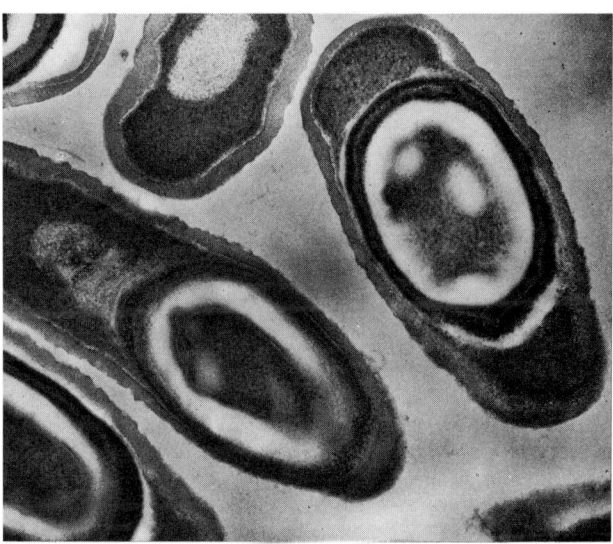

Institut Pasteur, Paris

Reproduction in bacteria. Upper left: bacteria reproducing asexually by fission. Upper right: sexual reproduction in bacteria. Lower right: bacteria with spores in their protoplasm. Under favorable conditions, the spores will germinate, developing into active bacteria.

energy they derive from this oxidation is used in building foods from carbon dioxide. The compounds formed by iron bacteria are deposited in ropy, rust-colored masses, which often appear in springs, water pipes, and reservoirs. Nitrifying bacteria are chemosynthetic bacteria that live in soils and that make nitrogen available to higher plants.

RESPIRATION IN BACTERIA

Bacteria resemble all other living organisms in their ability to carry on respiration—that is, the chemical breakdown of foods and the release from these foods of the energy used in growth, reproduction, and other vital activities.

Most bacteria, like most plants and animals, use free oxygen from the atmosphere in respiration and produce carbon dioxide and water as a result of the process. Such bacteria are called *aerobic,* which means "living in the presence of oxygen." They can live only when they have access to free oxygen—for example, in aerated soil and water, on the surface of other living organisms, or on the surface of foodstuffs.

Other bacteria maintain respiration in the absence of free oxygen. They are called *anaerobic* (living away from oxygen). They thrive in sealed, imperfectly sterilized cans of food, in the bodies of other organisms, and in poorly aerated soils and water. The respiration of such bacteria is commonly called *fermentation*.

The products of anaerobic respiration, or fermentation, are carbon dioxide or other gases and a variety of organic compounds. These compounds include ethyl alcohol, lactic acid (produced, for example, in sour milk), formic acid (responsible for various types of food poisoning), and butyr-

BACTERIA 25

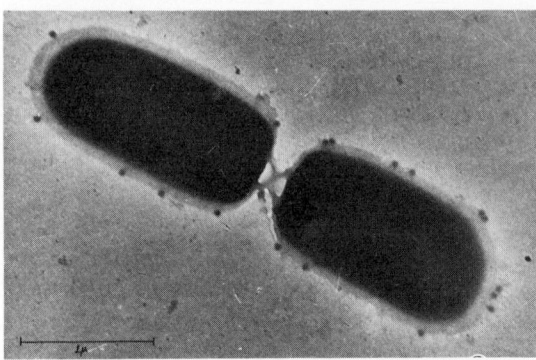

Escherichia coli dividing. Many virus particles, known as phages, are attached to the bacterial wall.

ic acid (which causes the disagreeable odor of rancid butter). The swelling of cans containing imperfectly sterilized foods is the result of the pressure of gases formed by bacterial fermentation. Foods in such swollen cans should never be eaten.

Some bacteria are able to respire either aerobically or anaerobically, depending upon whether or not free oxygen is available. The energy released from foods in both aerobic and anaerobic respiration is chiefly in the form of chemical energy. A certain amount of heat energy is also released. When damp hay is stored in a poorly ventilated barn, the respiration of the bacteria growing on the hay may cause heat energy to accumulate to a dangerous extent. Under such circumstances the hay may suddenly burst into flames. Many apparently mysterious fires may be traced to heat energy released by bacteria.

HARMFUL BACTERIA

Certain bacteria rank high among our deadliest enemies. These are the disease-causing, or *pathogenic,* bacteria. They are responsible for such human diseases as tuberculosis, typhoid fever, some types of pneumonia, meningitis, tetanus, cholera, diphtheria, leprosy, several types of dysentery, and various wound infections. They cause diseases among domesticated animals—tuberculosis, anthrax, fowl cholera, pneumonia, and glanders, among others. They are responsible for various diseases of crop plants, such as fire blight of pears, citrus canker, tomato and potato wilts, potato blackleg, and soft rot of celery. These diseases result in tremendous crop losses throughout the world's agricultural areas.

Bacteria also ruin great quantities of foodstuffs. They cause souring of milk, rancidity of butter, and spoilage of both fresh and canned fruits and vegetables. The *toxins,* or poisons, of bacteria in certain spoiled foods may result in ptomaine poisoning, botulism, and other types of food poisoning. Fortunately there are various methods of preventing or at least reducing bacterial activity in foods. Heat sterilization, refrigeration, and deep-freezing are all effective. So is desiccation, or the drying of foods. Since bacteria usually require considerable quantities of water for their activity, they cannot grow, reproduce, and respire in dried foods. Food may also be protected against bacteria by adding chemicals that are harmless to human beings but poisonous to bacteria.

Some bacteria cause the decomposition of fabrics, wood, and other products of organic origin. This is most common in tropical regions, where excessive heat is combined with high relative humidity.

BENEFICIAL BACTERIA

The harmful activities of bacteria (harmful, that is, from our viewpoint) are more than offset by other activities that are directly or indirectly beneficial to mankind.

Bacteria play an important part in the production of many foodstuffs. Among those foods are vinegar, certain types of cheese, and sauerkraut. Fermented plant tissues produced through bacterial action in silos form ensilage, which is fed to livestock. Various valuable organic chemicals are also the product, at least in part, of bacterial action. Among these are acetone, which is widely used in industry; butanol, which serves as a solvent for lacquers; and vitamins.

The activities of certain bacteria result in the retting, or rotting, of pectins and other carbohydrate materials that cement together the fibers in the stems of flax and hemp and in coconut husks. The retting process causes these valuable fibers to separate from the tissues that surround them. The fibers are later dried, spun, and woven. The retting is accomplished by submerging bundles of cut stems or coconut husks in water, or by allowing bacterial action to proceed in piles of dew-soaked stems in the fields.

Bacteria cause the decomposition of dead plant and animal bodies and of the wastes of living organisms. They break down the proteins, fats, carbohydrates, and other complex organic substances of these materials and transform them into carbon dioxide, ammonia, and other simple inorganic compounds. In this way, bacteria rid the earth of organic debris. They also restore to soil and air the simpler compounds that green plants require for food manufacture and, hence, for their existence.

Green plants make organic foods out of these inorganic substances of air and soil. They build these foods into their own protoplasm and cell walls. Animals obtain such foods by eating plant tissues, by devouring plant-eating animals, or by feeding on other carnivores that prey on plant-eating animals. Without bacteria, the supplies of inorganic raw materials required for the synthesis of organic foods would soon become exhausted, for these raw materials would be locked up in dead plant and animal bodies. Both plant and animal life would come to an end.

Certain exceedingly important nitrogen transformations are due to bacteria that live in the soil. These transformations ensure a continuing supply of nitrogen, in the form of nitrates, to plants. This is a vital matter, since nitrogen is essential to plant life. Ammonifying, nitrifying, and nitrogen-fixing bacteria are involved in nitrogen transformations.

Ammonifying bacteria transform various organic compounds, derived from dead plants and animals and their wastes, into ammonia. There are two kinds of nitrifying bacteria: nitrite bacteria, which convert ammonia into nitrites, and nitrate bacteria, which transform nitrites into nitrates — the nitrogenous, or nitrogen-containing, compounds most readily absorbed and utilized by most green plants.

Nitrogen-fixing bacteria convert nitrogen gas into compounds that the plants can absorb. The nitrogen-fixing bacteria include the so-called nodule bacteria, which inhabit the roots of legumes (clovers, alfalfa, soybeans, peas, cowpeas, and so on) and a few non-leguminous plants. The bacteria enter the roots of the plants through root hairs in the soil. As they grow and develop, nodules are formed. Nodules are more-or-less globular lumps that are large enough to be seen with the unaided eye when the roots of legumes are carefully freed of adhering soil.

Tetanus bacillus, an example of a bacterium with threadlike bodies extending from it.

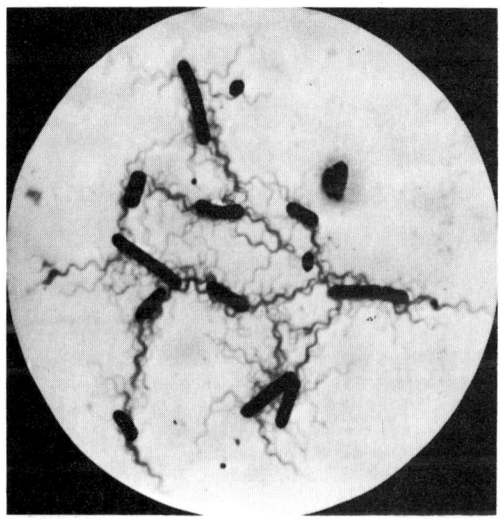

Institut Pasteur

The tuberculosis bacillus, *Mycobacterium tuberculosis,* magnified something like 17,000 times.

E. R. Squibb and Sons

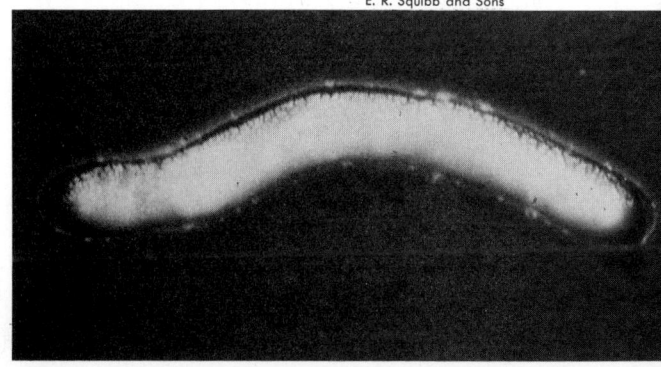

BACTERIA 27

AN EXAMPLE OF SYMBIOSIS

A nodule on a legume root consists in part of bacteria, in part of root tissues. The association between roots and nodule-forming bacteria is an example of *symbiosis* —a state in which two organisms, living together, are mutually beneficial. The roots of legumes supply carbohydrates and other foods to the bacteria. These organisms, in turn, through their ability to fix nitrogen gas, provide the roots with nitrogenous compounds. If the bacteria are separated from the roots, they cannot fix nitrogen at all, or else fix it ineffectively.

Root nodules excrete organic nitrogen compounds into the soils in which nodule-bearing plants grow. Thus they increase the nitrogen content of the soil and increase its fertility. The practice of crop rotation, in which one of the crops consists of legumes, is followed in part to increase the nitrogen content of agricultural soils.

The different varieties of nodule bacteria seem to be widely distributed in the soil. When a farmer plants alfalfa, or sweet clover, or cowpeas in a field, the roots of these plants usually develop nodules, because their roots come in contact with the naturally occurring nodule bacteria of the proper variety.

Commercial bacteriological laboratories sell preparations containing different varieties of nodule bacteria. Using such preparations, farmers inoculate seeds of leguminous crops before planting them. They can then be reasonably sure that there will be a high degree of nodule formation in the roots of these crops.

INTESTINAL BACTERIA

Bacteria play a vital part in the breakdown of foods in the digestive organs of various animals. Cattle and horses, whose diet consists entirely of plant tissue rich in cellulose, do not produce cellulose-digesting enzymes. Therefore they cannot digest cellulose without the help of bacteria. The anaerobic bacteria that live in the intestinal tracts of such animals digest cellulose material, which may be as much as one-third the total mass of the grasses eaten, and produce simpler substances. These can be absorbed through the intestinal walls and thus used as a source of nourishment. Bacteria occur in enormous numbers in the intestinal tracts of cattle, horses, and sheep, and are dropped in great numbers with the feces.

Certain intestinal bacteria are important to the animals in which they live for another reason. They produce and secrete certain essential vitamins, notably vitamin K and vitamin B_{12}.

Bacteria also occur in the human digestive tract, promoting the digestion of certain foods and furnishing vitamins. On the average, about 40 per cent of the mass of human feces consists of these bacteria, chiefly a rod-shaped bacterium named *Escherichia coli*. When *Escherichia coli* appears in water, milk, and foodstuffs, it is usually assumed that these substances have been contaminated with human feces or sewage.

ANTIBIOTICS AND VACCINES

Certain bacteria produce antibiotic drugs that are extremely valuable in treating various diseases of human beings and other mammals. Among the important antibiotics produced by bacteria are streptomycin, which is especially effective in treating tuberculosis and tularemia; and aureomycin and terramycin, both highly effective in cases of intestinal, urinary, and other internal infections, and certain types of pneumonia and influenza. It should be noted that the most widely publicized antibiotic, penicillin, is produced by a fungus, not by a bacterium.

Some kinds of pathogenic bacteria are used to make vaccines and serums. These are then used to prevent or treat diseases caused by the same bacteria.

A bacterial *vaccine* is a preparation of dead or weakened bacteria or bacterial products. It is injected into an animal body. The vaccine stimulates the animal to produce substances called *antibodies* in its blood. If active bacteria of the same kind as the injected bacteria enter the body at a later date, they are held in check or destroyed by the antibodies. The antibodies may persist in the blood for long periods of time,

Retting kenaf fibers. Bacterial action is involved in this process.

conferring upon the animal a type of disease resistance called *immunity*. This action is similar to that which occurs when an animal has had a disease and has recovered from it. During its diseased state, the animal produces and accumulates antibodies, which then prevent the later development of bacteria of the same type in its body. Immunity following an attack of typhoid fever or cholera, for example, usually persists throughout the life of the individual.

Bacterial vaccines are used chiefly in immunizing human beings and domesticated animals against such diseases as diphtheria, cholera, and typhoid fever. When a vaccine is injected into an animal it gives it a very mild form of the disease that these same bacteria would cause if they entered the animal's body in an active and unweakened condition. For example, the fever, digestive upsets, and headaches that often result when typhoid vaccine is given are actually symptoms of a very light attack of typhoid fever.

A *serum* is a preparation from the blood of an animal that has been inoculated with bacteria (or other disease-producing agents) and has recovered from the disease that these bacteria or other agents cause. Blood is removed from the animal and is cleaned and sterilized; the serum is then separated. This contains antibodies that the animal formed as a consequence of the disease that attacked it. The serum, injected into another animal, confers immunity upon that animal, should disease bacteria enter it. Serums are especially effective in treating or preventing tetanus, diphtheria, meningitis, and some forms of pneumonia.

Since the antibodies that develop in the blood following inoculation with a vaccine are produced by the animal thus inoculated, the type of immunity that results is called *active immunity*. In the use of serums, the animal that receives the serum acquires its immunity from the antibodies produced in the blood of another animal. Immunity of this type is called *passive immunity*.

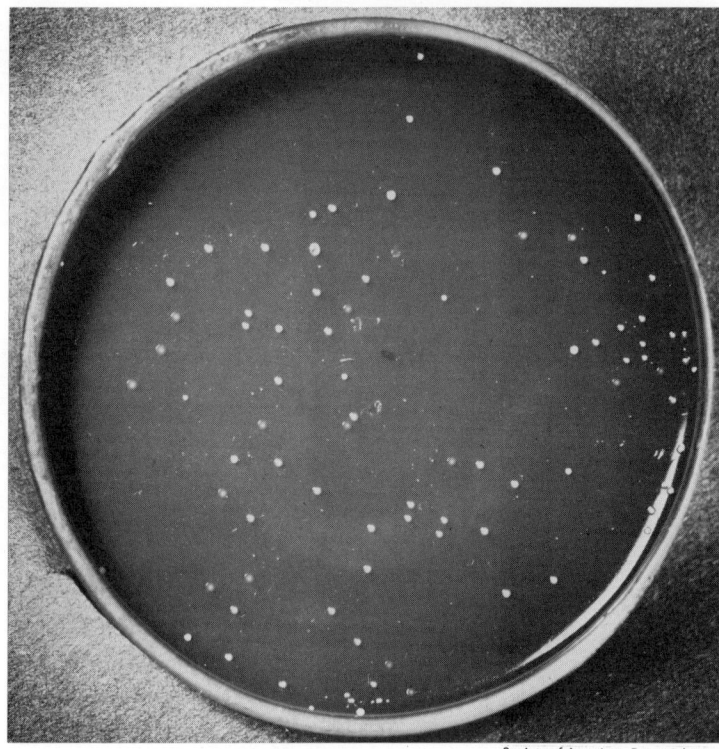

A photograph of colonies of *Lactobacilli,* which are derived from human saliva. They grow in a shallow Petri dish, in which their food is mixed with gelatin or agar.

Society of American Bacteriologists

Disease-producing bacteria, of course, may be held in check by substances other than vaccines and serums. Sulfa drugs and antibiotics destroy many disease-carrying bacteria within animal bodies. Certain highly toxic chemicals, such as mercuric chloride, iodine, and carbolic acid, may be used on body surfaces to kill bacteria. Such chemicals cannot be taken internally because they are poisonous to body tissues.

CULTIVATING BACTERIA

Bacteria are extensively cultivated in the laboratories of medical schools, hospitals, medical research institutes, and universities, as well as in commercial laboratories and factories that produce antibiotics, useful organic chemicals, cheese, preparations of nodule bacteria, and other economically valuable bacterial products.

The careful cultivation of bacteria makes it possible for researchers to carry on the fight against bacterial diseases and for manufacturers to make the products in which bacteria play a part.

Great care must be taken in cultivating bacteria. There is danger involved in handling disease-producing species. Also, for many purposes, bacterial cultures must be kept pure. A *pure culture* contains only the desired species of bacterium that is being studied or utilized. It is not contaminated by other types of bacteria.

Most bacteria that are important in medicine and commerce are heterotrophic and therefore cannot synthesize their own food. They must be supplied with organic nutrients, such as carbohydrates, proteins, fats, and vitamins.

The solid-medium technique is used more widely than any other in cultivating bacteria. In this method, the requisite foods are mixed with gelatin or agar, a gelatinlike material produced by some species of red algae, which has been stirred into boiling water to form a viscous liquid mixture. The mixture is sometimes poured into shallow dishes called Petri dishes. These are equipped with covers that admit air but prevent entry of dust and bacteria from the atmosphere. Sometimes the mixture is put into glass flasks or other types of containers.

The containers are sterilized for 15 to 30 minutes under pressure in order to kill any bacteria or bacterial spores that may have fallen into the mixture or the containers during the preparation of the medium. The containers are then removed from the sterilizers and allowed to cool. During the cooling process, the gelatin or agar becomes firm in the same way as a gelatin dessert. It is then ready to serve as a "garden" in which the growth of bacteria can take place.

The bacteria are placed on the surface of the solid medium by means of a sterile needle. The bacteria are obtained from a variety of sources: older cultures, soil, foods, infected plant or animal tissues, and the like. Despite all precautions, foreign, unwanted bacteria sometimes enter a culture. In that case the culture is useless because it is not pure and it has to be destroyed.

Different species of bacteria require different cultural treatment. For example, anaerobic bacteria must be grown in containers from which oxygen can be excluded. Some bacteria require cellulose in the growth medium; others, blood proteins; still others, amino acids, vitamins, or other organic compounds. Certain bacteria are grown in liquid solutions of nutrients, rather than in semisolid media of the gelatin type.

Typhoid fever bacilli

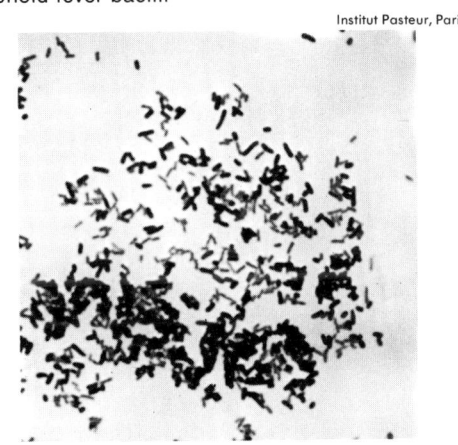

Institut Pasteur, Paris

Bacteria that cause pneumonia

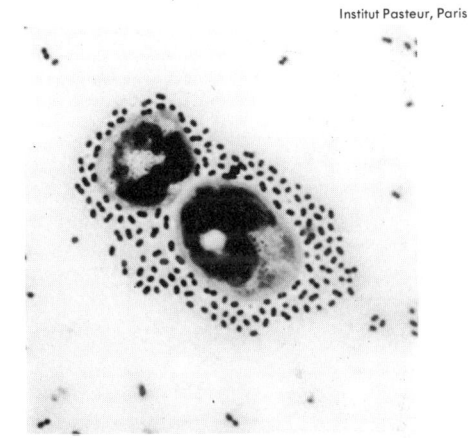

Institut Pasteur, Paris

Streptococcus bacteria responsible for many diseases

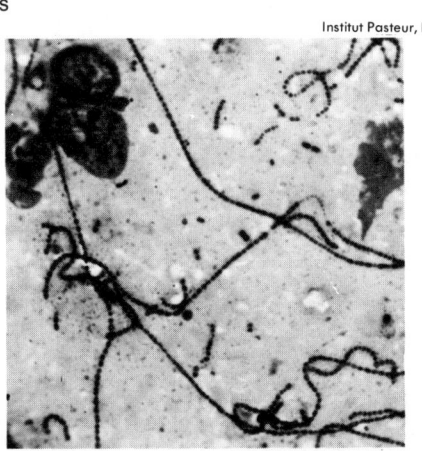

Institut Pasteur, Paris

Cholera bacteria

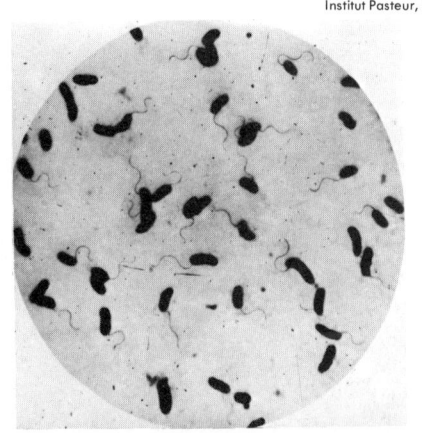

Institut Pasteur, Paris

ALGAE

by M. H. Berry

Microscopic glasslike cells found in almost all the waters of the world; scum on the surface of stagnant pools; bluish-green streamers attached to rocks in the vicinity of waterfalls; immense ribbons of seaweed found on offshore rocks—these are representative forms of the plants known as algae. They belong to the subkingdom of the Thallophyta. Members of this group lack roots, stems, and leaves and are by all odds the most primitive of all plants. In one important respect, however, the algae resemble higher plants. They possess the green pigment called chlorophyll. With this pigment, which absorbs the radiant energy of the sun, they can manufacture food by the process of photosynthesis.

Algae are chiefly water plants, dwelling in oceans, seas, lakes, ponds, rivers, ditches, and other bodies of water, large and small. Some species, however, are found on stones, the bark of trees, fences, and so on—generally moist environments that are not subjected to direct sunlight. The algae are infinitely varied in size and shape. Some consist of individual microscopic cells. Others form flat sheets, narrow filaments, or immense stemlike structures that may be more than 30 meters long. Certain algae have growths that strikingly resemble the leaves of higher plants.

These primitive plants are divided into seven phyla, or primary divisions.

BLUE-GREEN ALGAE

These slimy algae make up the phylum Cyanophyta. They contain a blue pigment, phycocyanin, in addition to chlorophyll and other pigments. The blue masks out the other hues and gives most species a dark blue color. Some species are other colors, however, ranging from orange to black.

Many species of blue-green algae are one-celled. The cells are very simple and show no definite nuclei. Other species form colonies. In these, one-celled plants are joined together to form threadlike fila-

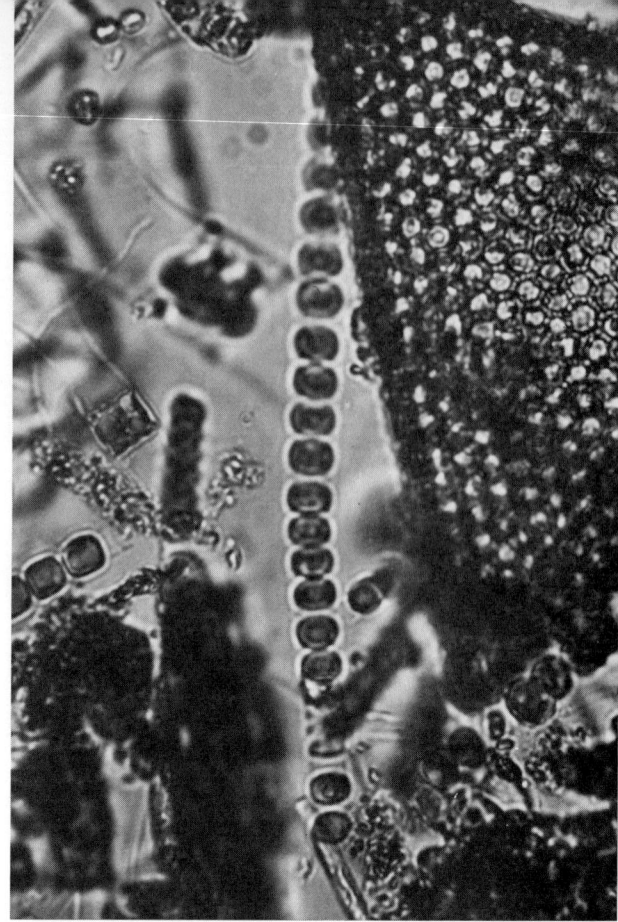

Oscillatoria, a primitive filamentous blue-green alga that can move about in a wavelike manner.

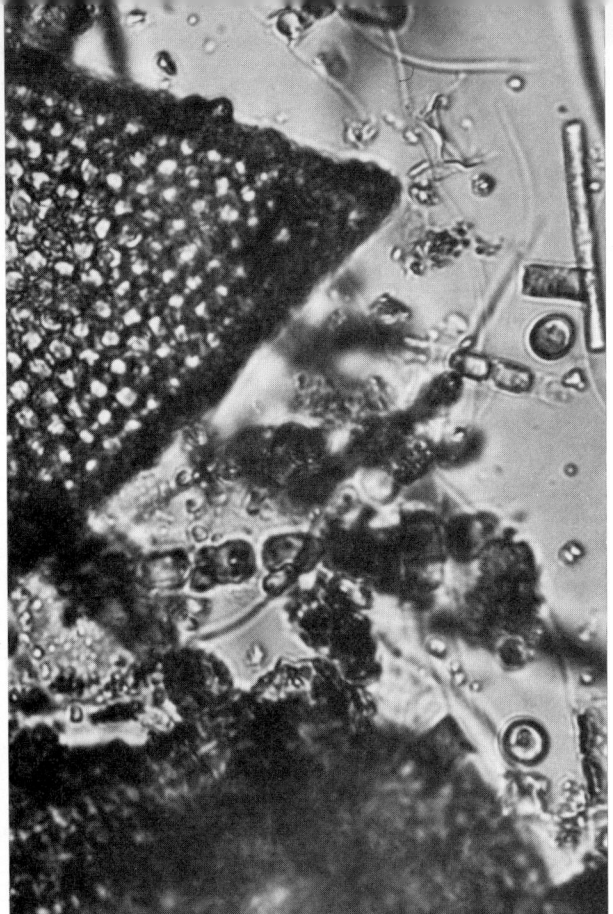

Left: diatoms, a particularly large one in the center. Deposits of these golden-brown algae have many commercial uses.

ments, sheets, or balls. The cells of the colonial plants are generally held together in a mass of sticky slime secreted by the individual cells. The blue-green algae reproduce only by asexual means.

Blue-green algae are found wherever there is ample moisture, in almost all parts of the world. They often contaminate drinking water, causing a very disagreeable odor and taste. The hot springs of Yellowstone National Park, in the western United States, are highly colored because of their presence. Curiously enough, they bring about the characteristic reddish color of the Red Sea. Blue-green algae are found in the snows of the Arctic. Some species thrive in the digestive tract of human beings and the lower animals, apparently without causing ill effects. Certain members of the group, including *Gloeocapsa* and *Nostoc*, have formed a partnership with fungi, making up the separate group of plants known as lichens.

Gloeocapsa, however, is generally found in the form of single cells. It sometimes forms colonies consisting of three or four cells enclosed in a gelatinous sheet. Eventually, these colonies break up into single individuals. *Gloeocapsa* usually occurs as a slimy coating on damp walls and rocks.

Oscillatoria is perhaps the commonest of the filamentous blue-green algae. It consists of a long, threadlike filament made up of separate rectangular cells, like so many dominoes placed side by side. Cell division at either end of the filament causes it to become longer. Often, gelatinous partitions are formed at several places within the filament, and the chain is ultimately broken up into several shorter ones. These continue to produce new cells until each new chain reaches the proportions of its parent chain.

Oscillatoria is one of the few plants that can move about. It has a wavelike, gliding motion. Botanists cannot account for this phenomenon, since the plant does

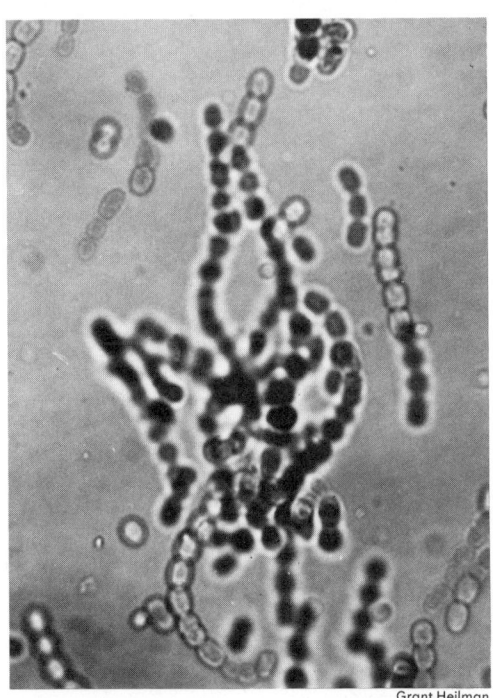

The blue-green alga, *Anabaena*, showing colonial organization. The chains are called *trichomes*.

not have the structures that often bring about motion in primitive plants and animals. It has no whiplike flagella, cilia, or hairlike structures.

Oscillatoria is found wherever there is ample moisture. Often it can be seen floating in great profusion on the surface of lakes and ponds, forming the so-called "dog-day waterblooms." One species is extremely abundant in moist soil and on flower pots in greenhouses.

Anabaena is a blue-green filamentous alga. It differs from *Oscillatoria* in that it cannot move. It is perhaps the most abundant alga in North America. Its cells are beadlike in form, and under the microscope the plant looks like an irregular strand of beautiful greenish beads. *Anabaena* is almost wholly aquatic and is generally restricted to fresh water. It is a favorite food of fishes. This alga is often the chief offender when the water supply of small villages becomes foul, with an unpleasant taste and a disagreeable odor.

EUGLENA AND ITS RELATIVES

These one-celled organisms make up the Euglenophyta. They are generally green in color and swim about in the water by means of flagella. (One group has no flagella.) They vary in form. Some are spherical. Others are egg-shaped or pear-shaped. Reproduction is generally by cell division. Sometimes considerable numbers of one-celled Euglenophyta form colonies. These algae, which are considered by botanists to be plants, are classified as one-celled animals, or Protozoa, by zoologists.

The best-known of the Euglenophyta is *Euglena*, which is found in stagnant ponds, swimming pools, and aquaria. This minute green organism has a single flagellum. It also has a minute red "eyespot," which seems to be sensitive to light. *Euglena* causes water to become greenish and cloudy, and often imparts an unpleasant flavor. Water containing excessive quantities of this alga is considered to be undesirable for drinking or swimming purposes.

THE PYRROPHYTES

These algae make up the phylum Pyrrophyta. They have pigments ranging from yellowish to brownish. Not only can they manufacture food in the presence of sunlight, but they also are able to store reserve supplies of food in the form of starch or compounds similar to starch.

The Pyrrophyta are found in both ocean and fresh waters. Most species consist of a single cell and move about by means of one or two flagella. Some one-celled species, however, have no flagella and do not move. A certain number of the Pyrrophyta are colonial, forming filaments. Like the Euglenophyta, the Pyrrophyta are included by zoologists among the protozoans.

GREEN ALGAE

The green algae, or Chlorophyta, are

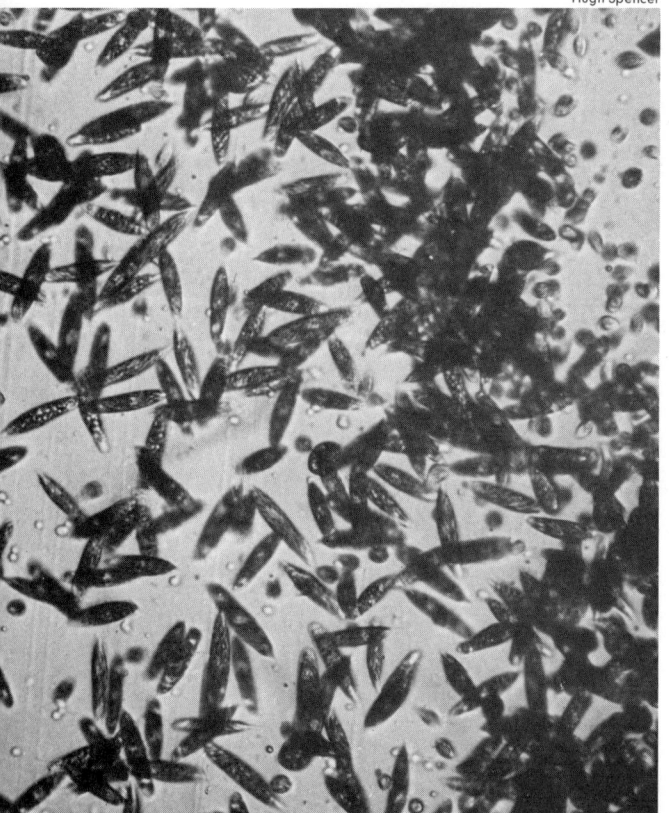

Euglena has a mixture of plant and animal features. They are motile, yet they possess chloroplasts.

Hugh Spencer

chiefly found in fresh water, though there are some marine representatives of the group. Certain forms have adapted to land life. They grow attached to moss, rocks, trees, and soil in places where the environment is not too dry. Occasionally these algae are found at high altitudes in patches of snow. Some species are one-celled. Others form colonies. Still others are multicellular, in the form of filaments (sometimes with numerous branches) or flat sheets. The green algae reproduce in various ways: by cell division, fragmentation (breaking off into fragments), or sexual reproduction. Certain species combine with fungi to form lichens.

The green algae have an important bearing upon our lives, both for good and for evil. In the course of their food-making activity they add oxygen to the water, thus making more of this essential gas available for fish and other organisms that form an important part of our food supply. They also serve as food for these creatures. On the debit side of the ledger, the green algae often cause water pollution in lakes, tanks, aquaria, and the like. They may impart unpleasant flavors and odors. If they grow too thickly, their respiration may seriously lower the oxygen content of the water. As a consequence, fish in the area may die of suffocation. Fortunately, it is possible to eradicate unwanted green algae from swimming pools and tanks by adding minute quantities of copper sulfate ($CuSO_4$) to the water; generally one part of $CuSO_4$ to several million parts of water suffices.

Among the most interesting of the green algae is *Chlamydomonas*. It is very common in ditches, pools, and lakes, and is often so abundant that the water appears to be green. It has also been found in the Alps and the Arctic, where it covers entire snowbanks. *Chlamydomonas* is a very primitive plant. Like so many other lowly forms of life, it can stand long periods of unfavorable environment.

Chlamydomonas often occurs as a single cell that moves about by means of flagella at the anterior, or front, end. The motion of this alga seems to be a pretty definite response to stimuli. For example, the plant moves toward moderate light and away from too much light.

Single individuals of *Chlamydomonas* sometimes withdraw their flagella and divide into two cells, which remain within a single cell wall. In many instances this cell division will go on until there are a number of cells, all included within the same cell wall. The arrangement is temporary, however. The individual cells ultimately break out of the wall, develop flagella, and become free swimmers.

Each flagellated cell then swims about until it comes in contact with a free-swimming cell from another colony. The two of them fuse and their flagella disappear. We now have a single cell called a *zygote*, representing the union of two reproductive cells called *gametes*. If the environment is

Green algae can reproduce sexually. This photo of *Oedogonium* shows the fertilized egg. The empty tube held the male gametes.

Grant Heilman

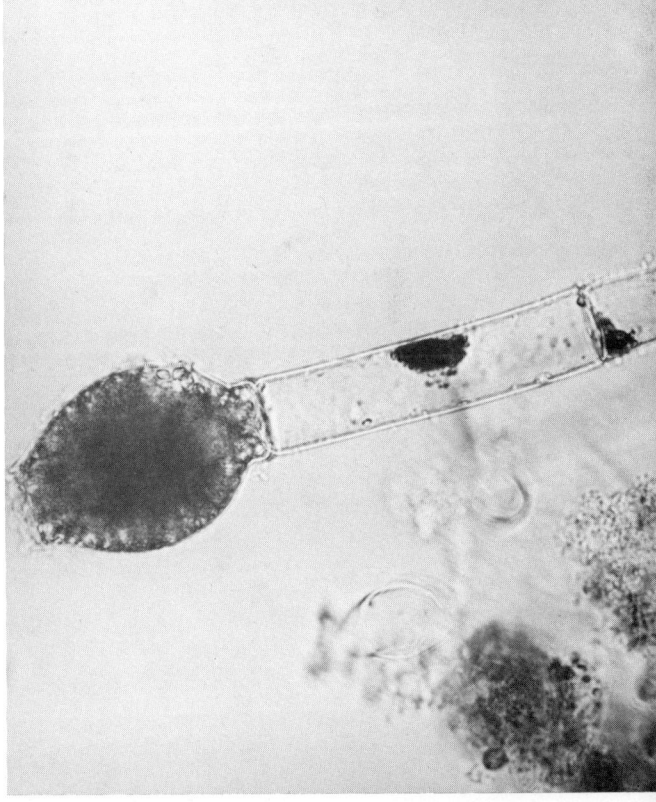

favorable—for example, following a spring shower—the zygote will divide and form four free-swimming cells.

Ulothrix is a green alga that thrives in fresh-water streams, particularly in places where the water does not get too warm. This plant forms colonies in which its cylindrical cells are set end to end in a green filament. At the bottom of each filament is a specialized cell called a *holdfast*. As the name indicates, this helps to anchor the plant to sticks and stones in the water. Sometimes the filaments are unable to attach themselves to anything, or else they fail to remain attached. In that case they are free floating.

Reproduction is often carried on by asexual reproduction bodies called *spores*. Under certain conditions the protoplasm of the cell divides and forms spores. These are later freed from the cell, and swim about in the water by means of flagella. After a time they become anchored to solid objects in the water and the flagella are lost. Each single-celled spore then divides and divides again, until a new filament is formed.

In *Ulothrix* we also find *alternation of generations,* in which one generation of plants produces spores and the following generation produces sex cells. One or more of the cells along a filament may produce gametes, or sex cells, in much the same way as spores are produced. After they are freed from the cell, the gametes swim about in the water. When a gamete comes into contact with a gamete from another colony, the two fuse and form a zygote. This later produces spores, each of which is capable of starting a new colony.

Ulva, or sea luttuce, reproduces in the same way as *Ulothrix.* It differs from the latter chiefly in the size of its colony. It

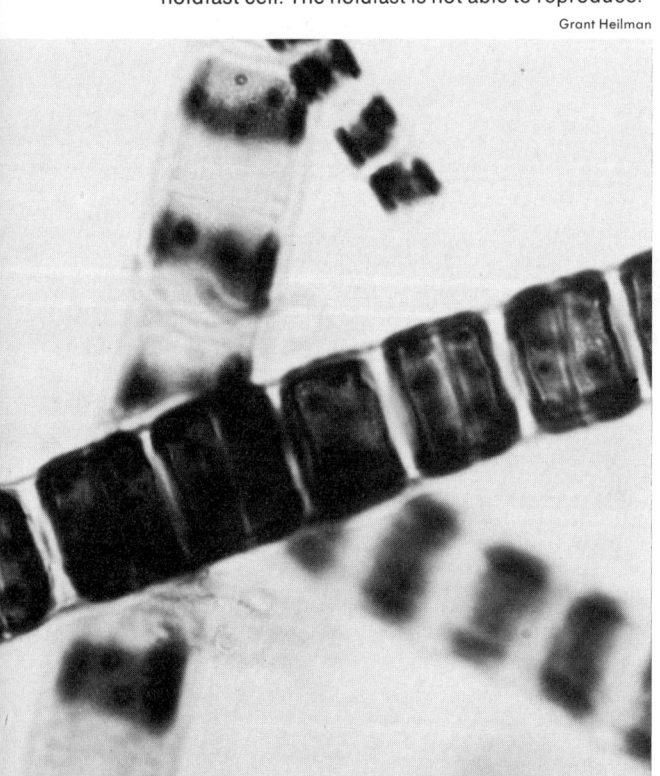

Ulothrix, a filamentous green alga. *Ulothrix* anchors itself to submerged objects by means of a holdfast cell. The holdfast is not able to reproduce.

Grant Heilman

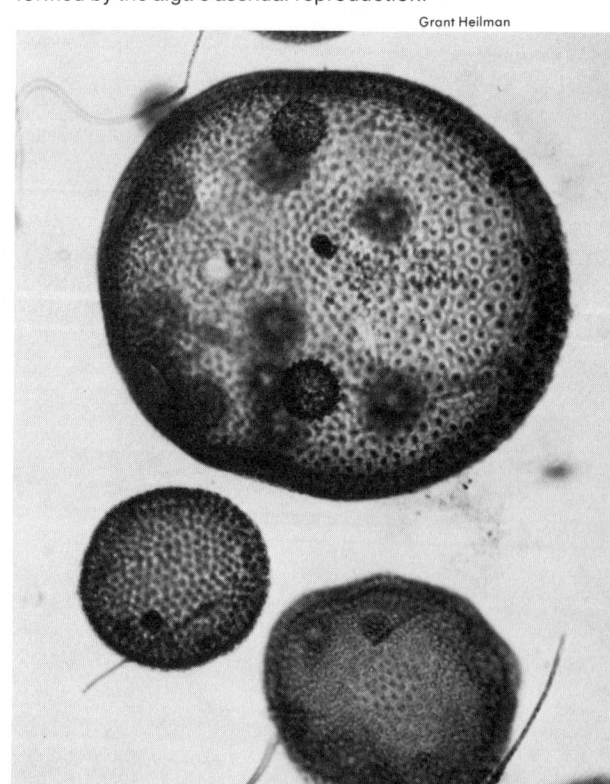

Volvox, a colonial green alga. The smaller spheres within the large colonies are daughter colonies formed by the alga's asexual reproduction.

Grant Heilman

forms a sheet two cells thick, several centimeters wide, and perhaps 30 centimeters or more in length. *Ulva* is usually found in salt water or on piling and rocks along the shore. It often forms part of the debris along the shore after a storm. It is used for food in the Far East.

Oedogonium is a fresh-water green alga that forms filaments. Some of these filaments are free floating. Others are attached to solid objects in the water. *Oedogonium* reproduces asexually or sexually. In sexual reproduction, certain cells produce large egg cells. Others give rise to lively little male cells, provided with many cilia. These male cells ultimately find their way to the egg cells, which they fertilize.

The green alga known as *Volvox* forms a curious sort of colony—a hollow sphere made up of thousands of cells arranged side by side. The tiny cells are provided with flagella, which project beyond the outer surface of the sphere like fuzz on a peach. As the flagella whip about in the water they give the whole colony a rather definite rotating movement. There are often many smaller colonies within the parent colony. They remain enclosed until the older colony ruptures. *Volvox* is a common alga of fresh-water ponds. It often serves as a host to microscopic animals called rotifers.

Protococcus, often also called *Pleurococcus,* forms striking green layers on bark on the shady side of trees. It is definitely a land plant. It is often called an aerial alga. *Protococcus* has marvelous powers of resistance. It can survive prolonged periods of drought and temperatures as low as $-40°$ Celsius. The cells are tiny, round, brilliant green objects, with one or two irregularly shaped chloroplasts (bodies containing chlorophyll). *Protococcus* is not equipped with flagella. It reproduces by asexual means. The daughter cells may remain attached to each other, forming colonies of two to four cells.

In *Cladophora,* the filaments are divided to form branches not unlike those found in some higher plants. *Cladophora* can be found in salt water and fresh water

Fucus, a brown alga. Members of this group are commonly called seaweed. The bloated sacs are reproductive bodies.

Bladder kelp, *Nereocystis leutkeana*, a brown alga.

alike at almost any time of the year. It seems to thrive under especially difficult conditions. It can be seen along rocky shores where it is constantly battered by the waves, or attached to boulders at the base of waterfalls, with the plant body extending downstream for a meter or more. It has been found at the bottom of lakes and under ice on rivers. The cells of *Cladophora* usually contain more than one nucleus.

The green scum seen on quiet pools, ponds, and lakes often consists largely of the green alga *Spirogyra*. This plant gets its name from the peculiar arrangement of its chlorophyll, which extends like a spirally twisted ribbon from one end of the cell to the other. *Spirogyra* is the genus of green algae commonly studied in elementary botany classes because the plant is easy to obtain and its cells are fairly large. It is used for food in some parts of the world, and it is a constituent of some lens paper.

Spirogyra is a filamentous plant, in which the cells are quite elongated. The colony is enlarged by cell division but the number of colonies is increased by a rudimentary form of sexual reproduction. When the filaments are lying close together, the adjacent cells develop protuberances. As these come into contact, the cell walls dissolve. The cell material from one cell (sometimes called the male cell) flows into the other cell (usually called the female cell). The two nuclei merge, and a zygote is formed. This secretes a thick wall and drops to the bottom of the pool. The zygote later develops into a new filament. Sometimes a *Spirogyra* cell produces a cell similar to a zygote without uniting with another cell. This zygotelike cell can produce a new filament in much the same way as a normal zygote.

The desmids are generally considered to be one-celled relatives of *Spirogyra*, because they reproduce in the same way. They are among the most beautiful of all microscopic plants. To the unaided eye they are only tiny green specks. But under the microscope they rival snowflakes in their infinite variety and their beauty of design. There are many different kinds of desmids. Perhaps the most striking of all is the jewellike *Micrasterias*.

Desmids are found in almost any sunny area in fresh water. Fine-meshed towing nets are often used to collect them. Also, it is often possible to obtain them by scraping the bottom mud and debris of almost any pool. Each desmid cell has a constriction in the middle, which makes it look at all times like a cell on the point of dividing. Some desmids are united to form chainlike colonies. Each cell of these colonies is sharply differentiated from all the others. Certain desmids can move, after a fashion.

An unusual green alga is *Vaucheria*, which is found in shallow water or in clumps of damp soil. This plant has no distinct cell walls. Many nuclei are scattered throughout the protoplasm, just as in the case of certain molds. It differs from the molds, however, in that it has chlorophyll. Like so many other green algae, *Vaucheria* reproduces both asexually and sexually.

BROWN ALGAE

The characteristic color of the brown algae, or Phaeophyta, is due to a brown pigment, fucoxanthin, which under normal conditions masks the green color of the chlorophyll that is present in the tissues.

Almost all brown algae are marine plants. They show considerable variety of structure. Some are in the form of filaments. Others are sheetlike or ribbonlike. Certain members of the group have structures resembling the leaves and stems of higher plants. They are generally attached by means of holdfasts to rocks and other solid objects in the water. Many of them have hollow structures that hold air and that help the upper part of the plant to float on or not far below the water's surface. Some brown algae are found only on the seashore, where they are alternately submerged and exposed to the air. These species are generally covered with a jellylike substance that holds water and prevents the plants from drying out while exposed to the air.

In various brown algae there is alternation of generations. At certain times of the year, parts of the plant develop spore-producing organs called *sporangia*. The spores that develop from these organs develop into plants that produce sex cells. These unite to form spore-producing plants. Other brown algae form spores that function as sex cells.

The brown algae are of considerable commercial importance to man. They are a source of food for fish and other animals living in the sea. When removed from the sea they are sometimes used as cattle feed. In the Far East and Europe they often

Laminaria (center), a kelp widely used as food, and bladder wrack, a species of *Fucus*.

Jeanne White, Audubon/PR

serve as food for human beings. They are eaten fresh or dried, or made into soups and broths. Some species yield iodine. Others make excellent fertilizer.

The brown algae known as kelps include the largest members of the group. The giant kelp, *Macrocystis pyrifera,* is reported to be the longest plant in the world. Thriving at ocean depths of 15 meters or more, it often grows to be many meters wide and well over 30 meters in length. It is harvested in California as a source of algin, a gelatinous material often used to give body to ice cream and other desserts.

Ribbon kelp, *Nereocystis leutkeana,* is often found along rocky shores and in shoals, where it is sometimes bothersome to small craft. The sea palm, *Postelsia palmaeformis,* is confined to the northern part of the Pacific coast. *Laminaria,* or the common kelp, is a comparatively small member of the group. It generally does not exceed 2.5 meters in length.

The brown algae that are alternately submerged and exposed to the air are called rockweeds. One of the best known is *Fucus.* Members of this genus are flat, much-branched plants that attach to rocks by means of holdfasts and are buoyed up by bladderlike structures. Because of these structures, the plants are sometimes called bladder wracks. *Fucus* is widely presented in elementary biology classes as a typical representative of the brown algae. It does not have alternation of generations. Gametes develop at the ends of the plant, and these gametes reproduce their kind.

Gulfweed, or *Sargassum,* is a brown alga provided with berrylike bladders. It has leaflike growths set on stemlike structures that sometimes reach great length. Masses of *Sargassum,* torn away from their moorings, are often carried along by ocean currents and collect in floating mats. They are found particularly in an area of the Atlantic Ocean called the Sargasso Sea. This area extends from the West Indies to the Azores. Columbus was the first to report the existence of this floating seaweed formation. It was widely believed at one time that ships were sometimes so enmeshed in the seaweed in the Sargasso Sea that they could not work their way free. But this is now held to be one of the innumerable tall yarns of the sea.

Robert C. Hermes, Audubon/PR

Sargassum, a seaweed. The white growths are colonies of wormlike bryozoans.

DIATOMS AND THEIR RELATIVES

These algae make up the phylum Chrysophyta. Their color ranges from yellowish green to yellowish brown. That is how the name Chrysophyta is derived: *chrysos* means "gold" in Greek. The best known members of this group are the minute plants known as diatoms, which belong to the class Bacillariophyceae.

Diatoms are generally one-celled organisms, though they are sometimes found in colonies. They are brownish in color because of the presence of the pigment diatomin, which resembles the pigment of the brown algae. The diatom cell wall contains silica. It is a glasslike covering, consisting of two sections that fit together like the top and bottom parts of a box. The cell walls are beautifully and delicately sculptured. Some are round. Others are oval or triangular. Still others are simple bars with hundreds of crosswise markings. When the diatoms die, the cell walls become skeletons which retain their shape for amazingly long periods of time. Some species of diatoms are able to move by swimming, gliding, or twisting. Diatoms generally reproduce by cell division. But in some species a simple form of sexual reproduction takes place.

These tiny plants are found almost everywhere where there is moisture and light: in the sea, in lakes and ponds, in flowing streams, in pools, on moist rocks, and in cultivated soil. They are sometimes found even in purified drinking water; fortunately, they are harmless to man. They often make up the bulk of the plankton, the passively floating or weakly swimming plant and animal life of the ocean. The diatoms thus form a sort of ocean pasture on which countless sea animals feed.

When diatoms die, their skeletons drop to the bottom of the sea or lake where they lived. The constant rain of these minute particles over countless centuries has resulted in the formation of large deposits of diatom-containing, or *diatomaceous earth,* up to a hundred or more meters in thickness. Diatomaceous earth is used in different ways. It serves to filter and clarify many liquids. It is an excellent insulating material for boilers, blast furnaces, and refrigerators. It is also used as a mild abrasive in metal polishes, scouring powders, and tooth pastes.

Hugh Spencer

Rockweed and various other marine algae exposed along a rocky shore at low tide.

RED ALGAE

The red algae, or Rhodophyta, derive their color from the red pigment phy-

ALGAE 41

coerythrin. They are found chiefly in ocean waters. Those growing in fresh water occur principally in cold, swiftly flowing streams.

Although called red algae, they actually exhibit a wide range of colors, including different shades of red, brown, and violet. Some red algae, such as genus *Porphyra,* are nearly black, as well, while others, such as genus *Bangia,* are almost without color. The striking coloration of many species results from the presence of the blue pigment, phycocyanin, in varying proportions, in addition to phycoerythrin.

These plants are multicellular. They occur in the form of filaments, ribbons, or sheets of fernlike or featherlike growths. Generally they range up to one third of a meter or so in length. Most marine species grow attached to solid objects in the water by means of holdfasts or special filaments. In practically all species there is alternation of generations. Neither the spores nor the sex cells are provided with flagella. The male sex cells float passively in the water until they come into contact with an egg cell, which they proceed to fertilize.

Like other algal groups, the red algae supply abundant food for fish and other animals living in the sea. They also serve as food for humans, particularly in Europe and the Far East. Among the edible varieties are Irish moss *(Chondrus crispus)* and laver (several species of the genus *Porphyra).* Irish moss is also used for curing leather and for shoe polish, as well as an ingredient in the manufacture of creams and shampoos.

Certain red algae, including Ceylon moss, *Gracilaria lichenoides,* yield a gelatinous material known as agar-agar. This substance absorbs a great deal of water. When it sets, it has a consistency like that of gelatin. It is used by researchers as growth material for bacteria. It also serves to thicken soups and broths, as a sizing material for textiles, as a mild laxative, and to provide body for puddings, pastries, ice creams, and other preparations.

Some of the red algae secrete lime and therefore have helped to build numerous coral reefs in the Indian Ocean and in various other parts of the world. Red algae have contributed to reef-building in the geological past, as well, probably dating back to Ordovician times.

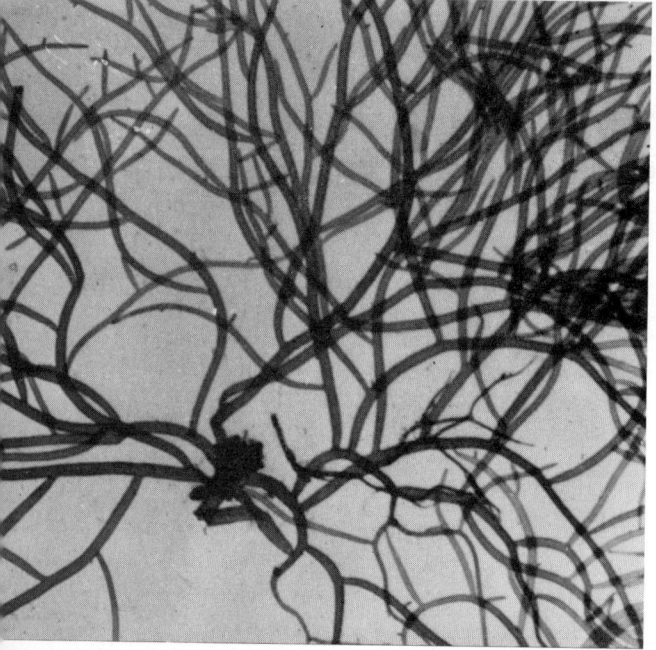

Red algae are commercially important. They are used in a variety of products. Top: *Chondrus crispus;* bottom: *Lomentoria.*

both photos, Hugh Spencer

Fungi come in many shapes and sizes. This one is a club fungus, *Polyporus versicolor*.

Grant Heilman

FUNGI

by Lindsay S. Olive

Mushrooms growing on a forest floor or displayed for sale in a supermarket. Yeasts used to make cakes rise or to brew rum and ginger beer. Molds that discolor bread and fruit. These are fungi with which everyone is familiar. Other fungi are less well known but just as important. A great many fungi are beneficial to mankind. The antibiotic penicillin, for example, derived from a small blue-green fungus called *Penicillium*, has saved countless lives and relieved much suffering. Other fungi are harmful to human beings and to our crops and other products. Hence the branch of botany that deals with the fungi is of particular interest. It is called *mycology*, from the Greek words *mykes*, meaning mushroom, and *logos*, meaning reason or science. Specialists in this field are known as mycologists.

The fungi are among the most primitive members of the plant kingdom. They are included among the thallophytes, or plants with a thallus. A *thallus* is a simple plant body that has no roots, stems, flowers, and seeds—structures we commonly associate with the higher plants. The thallus of a fungus is usually made up of branching threads called *hyphae*. The name *mycelium* is given to the sum total of the hyphae. The fluffy growth of a mold on bread is the part of the mycelium that extends above the surface of the bread. The hyphae that obtain nourishment for the mold are within the bread or upon its surface.

HOW FUNGI REPRODUCE

Most fungi reproduce in two different ways—sexually and asexually. The mycelium of many species produces both asexual spores and sexual structures. The asexual spores form a sort of dust made up of tiny particles. They are called *conidia* (singular: conidium), from the Greek word *konis*, meaning dust. When these spores germinate, they give rise to a mycelium similar to the one from which they themselves were derived. The conidia are commonly scattered far and wide by wind, rain, insects, birds, people, and other agents.

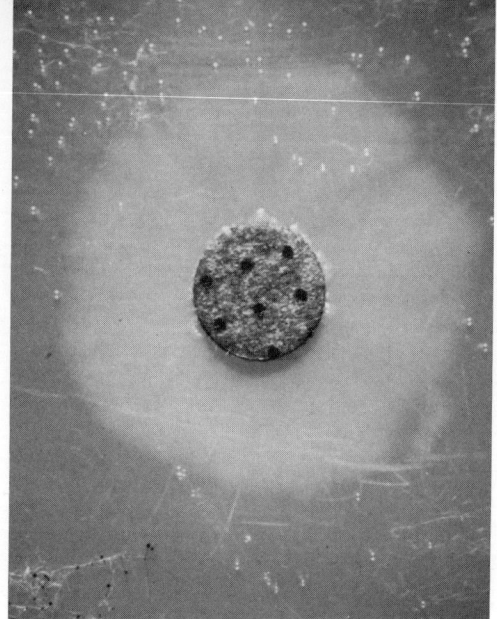

Penicillin (central disk), itself derived from a mold, can stop the growth of other molds.

Many conidia are so resistant that they can be blown for hundreds or even thousands of kilometers without losing their ability to germinate. This is one of the reasons why many species of fungi are so widely distributed over the earth.

Fungi show a great deal of variation in sexual reproduction. In all cases, however, the essential feature is the bringing together and fusing of two cell nuclei. In many of the higher fungi, the sexual spores are produced in conspicuous fruiting bodies, of which the familiar mushrooms

are a good example. The fruiting body is only a part of the complete plant. It arises from hyphae that are underground or in the wood of a decaying log.

FOODS AND ENZYMES

Algae and most higher plants contain a green pigment called chlorophyll. This substance enables the plants to manufacture food. Fungi lack chlorophyll. Thus they cannot make their own food, but are dependent on other living things for nourishment. Some fungi exist as *parasites,* growing on other living organisms and obtaining nourishment directly from them. Others feed on dead organic matter, such as decaying leaves or dead wood, on which they grow. Such fungi are called *saprophytes* ("plants feeding on rotting things").

True fungi are unable to take in solid foods as animals do. They must change such foods before they can use them. They do this by secreting enzymes—substances that speed up chemical transformations—into their surroundings. The enzymes break down foods consisting of complex organic materials into simpler substances that can be dissolved in water. When these substances are in dissolved form, the fungi can absorb them.

Some species of fungi are very restricted in their choice of foods—that is, in the organic materials upon which their enzymes can act. One small fungus group, for example, is known to occur only on the castoff horns and hooves of animals. Other species produce many different kinds of enzymes, which enable them to grow on a wide variety of substances.

The enzyme-forming activities of fungi are a vital factor in the process of *decay,* a process upon which the very existence of life on earth depends. These activities convert the complex organic matter present in the dead bodies of plants and animals into simpler substances that plants can absorb. In the plants, these substances are used in the manufacture of food elements. Animals obtain these substances by eating plants or by devouring plant-eating animals.

Human beings put the enzyme-forming activities of certain fungus species to

A green mold is growing on these oranges. Green molds are members of the class Ascomycetes.

work in brewing, baking, and cheese-making. The enzymes of other species bring about the decay of lumber and textiles and cause numerous diseases of animals and plants.

A great many trees, shrubs, and other seed plants have a remarkable relationship with certain fungi. The mycelium of these fungi invades the roots of the plants, but instead of harming the seed plants it helps them. It assists in transporting water and minerals from the soil to the roots of the plants. The seed plant, in turn, supplies food to the fungi. Fungi of this kind are called *mycorrhizal* (fungus-root). The relationship between root and fungus in this case is a good example of *symbiosis*—an association beneficial to both partners.

EVOLUTION OF FUNGI

No one is sure how the fungi originated. Since certain lower forms resemble the green algae in some respects, especially in form and mode of reproduction, some mycologists believe that the fungi arose from the algae. They argue that the fungi became parasites or saprophytes as they lost their chlorophyll and the ability to manufacture their own food.

Other botanists hold a different view. They point out that certain species of the lower fungi are animal-like in many ways. They believe, therefore, that the fungi evolved from the protozoans, the most primitive group of animals. To have a clearer idea of the origin and evolution of the fungi, we shall have to learn more about the species already known to us. Perhaps the discovery of new species will also help fill the gaps that now exist in our knowledge.

Nearly 50,000 species of fungi have been described. Some authorities estimate that this figure represents less than one-half the total number likely to be in existence. It is probable, therefore, that we shall discover many new and interesting species in the course of time. Large parts of the earth, including Africa, much of South America, the South Pacific, and Asia, remain comparatively unexplored as far as fungi are concerned.

The following discussion of the different kinds of fungi begins with the slime molds, which are recognized as the most primitive and animal-like. You will note that the scientific names for the different classes of fungi all end in *mycetes,* a combining form meaning "fungi." The first part of the name indicates the kind of fungus.

SLIME MOLDS

Most slime molds are classified in the class Myxomycetes (*myxa* means "slime" in Greek). They are commonly found on rotting logs, old stumps, and decaying humus.

When the spore of a slime mold germi-

Slime molds are primitive types of fungi. The stalks in the photograph are the sporangia, or spore cases. In the lower photo, the trails are really streams of myxamebas, or plasmodia.

both photos, Hugh Spencer, Audubon/PR

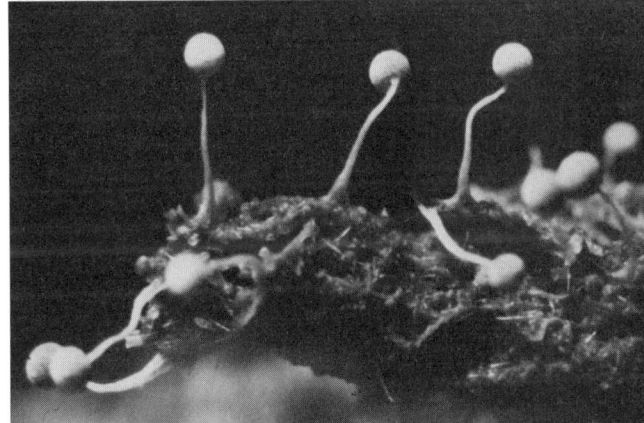

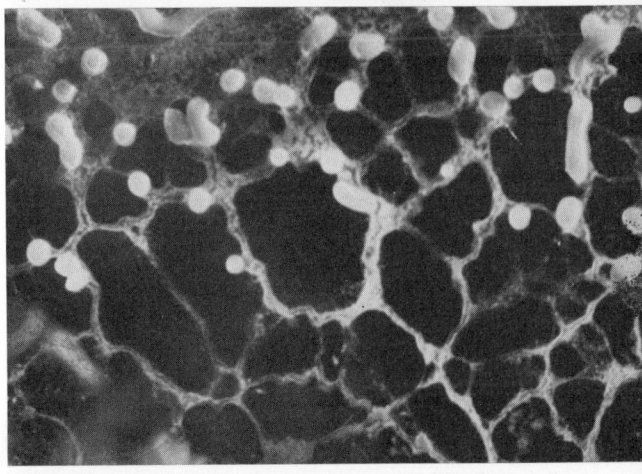

FUNGI 45

nates, it produces a cell that propels itself through the water by means of one or two whiplike projections called *flagella*. The flagella soon disappear and the cell begins creeping around like an ameba, a primitive one-celled animal. These cells are called *myxamebas*. Like the ameba, the slime mold changes its shape constantly and takes in solid particles.

When a slime-mold cell comes in contact with another cell of the same species, the two may fuse. The product of this fusion gradually enlarges while creeping about and feeding on bacteria and fungus spores. This feeding stage is called the *plasmodium*. It may grow to a comparatively large size on decaying logs or humus. Finally, it flows out onto an exposed surface. Here it appears as a soft thin layer or a hemispherical mass. It is a mass of protoplasm containing hundreds of nuclei but no cell walls. It varies in color—white, yellow, orange, red—according to the species.

The adult plasmodium now produces upright extensions. These develop into small stalked *sporangia,* or spore cases, containing numerous spores. In some species the entire plasmodial mass may develop into a single fruiting body.

In some types of slime molds, known as cellular slime molds and net slime molds, the adult form is really multicellular with a *pseudoplasmodium* consisting of many separate cells.

Slime molds have been used in many biological studies. The unicellular plasmodium is an excellent source of pure protoplasm for laboratory study.

ALGALIKE FUNGI

These fungi are the Phycomycetes, or "seaweed fungi." They include both saprophytes and parasites. Many species cause diseases of crop plants and of plants that are cultivated for decorative purposes. They are often serious pests.

Downy mildews. These fungi are so named because they produce a downy growth on the surface of infected plant parts, usually the leaves. The infection cells are blown about by the wind and cause other plants to be infected.

One of the best known downy mildews is the late blight of potatoes. When the white potato was first carried to Europe from South America many years ago, it was a botanical curiosity. Later, it became a staple crop throughout the European continent and the British Isles, especially in Ireland. In 1845, a devastating blight struck the potato plants. The potatoes, whether harvested or left in the field, rotted in large numbers. The famine that resulted left about half a million people dead in Ireland and brought about the migration of nearly two million Irish, principally to North America. It was with some difficulty that mycologists convinced the authorities that the disease was caused by a parasitic fungus—a downy mildew.

About 1865, another downy mildew was accidentally introduced from North America into Europe. Grapevines in France had been attacked by the root louse. It was hoped that this pest might be foiled if the French grape were grafted onto the American grape rootstock, since the latter was resistant to the insect. Unfortunately, the downy mildew of grape was brought into France together with the rootstock. The French grape proved highly susceptible to the fungus and the disastrous epidemic that resulted nearly ruined the French wine industry.

The discovery of the cure for grape downy mildew was as much of an accident as was the introduction of the disease. One day, the 19th-century French mycologist Alexis Millardet was walking in the country near Bordeaux when he made a significant observation. He noticed that certain vines, sprayed with a white substance in order to make the grapes unpalatable to marauders, were much healthier than unsprayed vines nearby. Millardet learned that the spray consisted of copper sulfate and lime. Obviously, it had held the downy mildew of grape in check and was an effective *fungicide,* or fungus killer. Under the name of "Bordeaux mixture," this spray was used extensively in the years that followed to protect plants from various kinds of fungi. It is still considered to be one of the most useful fungicides.

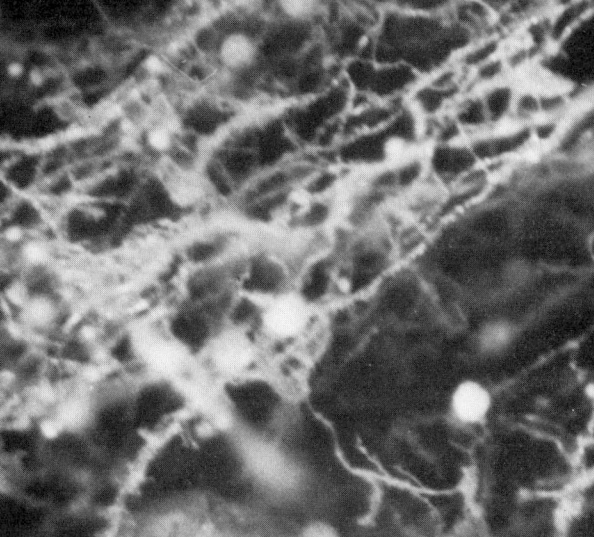

Left: 10-days growth of black bread mold on a slice of bread. Right: close-up of the mold, showing sporangia.

Both photos, Grant Heilman

Downy mildews also cause other plant diseases. They attack tobacco, cabbage, cucumbers, squash, and various other plants.

Black bread-mold group. The molds of this group are generally called *mucors*. There are many species. Most of them occur as fluffy growths on bread, fruits, vegetables, and preserved foods. They are also found in the soil and in decaying vegetation. They are white at first, but they soon become dark as their spores mature in large numbers.

Black bread mold is probably the best known example of these fungi. It develops on stale bread—not so frequently as in the past, however, because of the use of preservatives in bread. Great quantities of tiny black bread-mold spores are produced in spherical cases called *sporangia*. Any one of these spores, when carried by wind or air currents to a suitable substance, can germinate. It then produces a fluffy mycelium similar to the one from which it came. This same fungus is destructive to market vegetables and fruits, and causes a rot of sweet potatoes.

SAC FUNGI

These fungi make up the class Ascomycetes. "Asco" comes from the Greek *askos,* meaning "sac" or "bladder." All the plants belonging to this group have a number of *asci* (singular: ascus), or spore sacs. These are a product of sexual reproduction. In the higher sac fungi, such as the cup fungi, the asci are produced in various kinds of fruiting bodies.

Yeasts. The life cycle of a yeast is relatively simple. The typical thallus consists of a single small cell that reproduces by budding. The budding mass of cells that develops on the surface of a culture plate usually appears as a soft, creamy-white colony. Sexual reproduction also occurs in many species. It begins with the fusion of two similar cells. The end result is the produc-

Saccharomyces, a yeast. This photo of live cells shows several budding.

Grant Heilman

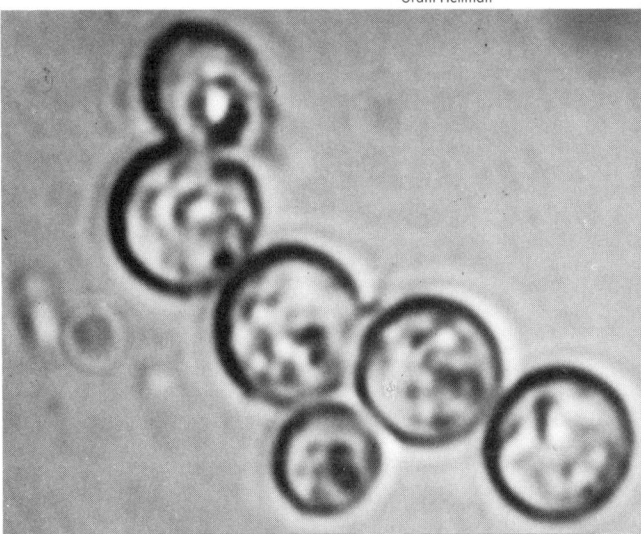

FUNGI 47

tion of several spores (usually four) within the characteristic sac, or ascus. These spores are very resistant.

There are many kinds of yeasts. They commonly occur in nature wherever simple sugars are available. They are especially common on fruits, in the sap from trees, and in soil, particularly in orchards and vineyards. The best-known are baker's and brewer's yeasts, both belonging to the species *Saccharomyces cerevisiae*.

Brewer's yeast has been bred and selected for high alcoholic yield. When the yeast is placed in a sugar solution and deprived of free oxygen, it carries on a special type of respiration called *alcoholic fermentation*. During this process, carbon-dioxide gas is given off and alcohol is produced. Brewer's yeast is the principal organism used in making beer, liquors, and commercial alcohol. A different species is used in wine making.

When baker's yeast is placed in bread dough, it causes the dough to "rise," or expand, by producing carbon-dioxide gas. Upon being heated, the gas pockets in the dough expand still more, making the final baked product light and porous. The alcohol produced is driven off by the heat.

Not all yeasts are beneficial. A few species cause serious diseases of man. *Candida albicans* is responsible for thrush, a disease of the mouth and throat. Other yeasts produce respiratory infections.

Blue and green molds. These are among the most common and widespread of all fungi. They occur on many different kinds of organic substances, including foods and textiles. In humid areas, they are frequently responsible for the mildewing of clothing. Although some blue and green molds reproduce sexually, the majority lack the sexual stage. They reproduce only by means of asexual spores, or conidia. These are produced in chains on upright branches of the mycelium. The large numbers of mature spores give these fungi their characteristic colors—greenish, blue-green or blue, and sometimes yellow, tan, or black. They differ from the black bread-mold group in that the mycelium is not fluffy and does not spread so rapidly.

A well-known genus is *Aspergillus*. Another is *Penicillium,* from which the antibiotic penicillin is derived. *Penicillium* has been put to various other uses by man. Roquefort cheese is made with one species of *Penicillium;* Camembert cheese, with another. The blue color of Roquefort cheese is due to the presence of great numbers of spores. A species of *Aspergillus* is the primary source of citric acid, widely used in flavoring candies and fruits. Various other species of *Aspergillus* and certain *Penicillium* species also produce this acid.

Powdery mildews. As the name of these fungi indicates, they give a powdery white appearance to the infected parts of plants, usually the leaves. This whiteness is caused by the presence of large numbers of asexual spores (conidia), produced in chains and distributed mainly by wind. These spores develop throughout the summer and cause the host plants to be reinfected during this period. In late summer and fall, the sexual stage appears in the form of little brown dots over the infected areas. These structures contain the ascospores, or sexual spores, which survive during the winter and cause new infections in the spring. The powdery mildews are plant parasites. They cause diseases of crop plants and ornamentals, including grape, apple, rose, and lilac plants.

Sphere fungi. This is the largest group of sac fungi. The fruiting bodies that result from sexual reproduction are small and spherical. They have necklike extensions through which the spores escape. Asexual spores (conidia) are also commonly produced in this group. Some of the species are parasitic, while others live on dead organic matter. Among the parasitic varieties is the chestnut-blight organism. Introduced into New York from Asia about 1900, it has completely destroyed the American chestnut. Other species of the sphere fungi cause Dutch elm disease, oak wilt, apple scab, and ergot of cereals.

One of the most useful species from the standpoint of basic science has been a pink bread mold known as *Neurospora*. A great deal of research has been done on the heredity of this organism. These studies have contributed greatly to our present understanding of genes, the hereditary units of all organisms.

Cup fungi. This is another large group of sac fungi. Its members are called cup fungi because the fruiting bodies are generally cup-shaped. They vary greatly in size from tiny, barely visible structures to conspicuous and sometimes brightly colored fleshy cups or saucer-shaped bodies. The larger ones are often edible. Some related forms, such as the morels, are not cup-shaped but have stalks and conical, pitted caps. Most mushroom-gatherers consider them a delicacy.

The majority of cup fungi live on dead wood, humus, and soil. Some of them cause destructive diseases. The brown-rot fungi, for example, attack peaches and other stone fruits, as well as apples and pears.

An odd group of fungi frequently classified with the cup fungi are the *truffles*. Their fruiting bodies are usually spherical rather than cup-shaped and what is particularly notable, they occur underground. They are mycorrhizal. As we pointed out, this means they are associated with the roots of higher plants in such a way that the relationship is beneficial to both organisms.

Truffles are considered a prime delicacy. In Europe, they are sought out by means of specially trained pigs and dogs, which can locate them by smell. Most truffles gathered for the market are two to five centimeters in diameter and show a dark, roughened exterior. When cut open, they have a distinctive aromatic odor. Their flavor is imparted to foods that are seasoned with them.

CLUB FUNGI

The club fungi, or Basidiomycetes, are the most highly developed group in the evolution of the fungi. The first part of the scientific name is derived from the characteristic spore-producing organ, which is called the basidium and is club-shaped.

Rusts and smuts. These are parasitic groups belonging to the lower orders of club fungi. The rusts are so called because one of their spore stages has a bright rust-brown color. There are a large number of

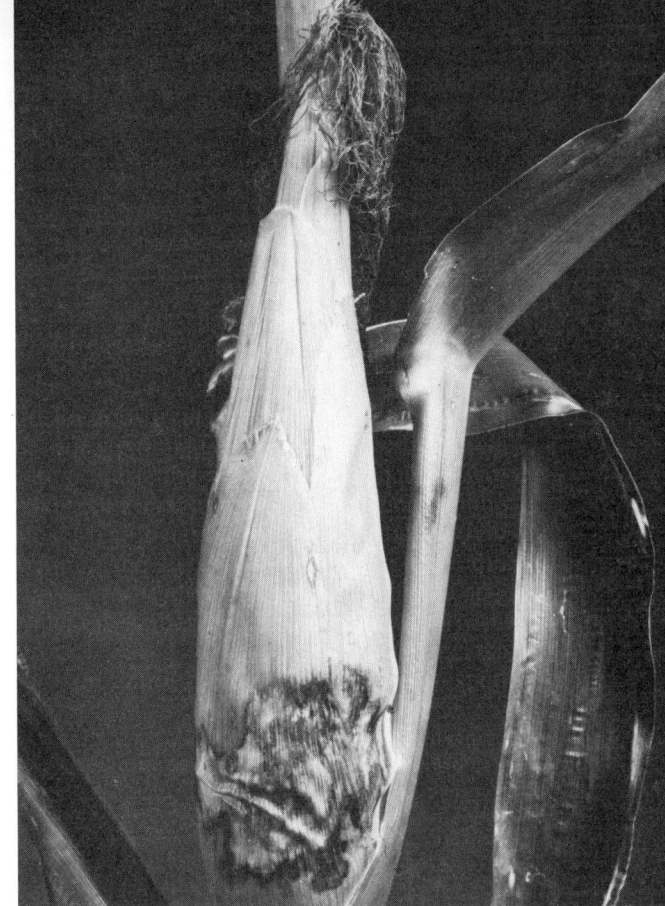

both photos, Grant Heilman

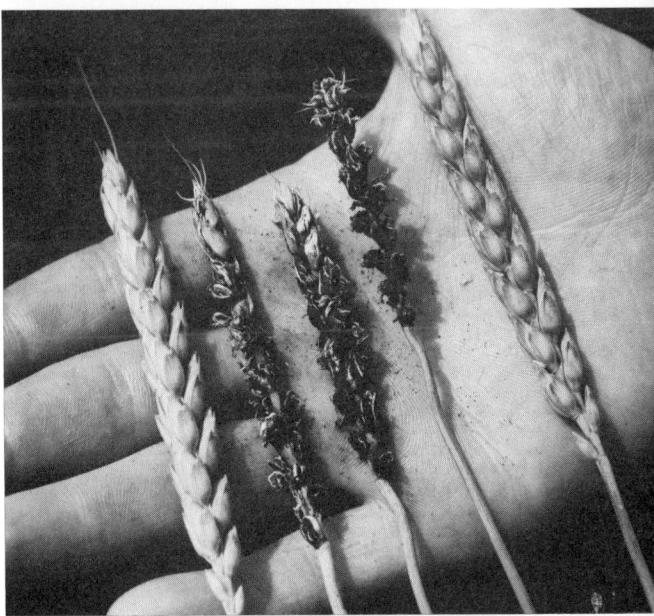

Fungi can cause serious damage to crops. Top: Southern corn leaf blight, caused by *Helminthosporium maydis*. Bottom: brown loose smut in wheat. Rusts, blights, and smuts belong to class of club fungi, the most advanced among the fungi.

FUNGI 49

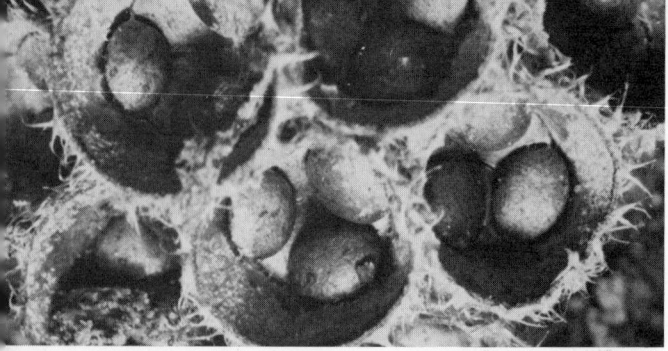

Club fungi vary greatly in appearance. From top to bottom: bird's-nest fungi; stinkhorn mushroom; coral fungus; puffballs.

species and they attack a wide variety of higher plants, many of which are valuable to mankind. One of the most destructive of the rusts, from our viewpoint, is the wheat-barberry rust. In order to complete its life cycle, this fungus must alternate between two host plants. It produces certain spore forms on wheat and different forms on barberry. Many rusts have this complicated life cycle. They cause much damage to cereals, vegetables, fruit trees, and some forest trees. Careful preventive measures help keep them in check.

The smuts have received their name because of the black masses of spores they produce on infected plant parts. Their life cycle is much simpler than that of the rusts; they complete their development on a single host plant. Smut spores are dry and easily blown about from one plant to another, or from one field to the next. They are quite resistant; those of some species pass the winter on the ground. One well-known smut fungus is corn smut, which turns corn kernels into tumorlike structures full of black spores. Other important smuts attack such plants as wheat, oats, barley, and rye.

Gill fungi. As we shall see, these highly evolved forms are often referred to as mushrooms. They are among the commonest and most conspicuous of all fungi, occurring abundantly in woods and meadows. The vegetative portion of a gill fungus—the mycelium—occurs in the ground and lives on organic wastes. At certain places in the mycelium swellings are produced. These develop into small spore cases called *sporophores*. In some of the more common species, these are enclosed in wrappers called *volvae*.

Under proper conditions of warmth and moisture, a sporophore bursts through the earth, rupturing the volva, and expands in umbrella fashion. The fully developed fruiting body consists of a stalk and cap. Under the cap are attached a number of thin gill-like plates, called *lamellae,* which radiate from the center. It is on these lamellae, or "gills," that the spores are produced. Gill fungi are extremely varied in size and color. Many are among the brightest and most attractive of the fungi.

Coral fungi. The fruiting bodies of these plants are much branched and sometimes strikingly resemble coral in appearance. They are creamy, tan, yellow, brown, or even purplish in color and fleshy in texture. Coral fungi can be found most often in temperate areas in woods made up of trees that lose their leaves every autumn. Most are edible.

Bracket or leather fungi. Their fruiting bodies often have a leathery texture. They are commonly found on the sides of dead trees and logs. They grow in shelflike fashion, one shelf rising above another. Bracket fungi are an important factor in wood decay.

Pore fungi. These plants are also important as wood-decay organisms. The spores are produced in pores on the underside of the fruiting bodies (sporophores). These bodies vary in shape. Some are thin layers. Others resemble gill fungi or bracket fungi. Pore fungi may cause great damage to lumber and to wooden structures; certain species attack living trees.

Puffballs. Most of these forms occur as saprophytes on the ground, on humus, or on rotting wood. The fruiting body of a typical puffball such as *Lycoperdon,* ranges in shape from spherical to pearlike. It is whitish at first and turns brown at maturity. The millions of tiny spores inside the mature fruiting body are expelled through a small opening at the top by pressure from wind, raindrops, or animals. Most puffballs are several centimeters high, but the giant puffball may be more than a meter in diameter. There are many edible forms.

Stinkhorns. These fungi, which are related to the puffballs, are well named for their carrionlike odor. Most stinkhorns have a stalk, usually with a cup at the base, and a cap covered with a dark, slimy mass of spores. It is this spore mass that causes the unpleasant odor.

Bird's-nest fungi. They are also related to the puffballs. They produce small cups, containing several tiny bodies in which the spores are borne. The entire cup structure resembles a miniature bird's nest with several tiny eggs. These fungi may be found on humus, dead wood, and the dung of herbivorous animals.

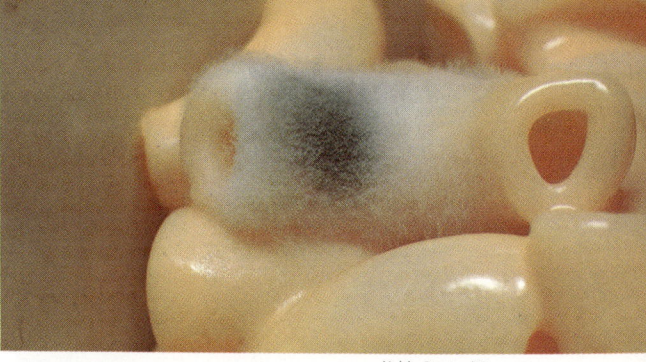
Noble Proctor/PR

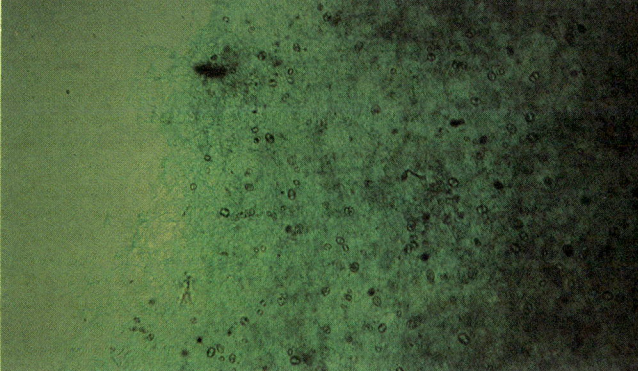
Russ Kinne/PR

Penicillium is an economically important mold. Top, the mold as it appears spoiling food. Lower photo: detail showing the structure of the mold.

Mushrooms and toadstools. The name "mushroom" refers to various fungi that have fleshy fruiting bodies and that grow on the ground, dead wood, or trees. The name is used most often in connection with the gill fungi described above. However, it is also applied to various other club fungi and even to certain sac fungi, such as the morels. Some persons reserve the name "mushrooms" for the edible species of these fungi; they call the poisonous species toadstools. Most students of the fungi, however, speak of poisonous and edible mushrooms.

Contrary to the belief of those unfamiliar with mushrooms, the great majority are harmless or edible. Relatively few are dangerous; some are mildly poisonous, while others are unpleasant tasting or have little flavor. However, since some deadly species are widespread, amateur mushroom hunters should be careful in choosing specimens for the dinner table. They should obtain a mushroom guide, of which many have been published.

After identifying a few common edible species, the mushroom picker should gath-

Top: the poisonous mushroom, *Amanita muscaria,* or fly mushroom, in several developmental stages. Bottom: A related mushroom, Caesar's mushroom.

er these and avoid all others. Only when a variety can positively be identified as edible should it be picked. Unfortunately, there are no other foolproof methods in distinguishing edible from unwholesome mushrooms.

The safest and certainly one of the most palatable is the common edible mushroom, *Agaricus campestris,* which may be bought at the grocery store. Its wild variety is common in nature, especially in meadows and lawns. Other wild edible mushrooms that are easy to identify are the morels, the coral fungi, and the oyster mushroom, which forms whitish fruiting bodies with stubby stalks on dead logs. The puffballs should also be included. One of the most delicious of all mushrooms is the bright orange-yellow Caesar's amanita, *Amanita caesaria,* which was a favorite of the Roman emperor Nero.

Unfortunately, other species of Amanita are among the most poisonous varieties. The fly amanita, *Amanita muscaria,* is quite similar in color to *Amanita caesaria*. It has a bright orange-red cap that is covered with whitish flecks. Like most species of *Amanita,* it has a veil around the stem, just below the cap. The base of the stalk is enlarged

Reproduction in higher fungi. The spore is released. It finds a suitable spot for growth, and then it germinates, grows, and reproduces.

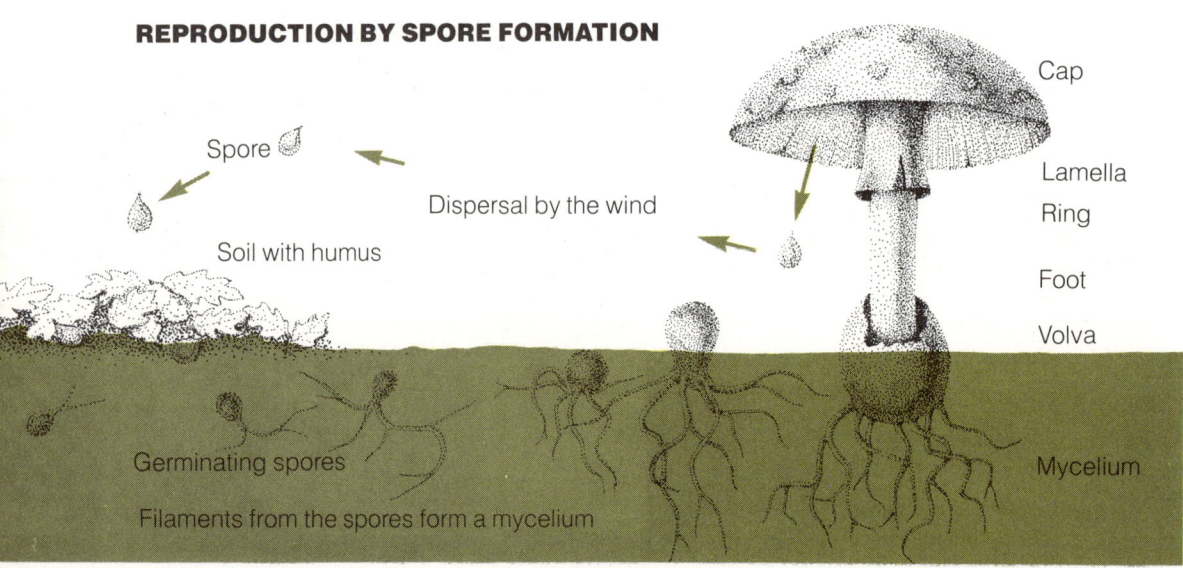

52 FUNGI

and bulbous. The deadliest of all mushrooms is the destroying angel, *Amanita phalloides,* which ranges in color from white to grayish. It has a veil and also a prominent cup at the base of the stalk. Although it is easy to recognize, it is probably the commonest cause of serious mushroom poisoning in North America. A single forkful is sometimes sufficient to cause death.

Symptoms of poisoning by mushrooms are often delayed for hours after the fungi have been eaten. If there is any suspicion of mushroom poisoning, a doctor should be sent for immediately. While waiting for the doctor, any effective means of emptying the stomach of its contents should be used.

It is interesting to note that certain species of mushrooms may cause nausea in some persons, usually without serious results, while they may be eaten by others with no ill effects whatsoever. It is possible that a food allergy is involved here.

Mushrooms may be commercially grown in caves, abandoned mines, dark cellars, and similar places. In the United States, most commercial growing is done in specially constructed houses in which the temperature and humidity can be controlled. The mushroom beds are generally made by spreading a thin layer of loam over a deep layer of well-rotted manure. After fermentation has taken place, the mushroom "spawn" is planted. This consists of the mycelium of the fungus, encased in bricks made of manure and loam. The mushrooms are harvested while firm and before the spore dust has started falling.

IMPERFECT FUNGI

These fungi form the class Deuteromycetes. Their name refers to the fact that the sexual stage is lacking. Reproduction is by means of asexual spores (conidia) only. Most imperfect fungi seem to have evolved from the sac fungi through the loss of the ability to reproduce sexually.

The imperfect fungi form a very large and important group. They occur widely, either as saprophytes on many different kinds of organic matter or as parasites of higher plants. Some cause mildewing of fabrics; others, molds of foods. The parasitic species are often very destructive to crop and ornamental plants. Most fungi that cause disease in human beings belong in the group of the imperfect fungi. They are responsible for the ringworm diseases.

LICHENS

A lichen is a combination of a fungus and an alga. Most often, the fungus is a member of the sac fungi, while the alga is a green or blue-green species. The thallus, or plant body, that results from the combination of the two is quite different from anything that could be produced by the fungus or the alga growing alone. The hyphae of the fungus make up the framework of the plant body. The algal cells or filaments occur within this framework.

The two organisms—fungus and alga—live together in a mutually beneficial relationship or symbiosis. By the process of photosynthesis, the alga manufactures all the organic food that they both require. The fungus brings in water and minerals and offers protection to the alga.

Lichens are most commonly found on trees. They also occur on rocks and on barren ground. They furnish fodder for reindeer and, in some places, for cattle. Certain kinds of lichens yield litmus, a dye used as a chemical indicator.

Lichens are made up of an alga and a fungus. The two very different organisms live in symbiosis.

C. Nuridsany

MOSSES AND FERNS

by Thomas Gordon Lawrence

As you walk beside a forest stream, you probably notice two kinds of plants growing in abundance: mosses and ferns. Screened from the strong sun by the tall trees and near a plentiful water supply, which most species need or prefer, the plants form a lush and beautiful covering on the rocks and soil.

Mosses and ferns, as well as all higher plants, belong to the subkingdom Embryophyta. This group, which is divided into several phyla, includes some 300,000 species of plants.

All mosses and ferns contain chlorophyll and thus can make their own food by the process of photosynthesis. All undergo *alternation of generations* in their life cycle, in which a spore-producing, or asexual, generation alternates with a gamete-producing, or sexual, generation.

The maidenhair fern, *Adiantum pedatum,* is a species common to North America.

Documents C. Nuridsany

MOSSES

A delicate, intensely green carpeting of moss is nearly always to be found in woodlands, along streams, and in many meadows. Even in well kept lawns there frequently are glistening moss colonies, more or less shaded by grass blades. Moss often thrives in the cracks of cement walks and old stumps, damp fences, and the bark of old trees. It sometimes grows on thatched roofs, which look as if they were covered with green velvet.

Mosses belong to the phylum Bryophyta, which is derived from Greek words meaning "moss plants." This phylum, which contains about 25,000 species, also includes the liverworts and hornworts.

REPRODUCTION IN MOSSES

The moss plants with which we are most familiar develop from spores. Each spore gives rise to a green, branching filament called a *protonema,* which looks much like a green alga. After a time, the protonema produces buds. Each bud grows into an erect shoot, with leaflike structures branching off from it. Instead of roots, the shoot has small threadlike bodies called *rhizoids,* which draw water and mineral salts from the soil. The leaflike structures of the plant contain chlorophyll and are capable of photosynthesis.

In the course of time, sex organs develop at the tips of the shoots. The male sex organs, which produce sperm, are called *antheridia* (singular: antheridium). They are spherical or egg-shaped. The flask-shaped female organs are known as *archegonia* (singular: archegonium). A single egg is formed at the base of each female organ. In time a canal extends from the base to the open mouth of the organ. In some mosses the antheridia and archegonia are produced at the tip of the same shoot. In others, they are produced on separate plants that are called *gametophytes,* or "sex-cell plants."

Fertilization can take place only in water, which may be in the form of rain, running water, or even a thin film of dew. As the antheridium becomes thoroughly wet, the sperm are released. They swim through the water by means of long whiplike structures called flagella. In due time, the sperm make their way to the archegonium. Several sperm may go down the canal to the egg at the base of the female organ. But only one of the sperm fertilizes the egg.

Once the egg is fertilized, an entirely different generation is launched. A new plant, called a *sporophyte,* which means "spore plant," grows out of the tip of the old plant. This sporophyte consists of a foot, the base of which is implanted into the shoot of the gametophyte, and a long, thin stalk with a capsule, containing spores, at the end. It receives its nourishment from the old plant. When the capsule is mature, it splits open. As the plant sways in the wind or is shaken by animals, spores are thrown from the capsule like salt from a shaker.

Each spore gives rise to a new gametophyte and the cycle begins anew. Thus there is what the botanist calls alternation of generations: from gametophyte to sporophyte to gametophyte again, and so on through countless generations of mosses.

MANY DIFFERENT KINDS

There are many different kinds of mosses. They are small plants, never standing more than several centimeters high. Some are so low and hug the ground so closely that they look like green turf. Some genera, such as plume moss *(Hypnum)* and fern moss *(Thuidium),* which looks somewhat like a midget fern, are more conspicuous. They thrive in shady woods, often completely concealing the logs and stones on which they grow.

Perhaps the most beautiful of all the mosses are *Bryum* and *Mnium,* which have broad, thin, translucent leaves. More than any of the others, these are the mosses that sparkle like gems when a shaft of sunlight

both, C. Nuridsany

Two species of moss. The top photo shows a moss with hanging spore cases. The bottom photo shows the capsules erect.

A dense growth of sphagnum moss forming a thick carpet in marshlands.

finds its way to the ground through the forest canopy.

A few mosses are dark brown or black. The pincushion moss (Leucobryum), on the other hand, is whitish green; its cushionlike masses look ghostly at the base of trees in damp woods. A moss that makes a satisfactory inhabitant for a home aquarium is the water moss (Fontinalis), which lives entirely submerged. In an aquarium it soon forms a dense underwater jungle where young guppies can escape the eager jaws of their parents and other fish.

Some mosses grow in caves where the light is only $1/500$ as bright as full sunlight. Luminous moss (Schistostega) is found in dark places in woods. It reflects light rays from its lenslike cells. The greenish-golden glow of this curious plant may have been responsible for fairy tales concerning goblin gold.

PEAT MOSS

The peat mosses make up the genus Sphagnum. They are quite different from other mosses. For one thing, peat mosses are far larger. They grow in boggy places, especially in temperate and subarctic regions. A peat moss leaf is generally only one cell thick and contains many dead cells. These absorb great quantities of water and give a pallid appearance to the plant.

The bogs in which peat mosses flourish are called peat bogs. In addition to peat mosses, these bogs often contain other mosses, as well as reeds, sedges, and rushes. The tannins and other organic acids and the salts that are found in abundance in peat bogs serve as natural preservatives. Consequently, the remains of animals and trees buried in bogs for hundreds of years have been dug up in excellent condition. The peat bogs of Ireland are perhaps the most famous. There are also extensive formations in Great Britain, the Soviet Union, Finland, Sweden, Canada, and the United States.

Florists use dried peat moss as packing for flowers, since the leaves absorb and hold moisture. Peat moss that has been ground up is often added to the soil of gardens. Not only does it hold water effectively but it prevents the soil from caking. Sterilized peat moss has been used for surgical dressings.

Dead plants in a peat bog sink to the bottom. As one layer piles up on another, the lower layers are gradually compressed and form the substance called *peat*. The element carbon makes up about 60 per cent of the peat—a much greater proportion than it was in the living plants. Peat is really

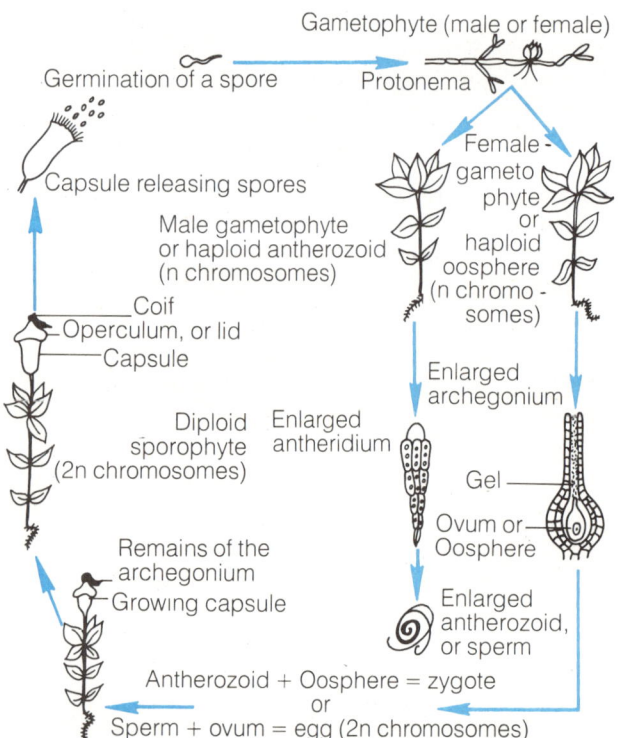

The life cycle of a moss. Sexual and asexual generations alternate.

the first step in nature's production of coal.

Peat is used chiefly as a fuel. To dig it out, the bog is first partly drained. The loosened surface is then removed, and rectangular sections of peat, called turfs, are cut with specially shaped spades or by machines. The turfs are stacked in the open air for drying, a process that takes about six weeks. Dried turfs are used as fuel in Ireland and other countries. Sometimes peat is pressed into the form of briquettes. These make better fuel than the turfs, since they are more compact and contain less moisture.

LIVERWORTS

The liverworts are among the simplest of all land plants. Most of them have a simple type of plant body called a *thallus*. This is a flattened, ribbonlike structure. Rhizoids grow from the underside of the thallus. They anchor the plant and supply it with water and minerals. Many liverworts have small, cuplike growths known as *gemmae cups* on their upper surface. Budlike bodies called *gemmae* are produced in the cups. When the gemmae are separated from the plant, they are capable of growing into new thalli.

One of the commonest and most interesting liverworts is *Marchantia*. As it grows, it splits again and again at the tip, so that presently it is as if green ribbons were growing in different directions from the original starting point. If the plant prospers, it soon sends up a group of what look like miniature crudely designed palm trees. These growths bear the sex organs. When the egg is fertilized it grows into a spore plant so small that only a specialist would notice it.

Another familiar liverwort is *Riccia*, which is frequently sold to tropical fish breeders for their aquaria. Grown in water, it repeatedly splits into very thin branches. Grown on land, it divides in much the same way but the branches are thicker.

In the so-called leafy liverworts, the thallus is so divided as to suggest leaves. These plants have a strikingly delicate appearance. They are at their best on dripping rocks, although they are most likely to be

both photos, C. Nuridsany

Liverworts. Top: photo showing well-developed aerial chambers. Bottom: *Marchantia,* one of the commonest liverworts.

found in the dimmest recesses of woods. They are especially abundant in moist tropical forests. Here they frequently cover with luxurious growth not only the ground but also the stems and leaves of other plants.

HORNWORTS

The hornworts, or horned liverworts, generally resemble the true mosses and liverworts, except that the hornworts have a simple, undifferentiated thallus and a complex sporophyte structure. The latter is able to produce spores over a prolonged period of time. In regard to number of genera and species, the hornworts are a relatively small group. However, they are often abundant in wet places.

MOSSES AND FERNS

FERNS

Ferns and their distant relatives, horsetails and club mosses, are the more primitive of the vascular plants, or tracheophytes. They were formerly classified together as pteridophytes ("fern plants"). Nowadays, however, the ferns are considered by botanists to be more nearly akin to the conifers and flowering plants.

Top: reproductive cycle of a fern. Bottom: Sori on a fern leaf. The spores contained in these cases will give rise to new fern plants.

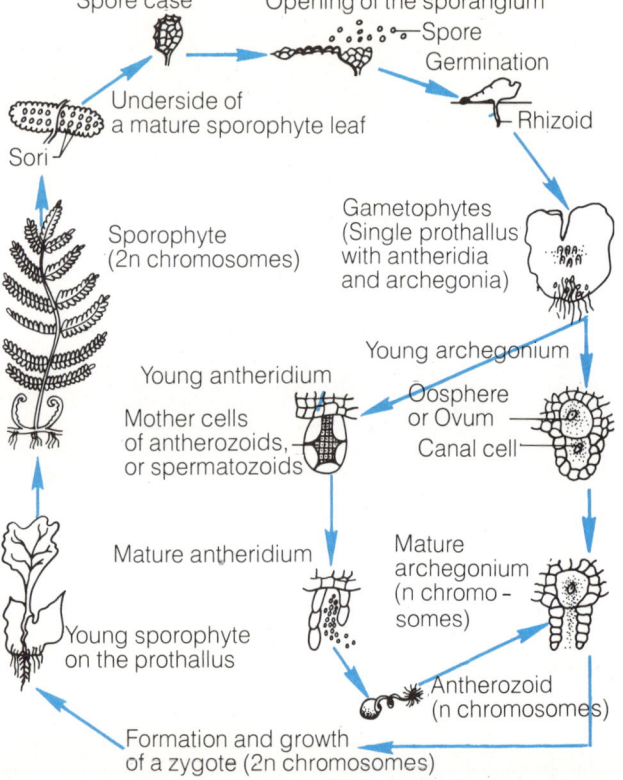

The club mosses are now placed in one phylum, the Lycopsida. The horsetails make up the phylum Sphenopsida. And the ferns make up the phylum Pteropsida, which comes from the Greek words meaning "fern-resembling plants." The Pteropsida also include all the true seed plants, which are the most highly developed plants of all. The ferns are segregated in a separate class, the Filicineae.

As in the mosses and liverworts, these plants undergo an alternation of generations. There is one important difference, however. In the ferns and their kin, it is the sporophyte, or spore-producing plant, that is the conspicuous form. The gametophyte, or sex-cell-bearing plant, is much simpler in structure.

CHARACTERISTICS OF FERNS

The ferns are more widespread and better known than the club mosses and horsetails. The roots, stems, and leaves of the fern sporophytes are like those of seed plants. Ferns absorb water and dissolved mineral salts by means of root hairs, just as seed plants do. The stems of most species remain underground. These hidden growths are known as *rhizomes*. The tips of the rhizomes slowly push their way forward through the soil as they grow. Thus they constantly tend to invade more and more territory. In certain cases, the stems of ferns grow vertically, forming lofty plants called tree ferns.

The leaves of the spore-bearing fern plants are called *fronds*. They carry on photosynthesis in exactly the same way as seed-plant leaves do. When young, the fronds of certain species are called *fiddleheads,* because the tips are curled up and resemble the scroll, or curved head, of a violin. As the frond grows longer, the tip uncoils. Most of our common ferns have large and usually very decorative leaves.

The underside of some fern leaves is dotted with structures called *sori* (singular: sorus). Each sorus is made up of a cluster of spore-containing cases known as sporangia, each mounted on a flexible stalk. When the sporangia dry out, the spores are released. If a fern leaf containing sori is kept in position over a glass slide, the slide will be covered after a time with a fine dust consisting of innumerable spores. Fern spores are provided with a strong protective covering. They are so light that the wind may carry them for some distance.

Many millions of spores may be produced by a fern plant in a single growing season. Comparatively few of the spores, however, develop into new plants. When a spore germinates, it gives rise to a small, flat, heart-shaped structure called a *prothallus,* which looks a good deal like the thallus of a liverwort. The undersurface of the prothallus has rhizoids instead of roots. The sex organs also develop on the undersurface of the plant. As in the case of the mosses, fertilization can take place only when water is present. The fertilized egg grows into a spore-bearing fern plant of the familiar type.

REPRESENTATIVE GENERA

The best place to look for ferns is on the floor of a forest. Here they are often more numerous than any other low-growing plants. They reach their fullest development in the tropics. The richest fern growth occurs not in dank, lowland woods nor in impassable jungles but in mountain forests where moisture is abundant.

In some regions, as in Hawaii, New Zealand, and parts of Australia, tree ferns ranging up to 12 meters and more in height and a meter in diameter are a familiar feature of the landscape. Many tropical ferns grow on the trunks or branches of trees. They are not parasites, but simply use trees as convenient anchorages. A few ferns are water plants. One of these, *Marsilea,* has floating leaves that suggest the four-leaf clover and therefore is sometimes called the clover-leaf water fern.

Among the best-known of all ferns is the brake, or bracken *(Pteridium).* It is particularly conspicuous in the heaths and moors of Great Britain and Ireland, though it is found in almost every part of the world. In Ireland the leaves of this plant reach a height of four meters.

The Christmas fern *(Polystichum)* has tough leaves that shine bright and green

both photos, C. Nuridsany

Top: the polypod is a fern commonly found in temperate regions. Bottom: new leaf of a polypod, still coiled in a fiddlehead.

even when covered with ice. The long, simple leaves of the walking fern *(Camptosorus)* take root at the tip and bud into new plants. The beautiful maidenhair fern *(Adiantum)* has small, lobed leaflets and shining, purplish-brown leaf stalks.

The leaves of certain ferns are used as packing for fruits and vegetables. The Japanese use the young fronds of some varieties as food. The trunks of some of the sturdier tree ferns serve for construction purposes, particularly in tropical lands. In the Hawaiian Islands, the soft hairlike scales of certain tree ferns *(Cibotium)* are used as a filling for mattresses. The male fern *(Dryopteris)* is known to pharmacists as the source of an effective vermifuge—a remedy used to expel or destroy intestinal worms.

The bracken fern, *Pteridium aquilinum*. This fern is very common in wild wooded areas. In some areas the fronds may reach four meters.

CLUB MOSSES

If you wander through deep pine or spruce forests, you will often find the ground covered with dense, bright growths of creeping ground pine. These plants are not related to pines but are club mosses (*Lycopodium*). The name is apt enough. These plants look very much like extremely large mosses. And, in many species the spores are produced in clublike cones. Most club mosses reproduce in much the same manner as the ferns do. They are more abundant in the tropics than they are in temperate regions.

These plants are not especially useful to mankind. Certain varieties are used in wreaths and other decorations. Lycopodium powder, consisting of the spores of a club moss, is used in fireworks and is sometimes set off in general-science classes to illustrate rapid combustion. It has also been employed in the treatment of certain skin diseases.

HORSETAILS

Horsetails belong to the genus *Equisetum*. They grow thickly in the gravel of railway roadbeds, in clumps along the banks of streams, and in swamps. Their jointed structure suggests that of sugar cane; the leaves have been reduced to small scales. In some species, the main stem bears a circle of smaller branches at each joint. In others, the stem usually does not branch, so that a thicket of this type resembles a fanciful green colonnade.

Horsetails contain a considerable amount of the hard substance called silica, or silicon dioxide (SiO_2), most familiar to us in the form of quartz. As a result, the texture of these plants is very rough. The dried stems have been used to scour pots and pans. The plants are sometimes known, therefore, as scouring rushes. Hay containing an abundance of horsetail plants is harmful to livestock.

Most species of horsetails are small plants, from one third meter to two meters in height. However, a vinelike tropical variety found in South America sometimes reaches a height of 12 meters. Most species send up both unbranched, fertile stems, each bearing a cone at the tip, and green, bushy stems. The cones contain the spore cases, or sporangia. Green gametophytes arising from the spores are small and ribbonlike.

Today, horsetails and club mosses form only a small part of the earth's vegetation. Even ferns are rarely the dominant plant in any locality. Yet millions of years ago, in the Carboniferous period, the forests of the world consisted largely of horsetails, club mosses, and tree ferns, all of enormous size. With the passage of geological time, countless generations of such forests grew and died. The plant debris accumulated in the swampy water of these forests until it was several meters thick. Later it was buried beneath sedimentary materials. The remains of the plants, completely transformed into a dark mineral, are now widely used in the form of coal.

SEED PLANTS

by Harry J. Fuller

In almost every type of landscape, most of the plants you see are seed plants, so called because they reproduce by means of seeds. There are more than 250,000 known species of seed plants—more than those of all the other plant groups put together. Botanists have been particularly interested in seed plants not only because of their numbers but also because of their variety and their outstanding importance in our lives. As a result, they have been more thoroughly investigated than any other plant group.

The seed plants belong, according to a widely accepted classification, to the phylum Pteropsida. This includes not only seed-bearers but also the ferns, which do not have seeds.

GENERAL CHARACTERISTICS

The seed plants show truly amazing variety in structure, growth, and reproductive processes. Yet they all have certain features in common:

1. Their characteristic reproductive structures are seeds, which are produced by *flowers* or *cones*. Each seed contains a tiny plant, the embryo, which results from a process of sexual reproduction. Upon sprouting, the embryo grows into a mature plant.

2. The sperm, or male sex cell, makes its way to the egg, or female sex cell, by way of a structure called a *pollen tube*, which is found only in seed plants. In most other plant groups—algae, ferns, mosses, liverworts, and many fungi—sperms can reach eggs only by swimming through water. Water is not necessary for fertilization in seed plants, because of the pollen tube. This fact accounts, at least in part, for the dominance of seed plants in the earth's vegetation. They can complete their reproduction under a much greater variety of environmental conditions than lower plants.

3. Seed plants possess complex *vascular tissues*. These are conducting tissues that transport water, minerals, foods, and other substances within the plant. Many lower plants also have vascular systems, but these are much less complex and less varied than the vascular tissues of seed plants.

4. Virtually all seed plants possess the green pigment *chlorophyll*. This is essential in *photosynthesis,* the basic process of food manufacture in plants. Only a few species of seed plants lack chlorophyll and therefore cannot carry on photosynthesis. Some of these plants are parasites, obtaining their food from the tissues of living plants. Others are saprophytes and derive their food from decaying organic materials in soils, the dead stumps of trees, and similar sources.

STRUCTURE AND PHYSIOLOGY

The bodies of practically all seed plants consist of stems, roots, leaves, and cones or flowers. One or more of these parts may be absent in a few species. For example, the asparagus plant and certain parasitic flowering plants lack leaves. In asparagus, the functions of leaves have been taken over by small, green stem branches.

Roots, stems, and leaves are called *vegetative organs*. They are concerned with maintaining the life, or vegetation, of the individual plant, rather than with the process of reproduction. Roots anchor the plant body firmly in the soil. They absorb water and mineral nutrients from the soil, conduct these materials upward into stems, and transport food downward from stems. Stems produce and support leaves and flowers or cones. They conduct water and nutrients upward and foods downward. They also store food. Leaves manufacture food in the process of photosynthesis.

Photosynthesis. In photosynthesis, carbon dioxide reacts with water, utilizing light energy absorbed by chlorophyll. As a

GYMNOSPERMS

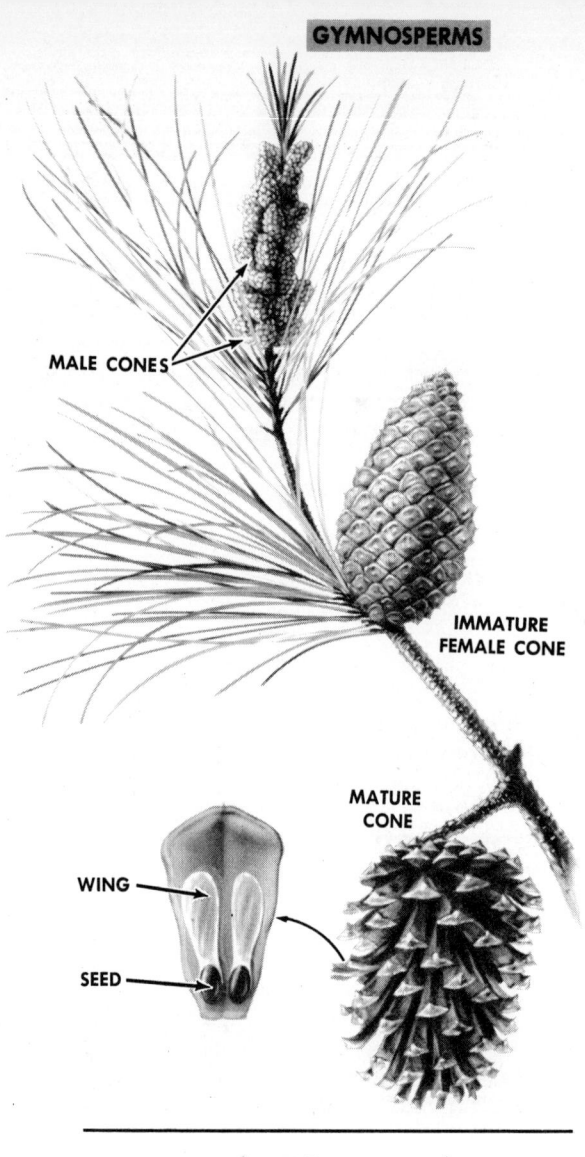

The reproductive processes of gymnosperms occur in cones. After fertilization has taken place, the seeds develop on the surface of cone scales. We show here a branch of the loblolly pine, with male and female cones. We also show two fully developed seeds, with winglike structures, attached to a scale. In time, the seeds will drop off.

result of this reaction, a simple sugar—glucose—is synthesized and oxygen is released. The carbon dioxide used in photosynthesis is absorbed by leaves from the atmosphere. The water required for the process is absorbed by roots from the soil. It travels upward through the conducting tissue of roots and stems into leaf veins, which are continuations of the conducting tissues of stems.

Since most of the chlorophyll of plants occurs in leaves, most photosynthesis obviously takes place in these plant structures. In some plants with green stems, such as tomatoes and petunias, the process also occurs in green stem tissues. Photosynthesis is the only major process for converting inorganic substances into food. It is also the only major source of oxygen in the earth's atmosphere. Without photosynthesis, all animal life, as well as almost all plant life, would come to an end. Animals depend upon this process for their food, since they derive various essential nutrients from plants or by devouring animals that eat

The reproductive processes of angiosperms occur in flowers. At the left is a cherry blossom, in cross section. The stamens are the male elements; the pistil, the female element. Unlike cherry blossoms, some flowers have only stamens or pistils. The seeds of angiosperms develop in structures that are known as fruits.

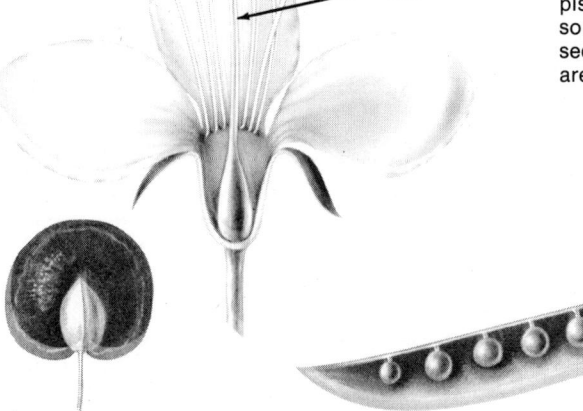

ANGIOSPERMS

62 SEED PLANTS

plants. They also depend upon photosynthesis to maintain the supply of oxygen, essential for respiration.

Respiration. Like all other living organisms, seed plants carry on respiration. In this process, foods combine chemically with oxygen and release their stored energy, chiefly as chemical energy. The energy derived from respiration is needed for growth, reproduction, and other processes.

Respiration is essentially the reverse of photosynthesis. Photosynthesis converts light energy into the stored chemical energy of foods. Respiration releases this stored energy for active use. Respiration differs from photosynthesis in other respects. It takes place in all living cells and at all times, while photosynthesis occurs only in chlorophyll-containing cells and only in the presence of light.

During the day, the rate of photosynthesis in green leaves exceeds the rate of respiration. Hence the release of oxygen and the intake of carbon dioxide in photosynthesis overbalance the use of oxygen and release of carbon dioxide in respiration. At night, when photosynthesis ceases, respiration continues. The result is that green plants absorb oxygen and release carbon dioxide. This characteristic gas exchange of green plants is responsible for the superstition that green plants "purify" air during the day and "poison" it during the night.

Growth. All seed plants obviously grow. The process of growth, which results in the formation of new tissues, occurs chiefly at the tips of roots and in the buds of stems. In many seed plants a growth tissue called *cambium* is located in stems and roots. The cambium lies between the two conducting tissues: the *xylem,* which conducts water and soil nutrients upward, and the *phloem,* which conducts foods manufactured in leaves downward. The growth activity of cambium leads to the production of new tissues transversely, or crosswise, in stems and roots. Therefore it causes these organs to grow in diameter. Growth processes are influenced by many factors, including the hereditary nature of the individual plant, growth hormones, water supply, nutrient supply, temperature, and light.

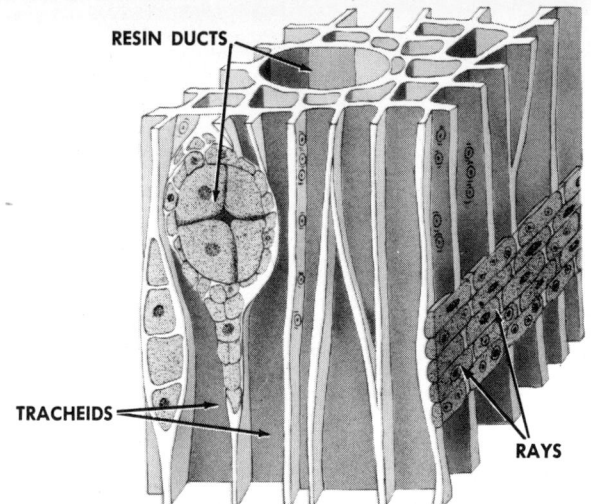

GYMNOSPERM XYLEM (PINE)

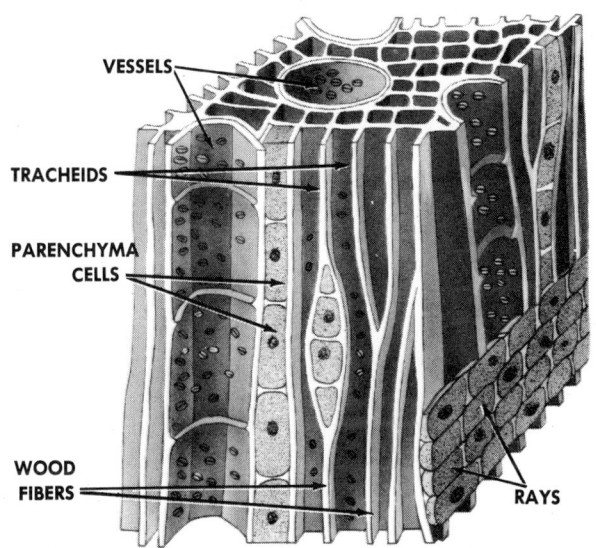

ANGIOSPERM XYLEM (OAK)

The two diagrams above show three-dimensional sections of gymnosperm and angiosperm xylem. Note the comparative simplicity of xylem in gymnosperms; angiosperm xylem is considerably more complex in structure.

CLASSIFICATION

Living seed plants are classified in two main groups: Gymnospermae, or gymnosperms, and Angiospermae, or angiosperms. Gymnospermae, taken from two Greek words meaning "naked seeds," are so called because their seeds are produced on the surface of cone scales. In the Angiospermae, which means "seeds in a vessel," seeds develop within structures called

SEED PLANTS 63

Left: cone and the fernlike leaves of *Encephalartos villosus*, which belongs to the order Cycadales. Right: a fine specimen of *Ginkgo biloba*, the Chinese maidenhair tree—the only living species of the order Ginkgoales. Middle: close-up of the leaves and fruits of this tree.

fruits. The gymnosperms are represented by comparatively few species—about 750—as against the approximately 250,000 known species of angiosperms.

In addition to the manner of seed production, there are other differences between these two plant groups. The reproductive processes of gymnosperms occur in cones. In angiosperms, they occur in flowers. Embryos of gymnosperms have more than two seed leaves, or *cotyledons*. Those of angiosperms have either one or two seed leaves. The ovules (structures that develop into seeds) of gymnosperms bear their eggs in tiny female sex organs called *archegonia*. Archegonia are lacking in angiosperm ovules. The xylem tissue of most gymnosperms is relatively simple. It consists mostly of vertically arranged cells, called *tracheids*, which conduct sap and serve as support for the stem; horizontally arranged cells, called *rays;* and an extensive system of *resin ducts*. The xylem of angiosperms is more complex. It is usually made up of several types of cells and cell groups: tracheids, vessels (long, continuous conducting tubes), fibers, and *parenchyma* (storage) cells and rays.

CONE-BEARING PLANTS

The gymnosperms are woody plants—that is, trees or shrubs. They possess roots, stems, and leaves, as well as the reproductive structures called cones. In most gymnosperms, there is a dominant main stem, with conspicuously smaller branches. In some species there are no branches at all. The main stems of certain gymnosperms, such as the redwoods of California, may reach a height of more than 90 meters. The diameters of their trunks are sometimes enormous.

The leaves of most gymnosperms are needlelike or scalelike. They contain chlorophyll and therefore carry on photosynthesis. In certain gymnosperms, such as cycads and ginkgoes, the leaves are broad and thin, rather than needlelike. Most gymnosperms are evergreen. That is, they retain their leaves throughout the year and often, as in pines and firs, for several years. A few species, including larches, bald cy-

Left: Ponderosa pine *(Pinus ponderosa),* a member of the order Coniferales. Middle: two cones of this tree. Right: *Welwitschia bainesii*, with its long, leathery leaves and its numerous small cones. A native of South-West Africa, this plant belongs to the order Gnetales.

presses, and ginkgoes, are deciduous. They lose their leaves in autumn, remain leafless throughout the winter, and produce a new crop of leaves in the spring.

Many species of gymnosperms produce essential oils that are volatile and highly scented. These oils, which appear to be waste products, are responsible for the characteristic odors of the needles and wood of many gymnosperms, including pines and cedars. When essential oils combine with oxygen, they form extremely viscous liquids or brittle solids called *resins*. Resins are important raw materials of industry. They are obtained by cutting gashes into the sapwood of tree trunks. The resins ooze out of the cuts made in this way and are caught in containers. Among the most valuable gymnosperm resins are the turpentines, derived chiefly from several species of pines and used principally in the preparation of paints and varnishes.

The living species of gymnosperms are generally divided into four orders: the Cycadales, or cycads; the Ginkgoales, or maidenhair trees; the Coniferales, or conifers; and the Gnetales. It is believed that the gymnosperms evolved from some group or groups of extinct ferns, probably by way of an order of curious plants called seed ferns. These plants had fernlike leaves and stems, but reproduced by means of seeds; the seeds were produced, not in cones, but directly on the leaves. The study of rocks of great geological age reveals the fossils of many gymnosperm species now extinct.

THE CYCADS

The order Cycadales consists of about 100 species. The plants belonging to this group have unbranched, somewhat woody stems and compound, fernlike leaves that grow in a tuft from the tip of the stem. The leaves are subdivided into flat, thin structures called *leaflets*. Many cycads resemble huge pineapples. They have a squat, thick stem, surmounted by a crown of leaves.

Cycads are chiefly tropical and sub-

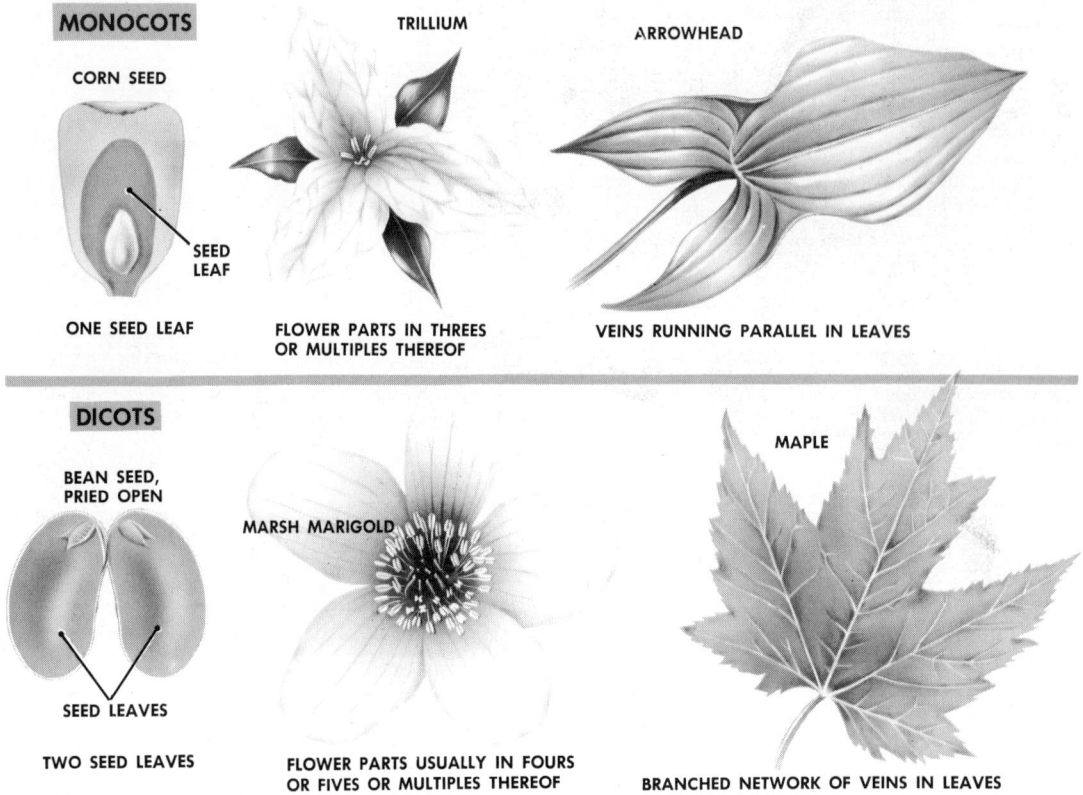

tropical. They thrive in southern Florida, Mexico, the West Indies, tropical Asia, Australia, and South Africa. The cycads are often cultivated as ornamental plants in display greenhouses and in tropical gardens. One Asian cycad produces a valuable gum. Another species, the sago palm, is extensively cultivated in Malaysia and Indonesia, and is the source of sago starch, or pearl sago, which is widely used as a food.

The reproductive processes of cycads occur in cones. There are two types of cones: large and usually woody female cones and smaller, more delicate male cones. The male cone bears scales that produce pollen grains. These grains are carried by wind to the ovules borne on scales of the female cone. Pollen tubes, produced by the grains of pollen, grow into the ovules. Sperms from the pollen tubes then fertilize the eggs contained in the ovules. A fertilized egg develops into the embryo of the seed. When it is mature, the seed falls to the ground, where it sprouts.

MAIDENHAIR TREES

The Ginkgoales, or maidenhair trees, numbered many species in past geological ages. Today there is only one living species—*Ginkgo biloba,* the Chinese maidenhair tree. This tree sometimes attains a height of 28 to 30 meters; its trunk may have a diameter of more than a meter. The leaves are crowded on short spur branches. They are 5 centimeters or more in length and are broad and fan-shaped, with prominent riblike veins. Some trees produce small male cones; others produce female cones of simple structure. A female cone bears two tiny cuplike scales, each containing one ovule. Pollination is by wind. After fertilization, each ovule develops into a seed with a single embryo. The surface tissues of the seed become fleshy at maturity and give off a very unpleasant odor.

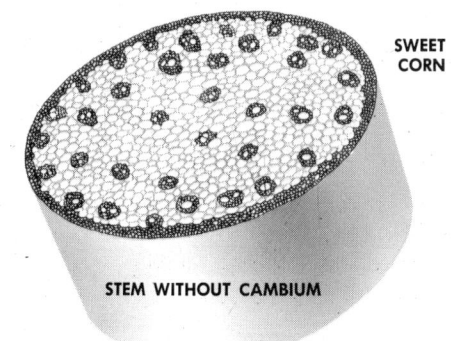

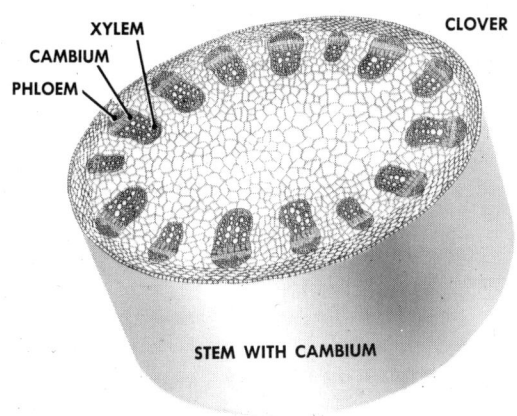

times they are produced in clusters of from two to five needles. The majority of conifers are evergreen, though a few are deciduous.

The reproduction processes of most coniferous trees are quite similar. A pine tree, for example, produces two kinds of cones. There are small, rather soft male cones, each bearing a number of diminutive pollen-producing scales. There are also large, quite woody female cones, which have numerous cone scales. Each female cone scale bears two ovules on its upper surface. Pollen grains, shed by the mature scales of male cones, are carried by wind to female cones. The grains of pollen land on an ovule and form pollen tubes. These penetrate the ovule through a tiny pore, the *micropyle*, in the ovule coat. The pollen tubes carry sperms into the ovule, and the eggs in the archegonia of the ovule are fertilized by the sperms. Since several archegonia and eggs are present in an ovule, multiple fertilization may occur. Usually, however, only one fertilized egg develops into an embryo in the ovule. The embryo and the surrounding parts of the ovule grow, and ultimately a mature seed is formed. This falls from the female cone scale to the soil.

Coniferous trees are very widely distributed on the surface of the earth, particularly in the cooler parts of the temperate zones and in elevated areas of the tropics.

The conifers are the most valuable economically of all the gymnosperms. They furnish large quantities of softwood lumber (pine, redwood, fir, spruce, and so on). They produce resins used in paints and varnishes. They are also widely used as ornamentals. A few species, including the piñon pine of the southwestern United States, produce edible seeds. The bark of some species yields tannins, used in tanning hides. Conifers growing in extensive forests break the force of falling rain and bind the soil, thus preventing soil erosion.

Ginkgo trees have no economic value. They have been widely planted, however, in many parts of the world in parks and gardens and along boulevards as ornamental trees.

CONIFERS

The nearly 600 living species of the Coniferales are much more familiar than the members of the other orders of gymnosperms. Among the better-known conifers are pines, spruces, Douglas firs, true firs, yews, incense cedars, true cedars, junipers, cypresses, sequoias (redwoods and California big trees), and larches, or tamaracks. Pines, which number about 80 species, make up the largest subgroup of the order.

Most conifers are trees, but a few species are shrubs. They usually have straight, dominant trunks, often reaching heights of more than 30 meters. The leaves of most conifers are needles or small scales. Often they are borne singly on the stems; some-

THE GNETALES

This order, numbering about 65 species, is divided into three quite distinct genera: *Ephedra, Welwitschia,* and *Gnet-*

um. *Ephedra* is found in dry or rocky regions; tiny leaves grow on its green stem. *Welwitschia,* which has but one species, is a rather curious plant of southwest Africa. It has a huge taproot and two thick and tough leaves. The plants belonging to the genus *Gnetum* are natives of the tropics. They are found in Asia, Africa, South America, and the West Indies. They are mostly vines. Sometimes, however, they occur as shrubs.

FLOWERING PLANTS

As already pointed out, the 250,000 known species of angiosperms produce their seeds within fruits. The fruits may become dry when they are mature, forming structures such as pea pods, bean pods, and lily capsules. Or they may become soft and fleshy, as in the case of tomatoes, grapes, watermelons, and peaches.

MONOCOTS AND DICOTS

There are two subclasses of angiosperms. The Monocotyledonae are commonly called monocotyledons or monocots. The Dicotyledonae are called dicotyledons or dicots. There are various differences between the two groups. Monocots have only one seed leaf, or cotyledon, in the embryo. Dicots have two seed leaves. The flower parts of monocots are in threes or in multiples of three. Those of dicots are most often in fives, or multiples of five, less frequently in fours, rarely in twos and threes. The leaves of most monocots, including grasses and irises, are much longer than they are wide, and the veins usually run in the same direction lengthwise in the leaf. Leaves of most dicots—elms, lettuce, maples, and lilacs, for example—are about as long as they are broad or usually not more than two or three times longer than they are broad. The veins form an extensively branched network in the leaf. The stems of monocots generally lack cambium. Consequently they increase but little in diameter. (The palms form a notable exception.) Most species of dicots have cambium in their stems and roots. As a result, their girth, especially in trees, is likely to be very impressive.

How did the angiosperms originate? The prevailing view, based largely on speculation, is that the earliest angiosperms evolved from the seed ferns or some other

The plants on the opposite page are examples of monocots. Left to right: awns of wheat, narcissus blossom, stem and foliage of a coconut palm. Plants shown above are dicots. Left to right: lilac blossoms, petunia blossoms, sugar maple tree.

group of primitive and now extinct plant forms. It is generally agreed that the most primitive living angiosperms are probably the magnolias, tulip trees, and their relatives. Other types of angiosperms may have evolved from primitive magnolialike ancestors. Most botanists regard the monocots as derivitives of primitive dicots of the magnolia type. The older view that monocots are more primitive than dicots is no longer accepted.

There are almost 50 orders of angiosperms. These orders are subdivided into about 300 families, which cannot be described in a brief account of this type. The monocot orders include cattails, grasses, sedges, lilies, tulips, narcissuses, irises, amaryllids, palms, bananas, and orchids. Among the dicot orders are almost all broad-leaf trees, such as willows, elms, oaks, maples, apples, walnuts, mahoganies, locusts, plum trees, and the like; most shrubs, including lilacs, privets, roses, forsythias, and snowballs; most woody vines, such as wisterias; and thousands of species of plants that generally have herbaceous stems, such as sweet potatoes, bindweed, morning glories, petunias, marigolds, zinnias, celery, soybeans, sweet peas, and larkspurs. The monocots number about 50,000 species; the dicots, nearly 200,000.

PLANT STRUCTURE

All angiosperms have well differentiated conducting, food-making, storage, and strengthening tissues. The typical angiosperm body consists of four major parts: roots, stems, leaves, and flowers.

There are notable variations among roots. For example, grass roots are slender, fibrous structures. Those of carrots, sugar beets, and turnips become fleshy and enlarged as food and water are stored in them.

Stems are just as varied. Certain stems are modified into tendrils or into organs of food storage. The bulbs of onions and Irish-potato tubers are examples of the latter. Some stems are modified for vegetative reproduction. The runners of strawberry plants, and rhizomes of quack grass are examples.

The leaf forms of angiosperms are almost infinitely varied. The leaves of most species are broad and thin. In some flowering plants, they are modified into tendrils. In still others, they form protective scales over buds. Leaves of certain species, such as the Venus's-flytrap, are specialized for

SEED PLANTS

trapping and digesting the bodies of insects. Century-plant leaves serve primarily for food or water storage.

The flowers are where the reproductive processes that lead to seed formation take place. Like the other plant structures, flowers show great variety. Some flowers, such as peonies, roses, and carnations, have brightly colored, showy petals. Other flowers, such as those of oak and willow, have no petals at all. Most flowers contain both stamens, or male parts, and a pistil, or female part. But in oaks, willows, and walnuts, some flowers have stamens while others have pistils.

The mode of pollination—the transfer of pollen grains from stamens to the stigmas, or tips, of pistils—depends upon the type of flowers found on a plant. In some, particularly if the flowers are conspicuous because of the size or color of their petals, or are fragrant, or if they produce nectar, insect pollination takes place. In others, wind carries the pollen.

Pollination is the first stage in the reproductive process. When a pollen grain, carried by insects or other animals (such as hummingbirds) or by the wind, lands on a stigma, it begins to grow, forming a long, slender pollen tube. This grows downward into the ovary of the pistil. Within the ovary are the ovules. Each ovule contains a central structure—the embryo sac—within which a single female sex structure, the egg, is produced. A pollen tube enters a pore (micropyle) in the surface tissues of the ovule, and discharges sperms into the embryo sac. A sperm fertilizes the egg, and the fertilized egg, or zygote, begins developing.

Foods are transported into the ovule. It grows and its surface cell layers develop into the seed coat. In the meantime foods and often water move into the tissues of the surrounding ovary, which ultimately forms the mature fruit, which releases seeds. When a seed reaches the soil and finds suitable conditions of temperature, moisture, and oxygen supply, its embryo sprouts and in time grows into a mature flowering plant. In some angiosperms, including navel oranges, pineapples and bananas, fertilization does not occur; the fruits are seedless.

IN MANY ENVIRONMENTS

Angiosperms thrive in just about every type of climate and environment. And although the angiosperms are primarily land plants, they include some species of water plants. These grow principally in the fresh water of lakes, ponds, and streams. Among aquatic or semi-aquatic freshwater angiosperms are water lilies, cattails, duckweed, pondweed, tape grass, and frogbit. Very few angiosperms grow in salt water.

PEOPLE USE ANGIOSPERMS

Angiosperms play a more important part in our lives than any other plant group. Generally they are beneficial to mankind. We derive some of our most useful plant products from angiosperms. They are the source of such foods as fruits, grains, vegetables, and nuts.

Angiosperms supply us with hardwoods such as oak, hickory, maple, cherry, and mahogany. We get nonalcoholic beverages from coffee, tea, and cocoa plants; and fibers from cotton, flax, and hemp plants. Angiosperms are also responsible for the flavors of spices; for latex products such as rubber, chicle, and gutta-percha; for essential oils used in perfumes, colognes, soaps, and cosmetics; for drugs such as quinine, digitalis, cocaine, and belladonna; and for tannins and dyes.

Angiosperms are beneficial to us in less direct ways. Their roots check soil erosion and thus aid in preventing floods. They provide shelter and food for many wild animals. They lend beauty to landscapes.

On the other hand, certain angiosperms are harmful or disadvantageous to human beings. Some species are weeds that compete with crop plants for space, light, water, and soil nutrients. Certain angiosperms liberate wind-borne pollens that cause various kinds of allergies. Other species produce toxic substances that cause surface or internal poisoning. Still other species, such as the opium poppy, *Erythroxylon coca,* and the hemp plant produce narcotic substances that cause human misery and economic loss.

VITAL PROCESSES OF PLANTS

There are certain important differences between plants and animals. First, most plants are able to manufacture their own food supply by a series of chemical reactions known as *photosynthesis*. No animal can do so, unless certain lowly organisms, called phytoflagellates, are animals, as many zoologists believe. Second, animals generally have a maximum size and a rather definite form. These do not change to any marked extent after the animals become mature. In plants, both size and form are quite variable and depend on the nature of the environment. Third, the cell walls of most plants are made up primarily of a carbohydrate compound called cellulose. Practically all animals lack cellulose. The only exceptions are the tunicates, whose coat is almost identical with cellulose. Finally, most plants are firmly fixed in the soil in which they grow, while most animals can move from one place to another.

The differences between plants and animals, then, are quite striking. Yet the similarities between them, particularly when we consider their vital processes, are perhaps even more remarkable. Both plants and animals use foods as sources of energy and as materials for the building of their living substance. Both are able to digest complex foods. Both produce regulating substances called *hormones,* which influence various processes of development. Both grow as new cells are formed from old cells or as cells are enlarged. In both plants and animals, previously similar cells become differentiated, in accordance with an orderly pattern, to yield the final body form. Both plants and animals take part in reproductive processes that lead to the development of offspring. Both are composed mainly of water and require water, in varying quantities, for the normal processes of life. Both transport materials within their bodies. Both are sensitive to the environment and are able to react to these conditions.

When we deal with the vital processes of plants, therefore, we must not think of these processes as characteristic of plants alone. We must bear in mind that most of them are found, to a greater or lesser extent, in all living things. A rose plant assimilates and digests food, respires, grows, reproduces its kind, and responds to the varying conditions of its environment. So do earthworms, lobsters, and people.

PLANT METABOLISM

The word "metabolism," as applied to plants, has a wide range of meaning, since it includes all the chemical reactions that go on within the living system. All phases of metabolism are derived ultimately from the energy supplied by the sun. Solar energy, as we shall see, plays a direct role in certain vital metabolic reactions. Through these reactions, such simple substances as water and carbon dioxide are built up into large molecules of sugar, starch, cellulose, fat, protein, and nucleic acids. Locked within these molecules is energy obtained from the sun—energy needed by the plant for the activities of life. In other metabolic reactions, the large molecules break down into simpler substances, often with the release of great quantities of energy.

Some of this energy is lost to the plant, since it is given off in the form of heat. The rest is not immediately available. It is conserved at first, in the form of chemical bonds, in the substance called adenosine triphosphate (ATP). It is when the ATP splits up into adenosine diphosphate (ADP) and free phosphate that the energy stored up in the ATP is finally released to the plant. It is then used for such activities as growth, movement, reproduction, and the repair of bodily tissues.

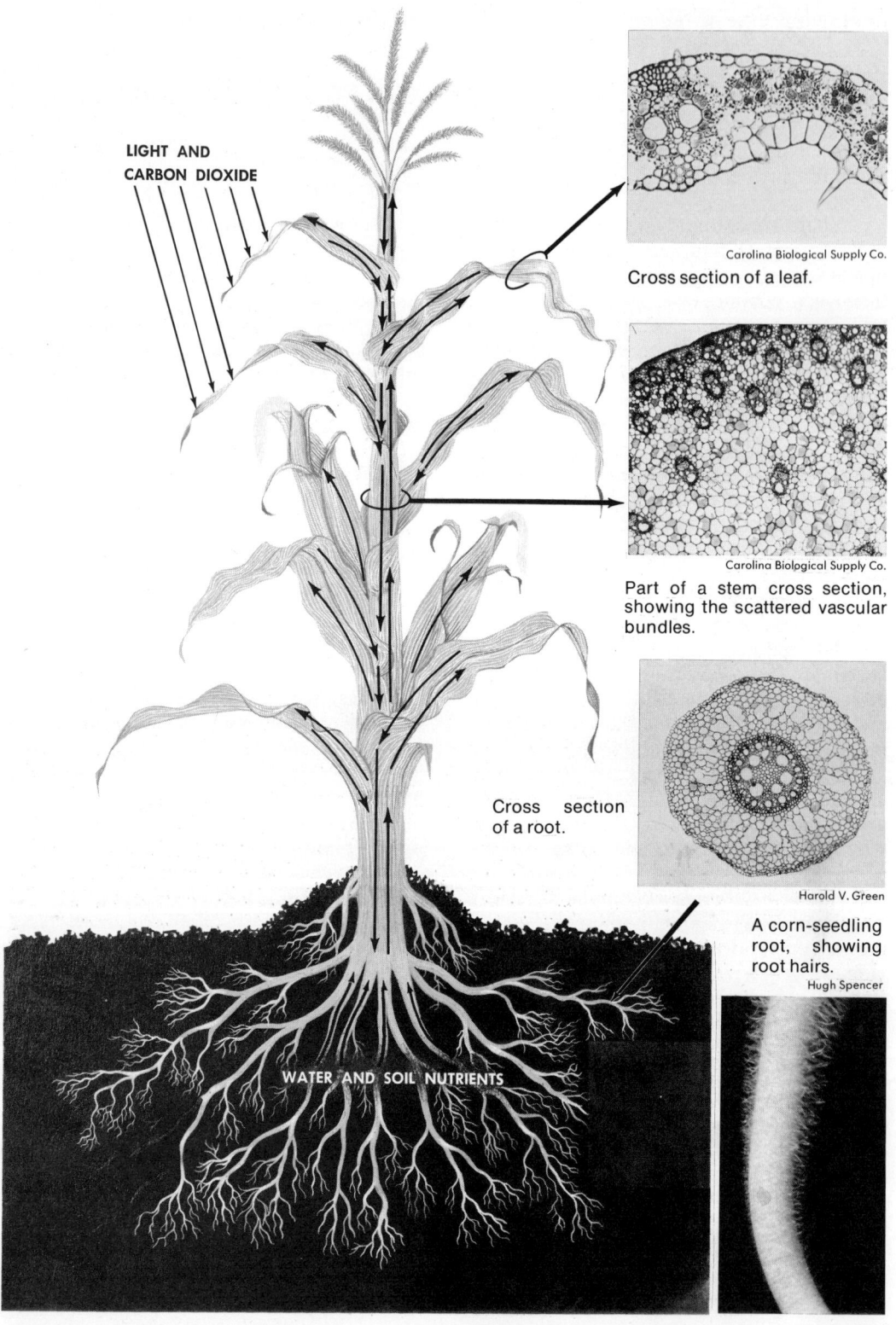

Cross section of a leaf.

Part of a stem cross section, showing the scattered vascular bundles.

Cross section of a root.

A corn-seedling root, showing root hairs.

72 VITAL PROCESSES OF PLANTS

ESSENTIAL ELEMENTS

The raw materials needed for the complicated chemical reactions of metabolism are water, carbon dioxide, oxygen, and various nutrients absorbed from the soil. They are introduced into the plant in different ways.

Water is absorbed by the surface cells of roots and is transported into the central portions of the roots. From there it passes through a conducting tissue called *xylem,* which extends upward from the roots into the stems and then branches from the stems into the buds, leaves, flowers, and fruits.

Carbon dioxide, which usually makes up about 0.03 per cent of the atmosphere, enters plant bodies chiefly through pores in the surfaces of leaves. The opening and closing of these pores, which are called *stomata,* are regulated by the expanding and contracting movements of the special cells that surround them. The stomata are usually open in the light and are closed during the hours of darkness.

The soil nutrients that seem to be necessary for the normal metabolic activities of all green plants include relatively large quantities of nitrogen, sulfur, phosphorus, potassium, calcium, and magnesium, and smaller quantities of iron, copper, manganese, zinc, molybdenum, boron, cobalt, and chlorine. These 14 elements, together with the carbon, hydrogen, and oxygen coming from atmospheric sources, make up the *essential elements* of plant nutrition. The lack of any one of these elements can prevent the normal functioning and development of a plant.

The soil elements are absorbed mainly as electrically charged particles, or ions, which come from various chemical compounds found in the soil. Like water, the nutrient ions are absorbed by the roots and then move upward in conducting tissue to the leaves and other aerial parts of plants.

We know something about these essential elements, though much remains to be learned. Nitrogen, for example, forms part of the molecules of all nucleic acids and proteins, as well as of chlorophyll, a green pigment that plays a vital part in photosynthesis. Phosphorus, too, is found in nucleic acids and proteins, and it is intimately involved in energy transformations. Magnesium is present in chlorophyll molecules. Zinc is essential in the formation of growth hormones. Though iron is not found in chlorophyll molecules, it plays an important role in their manufacture. It is also a part of certain active proteins, called enzymes, that are involved in the release of energy from foods.

PHOTOSYNTHESIS

The substances drawn from the air and the soil are the raw materials that the plant uses in the manufacture of food. The first step is photosynthesis ("putting together in light"). In this process, water, absorbed by roots from the soil, and carbon dioxide, obtained from the atmosphere, react in the green tissues of plants in the presence of sunlight to form a simple sugar, glucose. This substance serves as a basic material from which the plant manufactures other organic compounds, such as sucrose, starch, fats, and oils.

Photosynthesis takes place only in cells that contain the green pigment chlorophyll. In many plants, including most trees and shrubs, chlorophyll is present in the leaves and nowhere else. It is, therefore, only in the leaves of such plants that photosynthesis can occur. In other plants, such as corn, zinnia, and tobacco, not only the leaves but the stem tissues as well contain chlorophyll and are therefore capable of

A plant such as the corn plant *(Zea mays),* shown at left, draws in through its roots water and various nutrients. These pass upwards through conducting tissues (vascular bundles) into the stems and leaves. Carbon dioxide enters the leaves from the outer air through pores. In the leaves (and to a certain extent in the stem, in corn plants), carbon dioxide reacts with water in the presence of sunlight to form a simple sugar. Part of this is used by the plant as a source of chemical energy. Another part serves as a basic material from which the plant manufactures other compounds. In the photographs to the right of the diagram are close-ups of parts of the corn plant.

There would seem to be nothing in common between the giant redwood tree (far left), the microscopic *Spirogyra* (upper left), and the maiden-hair fern (lower left). Yet they all manufacture food from simple materials, in the presence of light, by the process known as photosynthesis.

(far left) Save-the-Redwoods League; (upper left) Hugh Spencer; (lower left) Hugh Spencer.

photosynthesis. In still other plants, including tomatoes and grapes, there is chlorophyll in young fruits, which take part in the food-manufacturing process.

Certain plant species lack chlorophyll and so cannot carry on photosynthesis. These species, which are mostly fungi and bacteria, cannot make organic compounds from simple inorganic substances of air and soil, as the green plants can. Therefore, like animals, they must obtain from green plants, directly or indirectly, the food substances that these green plants manufacture.

The process of photosynthesis is described in greater detail in the article "Photosynthesis."

DIGESTION

Many of the foods stored in plant tissues have a complex chemical structure or are insoluble in water. Therefore they cannot be quickly or easily used to provide energy or to build new tissue. They must first be converted into simpler foods or into water-soluble foods by metabolic processes called digestion.

These processes are controlled by special proteins called *enzymes,* which are manufactured by living cells. There are many different kinds of digestive enzymes but all have certain features in common. They are so efficient that a small amount can act on a large amount of food very quickly. They are also highly specialized.

Each type of enzyme ordinarily digests only one type of food or several types of chemically related foods. None of them are used up in the digestive processes that they promote.

There are three major groups of digestive enzymes in plants: carbohydrate-digesting, fat-digesting, and protein-digesting. One of the most widely occurring plant enzymes is amylase, or diastase, which digests starch to malt sugar. Another is invertase, or sucrase, which digests cane sugar (sucrose) to glucose (grape sugar) and fructose (fruit sugar).

Digestion is especially active in plants when seeds sprout. The reason is that the stored food of seeds is chiefly in the form of starch, fats, and complex proteins. These must be converted into simpler substances before they can be made to yield energy or before they can be transformed into other substances. Digestion also takes place when various fruits, such as bananas, ripen, and when potatoes and other tubers sprout. In these cases, starch has been converted into sugars.

RESPIRATION

Living plants require large amounts of chemical energy for growth, reproduction, reactions to external stimuli, and other vital processes. This energy is made available through the process of respiration. During respiration, foods—chiefly glucose—are chemically broken down and their stored energy released. A part of the energy liberated by respiration is heat energy, which is given off into the air from the body surfaces of plants. Most of it, however, is the chemical energy of ATP.

The reacting substances and products of respiration, as it occurs in most plants, may be represented by this simplified chemical equation:

$$C_6H_{12}O_6 + 6O_2 \longrightarrow$$
One Molecule + Six Molecules   Yield
of Glucose   of Oxygen

$$6H_2O + 6CO_2 + \text{Energy (Stored in ATP)}$$
Six Molecules + Six Molecules of + Energy
of Water   Carbon Dioxide

This equation shows none of the intermediate chemical reactions of the process. The respiration described by this equation is called *aerobic respiration*, because it occurs in the presence of free gaseous oxygen, found in the atmosphere. This sort of respiration occurs in most plants.

Some lower plants, however, including yeasts and certain bacteria that normally live in the absence of free oxygen, carry on another kind of respiration. This is known as *anaerobic respiration,* or *fermentation.* In this process, sugars are broken down in the absence of oxygen. The products that are formed are carbon dioxide and alcohol or some other compound, such as lactic acid. Anaerobic respiration is not limited to the lower plants. It may also occur in higher forms, if these are deprived of free oxygen.

During daylight hours, the sugar glucose is manufactured in the leaf more rapidly than it can be used or transported to other tissues. The excess glucose is converted into starch, which is stored in the leaf for a time. Later it is changed into a soluble sugar, which moves from the leaf. Top right: a starch test, applied to a bean leaflet after it had been exposed to sunlight. A stencil of the word "starch" had been set on the leaflet in the early morning. In the late afternoon, the stencil was removed and the chlorophyll of the leaflet was extracted with alcohol. The leaflet was then treated with iodine and photographed. The areas that had been exposed to the sun, including the stenciled letters, stained blue, indicating the presence of starch. Since glucose is quickly converted into starch in many plants, the presence of starch in a leaf shows that photosynthesis is going on. Starch, built up by the plant from glucose, is a form of stored food. It is found in great quantities, for example, in corn kernels, like the one at bottom right.

It was formerly believed that aerobic and anaerobic respiration had very little in common. Actually, they are closely related. They both begin with the chemical conversion of glucose into an intermediate organic compound, pyruvic acid. In the presence of oxygen, pyruvic acid is converted into carbon dioxide and water. This happens, of course, in the case of the higher plants, which carry on respiration in the presence of oxygen. If oxygen is absent, as is generally the case with yeasts and bacteria, the

Photo by Dr. Carl Wilson, from "Botany," 4th ed., by C. Wilson and W. Loomis, copyright © Holt, Rinehart and Winston, Inc. By permission.

Corn Industries Research Foundation, Inc.

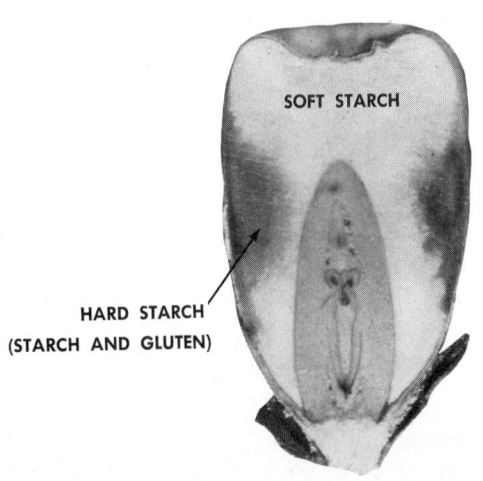

SOFT STARCH

HARD STARCH
(STARCH AND GLUTEN)

pyruvic acid is changed into carbon dioxide and alcohol, or lactic acid, or some other compound.

The yeasts, which are one-celled fungi, are responsible for industrial alcoholic fermentation. They derive the energy for their growth and reproduction from their anaerobic breakdown of sugar. The brewer is primarily concerned with the alcohol that they form in this process. From the viewpoint of the yeasts, however, the alcohol is only a waste product—it is the energy released in respiration that is vital to them. The carbon dioxide that is produced in alcoholic fermentation is given off in bubbles, which escape from the liquid or other medium in which the yeasts live. Carbon dioxide bubbles liberated by yeasts in dough cause the dough to rise.

The various processes that make up respiration are controlled by a group of respiratory enzymes. They operate as controlling mechanisms through a series of very complex chemical reactions. Some of the lower plants, which normally live under anaerobic conditions—that is, in the absence of oxygen—are unable to carry on aerobic respiration when free oxygen is available. The reason is that they lack certain enzymes required for the completion of aerobic respiration.

COMPARING PHOTOSYNTHESIS AND RESPIRATION

It is interesting to note that respiration is the reverse of photosynthesis. Respiration releases energy from foods, whereas photosynthesis stores energy in food. The products of respiration are carbon dioxide and water. The products of photosynthesis are glucose and oxygen. Respiration, however, occurs in all living cells and occurs in both light and dark. Photosynthesis takes place only in green cells and only in light.

Photosynthesis and respiration account for various interesting phenomena involving leaves and other plant parts. Among other things, they explain why green plants give off oxygen during the day and carbon dioxide at night. Leaves and other green parts of plants carry on both photosynthesis and respiration during the day, when they receive light. Under these conditions, photosynthesis proceeds more rapidly than respiration.

The plant draws in much more carbon dioxide than it releases, and much less oxygen than it releases. The result, therefore, is that the plant gives off oxygen during the daytime.

At night, with the disappearance of light, photosynthesis comes to an end. Respiration, however, continues, since it is independent of light. The result is that the plant gets oxygen from the atmosphere. It releases carbon dioxide (an end product of respiration) to the atmosphere, but does not release oxygen. At night, therefore, plants deplete the supply of oxygen and add to the supply of carbon dioxide.

PLANT GROWTH

In plants, as in animals, growth is a cellular process. It involves the production of new cells from pre-existing ones, the enlargement of newly formed cells, and, finally, the differentiation of enlarging cells into the mature tissues of the body. Obviously, increase in size is only one phase of the total growth process.

Various requirements must be met if normal growth is to take place. There must be abundant chemical energy, which is provided by respiration. The growing plant must have proteins and other organic compounds for the building up of protoplasm. It must have cellulose and various related compounds for the construction of cell walls. Conditions that interfere with photosynthesis or that cause stored food to be depleted are quickly reflected in diminished growth.

Higher plants have special growth zones known as *meristems*. These are located in the buds of stems, in young flower parts and young fruits, and immediately behind the tips of young roots. Lateral, or side, meristems occur in the stems and roots of many plants as a growth layer called the *cambium*. The cambium is located just outside the xylem tissue. In the outer portion of tree bark, a meristematic tissue called *cork cambium* makes up the surface of woody stems and roots. Another

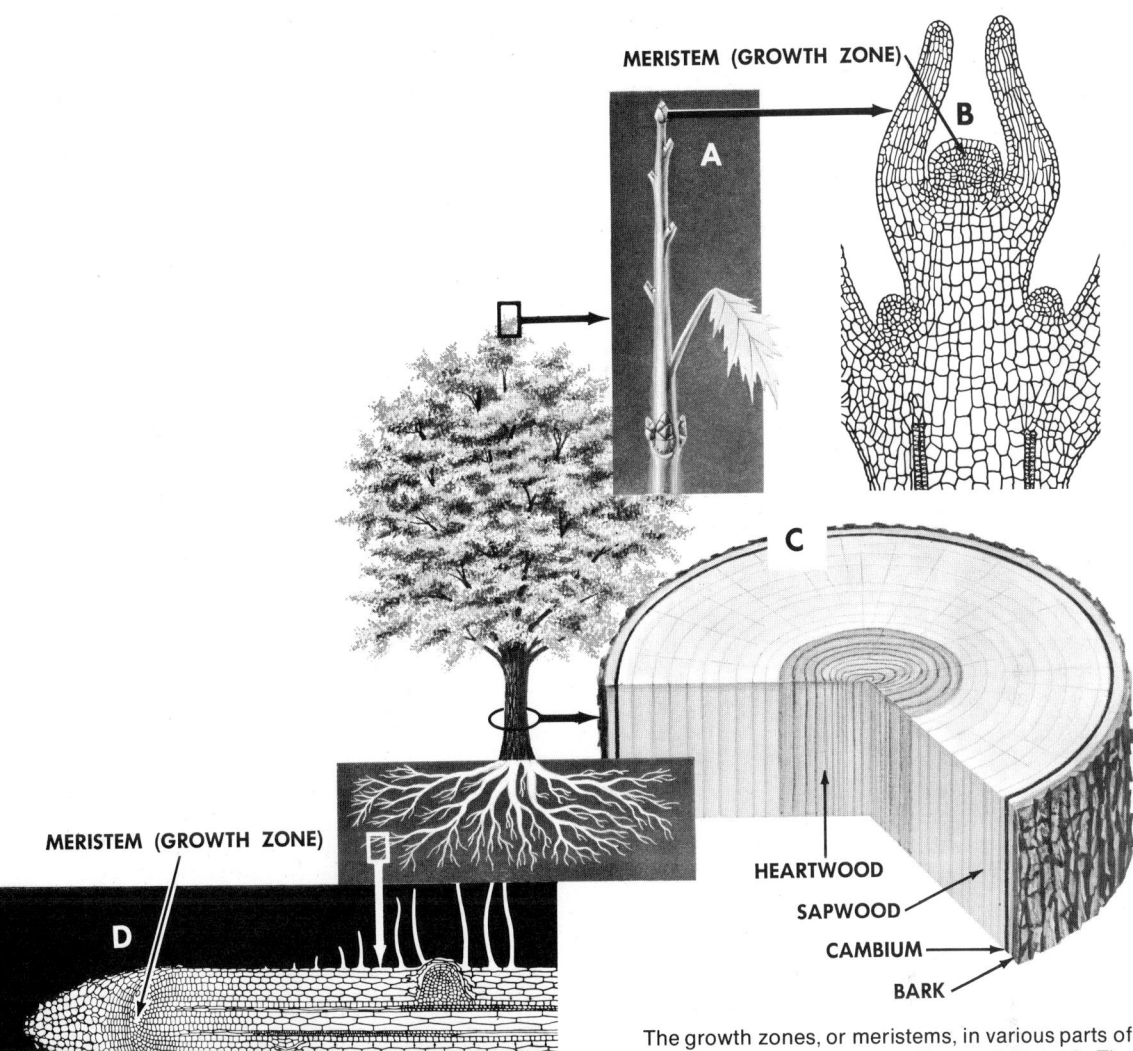

The growth zones, or meristems, in various parts of a tree. A shows a stem with terminal bud. The growth zone is shown close-up in B. C is a cross section of a tree trunk. The growth zone is indicated by the dark area (cambium). D shows the growth zone in a root tip.

type of meristematic tissue, the *pericycle*, lies in the central mass of root tissues. It gives rise to new cells that develop into root branches.

Meristematic cells have thin, elastic walls, abundant protoplasm, and a high level of metabolic activity. The creation of new cells begins with the formation of large quantities of deoxyribonucleic acids (DNA). These substances, found in the chromosomes of the cell nucleus, are intimately connected with development according to a hereditary pattern. DNA consists of four kinds of building blocks. The order in which these building blocks occur determines the order of the amino acids that make up the proteins of cells. Thus it controls the nature of the cellular enzymes and the general development of the cell.

When the DNA in the cell has doubled in quantity, two nuclei form from the original nucleus of the meristematic cell. As these two nuclei, both identical to the parent nucleus, are produced, a new wall is

VITAL PROCESSES OF PLANTS

formed across the cell and two cells arise, each with its own nucleus. Millions of new cells are created in this way by cell division, or *mitosis,* during the growth of a plant.

The newly formed cells increase rapidly in size, partly through the assimilation of food into their protoplasm, partly through the accumulation of water in the cells. The cell walls are stretched in the process.

The next phase of growth is *cell differentiation.* Division of labor occurs among the newly formed, enlarging, and maturing cells. As a result, the various permanent tissues are formed. These include conducting, strengthening, food-making, and storage tissues. Ordinarily, mature tissue cells do not divide any more, but retain their characteristics and their functions throughout their lives. Under certain conditions, however, some cells may continue to undergo division, thus contributing to further growth. This often occurs in the formation of roots on stem and leaf cuttings.

The growth of plants is under the general control of *hormones,* which are synthesized by plants and which are present in all their growing tissues. One growth hormone, indole-3-acetic acid (IAA), has been chemically isolated from plants. It has been shown to influence cell enlargement, leaf fall and fruit fall, the growth of fruits from the ovaries of flowers, the mutual interactions of buds, and various other growth phenomena. Another substance of more complicated chemistry, called gibberellin, has also been isolated. Gibberellin controls the growth of stems and the onset of flowering and fruiting in certain plants. Substances called kinins also occur in plants. They regulate cell division and the processes of aging.

Certain synthetic organic compounds that are chemically related to IAA are now available in commercial preparations. They are used to hasten the formation of roots on stem cuttings, to reduce or prevent premature fruit fall from trees, to produce seedless fruit, and to overcome the dormancy of tubers, bulbs, and other underground stems. They serve to kill weeds by accelerating their respiration and thus depleting their food reserves. They promote the healing of wounds left by pruning, hasten tissue unions in grafts, and retard the growth of trees in nursery storage.

Gibberellin is also available commercially, being produced in a fermentation process by a fungus. It serves to promote germination and to increase the growth rate of yield, especially in grapes and other fruit crops. Germination may be stimulated by synthetic kinins. These substances also promote bud growth and increase the longevity of cut leaves and flowers.

ENVIRONMENT AND PLANT GROWTH

The growth of plants is influenced to a certain extent by the conditions of the environment. Among the most important environmental factors is the availability of water, oxygen, carbon dioxide, and soil nutrients. Soil and air temperatures are also vital factors. So are light intensity and light duration—that is, the number of hours of light received every day. Poisons, mechanical factors such as pressure and contact, and the attacks of parasites also affect growth.

Different species of plants are affected in various ways by these external influences. Some plants, such as water lilies, are able to grow only in water. Others, including most cacti, thrive in soils in which the water supply is extremely limited. The majority of plant species grow best in soils of moderate to abundant moisture content. Different species of plants also react differently to temperature conditions. Plants that are natives of far northerly latitudes are able to withstand low winter temperatures without injury but cannot endure high summer temperatures. Plants of tropical origin must have warmth throughout the year.

Plants also have different soil nutrient requirements. For example, some species require high concentrations of nitrogen, while others thrive only if relatively low concentrations are present. Since plants react in different ways to the food elements in the soil, they require different kinds of fertilizers for maximum production.

Plant growth is influenced to a marked degree by the gases present in the atmo-

The manner of growth in plants is affected by the environment. The broad leaves of the water lily (left) float on the surface of ponds and streams. These leaves give off considerable moisture because of their huge size. This allows the plant to live in water. The baobab (right) has a bottle-shaped stem that acts like a sponge to hold water for the dry season.

sphere. The concentrations of carbon dioxide and oxygen in the atmosphere fluctuate but little. Therefore these gases, required for plant metabolism, are usually available to plants in adequate quantities. Sometimes, however, unwanted gases accumulate. In heavily industrialized areas, where oil and coal are burned on a large scale, and in regions of hot springs and active volcanoes, gases that are not ordinarily found in the atmosphere may be present in harmful concentrations. They may hold back plant growth and may even kill unusually sensitive plants. These harmful gases include carbon monoxide, sulfur dioxide, hydrogen fluoride, hydrogen sulfide, and chlorine.

Since light is essential in photosynthesis, it is an all-important factor in plant growth. Most plants cannot live long in the absence of light, for under such conditions they manufacture no food and therefore starve. Plants that are kept for a time in darkness or in very dim light develop long, weak stems. Their leaves remain stunted and underdeveloped, and they fail to develop chlorophyll in their tissues. The intensity of light is also important. For example, most range grasses, sunflowers, corn, wheat, and milkweeds are sun-loving plants that grow best in bright sunlight and cannot survive in partial or general shade. On the other hand, mosses, ferns, cocoa trees, and many woodland wild flowers cannot endure full sunlight for a prolonged period of time. They can live and grow normally only when they are shaded during all or most of each day.

PLANT MOVEMENTS

Both plants and animals possess the property of *irritability*. That is, they are sensitive to various factors of the environment and they can react to these factors by a series of movements. In some ways, such reactions are quite similar in plants and animals. In both these forms of life, the reactions are brought about by such stimuli as temperature, light, contact, chemical agents, electricity, gravity, water, and gases. In both animals and plants, the stimuli are frequently transmitted from one part of the body—a receptive zone—to another part—a reactive zone.

In some important respects, however, plant reactions resulting in movements differ from the corresponding animal reactions. For one thing, the reactions of all higher plants involve only the movements of parts and never the movement of the entire body of the plant from one place to another. Again, the reactions of most animals result from the contraction and extension of muscles; those of plants, from shifts in growth rates or changes in internal water pressure. In the third place, the higher animals have specialized sense organs that

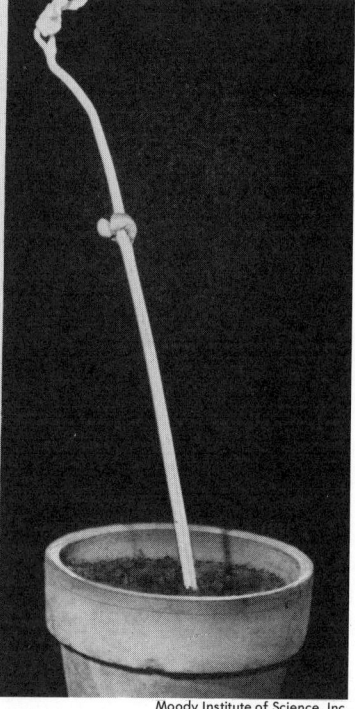

A striking illustration of the importance of light in the growth of plants. The bean on the left was given a normal amount of light, while the bean on the right grew in darkness.

Moody Institute of Science, Inc.

receive external stimuli. Plants, with few exceptions, have no such specialized structures. They receive stimuli from the outer world rather generally through their more newly formed tissues. Finally, in most animals the effects of stimuli are transmitted by specialized nerve tissue. In plants they are transmitted by diffusing chemical substances.

Plant movements are brought about either by changes in growth rates—that is, by growth reactions—or by changes in internal water pressure—the so-called turgor reactions, or movements. These two kinds of reactions differ in various respects. Growth reactions occur in all higher plants, wherever there are actively growing tissues. Turgor movements, however, are found only in certain families of flowering plants, including the legumes, wood sorrels, and several other groups. Growth reactions are the chief means by which the higher plants adjust to their environment. They are definitely beneficial to the plants. Turgor movements rarely benefit the plants in which they occur. Growth reactions require hours, days, or even weeks for their completion. Turgor movements are relatively rapid; they may be completed within a second or, at most, in 30 minutes.

Growth reactions. The most important growing reactions are known as *tropisms,* a word derived from the Greek *trope,* meaning "a turning." Tropisms of plants are bending movements of more or less cylindrical organs, such as stems, roots, and stalks. They are caused by different concentrations of growth hormones on the two sides of the stimulated organ.

There are different kinds of tropisms. These are indicated by an appropriate Greek word placed before "tropism." The prefix represents the type of stimulus. For example, sunlight brings about phototropism—*photos* is the Greek word for light. The earth's gravitational force causes geotropism—*geo* means "earth." A reaction to water is called hydrotropism—*hydro* means water. Contact brings about thigmotropism—*thigma* means "touch". And a reaction to shadow is called skototropism—*skoto* means "darkness." The normal growth hormone, indole-3-acetic acid, is involved in all tropisms.

Phototropism and geotropism have been more thoroughly investigated than the rest. Phototropism takes place when a plant grows in such a way that one side receives more illumination than the other, as in the case of a plant set upon a windowsill. Under such conditions, the plant will turn its stems and leaves toward the sun. If such a plant is to grow symmetrically, it should be turned regularly (if it is in a flowerpot), so that the different sides will face the sun in turn.

You can observe the action of geotropism, brought about by the stimulus of the earth's gravitational force, when you place a plant in a horizontal position. Within a few hours or days, the main stem of the plant will begin to bend upward, assuming the normal upward direction of growth. The main roots will bend downward.

In either phototropism or geotropism, there can be no question of purposeful activity. A plant growing on a windowsill does not turn its leaves and stems toward the sun because it is seeking a source of light. If a plant is set horizontally, it does not struggle to attain an erect position. In either case, the movements result from the unequal distribution of growth hormone.

In a plant growing vertically, the hormone is formed in the tip of the stem and ordinarily diffuses downward into rapidly enlarging cells. It diffuses equally on all sides. If, however, light intensity is greater on one side than on the other, an unequal distribution of growth hormone occurs. The less brightly illuminated side will then receive more of the growth hormone than the other side. The darker side will grow more rapidly than the opposite one, and, as a result, the stem will bend.

A shift in hormone distribution also occurs if a plant is placed in a horizontal position. The lower side of the stem will receive more hormone than the upper one. Hence growth will occur more rapidly on the lower side, and the stem will slowly bend upward.

Similar inequalities in distribution of growth hormone, corresponding to shifting external stimuli, are responsible for other tropisms, such as the skototropism exhibited by certain vines. When the vine seedlings first emerge, they grow toward the shadow of the nearest object up which they may be able to climb. This is not merely negative phototropism, since seedlings around a tree will approach it from all directions. Once the vines start to climb, they revert to phototropism. Botanists still do not understand how stimuli bring about growth-hormone movement.

Turgor movements. These occur in the flowers of some species of plants, such as moonvine, morning glory, and four-o'clock. They occur in the leaves of other species,

Top: phototropism. Successive stages in the phototropic bending of a bean seedling. Exposures were made on the same plate at 40 minute intervals. Middle: geotropism. The main stem of the plant bends upward, while the main roots bend vertically downward. Bottom: the bean roots grow downward, no matter in what position the bean is placed.

(top) From "The Plant World," by H. J. Fuller and A. B. Carothers, © Holt, Rinehart, and Winston, Inc. By permission. (middle) William G. Smith, Jr.—Boyce-Thompson Institute for Plant Research. (bottom) Moody Institute of Science, Inc.

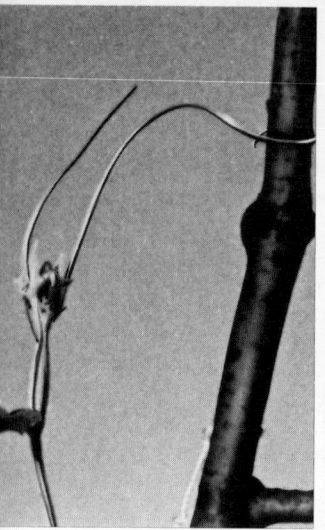 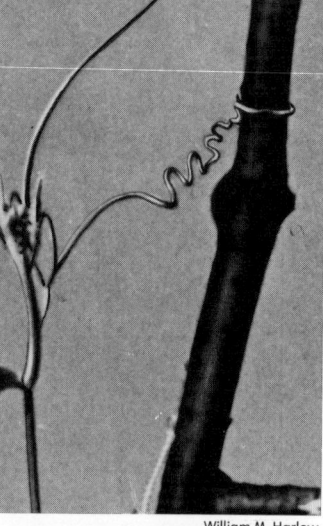

A wild cucumber tendril attaches itself to a support. When contact is made (left), the tendril twines around the support. The twining movement stops and a series of coils (right) is produced. These movements are turgor movements.

William M. Harlow

including clovers, beans, locusts, and various plants that trap and digest insects. Apparently hormones are involved in turgor movements, just as they are in tropisms. Little is known of these hormones. They seem to be chemically different from indole-3-acetic acid, the hormone that is involved in tropisms.

The turgor reactions of the flower petals and leaves of such plants as clovers and locusts are usually called sleep movements, because they are stimulated by the daily changes of light intensity as night alternates with day. The leaflets of white clover, for example, occupy a horizontal position between sunrise and sunset. At nightfall, they point upward from the stem and maintain that position during the night. At dawn, they return to their daytime horizontal position. In garden beans, the leaflets point downward from the stem at night and have a horizontal position during the day.

Leaf reactions such as these result from changes in the internal water pressure of small masses of tissue, located at the bases of the leaflets, in response to changing light conditions. Ordinarily, these reactions are rather rapid; sleep movements are usually completed within 30 minutes after a pronounced change in light intensity. The exact significance of these movements is unknown. Most authorities assume that they do not benefit the plant.

Some turgor movements, such as the daily sleep movements of leaves and flowers, will persist for many days even if the plant is removed to a totally darkened room. Such persistent rhythms, called endogenous rhythms, maintain a roughly 24-hour interval, and are believed to be controlled by internal timing mechanisms called biological clocks. The nature of such mechanisms is unknown.

The plant called the Venus's-flytrap ensnares insects in its leaves as a result of rapid turgor movements in the leaf halves. Each leaf has a hinged portion, whose inner surface contains hairs sensitive to contact. When an insect alights on these hairs, the contact stimulus causes the leaf halves to close rapidly. The reaction is usually completed within two seconds and effectively traps the insect. The leaf then digests the softer parts of the victim's body. The Venus's flytrap is one of the few plants with specialized sensory devices—the hairs on the inner surface of the leaf. Obviously, in this case the turgor movement is beneficial to the plant.

The leaves of the sensitive plant (*Mimosa*) fold in a striking fashion when they are touched or when they are subjected to electric shock or to changes in light intensity. These reactions, like those of the Venus's-flytrap leaves, are almost instantaneous. They result from rapid turgor changes in cushions of the cells at the bases of the leaflets. It is hard to see how these reactions benefit *Mimosa*.

REPRODUCTION

There are certain other vital processes that are of the utmost importance in the development and survival of plant species—those that have to do with reproduction.

Recent investigations have thrown much light on the formation of flowers and on other phases of reproduction in flowering plants. Flowers are really clusters of organs that have to do with the processes of reproduction. Their formation is influenced by various factors: the availability of stored food in the plant and temperature and light conditions—particularly the periods of light duration, or *photoperiods,* as they are called.

Large amounts of food are required for the formation of flowers, for the reproductive processes occurring within flowers, and for the development of fruits and seeds from flowers. This food is used partly to provide the necessary energy, partly to build up new protoplasm and cell walls. Every flowering plant, therefore, must pass through a vegetative period, which varies in length in different species. During this period, the plant absorbs raw materials, manufactures and stores food, and grows, before it can form flowers.

The plant does not necessarily begin to flower after a certain quantity of food reserves has accumulated. Many plants require exposure to certain definite temperatures or photoperiods, or both, before they can form flowers. For example, winter-wheat plants and apple trees must have at least a brief exposure to low winter temperatures before they can flower normally. That is why they cannot be successfully grown in the tropics. Some plants require short photoperiods; others, moderate ones; still others, long ones.

We do not know just what these critical temperatures and photoperiods have to do with the formation of flowers. However, it appears likely, on the basis of numerous experiments, that a specific flowering hormone is produced within plants under the proper conditions of temperature and light. This flower-inducing hormone has tentatively been named *florigen.* Its chemical composition has not yet been determined, but it appears to have some relation to the gibberellins. The influence of temperature and light on florigen production appears to depend upon the action of a recently discovered protein pigment called *phytochrome.* Phytochrome can exist in two forms, which we can think of as a "dark" form and a "light" form. The ratio of these two forms is controlled by heredity, temperature, and light, and controls many aspects of plant behavior.

After flowers have appeared and have opened, their reproductive processes begin. The first of these processes is *pollination.* This is the transfer of pollen grains by insects, birds, or wind from the stamens, or male parts, to the pistils, or female parts, of flowers.

Within the pistil, pollen grains form tubes. These reach the ovules, or immature seeds, located in the ovary, which is at the base of the pistil. Fertilization then occurs, as sperms from the pollen grains unite with eggs within the ovules. One sperm fertilizes one egg. The fertilized egg, or zygote, then begins to develop. Ultimately, it will become the embryo of the seed.

Under the stimulus of the growth hormone contained in the pollen, the ovary also begins to grow. During this period, large amounts of food and often water move into the ovaries and ovules. As the ovule grows, its surface cell layers develop into the seed coat. The ovary becomes a mature fruit.

If the fruit is a *dry fruit,* as in bean, tulip, and corn plants, it will split open, either along a definite seam or irregularly. If it is a *fleshy fruit,* as in tomato, grape, and peach plants, its tissues ultimately disintegrate. In either case, the seeds are liberated from the inside of the fruit.

If the seeds reach the soil, and if the temperature, moisture, and oxygen conditions are suitable, the seeds will germinate. The embryo, in time, will grow into a mature flowering plant and the cycle will begin anew.

In many plants, the development of fruits from ovaries and of seeds from ovules exerts such a drain upon the food stored in roots, stems, and leaves that these vegetative parts die. This commonly happens in annual plants, in biennials during the second year of their life, and in some perennials.

For instance, century plants have a vegetative period as long as 40 years. After that, they flower once, produce seeds, and die. There is sketchy evidence that, in some species at least, the repressive effects of fruiting upon vegetative growth may be due to hormonal action. It is becoming increasingly clear to scientists who study plants that hormones play important roles in plant reproduction.

VITAL PROCESSES OF PLANTS

PHOTOSYNTHESIS

by Beth Schultz

Sometimes, in a plot of grass, you can see a mushroom. The mushroom is easy to see because it is usually white, whereas the grass is green.

Grass is green because it contains a chemical known as *chlorophyll*. The name comes from two Greek words that mean "green leaf". Mushrooms do not contain chlorophyll. Nor do plants such as molds and bacteria. But most of the plants with which you are familiar—trees, flowering plants, ferns, mosses, algae—are green plants. At least some of their cells contain chlorophyll.

Chlorophyll enables a green plant to make food. Practically all other plants and most animals cannot make their own food. They depend on green plants for food, either directly or indirectly, and also, as we shall see, for oxygen.

The process by which green plants make food is called *photosynthesis*. The word "photosynthesis" is derived from two Greek words meaning "light" and "putting together." Green plants put certain chemical elements together to make food. To do this, they need light energy. Normally this energy comes from the sun. But green plants can also grow indoors, using light from electric lamps rather than from the sun to carry on photosynthesis.

In addition to light energy and the presence of chlorophyll, plants need carbon dioxide and water to make food. Through a series of complicated reactions, food in the form of sugar is made. This sugar is often converted into starch, which stores the chemical energy of sugar.

IMPORTANCE OF PHOTOSYNTHESIS

All living things need a constant supply of energy. One constant source of energy on earth comes from the sun, in the form of radiation. In photosynthesis, energy in the form of radiation, or light from the sun, is changed into chemical energy.

Swiss Travel Bureau

Cattle feeding on grass. By means of photosynthesis, the grass produces foodstuffs for its own benefit. But the cattle also get the benefit of the food in the grass: after eating it, they convert the grass into energy for life and into flesh. People indirectly "eat" the grass when they eat beef.

Green plants, of course, carry out photosynthesis to make food only for themselves. But animals and nongreen plants depend on green plants for their food, which contains the energy they need for life. They also need the oxygen given off by green plants in photosynthesis.

Some animals, such as cows, horses, and certain fish, eat the green plants. Other animals, such as lions, eagles, and sharks, eat the animals that ate the green plants. Nongreen plants get their food from either dead or living animals or plants.

THE STUDY OF PHOTOSYNTHESIS

The chemical activities of green plants did not begin to be scientifically recognized until some 300 years ago. In the 1640s, a Flemish doctor named Jan Baptista van Helmont put 90 kilograms of dry soil into an earthenware pot. He soaked the soil with water and planted a 2-kilogram willow in it. For five years Van Helmont protected the willow from dust. He also supplied it with pure water only, in the form of rain or distilled water.

At the end of this time, Van Helmont removed the willow from the pot and weighed the plant. He found that the willow had grown to weigh 76 kilograms, a gain of 74 kilograms. The soil, after it had been dried, weighed nearly 90 kilograms—nearly as much as it did at the beginning of the experiment.

The Flemish doctor concluded—wrongly, as we know now—that the extra 74 kilograms of willow plant developed only from the water he had supplied to it. He did not suspect how important the substances in air and soil also were for the growth of the plant.

The idea that plant matter comes from water alone lasted until the 1770s. In 1771, the English chemist Joseph Priestley observed that green plants, in daylight, "restore," or refresh, air that has been "injured," or exhausted, by candles burning or by animals breathing. That is, green plants somehow reverse the chemical effects of breathing and burning. In air refreshed by green plants, candles could burn and animals could breathe again. Priestley, however, did not then know about oxygen and the role it played in these experiments.

Modern understanding of photosynthesis really began with the discovery of oxygen and the work of a Dutch-Austrian physician named Jan Ingenhousz.

Beginning in the 1770s, Ingenhousz conducted experiments on plants and uncovered some basic facts about photosynthesis. He also showed that green plants do not "purify" the air at night. Instead of taking in carbon dioxide and water and giving off oxygen, green plants in darkness simply respire. That is, they take in oxygen and produce carbon dioxide and water as wastes. Ingenhousz, however, did not realize that plants also respire in sunlight.

Another great step in photosynthetic study came in 1837, when *chloroplasts* were discovered. These are tiny green bodies in the cells of most green plants. Later it was found that chloroplasts are green because they contain chlorophyll. These are the cell parts where photosynthesis takes place.

In the following years, scientists learned in more detail how light, chlorophyll, water, and carbon dioxide are needed to produce food and oxygen. In 1938, Robin Hill, an English chemist, discovered that oxygen released during photosynthesis comes from the molecules of water absorbed by green plants.

Until the late 1930s, most of the complicated reactions of photosynthesis were unknown or little understood. Since then, new techniques have opened the way to rapid progress. The electron microscope, for example, has helped in the study of the structure of chloroplasts. With a chemical method known as chromatography, scientists have been able to separate and analyze very small amounts of plant substances. Still other very sensitive methods have been used to detect plant chemicals.

To study the substances that are formed during photosynthesis, scientists also use atoms that give off radioactivity. By detecting the nuclear radiation from these atoms, scientists can follow the various chemical elements through the photosynthetic reactions. Thus, they identify the compounds that are formed.

THE CHEMISTRY OF PHOTOSYNTHESIS

PHASE ONE—THE LIGHT REACTION

When a chlorophyll molecule absorbs light, one of its electrons becomes "excited." That is, the electron acquires additional energy and leaves the molecule. This electron can supply the energy needed to form new compounds.

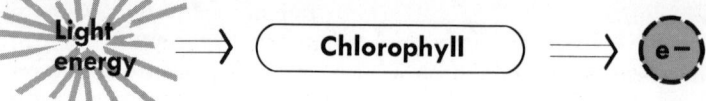

After electrons are released from the chlorophyll molecules, substances called electron carriers take them to energize the following chemical reaction. A phosphorus-containing compound called *adenosine diphosphate*, or ADP for short, reacts with other phosphate groups (P) to make *adenosine triphosphate*, or ATP.

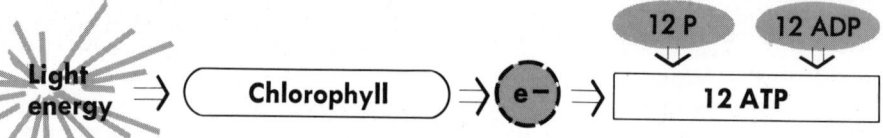

As these reactions occur, another also takes place. Molecules of water break up, or ionize.

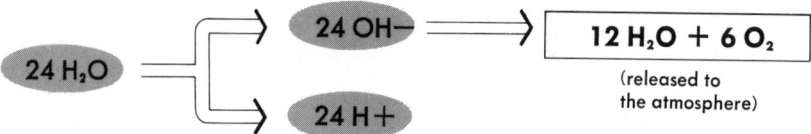

The energy of chlorophyll's excited electrons also sparks another reaction. It enables another phosphorus compound, *triphosphopyridine nucleotide*, or TPN, to accept hydrogen atoms from water molecules in the cell.

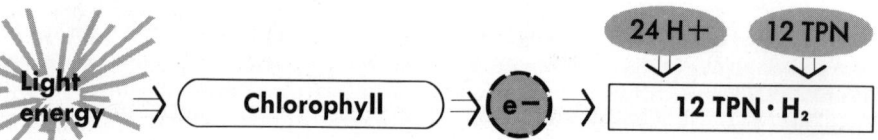

PHASE TWO—THE DARK REACTION

Chemical energy is released when ATP loses phosphate, becoming ADP. Energy is also released when TPN loses hydrogen. This hydrogen combines with carbon dioxide to form glucose. The energy released from ATP and the TPN-hydrogen compound is trapped in the glucose molecules.

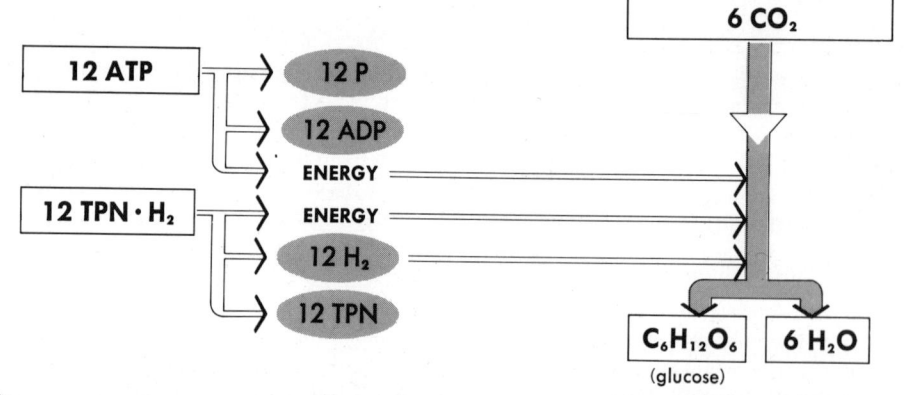

86   PHOTOSYNTHESIS

THE CHEMISTRY OF PHOTOSYNTHESIS

Here, in very brief form, is what happens during photosynthesis:

1. Light reaches the cells that contain chlorophyll.
2. The chlorophyll absorbs the light energy and changes it to chemical energy.
3. Water molecules in the cells break up. The oxygen from the water is set free.
4. The hydrogen from the water molecules combines with carbon dioxide from the environment to form glucose, a simple sugar.
5. The glucose molecule is held together by the chemical energy. In other words, the energy is trapped within the glucose molecule.

Some of these steps require light; others do not. This leads us to a discussion of the two phases of photosynthesis.

Phase one is sometimes called the *light reaction,* because light is needed to carry on the process. During this phase, light energy is trapped and converted into chemical energy. At the same time, oxygen is set free.

Phase two is sometimes called the *dark reaction,* because light energy is not needed. It is a strictly chemical process. In this phase the energy trapped in phase one is used to make the glucose.

The following chemical equation summarizes the steps in photosynthesis:

$$\text{Energy} + 6CO_2 + 6H_2O \rightarrow C_6H_{12}O_6 + 6O_2$$

Read this as "energy plus 6 molecules of carbon dioxide plus 6 molecules of water yield 1 molecule of glucose and 6 molecules of oxygen." The letters C, H, and O stand for the elements carbon, hydrogen, and oxygen. The large numbers stand for the numbers of molecules. The small numbers below and to the right of the letters represent the numbers of atoms in the molecules. Thus, O_2 means "2 atoms of oxygen."

GREEN PLANTS RECYCLE

If we read the photosynthesis equation backward—from right to left—we get the chemical reaction for *respiration.* Respiration is the chemical breakdown of food in an organism for energy. In many living things, respiration takes place when oxygen combines with, or "burns," sugar.

Thus, oxygen plus sugar yields energy, with water and carbon dioxide produced as wastes. But carbon dioxide and water are also the raw materials of photosynthesis. From them, a green plant makes food and oxygen. That is, it *recycles* these wastes into useful products.

CHLOROPHYLL IN THE CELLS

Photosynthesis takes place in plant cells containing chlorophyll. There is not just one form of chlorophyll, but four different chemical forms. However, they all react in the same way to light.

The organization of chlorophyll in a cell depends on the type of plant. The simplest plants that contain chlorophyll are one-celled algae, which live in water or on moist surfaces.

In contrast, higher plants—mosses, liverworts, ferns, and seed plants—which usually live on land, have specialized tissues. These tissues are composed of millions of cells of different types. The cells in certain plant parts, especially in stems and leaves, are full of chloroplasts, which contain chlorophyll. Other parts, such as the roots and blossoms, have no chlorophyll and so they do not have the ability to carry out photosynthesis.

Scientists once thought that chlorophyll is able to photosynthesize simply because of its chemical structure. But they have since learned that when chlorophyll is removed from chloroplasts, it loses much of its photosynthetic powers. Why?

The answer seems to lie in the way that chlorophyll is organized inside a chloroplast. This organization is chiefly responsible for the efficiency of chlorophyll in turning light energy into chemical energy.

Electron microscopes have shown that within a chloroplast the molecules of chlorophyll are arranged in definite layers, called *lamellae.* Other colored molecules, called *pigments,* which aid chlorophyll in its task, also form part of the lamellae. In the chloroplasts of most plants, the lamellae form orderly bundles called *grana.*

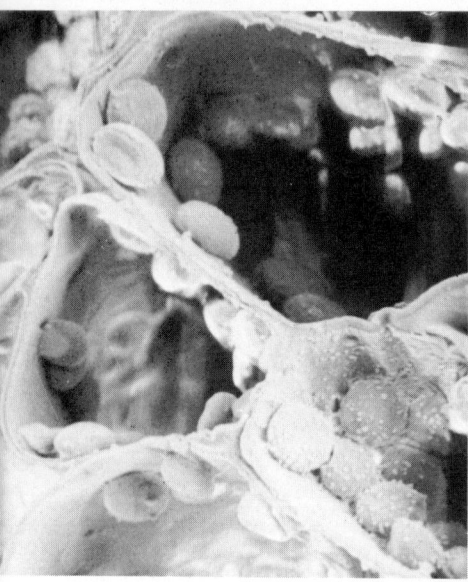
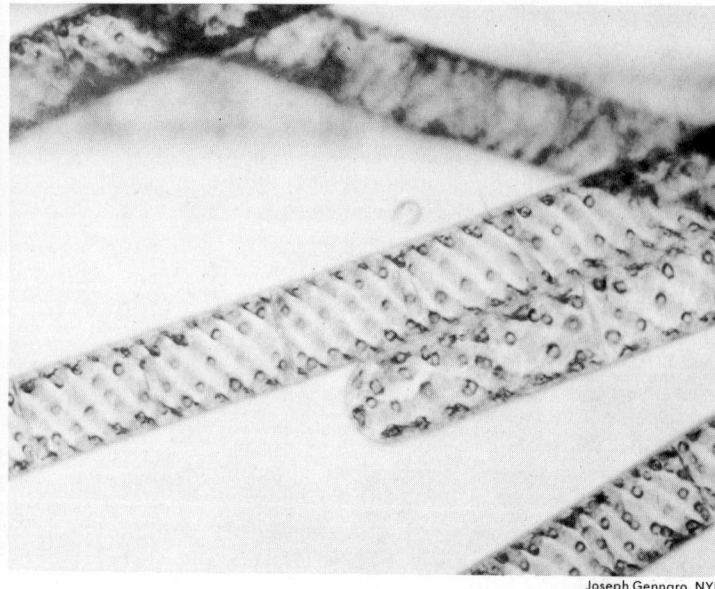

In most green plants, chlorophyll is contained in structures known as chloroplasts. The plant cells (above left) have been magnified 2,000 times. You can see the round-bodied chloroplasts attached to the walls of the plant cells. In the alga *Spirogyra* (above right), the chloroplasts are arranged in spiral structures. Algae are the simplest plants containing chlorophyll.

CONTROL OF PHOTOSYNTHESIS

Scientists can trace the paths of light, carbon dioxide, and water to the cells where photosynthesis takes place. For example, algae grow at or near the water surface, where sunlight penetrates. They absorb water and carbon dioxide dissolved in water directly through their thin cell walls.

On the other hand, land plants have leaves, which collectively expose a large surface to sunlight. They contain thousands of millions of cells with chloroplasts. Air containing carbon dioxide enters the leaves through openings called *stomata*. From these the air diffuses, or spreads, among the loosely arranged cells of the leaves.

The stomata have many functions. Not only do they admit carbon dioxide for photosynthesis, they also pass out the oxygen formed in this process. In addition, stomata absorb oxygen from the air for respiration and release carbon dioxide as a waste back into the atmosphere.

Stomata also control the amount of water passing out of the plant. They do so by adjusting their size. *Guard cells* surrounding the stomata shrink when they lose water, thus closing the stomata. Thus, too much water loss is prevented.

Water for photosynthesis and for other life processes is drawn by land plants from the soil, usually through roots. The water passes from the roots to the stem and then into the leaves. Excess water evaporates and escapes into the atmosphere.

The amount of photosynthesis that takes place depends on the supply of light, water, and carbon dioxide. It also depends on the temperature. If any one of these factors is unfavorable, photosynthesis slows down. For example, if an early summer season lacks sufficient rain, the soil and air will be dry. Plants will grow slowly or not at all, even though there is abundant sunshine and clean air containing carbon dioxide. The guard cells lose enough water to close the stomata, and little or no carbon dioxide enters the leaf. But the closing of the stomata prevents water loss; the plant will stay alive even though it cannot grow. In this case, water supply determines the rate of photosynthesis.

ROOTS AND STEMS

by Clyde M. Christensen

The outer appearance of an oak tree or rose bush or any other seed plant on a bright summer day gives no idea of the activity going on within. Certain raw materials are being absorbed by the roots and transported through the stems to the leaves. Other materials are drawn by the leaves from the outer air, are transformed into foods, and passed to the stems and from the stems to the roots. All these activities are called *vegetative*.

ROOTS

A scientist placed a small rye plant in a box of soil. After four months, the plant was dug up and its roots were studied. This single plant had nearly 14,000,000 roots. The combined length of the roots was 623 kilometers.

Why are there so many roots on a comparatively small plant? How do they serve the plant?

HOW ROOTS DEVELOP

The first root to develop in a plant is the *primary root*. It arises from the embryo contained in the seed. When the seed germinates, the tiny rootlet emerges and penetrates the soil. Within a few days it begins to put out branches from the tissues back of its growing point. These branches in turn give rise to other ones. All the roots that come from the primary one are known as *secondary roots*. The young primary root that pushes its way into the soil has several distinct regions. At the very tip is a protective covering called the *root cap*. This serves as a buffer when the rootlet pushes forward between and around jagged and sharp-edged soil particles. Immediately behind the root cap is a zone of actively dividing cells. Next comes a zone of cell elongation, in which the cells grow chiefly in length. Finally there is a zone of maturation. Here the different tissues found in mature roots are formed. Tiny hairlike projections, called *root hairs*, form the outermost layer—the *epidermis*—of this zone. A root hair is simply an elongated part of a cell in the epidermis.

A plant has many more root hairs than roots. That rye plant with the 14,000,000 roots had over 14,000,000,000 root hairs. They greatly increased the surface area of the roots. This is very important because the roots hairs are the chief nutrient-gathering parts of the root. They come in close contact with small particles of soil and attach themselves firmly to the particles. Soil water and minerals dissolved in this water

Young primary root. A root cap serves as a protective covering. Above this is the zone of dividing cells. Then there is a zone of elongating cells and above that a zone of mature cells. Root hairs form the outermost layer of this last zone.

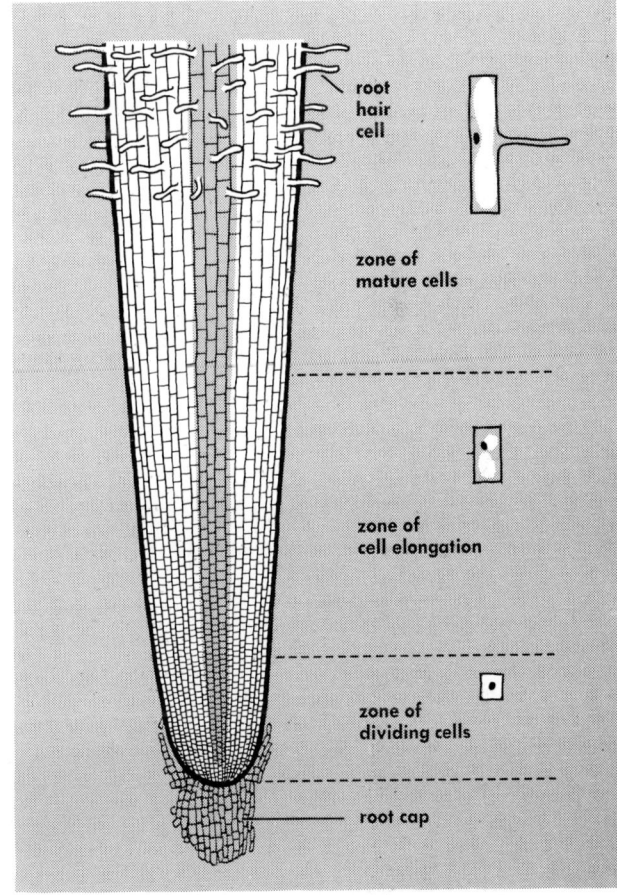

diffuse through the cell walls of the root hairs. They pass to conducting vessels in the roots and up through the stems to the leaves. Here they serve as raw materials for the manufacture of food. The root hairs are very delicate; if exposed to the outer air, they soon die.

MYCORRHIZAL ROOTS

Some roots absorb water and minerals from the soil by forming a partnership with a fungus. These are called *mycorrhizal roots,* or "fungus roots". They consist of the rootlet of a higher plant combined with the *mycelium*—an interwoven mass of threadlike filaments—of a fungus.

Mycorrhizal roots are usually short, stubby, thick, and much branched. They have few or no root hairs. The fuzzy mycelium may cover the outside of the rootlet or may flourish within its cells. In either case, strands of mycelium grow out into the soil.

Like root hairs, the fungus mycelium absorbs water and dissolved minerals from the soil and transports them to the root. It may also contribute to the root certain growth-promoting substances manufactured by the fungus. Mycorrhizal roots are found in many kinds of plants, including conifers and orchids.

CROSS SECTION OF A ROOT

The interior of a very young rootlet is made up mostly of unspecialized cells. As the root grows older, different types of tissues develop. If we examine the cross section of a root in the root hair zone, we see several clearly defined areas. As mentioned earlier, the outer layer of cells is the epidermis, from which root hairs arise. Inside this outer layer is the *cortex*. It is made up of thin-walled cells, called *parenchyma* cells, with numerous air spaces between them. Water and salts absorbed by the root hairs pass through the cortex to the center of the root. In many plants the cortex also serves not only for transportation but also to store food.

Next to the cortex we find a single layer of cells called the *endodermis*. The name comes from two Greek words that mean "inner skin". This is the outer part of the conducting portion of the root—the so-called *vascular* (conducting) *cylinder,* or *stele*. Inside the endodermis is a layer of small cells called the *pericycle*. Cells produced here penetrate the cortex and epidermis and form secondary roots.

Within the pericycle are the conducting tissues called xylem and phloem. They are found in every part of the plant, from the roots to the leaves. *Xylem* conducts materials upward from the roots to the leaves by way of the stems. There are two types of conducting cells in this tissue: tracheids and vessels. *Tracheids* are long, tapering cells, often with thin areas, called pits, in their walls. *Vessels* are made up of cylindrical cells placed end to end. The end walls of these cells have disappeared, allowing materials to pass through from one to another. Tracheids and vessels are not always found together. In addition to these conducting vessels, xylem has strengthening fibers and storage cells.

Materials are conducted in *phloem* tissue from the leaves to the stems and through the stems to the roots. The chief conducting cells of phloem are *sieve tubes*. Like xylem vessels, these consist of long cylindrical cells set end to end. In this case the end walls have perforations like those of a sieve. Smaller cylindrical cells, called *companion cells,* are sometimes found beside the sieve cells. Phloem also contains fiber cells for strength and other cells for storage.

When viewed in cross section, root xylem cells radiate like the spokes of a wheel from the center of the plant or from a "hub" made up of parenchyma cells. Small groups of phloem cells are found between the spokes of the xylem.

In perennial roots—those continuing to live from year to year—there is a *cambium zone* between the xylem and the phloem. This zone is made up of actively dividing cells. Xylem is formed on the inside of the cambium, phloem on the outside. As a result, the root grows in diameter. Just outside the phloem there is another cambium layer—the *cork cambium*. The cells that develop in this region have waxy, waterproof walls and are known as cork cells. The two outermost zones of the young root—the epidermis and the cortex—gradually disintegrate. The cork cells then form an outer, barklike layer. We find roots of this kind in most trees.

ROOT SYSTEMS

As the roots develop, they form a more or less complicated network. Generally speaking, the root systems of seed plants fall into two classes—fibrous and taproot. In the *fibrous system,* there are many branches that are about equal in diameter. A system of this kind may be very extensive, as we saw with the rye plant mentioned earlier.

In a *taproot system,* the primary root that develops from the embryo grows faster than the branches that arise from it. It becomes a main root that grows down through the soil, with much smaller roots branching from it. Usually a taproot is thick toward the top and tapers downward to a point. Among the common plants with taproots are carrots, radishes, beets, and dandelions. There are all sorts of variations of the two main types of root systems.

Some roots, including those of various grasses, do not extend more than 25 centimeters or so beneath the soil. The roots of trees may penetrate the ground to a depth of three meters or more. Such roots, of course, can make use of water sources that would not be available to plants whose root systems are near the surface. Deeply penetrating root systems also help to anchor the plant firmly to the soil.

Roots serve other purposes besides absorbing materials from the soil and anchoring the plant. Many roots store food produced during the growing season. This is especially true in certain biennial, or two-year, plants, such as carrots and beets. In these, food is accumulated in the roots during the first year; flowers and seeds are produced in the second.

SPECIALIZED ROOTS

Certain roots arise neither from primary nor secondary roots but from stems or leaves. They are called *adventitious roots.* A new plant grows from a tulip bulb—which is really an underground stem—because adventitious roots arise from the bulb. When a gardener sets a willow-stem cutting or a rubber leaf in the ground, adventitious roots are formed and a new plant develops. This is called *vegetative propagation.*

The stems of certain plants, such as the banyan tree, bear *aerial roots.* These grow downward until they reach the ground. They then penetrate the soil and firmly establish themselves. They are called *prop roots.* The banyan tree's prop roots will produce other roots, as well as new stems. In this way an entire colony of banyans, covering half a hectare, can arise from a single tree.

Mangrove trees, which grow along ocean shores in tropical regions, also put forth prop roots. They reach the water and penetrate the sandy bed. The numerous roots growing down from a mangrove tree form a tangled mass in time. Drifting sand, decaying leaves, and other materials are trapped here. In this way soil is built up.

The aerial roots of orchids play an important part in the life of the plant. For

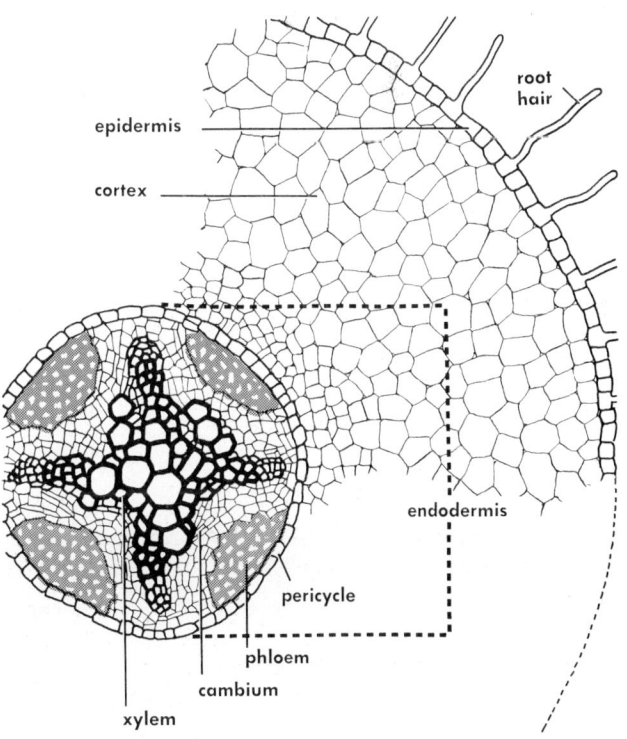

Cross section of a mature bean root in the zone of hair roots, showing the areas into which the root is divided. Cells have become specialized in some areas to carry out specific functions.

one thing, they serve to anchor the orchid to the tree or other support on which the orchid grows. They also absorb water from the moisture in the air or from rain. Food materials for the orchid are drawn from the debris that collects around the roots. An orchid that grows on a living plant is not a parasite, because it does not obtain its food from the host plant.

The mistletoe, which grows on plants such as oaks and maples, is parasitic. Its roots develop into suckers. These enter the tissues of the host plant and absorb food from them.

Among the most curious of specialized roots are those of the bald cypress tree, which grows in swamps. These roots, called *knees,* project upward from the water or waterlogged soil in which the cypress grows. They absorb air, which is then transmitted to the tissues of the plant.

STEMS

Like roots, stems serve several purposes. They produce and support leaves, which are the food-making centers of plants. They provide passageways along which various substances are transported within the plant. Stems also serve as a storage place for foods. And in certain cases, parts of stems propagate plants.

The first stem develops from the embryo of the seed. A miniature stem emerges as the seed germinates. Generally, it makes its way upward through the soil and into the air. As long as its growing point remains alive, it becomes longer. In time branches develop. If the growing point is killed or removed, one or more growing points may be formed below the cut end. Stems that grow above the ground in this way are called *aerial stems.* As we shall see, certain stems grow underground.

Stems show various kinds of growth patterns. They differ in size, in their method of branching, and in length of life. It is customary to divide them into two classes— herbaceous and woody.

TYPES OF STEMS

Herbaceous stems are generally slender, greenish, and comparatively soft. They are annual; that is, they live through only one growing season. In some plants, such as sweet peas and squash, the roots are also annual. The entire plant dies at the end of the growing season. The hollyhock, hibiscus, and certain other plants have annual stems but perennial roots. The stems die at the end of the first year. The next year, new stems arise from the roots. Plants with herbaceous stems are called *herbs.*

Woody stems are taller, thicker, and harder than herbaceous stems. When they are mature, their surface is made up of bark. Plants with woody stems may be either trees or shrubs. A *tree* usually has a thick main stem, called a *trunk,* which branches only at some distance above the ground. In a *shrub,* there are a number of comparatively slender main stems, which branch abundantly. Trees grow taller than shrubs.

Actually there is no hard and fast distinction between trees and shrubs. A plant in one place may grow as a tree. In another, it will be a shrub. A number of forest trees are large and tall on the lower slopes of mountains. In the timberline area they grow as shrubs. When elms and various other trees are grown in hedges and clipped regularly, they develop as shrubs.

BUDS, PORES, AND SCARS

The actively growing stem has a number of buds. These are particularly conspicuous in trees or shrubs that lose their leaves after the growing season has come to an end. Some buds are formed in the *leaf axil* —the acute angle formed where the leaf stalk joins the stem. These are called *lateral, or side, buds.* Generally each stem also has a bud at its tip—a *terminal bud.* Terminal buds continue the growth in length of a stem. Lateral buds give rise to new leaf-bearing twigs. Certain buds, called *floral buds,* develop into flowers.

In some plants, lateral buds grow spirally along the stem. In others they occur in pairs, one bud opposite the other on the stem. The twigs that develop from the buds will show the same arrangement—either spiral or opposite. Leaves, buds, and the branches originating from buds arise from

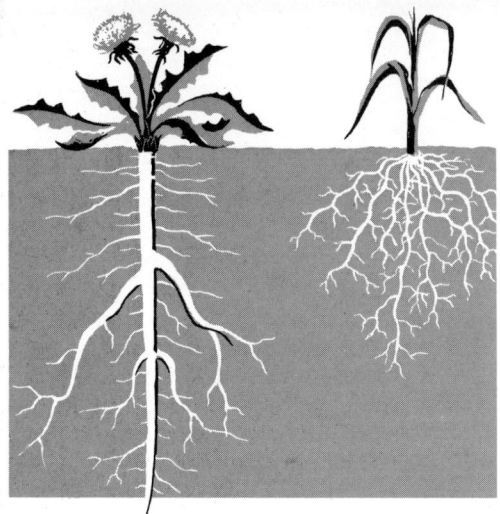

Left: taproot system of a dandelion. The primary root grows down through the soil, with smaller roots branching from it. Right: the fibrous root system of a grass. The different root branches are about equal in diameter.

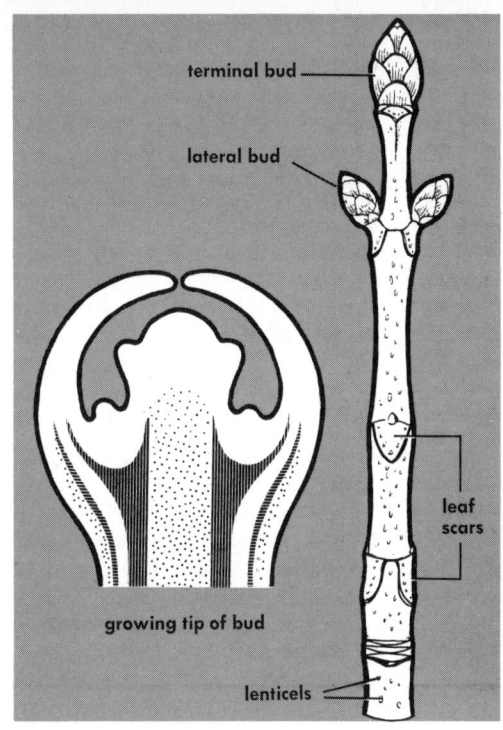

Left: cross-section of a bud tip. Right: an actively growing stem, showing buds, leaf scars, and lenticels, or pores.

slightly enlarged portions along the stem called *nodes*.

Buds are conical structures, made up of cambium tissue. They are really greatly shortened stems. In most herbaceous plants bud tissue is protected from the outer air only by embryonic leaves. Such buds are called naked. In woody stems the buds are usually covered with bud scales, which provide ample protection by overlapping.

Not all buds are to be found at the tips of shoots or the axils of leaves. Many plants develop *adventitious buds* at other places. Often these arise when the plant suffers injury. For example, if the leaves or twigs are eaten by animals, adventitious buds will appear. They will give rise to new twigs and leaves. In this way a plant can endure repeated grazing by animals or cutting by the hand of man.

A number of perennial plants have *dormant buds*, which are formed in the young stem. They continue to grow as the stem increases in diameter but do not become visible outside of the bark. If a part of the stem is cut off, these dormant buds will produce sprouts. When certain oak trees are felled, the dormant buds in the stump give rise to new shoots. Conifer trees and shrubs are often pruned to give them some desired shape. The new twigs that grow out back of the cut ends come from dormant buds, or adventitious buds, or both.

In the woody twigs of trees and shrubs we see other structures besides buds and leaves. Scattered over the surface are a great many tiny pores, called *lenticels*. Sometimes they are more or less circular. They may also take the form of slits. Gases pass through the lenticels from the plant tissues to the outer air, and from the outer air to the tissues.

The surface of twigs also shows *leaf scars,* which mark the places from which leaf stalks have broken off.

SPECIALIZED STEMS

Some stems do not look like stems at all. The climbing organs called *tendrils* are sometimes modified stems. We find them in such plants as grape and Boston ivy. A *twiner* is a stem that spirals around a solid support. The morning-glory shows such an arrangement. Some green stems, including those of various cacti, look like leaves. They manufacture food, just as leaves do.

The stems called *runners,* or *stolons,* grow horizontally just above the surface of the soil. They serve to spread and propagate the plant. The runners of strawberry plants, for example, develop new plants at their joints, or nodes, if these touch the ground.

The underground stems known as *rhizomes* look a good deal like roots. In fact, the name comes from a Greek word meaning "mass of roots." Buds, roots, and leaves arise at each node of a rhizome and form new plants. The rhizomes of herbs such as false Solomon's seal, bloodroot, and rhubarb are perennial. The rest of the plant dies each fall. Quack grass, a weed detested by gardeners, spreads rapidly by means of rhizomes. When these are cut or broken into sections, each section can give rise to one or more new plants.

In certain plants, such as potatoes and Jerusalem artichokes, we find fleshy, thickened underground stems known as *tubers*. They develop buds—the "eyes" of potatoes. Generally the rhizomes connecting the tuber to the main stem of a plant die after the growing season. The tubers remain alive, and in the next growing season they put forth shoots.

A *bulb* is a very short underground stem that bears a cluster of thick, overlapping leaves. These leaves store food. Onions, tulips, lilies, and other familiar plants develop bulbs, each of which can give rise to a new plant.

Gladiolus, crocus, and various other plants form underground stems called *corms*. These resemble bulbs somewhat. However, the stems are much more prominent and the leaves are generally smaller and thinner than those of bulbs.

STEM STRUCTURE

Stems also differ in internal structure. The young twigs of most woody plants show much the same type of cross section as the mature root. There is an outer layer of cells—the epidermis—which serves as a covering. Next to this is a cortex, made up of storage and supporting cells. The central part of the stem, the stele, contains a number of concentric layers.

The outermost layer of the stele is the pericycle. Within this is the phloem, which conducts food from the leaves to the lower parts of the stem and to the buds. A cambium layer comes next. As in the roots, it consists of cells that divide and cause the stem to grow in diameter. Inside the cambium is the xylem, which conducts water and dissolved substances upward to the leaves. The core of the plant is composed of *pith*. This serves chiefly to store food. Extending outward from the pith, usually through both xylem and phloem, are a series of vascular rays. They conduct foods and other materials from the inner part of the plant to the outer, and also store foods.

After the first season of growth, a given section of the stem does not increase in length. Rather, the stem becomes longer through the development of the terminal bud at its tip. There is, however, a gradual increase in diameter, as new xylem and phloem cells and new vascular rays are

Stems that propagate plants. Left: the runner of a black raspberry grows horizontally above the surface of the ground. New plants will develop at its joints if they touch the ground. Right: a sedge rhizome with new plants growing from the joints.

formed. A cork cambium layer outside the cortex now produces cork cells. These are heavily impregnated with suberin, a fatty substance that is impervious to water and resists attacks by insects and fungi. The cork cells come to form the external covering of the plant. The epidermis disappears. So do the outer part of the cortex and the pith at the center of the stem. All this sometimes happens after the first year's growth, sometimes after several years.

The fully developed woody stem, therefore, is quite different from the young twig in internal structure. There is a thick central core of xylem, surrounded by a thin layer of cambium. Outside the cambium are layers of phloem, pericycle fibers, and cortex. Beyond the cortex is a cork cambium layer, which produces cork cells. Cork cells make up the external layer.

The xylem that makes up the central core of mature woody stems is called *wood*. The name *"bark"* is given to the rest of the tissues: phloem, pericycle, inner cortex, cork cambium, and cork. The cork layer is often called the outer bark; the other bark tissues make up the inner bark. As the stem continues to grow in diameter, the outer bark often splits lengthwise, and its texture becomes rough. The bark of the mature woody stems of many plants contains lenticels. They account for the conspicuous horizontal markings of birches.

Both the wood and bark increase in girth as new layers of xylem and phloem are added. The bark, however, does not grow nearly so rapidly as the wood. In most trees it does not exceed five centimeters, while the central core of wood may have a diameter of 60 centimeters or more. The bark of the giant sequoia may be 30 centimeters thick at the base of the tree. However, we must remember that the total diameter of this huge tree may exceed six meters.

SEASONAL CHANGES

The cambium layer, which forms new xylem and phloem cells, is not equally active throughout the year in regions where cold and warm seasons alternate. There is one active period each year. During this period the cambium forms a new layer of

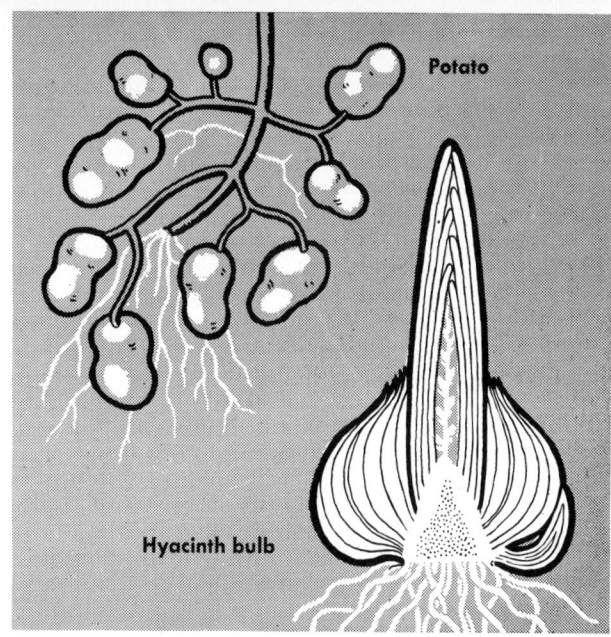

Among the stems that do not look like stems are tubers and bulbs. The tubers of the potato plant are well known. They are fleshy underground stems connected by rhizomes. Bulbs are short stems with overlapping leaves. Here we show a hyacinth in cross section.

xylem and one of phloem. So little phloem is produced, comparatively speaking, that it is hard to make out the different layers laid down over a period of years. Much more xylem is produced. Generally the layers formed each year make up a distinct series of concentric rings called *annual rings*. By counting the number of such rings in the cross section of a log, we can tell how old the tree was when it was felled. Trees in regions where rainy and dry seasons alternate may also show distinct annual rings. It is generally hard to make out rings in tropical trees that grow more or less uniformly throughout the year.

In most woody stems of the temperate zones, the conduction cells formed in the xylem in the spring are relatively large in diameter and thin-walled. Those produced during the summer are smaller and have thick walls. As a result each annual ring is made up of two bands of cells: an inner band called spring wood, and an outer band called summer wood.

The annual rings of a tree are not uniform throughout. Their width will depend on the type of weather that prevailed when the rings were formed. For example, if there has been more rain than usual dur-

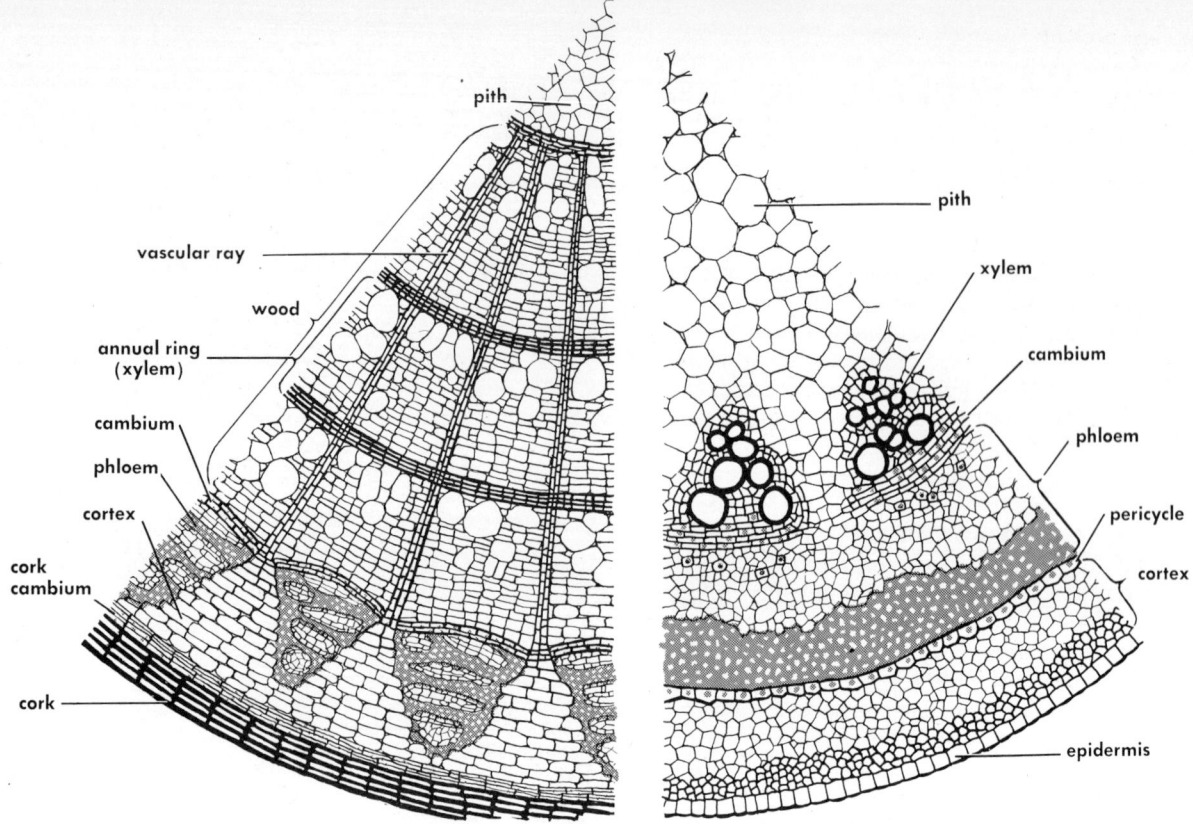

Left: A fully developed linden stem in cross section. The xylem (wood) is more prominent than the pith, and vascular rays are highly developed. It is quite different in structure from a young twig. Right: A young twig of a woody plant. The pith, serving as a food storehouse, is very prominent. The xylem is not highly developed, and there is no cork surrounding the cortex.

ing the growing season, the rings will be comparatively wide. They will be quite a bit narrower if the weather has been unusually dry. Generally speaking, annual rings seem to follow an 11-year cycle that corresponds to the sunspot cycle.

MATURE STEMS

The outer wood of most trees is light in color. It contains many living cells and actively transports water and dissolved minerals. This area is known as *sapwood*. After the cells are from 10 to 20 years old—that is, when they are 10 to 20 rings away from the cambium—they die. They are then transformed into *heartwood*. Of course, as the tree becomes older, its heartwood becomes more prominent.

In many woods, such as cypress, redwood, white oak, black locust, catalpa, and chestnut, heartwood is dark in color. This is due to the gums, resins, tannin, and other materials that have been deposited in the cells. Dark heartwood is generally very resistant to decay. The heartwood of true firs, some of the spruces, aspen, and basswood is faintly colored, or colorless. It decays rather quickly.

If the heartwood of a tree has rotted away, the only effect upon the tree may be that its structure is weakened. The xylem cells of the sapwood continue to supply material drawn from the soil to the leaves. A tree may be quite hollow and yet continue to put forth new twigs and abundant leaves.

The structure of mature herbaceous stems is quite different from that of woody stems. There are two distinct kinds of herbaceous stems—those of dicotyledons, or dicots, and of monocotyledons, or monocots. Dicots have two cotyledons, or seed leaves, in their seeds; monocots have only one.

Some dicots—violets, clover, and snapdragon, for example—have herbaceous stems. These stems are like those of young woody twigs. They have an epidermis, a cortex, and a central zone—the stele—which contains phloem, cambium, xylem, and pith. In some herbaceous dicots, the phloem, cambium, and xylem each make up a continuous layer. In others, they are concentrated in a number of bundles, called *vascular bundles*. These are arranged in a circle. The cambium they contain is not active. Therefore the stems do not grow much in diameter.

Most monocots have herbaceous stems. Those of monocots such as grasses, lilies, and orchids generally have no cambium tissue. The xylem and phloem always occur in vascular bundles. Often these are scattered at random throughout the stem, with supporting tissue between the bundles. In the grasses and their relatives, such as the bamboo, the vascular bundles and their supporting tissues are arranged in a circle. The stem is hollow, except for solid sections at the joints, or nodes. Since there is no cambium in monocot herbaceous stems, these stems can increase in diameter only as existing cells increase in size.

THE RISE OF SAP

Liquids generally flow from a higher place to a lower one because they are drawn by the force of gravity. In the stem, however, water drawn from the soil rises, sometimes for 30 meters or more, apparently in defiance of gravity. The rise of *sap*—that is, water and the mineral salts dissolved in it—has puzzled many generations of botanists. Many theories have been proposed to account for it. According to modern belief the rise of sap is due to a combination of various forces.

One force is known as *root pressure* because it originates in the roots of plants. Salts are more highly concentrated in the sap within the root hairs than in the soil solution. Therefore, water passes from the soil into the cells by the process of osmosis. Dissolved minerals, in the form of ions, also pass from the soil solution, through the cell membranes and into the cells. This passage may be by simple diffusion or by absorption, whereby the cells exert energy in order to "trap" the ions. The cells swell and become turgid. The water concentration is now higher within them than it is in the cells next in line. Hence the sap—water and dissolved minerals—passes into these adjacent cells and from these to the next ones. The root pressure brought about in this way is great enough to force sap some distance through the xylem.

When one end of a glass rod with a fine tube running lengthwise through it is placed in water, the water will rise some distance in the tube. This is due to the force called *capillary action*. It is believed that this force is also involved, to some extent, in the rise of sap. The tracheids and vessels of xylem are very thin tubes, and the water will rise in them just as it does in a thin glass tube. This force could account only for a part of the total rise of sap.

Transpiration, or evaporation of water, is chiefly responsible for the rise of sap. Large quantities of water evaporate from the leaves. When this happens, the leaf cells nearest the outer air become partly dried out. Water then diffuses into these cells from the adjacent ones by osmosis. These cells in turn lose part of their water content and receive water from the cells next in line. In this way an upward pull of water is transmitted all along the conducting cells of the plant.

Root pressure, capillarity, and evaporation are the known forces that account for the rise of sap in stems. Other forces, as yet unknown, may also be involved.

Cross section of a tree trunk. The trunk consists mostly of xylem (wood)—heartwood and sapwood. The bark occupies relatively little space. Note the annual growth rings.

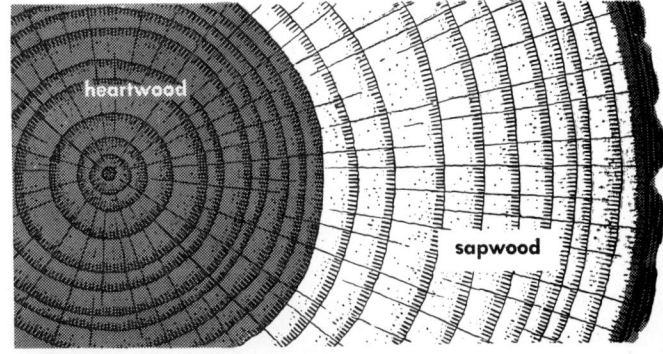

PEOPLE USE ROOTS AND STEMS

Roots and stems are indispensable to us. They support the growth of fruits and seeds, which we use as food and for various other purposes. But this is only part of the story. They also serve us in many other ways, only a small fraction of which we can list here.

We appropriate for ourselves the foods stored in various roots. Among the important root foods in temperate regions are carrots, beets, sugar beets, radishes, turnips, and sweet potatoes. In some tropical areas the roots of yams and cassava, from which tapioca is derived, form the bulk of the vegetable diet.

Horseradish and various other condiments and flavorings such as licorice and sarsaparilla are derived from roots. In the Far East, ginseng roots are consumed in some quantity.

Certain roots yield dyes and were long used as a source for coloring matter. Dyes derived from roots have now been replaced largely by synthetic dyes.

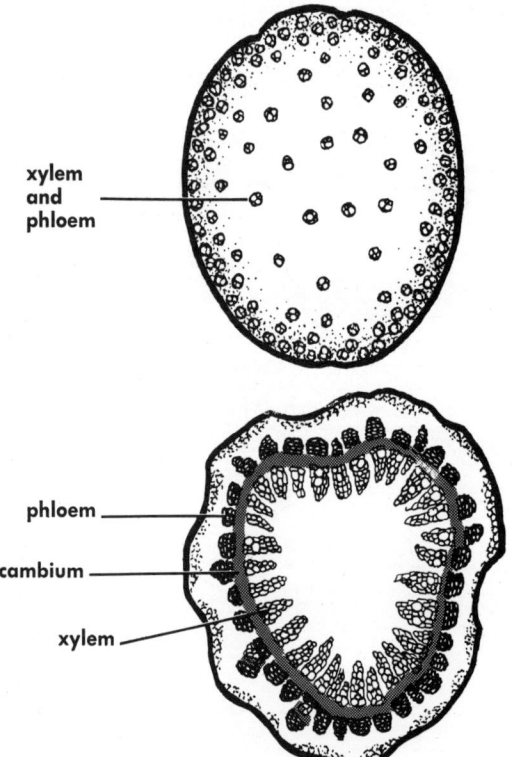

Two types of herbaceous stems shown in cross section. Top, the stem of *Zea mays,* a species of corn. Corn is a monocot. Below, the stem of a sunflower. The sunflower is a dicot.

Roots help to conserve the soil by clinging firmly to soil particles. In regions where erosion by wind or water is excessive, soil-binding plants are used to prevent productive lands from going to waste and also to reclaim wasteland.

The roots of most leguminous plants, such as beans, sweet clover, and alfalfa, have swellings called *nodules*. These contain nitrogen-fixing bacteria, which help make nitrogen available to the plant. There is plenty of free nitrogen in the air that circulates in the soil, since air is made up mostly of nitrogen. But plants cannot make use of the uncombined element. Nitrogen-fixing bacteria in the root nodules convert free nitrogen into nitrogen compounds, which are then used by the plant. In many agricultural regions, a legume of one kind or another is a regular part of the crop-rotation plan. The nitrogen compounds formed by the bacteria are added to the soil if the crop is plowed under. In this way fertilizer is provided for the crops that follow in the succeeding years of the rotation.

The stem product most useful to humans is wood—the xylem of trunks and branches. This substance consists mostly of cellulose. It also contains a considerable amount of lignin and smaller amounts of other substances, including resins, starch, minerals, and gums. Wood provides us with lumber, paper, and insulating boards and other compostion products. Chemicals such as tar, pitch, and turpentine are also derived from wood. It is raw material for synthetic substances such as cellophane and rayon.

Stem fibers give us textile materials such as linen (from the flax fiber), jute, and ramie. The sap, or latex, of the South American rubber tree, *Hevea brasiliensis,* yields natural rubber. Quinine, invaluable in treating malaria, is derived from the bark of the cinchona tree. Cinnamon comes from the bark of the cinnamon tree; cork, from the bark of the cork oat. Other commercial products obtained from stems include resins, chicle (used for chewing gum), and lacquer. Many important foods are obtained from aerial stems, including cane sugar and maple sugar. Underground stems provide potatoes, onions, and garlic.

Leaves are the food factories for most green plants. This autumn leaf contains several pigments, the most important of which is chlorophyll—a remarkable substance that keeps the food factory going.

Barbara G. Hallowell

LEAVES

by Clyde M. Christensen

"Nature made ferns for pure leaves, to show what she could do in that line," wrote Henry David Thoreau, the 19th-century U.S. writer. He was expressing an appreciation of leaves that is shared by many people. Some people so enjoy looking at leaves that they make leaf collections. Some women mix leaves with flowers and wear them as corsages. Other people use the colorful leaves of autumn to decorate their homes.

More important to us, however, are the foods and other products we get from leaves. The leaves of plants such as cabbage, lettuce, spinach, and watercress are vegetables. We often flavor our food by adding leaves of bay, marjoram, peppermint, or thyme.

The leaves of wild and cultivated grasses supply feed for many kinds of domestic animals and for wild animals that are hunted for game. Two valuable fibers—sisal and Manila hemp—are leaf products. Tea and tobacco are derived from leaves. So are drugs such as cocaine and digitalis, and dyes such as henna and indigo.

The leaves are the food-making centers of the plant. During the process known as *photosynthesis,* they put together carbon dioxide and water to produce the sugar glucose. Oxygen is also produced. An essential factor in this process is the green pigment *chlorophyll*. This substance is found in *chloroplasts,* small bodies located in most leaf cells. Chlorophyll acts as a trap to absorb some of the energy of the sun's rays. This energy is used to transform the water and carbon dioxide into sugar.

LEAF SIZE AND ARRANGEMENT

In size, leaves range from the tiny scales of the white cedar to the giant leaves of certain palm trees. There is also a great variety of leaf shapes. Leaves may look like needles (most conifers), or pitchers (pitcher plant), or floating platters (giant water lily), or bayonets (agave), or pincushions (sundew). The margins, or edges, of leaves are smooth or indented, according to the species. The indentations may be like the teeth of a saw or may show a wavelike pattern. Generally leaves are set spirally on the stem. In some cases they are arranged in pairs, one leaf being opposite the other. There may be more than two leaves, spaced more or less evenly, at a given point of the stem.

Leaves arise from tissue just back of

Left: oak leaf in autumn. Notice that the chlorophyll has receded to the area of the leaf veins. Veins carry nutrients and fluids. Middle: young horse chestnut leaves. Right top: the poinsettia's "flower" is really made of red leaves. Right, bottom: the pitcher plant leaves trap insects.

the growing points of buds. Generally a leaf consists of a leaf stalk, or *petiole,* and the leaf proper, called a *blade.* In some leaves there are small projections, called *stipules,* at the point where the petiole meets the stem. The petiole contains xylem and phloem and forms part of the conducting system of the plant. Certain leaves have no petioles; the blade grows out from the stem. The framework of all blades is made up of a network of *veins.*

In elms, oaks, and many other trees each leaf consists of a single blade and petiole. In other trees the leaf blade is divided into several *leaflets,* each attached to the same petiole. The leaflets may be arranged on opposite sides of the petiole, producing a featherlike effect. This arrangement is found in the rose, the ash, and most legumes, including peas and beans. In plants such as palms and horse chestnut trees, the leaflets arise from a common point at the end of the petiole. They radiate from this point much as fingers do from the palm of a human hand.

LEAF STRUCTURE

The leaves of most plants have a similar microscopic structure. There is a surface layer of cells, the *epidermis,* covered on the outside with a waxy coating, or *cuticle.* The hairs found on some leaves are outgrowths of epidermal cells. The epidermis has various openings, or pores, known as *stomata* (singular, "stoma"). Gases such as carbon dioxide, oxygen, and water vapor pass in or out through these pores.

Each stoma is surrounded by a pair of cells called *guard cells.* These change their shape as their water content increases or decreases. In this way they regulate the size of the stoma. The guard cells make up a relatively small part of the total number of cells on plant surfaces. Even so, there may be a million or more stomata in a single leaf.

Both sides of a leaf contain an epidermal layer. The upper epidermis has fewer stomata than the lower one and its cuticle is thicker.

The cell tissue between the upper and lower epidermis is known as *mesophyll.* It is made up of two distinct layers. The *palisade layer,* which is beneath the upper epidermis, consists of relatively long, narrow cells. These are compactly arranged, with the long axis vertical to the leaf surface. Under the palisade cells is a *spongy layer.* Its cells are loosely packed and there are many air spaces between them.

The cells of both the palisade and spongy layers contain chlorophyll. So do the guard cells. These cells, therefore, take part in the process of photosynthesis.

Through the mesophyll run the *leaf veins*. These contain the conducting tissues xylem and phloem, arranged in vascular bundles. *Xylem* conducts water and dissolved substances from the petiole throughout the leaf blade. The food manufactured in the blade is carried in the *phloem* to the *petiole,* from which it is transported to the stems and roots.

During the growing season the leaves of most plants are green. This color is due to the chlorophyll, which is really two related pigments—chlorophyll A and chlorophyll B. Two other pigments—xanthophyll, a yellow pigment, and carotene, which is yellowish-orange—are also present in most leaves. Generally, however, these pigments are masked by the chlorophyll pigments, which are far more plentiful. In certain plants, not only the leaves but the petals, stems, and other parts are reddish in color. This is due to the presence of pigments called anthocyanins, which are found in the sap.

LEAF FALL

The leaves of most perennial broadleaved trees are shed in the fall. In certain cases leaves are shed more or less continuously throughout the year. Each leaf falls after it reaches a certain age, which may range from one year to five years, depending on the species. This is true of most conifers and of certain other plants, including live oak and holly.

The first stage in leaf fall is the formation of an *abscission* ("cutting off") *layer* at the place where the leaf joins the twig. A layer of cork cells develops below the abscission layer. This will provide a protective covering over the spot from which the leaf will be detached.

Not long after the abscission layer is formed, it begins to loosen and dry out. After a while it ruptures. The leaf is held to the twig only by the vascular bundles. As the leaf is attacked by frost or sways in the wind, the vascular bundles begin to weaken. Finally they snap and the leaf falls.

The formation of the abscission layer in leaves accounts for their showy autumnal colors. Once this layer has been laid down, the flow of materials to and from the leaves is shut off. The cells of the leaf are deprived of life-giving elements and die; their pigments decompose. Chlorophyll breaks down more rapidly than the yellow xanthophyll and the yellowish-orange carotene. These now predominate and give the leaf its yellowish and golden hues. The red colors seen in leaves in the autumn are due to the anthocyanins—the pigments dissolved in the cell sap.

MANY DIFFERENT FUNCTIONS

Leaves serve many purposes in addition to food-making. The bud scales of woody plants are leaves that furnish protection to the tender tissues they enclose. Tendrils are sometimes modified leaves. Part of the leaves of the sweet pea are broad leaflets; the rest are tendrils, which attach the plants firmly to a support. Most plant spines, such as those of the cactus

The structure of leaves varies. Left: the arrangement of the leaves on this large fern suggests an ostrich plume. Middle: a magnolia leaf is very smooth and waxy. Right: fan-shaped leaves from a maidenhair, or gingko, tree.

John Lewis Stage/PR   Richard Parker, Audubon/PR   Leonard Lee Rue III, Audubon/PR

Upper left: underside of fern leaves, laden with reproductive structures called sporangia. Middle: agave has spiked, two-toned leaves. Right: scrub-oak leaf. The leaf has lobed edges and prominent veins. Bottom left: this Chinese ivy has leaves with serrated edges; the leaves are arranged in a spiral pattern.

plant, are modified leaves or leaf stipules. They serve to protect plants against excessive browsing by animals. Bulb leaves serve as food-storage organs.

The leaves of carnivorous plants trap, hold, and digest insects and other small animals. Carnivorous plants are found chiefly in bogs, sandy areas, and tropical rain forests. In such areas nitrogen and various other essential food elements are likely to be present only in small amounts. Such elements are provided by animals trapped in leaves. More than 400 species of carnivorous plants are known.

Pitcher plants are carnivorous plants that trap insects in a trumpet-shaped or pitcher-shaped leaf. At the base of the leaf is a reservoir in which water collects or in which sap is exuded by the plant. Insects are attracted to the leaf by its odor or color. They venture into the mouth of the pitcher and slide down into the water. Downward-pointing bristles in the interior walls prevent the insects from climbing up again. They are then slowly digested by fluids secreted by the leaf.

The leaves of the sundew plant bear numbers of long, sticky, glandular hairs. When an insect alights on a leaf, it touches several of the hairs and is held by them. Some of the surrounding hairs bend over and help hold the victim fast. Eventually it is digested.

The Venus's-flytrap is another carnivorous plant. Its leaves are nearly flat when they are open. Their margins bear spikelike structures and there are sensitive hairs on the inner surface of the leaves. When an insect touches some of the sensitive hairs, the leaf blades fold shut along the midrib like the two halves of a book. The spikes at the margins interlock and the insect is trapped. After the prey has been digested, the leaf opens, ready for the next victim.

A dahlia in full bloom. Dahlias are widely cultivated autumn-blooming flowers. New York Horticultural Society

FLOWERS

by Jenny Tesar

Flowers are found almost everywhere. In the tropics flowering vines grow on trees, competing with other lush vegetation for sunlight. Colorful flowers brighten dull expanses of hot desert. Lakes and even salty marshes can be home to flowers. Water hyacinths and water lilies float aimlessly in lakes and ponds, while pickerelweed and pond lilies stay anchored to the bottom but show off their flowers above the water line. Meadows and fields explode with color as springtime dandelions, bluebonnets, thistle, and cowslip give way to the wild roses, black-eyed susans, and goldenrod of summer. In woodland areas, violets and bloodroot fade away as azaleas and rhododendron bloom. Even mountainous areas with snow-capped peaks have flowers, with edelweiss and mountain heather following snowdrops that break through the snow. Many of the flowers that are familiar to us are large and brilliantly colored. Other flowers are tiny and inconspicuous—so small that you do not notice them unless you search for them. All, however, are built on the same basic pattern. And all have the same function: to reproduce the species.

A flower exists for only a short while. Then, parts of it develop into a fruit. Within the fruit are the seeds, which will produce a new generation of plants when they return to the soil.

THE PARTS OF A FLOWER

A typical flower has four main parts: the sepals, petals, stamen, and pistil. These

New York Horticultural Society

Derek Fell

Burpee Seed Co.

parts are attached to a stem tip, called the *receptacle,* which is slightly enlarged.

If you look at a typical flower bud, you find that it is covered by green leaflike structures. These are the *sepals*. Their function is to enclose and protect the other parts of the bud, before they are fully developed. In an open flower, the sepals can be found beneath, or outside, the petals.

Many flowers are known by their color. We talk of red roses, goldenrod, and purple asters. In some cases, the sepals provide the distinctive color. Usually, however, the bright, showy parts of a flower are the *petals*. Their main function is to attract insects, which play an important role in pollination.

Petals attract insects in several ways. The glistening white or bright colors of many flowers are inviting to insects and even to certain birds. The petals of some flowers have glands that secrete nectar. This sugary liquid is greedily sought by bees, butterflies, and other flower-visiting insects, which use the nectar as food.

The odors of oils and other substances produced by the petals of many plant species are still another way of tempting insects. People, too, often find these odors very pleasant. As a result, the oils secreted by flowers such as roses, lavender, and jasmine are used in making perfumes.

Some plants, such as skunk cabbage, have very unpleasant floral odors. Often, these plants also have purplish- or reddish-brown petals or other structures that resemble decaying animal flesh. Bees and other insects that visit pleasantly scented, brightly colored flowers are not attracted to these plants. But other insects are, especially those that frequent rotting flesh and other foul-smelling organic matter.

Within the enclosure formed by the flower petals are the *stamens*. These are the male organs of the flower. A stamen usually consists of a slim stalk, or filament, which bears at its tip a single, enlarged *anther*.

Flowers come in many shapes, sizes, and colors. Top: the Oriental poppy, an easy-to-grow, hardy species. Two well known spring-blooming flowers—Darwin hybrid tulips (middle) and forsythia (bottom).

Left: American cowslip *(Dodecathon),* a wild flower of woodlands, prairies, and mountainous areas. Center top: Flowering dogwood—the white petal-like structures are modified leaves surrounding a tiny greenish flower head. Center bottom: a Royal Highness rose. Right: a flowering plum in full bloom.

Pollen grains are produced in the anther. Later, these grains will form the sperm, or male reproductive cells.

In the center of the flower is the *pistil* —the female organ. A pistil often looks like a flask with three fairly distinct parts. At the base is a sphere-shaped *ovary*. This is where the eggs develop and, later, the seeds are formed. A slender stalk, the *style*, rises from the top of the ovary. At the top of the style is a slightly enlarged *stigma*.

Before seeds can form, pollen grains must pass from the anthers to the top of the stigma. To ensure that the pollen will stick to its surface, the stigma is often rough or sticky. Similarly, the outer wall of the pollen grain is usually covered with spinelike structures that help it adhere.

In roses and many other flowers, the floral organs are all separate and distinct from each other. In other flowers, some of the parts are joined together. In morning glories and petunias, for example, the petals are fused into trumpetlike structures.

The parts of a flower usually occur in multiples of 3, 4, or 5. A lily, for example, has 3 sepals, 3 petals, 6 (2 × 3) stamens, and 1 pistil with 3 sections in the ovary. The evening primrose is an example of a flower with 4 sepals, 4 petals, 8 stamens, and an ovary with 4 sections.

TYPES OF FLOWERS

Most of the flowers that you see in a garden have all the parts that we just described. They are *complete flowers*. Some flowers, however, do not have all four floral parts and are known as *incomplete flowers*. The tiny flower of a grass plant, for example, has neither sepals nor petals. The flowers of corn and other grains and of many trees are also incomplete, lacking at least one of the four parts of a complete flower. Sepals and petals are often termed *accessory parts* of a flower because they are not concerned with reproduction.

When a flower contains both stamens and pistils, it is said to be *perfect,* or bisexual. Orchids, lilies, roses, and sweet peas are all perfect flowers, containing both male and female reproductive organs in each flower. When a flower does not contain both stamens and pistils, it is said to be *imperfect,* or unisexual. It may be a pistil-

Flowers are found even in the desert. These yellow flowers belong to a fishhook cactus.

late, or female, flower or it may be a staminate, or male, flower.

In some species, such as corn, pumpkin, oak, and squash, the pistillate flowers and the staminate flowers are borne on the same individual plant. The plant is then said to be *monoecious*. In other cases, including the willow tree, asparagus, hemp, and date palm, the pistillate and staminate flowers are borne on separate plants. These are said to be *dioecious*.

A flower may be perfect but incomplete. A calla lily, for example, is perfect because it has both stamen and pistils, but it is incomplete, lacking sepals.

FORMING A FLOWER

A flower starts as a bud. In many plants the same stem tip that forms leaves will later form a flower. Small projections form along the sides of the stem tip. These develop into the various parts of the flower.

What causes flowering? What determines when a plant will produce flowers rather than leaves? Many factors seem to be involved. The plant must have an ample supply of food. Also, certain hormones must be produced.

Environmental factors can also affect the start of the flowering period. The relative length of day (light) and of night (dark) is one such factor.

The daily period of light to which a plant is exposed is called a *photoperiod*.

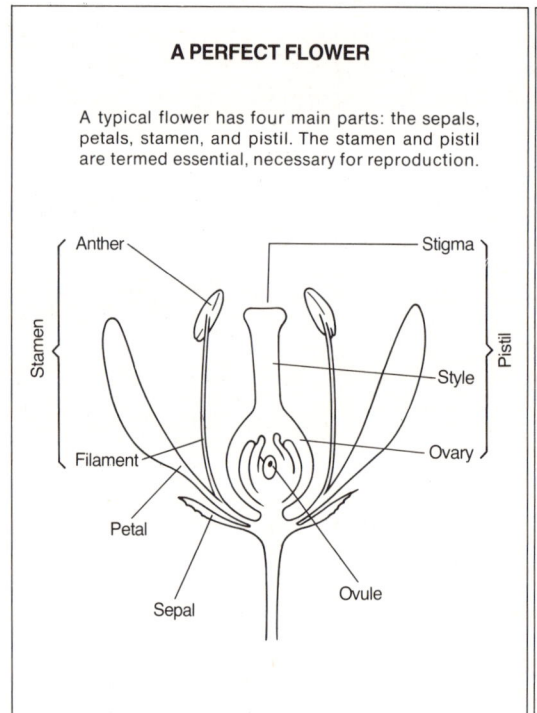

A PERFECT FLOWER

A typical flower has four main parts: the sepals, petals, stamen, and pistil. The stamen and pistil are termed essential, necessary for reproduction.

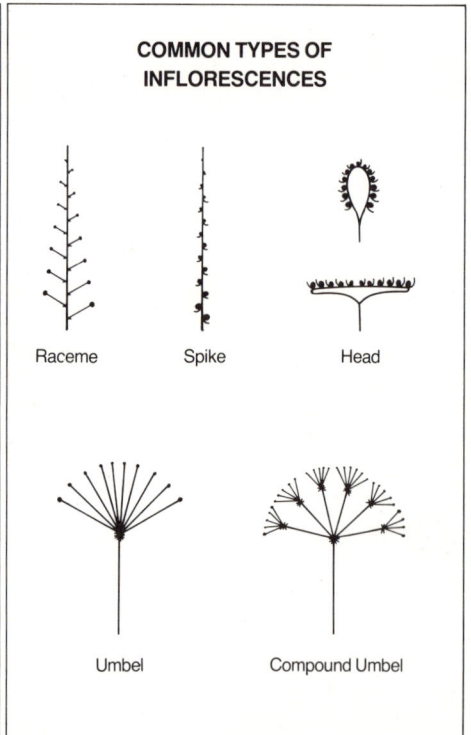

COMMON TYPES OF INFLORESCENCES

Raceme Spike Head

Umbel Compound Umbel

Plants that bloom in the spring or fall are short-day plants. Violets, poinsettias, and chrysanthemums are examples. Long-day plants, such as clover, potatoes, and most grains, usually bloom in the summer. They flower only when the day length is comparatively long.

Other plants bloom only when the day length is neither too long nor too short. And some, such as dandelions and tomatoes, will bloom during a wide range of day lengths.

Another environmental factor that influences flowering is temperature. Plants such as lettuce, cotton, and peppers flower only (or best) at relatively high temperatures. The flowering of celery, carrots, and onions is favored by low temperatures. Some plants flower best at intermediate temperatures. Still others blossom under a wide range of temperature conditions.

Knowledge of the effects of temperature and light on flowering is used by florists. By regulating environmental conditions in greenhouses, they can make plants such as chrysanthemums and carnations bloom on any chosen day of the year.

POLLINATION

We have been talking about many different kinds of flowers. Certainly there is much variety among them—in size, color, shape, number of parts, periods of blooming, and so on. In every case, however, the flower has just one function: to make seeds.

If seeds are to be made, pollen must get from the anther to the stigma on the tip of the pistil. This transfer of pollen is called *pollination*. In a perfect flower, pollen from its own stamen may reach the pistil. Or the pollen from one flower may land on the pistil of another flower on the same plant. This

Top left: Common garden petunia, often used for bedding and borders. Top right: an azalea. Bottom left: rock asters, wild flowers of mountainous areas. Bottom center: night-blooming cereus *(Hylocereus).* Bottom right: *Impatiens.*

is *self-pollination.* When pollen is carried from the stamen of one plant to the pistil of a flower on another plant, the transfer is called *cross-pollination.*

Cross-pollination depends on an outside agent to carry the pollen from one plant to another of the same species. The most common agents are insects and wind, though birds and water also sometimes carry pollen. Plants that depend on insects advertise their presence with bright colors or strong odors. Insects visit the flowers to feed on nectar and pollen. The pollen is sticky and easily adheres to an insect's hairy body. When the insect then moves on to another flower of the same species, some of the pollen rubs off.

In many other plants, the wind carries the pollen from one plant to another. In most of these plants the stamens and stigma are exposed to air currents. The pollen grains are lightweight and easily carried by the wind. Among the many plants that are wind-pollinated are grasses, corn, oak, pine, and fir.

In cross-pollination the characteristics of two plants are mixed. This often results in plants that are superior to those that result from self-pollination. This is true for certain varieties of apples, cherries, pears,

Top left: a giant sunflower. Sunflowers turn to follow the sun's path. Bottom left: iris, a popular ornamental flower. Top right: a garden of annuals that includes marigolds and petunias. Bottom right: a tree peony.

and other kinds of fruit. People who grow these fruit trees usually keep beehives in their orchards to ensure cross-pollination of the blossoms.

Pollination is the first step in making seeds. The next step is *fertilization*—the union of egg and sperm. This occurs in the ovary. The sperm, which is formed from a pollen grain, travels down the pistil through a pollen tube. When the sperm reaches the ovary, it fertilizes an egg, and seeds are formed.

After it is fertilized, the egg begins to grow. The ovary grows, too. In time, the ovary becomes a fruit, containing one or more seeds. When the fruit is ripe, it may drop to the ground and the seeds develop into new plants. In many cases, however, the fruits or seeds are carried by wind, water, insects, birds, and other animals before they take root. In some cases, the fruits "explode," scattering the seeds. Seeds sprout, or develop, if temperature, moisture, and other environmental conditions are favorable.

INFLORESCENCES

The many differences between flowers are useful to people who are interested in identifying plants. If you were given a rose and a dandelion, you could tell them apart because you have learned to identify these two flowers. What differences between the rose and the dandelion do you use to identify them?

One important characteristic used in identification is the relative position of the flower on the plant. In some plants, such as roses, magnolias, and violets, each flower is solitary, borne on a stem that is some distance from other flowers on the plant. In most flowering plants, however, the flowers are borne in clusters. These flower clusters are called *inflorescences*.

There are many types of inflorescences. We'll just look at some of the most common ones:

1. Raceme. The main axis of this inflorescence is elongated. The flowers are borne on short stalks known as *pedicels*. The pedicels are all approximately the same length and are more or less equally

Derek Fell

New York Horticultural Society

New York Horticultural Society

Top: red poinsettia, a favorite Christmas decoration. The showy part of the plant is the bright red leaves surrounding a small yellow flower. Middle: water lilies floating in a pond. Bottom: pansies started indoors. These do well outdoors and are commonly found in flower gardens.

FLOWERS 109

distributed along the main axis. This simple type of inflorescence is found in such plants as currant, radish, hyacinth, and snapdragon.

2. Spike. This is similar to a raceme but the individual flowers do not have stalks. They are sessile, attached directly to the main axis. Wheat, cattail, gladiolus, and common plantain have spiked clusters.

3. Simple umbel. Here, the main axis is very small. The stalks that bear the flowers all seem to arise at the same point and all the stalks are of about the same length. Thus an umbel looks rather like an umbrella. Examples include ginseng, bluelace flower, and onion.

4. Compound umbel. In a compound umbel, the branches grow from the tip of the main axis. Each branch then bears an umbrellalike cluster of flowers. Carrot, dill, and Queen Anne's lace have this type of inflorescence.

5. Head. In this disk-shaped inflorescence, both the main axis and the pedicels are very reduced in size. The flowers are crowded close together, forming a disk-shaped arrangement. Examples include the daisy, sunflower, and the thistle.

The head is one of the most unusual and interesting types of inflorescence. In a head, the flowers are usually very small and tightly crowded on a flattened receptacle. The daisy is a good example; it isn't a single flower but a whole cluster of flowers. Marginal flowers, called ray flowers, are large and brightly colored. Their function is to attract insects. The flowers in the center of the daisy head are much smaller and less conspicuous. These are the disk flowers. They produce the seeds.

Dandelions are also examples of head inflorescences. Here, however, the entire head is composed of ray flowers, each of which produces a tiny one-seeded fruit.

The next time you walk through a garden look at the flowers. Can you identify different inflorescences? Can you find flowers with fused parts? Can you tell which flowers are pollinated by insects and which by the wind?

You will find many floral designs in the garden. Each is a thing of beauty. But what is really remarkable is that nature has found so many ways in which to ensure the formation of seeds and, hence, of flowering plants on our planet.

Top left: crocuses. Bottom left: torch zinnia, a popular garden and cut flower. Top right: Easter lily. Bottom right: camelia-flowered balsam.

FRUITS AND SEEDS

by Sanford S. Tepfer

If you ask a friend to name some fruits, the reply would probably include such things as bananas, apples, grapes, and peaches. If your friend is a botanist, then beans, tomatoes, and cucumbers might also be on the list. All of these are fruits because all consist, at least in part, of a ripened ovary. Generally, they contain seeds that, under proper conditions, may develop into new plants.

The first step in the formation of a fruit is pollination. Ordinarily, if pollination fails and fertilization does not occur, no fruit develops. It is possible to make a plant bear fruit, however, even though its flowers have not been pollinated, by applying synthetic hormones.

Basically, a fruit results from a change in the ovary wall. Hormones provide the stimulus that is required. In some plants, parts other than ovaries also become part of the fruit.

Fruits have a very specific purpose: to protect and help disperse the seeds.

FRUITS WITHOUT SEEDS

Some fruits can develop naturally even though pollination has not taken place. Such fruits are called *parthenocarpic*. They generally are seedless.

Many cultivated varieties of plants have become seedless or practically so. In some instances there are only traces of seeds, such as the black specks of the banana. Plant breeders have deliberately developed seedless varieties of grapes and oranges, which are now naturally partheno-

Buzzini

Ministere de l'Agriculture

C. Rives

Fruits are ripened ovaries that usually contain seeds. Fruits come in many sizes and colors. Some familiar fruits are lemons (top), peaches (middle), and apples (bottom).

Tomatoes, peas, beans, and summer squash are vegetables. Yet to a botanist, they are also fruits, for they, too, consist of a ripened ovary.

carpic. Of course, since the fruits have no seeds, new plants can be grown only from stem or root cuttings, or by some other means of vegetative propagation.

Seedless fruits are exceptional forms. Generally speaking, every fruit has one or more seeds that have developed from ovules.

STRUCTURE OF FRUITS

When a fruit arises from the ovary wall of a single pistil, it is called a *simple fruit*. The ovary wall grows larger and becomes the covering of the fruit. It is then known as the *pericarp*. Simple fruits, each derived from a single pistil, are found in many plants, including dates, olives, cherries, and elms.

A fruit may develop from a number of pistils in the same flower. A fruit like this is called an *aggregate fruit*. The raspberry and blackberry are familiar examples.

Multiple fruits develop from the pistils of several different flowers, all more or less fused. A pineapple plant has a multiple fruit; so has a mulberry.

In plants such as apples and pears, only a part of the fruit is developed from the pistil. The rest is formed from other parts of the flower. This type is called an *accessory fruit*. In the apple fruit, for example, only the central part of the core develops from the ovary. Most of the fruit is derived from the receptacle and outer floral organs.

The strawberry is also an accessory fruit. There are many minute pistils in the flower of the strawberry. Each of these produces a dry, hard, seedlike fruit. The fleshy part of the strawberry is the receptacle, which has become swollen and juicy.

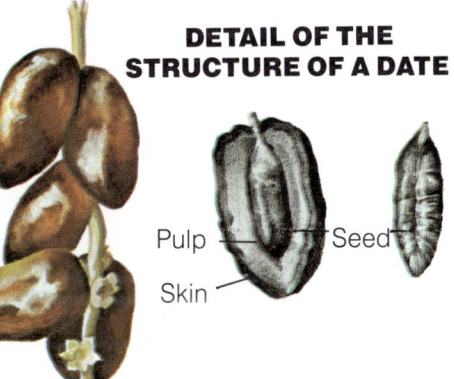

DETAIL OF THE STRUCTURE OF A DATE

Pulp — Seed
Skin

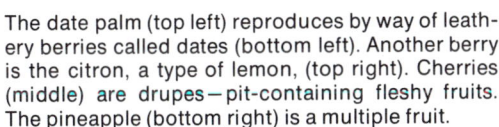
The date palm (top left) reproduces by way of leathery berries called dates (bottom left). Another berry is the citron, a type of lemon, (top right). Cherries (middle) are drupes—pit-containing fleshy fruits. The pineapple (bottom right) is a multiple fruit.

MATURE FRUITS

When fruits become mature, they are either fleshy or dry. In a *fleshy fruit,* such as an apple or an orange, the ovary and any flower parts that may be associated with it form a fleshy structure surrounding the seed. In *dry fruits,* the ovary forms a hard coat for the seed. Acorns and pea pods are examples of dry fruits.

Fleshy fruits. There are three different kinds of fleshy fruits: berries, drupes, and pomes. In a *berry,* the entire ovary wall be-

Walnut fruit. On the tree, the fruit is covered by the green skin. The inner wall of the ovary forms a hard, woody shell around the developing seed. The exposed walnut seed has two lobes of endosperm that supply food to the embryo.

comes fleshy or leathery. Dates, grapes, and tomatoes are all berries. The date has only one seed; the grape has several; the tomato has a great many. Some berries, such as the pumpkin, have a hard rind. Others, including oranges, have a tough outer skin.

In the *drupe,* the inside layer of the ovary wall is hard, forming a pit or stone. The middle layer is fleshy. The outer layer is the thin skin of the fruit. Plums, peaches, cherries, and almonds are all drupes. We cultivate the first three of these fruits for the fleshy part. In the almond, the fleshy layer is removed before the fruit is marketed.

A *pome,* such as an apple, pear, or quince, is an accessory fleshy fruit. Much of its pulp is derived from the receptacle and other parts of the flower. We can often see the remains of the calyx at the end of the fruit opposite the stem. As a matter of fact, apple growers refer to the "stem end" and the "calyx end" of an apple.

114   FRUITS AND SEEDS

Dry fruits. In dry fruits the pericarp is not clearly divided into layers. Some dry fruits, called *dehiscent,* open when they are mature and shed their seeds. A *pod,* such as that of the pea or bean, is a familiar type of dehiscent dry fruit. The pod is derived from a single pistil. The ovules, which later develop into seeds, are attached to one of the inner surfaces. When the pod is mature, it splits lengthwise along two margins, forming two valves, or halves, and thus releases its seeds.

The dry fruit called the *follicle* differs from the pod in that it splits along only one margin. The fruits of the milkweed and larkspur are follicles, though they are sometimes incorrectly called pods.

Some dehiscent fruits are derived from a compound ovary, made up of two or more united carpels. The *carpel* is the seed-bearing part of the pistil. In a *capsule,* such as the fruit of the poppy, violet, and snapdragon, the seeds escape through pores or slits that develop in each carpel.

Certain dry fruits, called *indehiscent,* never split open when they are mature. Generally they contain only one or two seeds. There are five different kinds of indehiscent fruits: achenes, samaras, schizocarps, grains, and nuts.

In an *achene,* such as the fruit of the sunflower and the buttercup, the pericarp remains separate from the solitary seed within it.

The *samara* has much the same type of structure as the achene. However, the pericarp of the samara develops a large wing, which aids in the dispersal of the fruit. Samaras may be single, as in the ash, or double, as in the maple.

A *schizocarp,* of which the fruits of carrot, dill, and celery are examples, is usually formed of two fused carpels, which split apart when mature. Each part usually contains a single seed.

In a *grain,* the seed is completely fused to the inner surface of the pericarp. This type of fruit is found in cereal crops, such as wheat, rice, and corn.

A *nut* has a thick, hard pericarp. Acorns, hazelnuts, pecans, and cashew nuts are well-known examples. In certain nuts, such as pecans, the outer part of the ovary wall becomes a husk, while the inner part forms a hard shell. Some of the "nuts" sold by grocers are not nuts in the botanical sense. The Brazil nut, for example, is really a seed. And, as mentioned earlier, the almond is the "pit" of a fleshy fruit.

THE FORMATION OF SEEDS

While the ovary and other flower parts are ripening into a fruit, the ovule

Soybean pods and beans. When the pods are mature, they split and release their seeds. Soybean is a nutritious food. Its spun protein fibers (bottom) are now being used as meat analogs.

General Mills

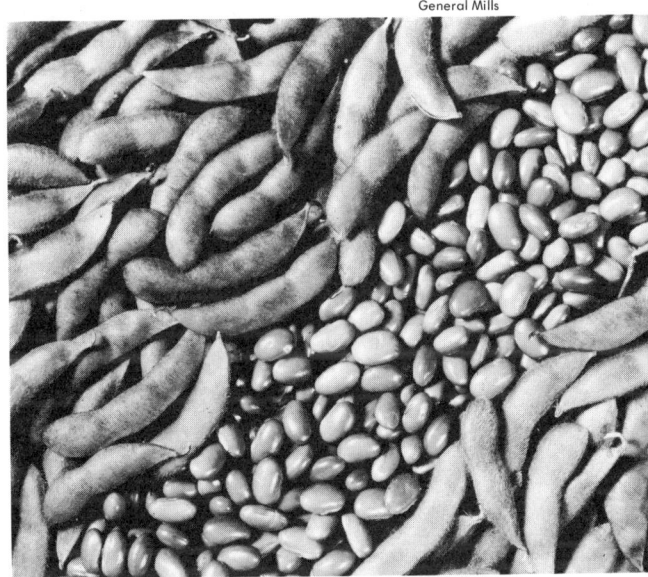

USDA

FRUITS AND SEEDS

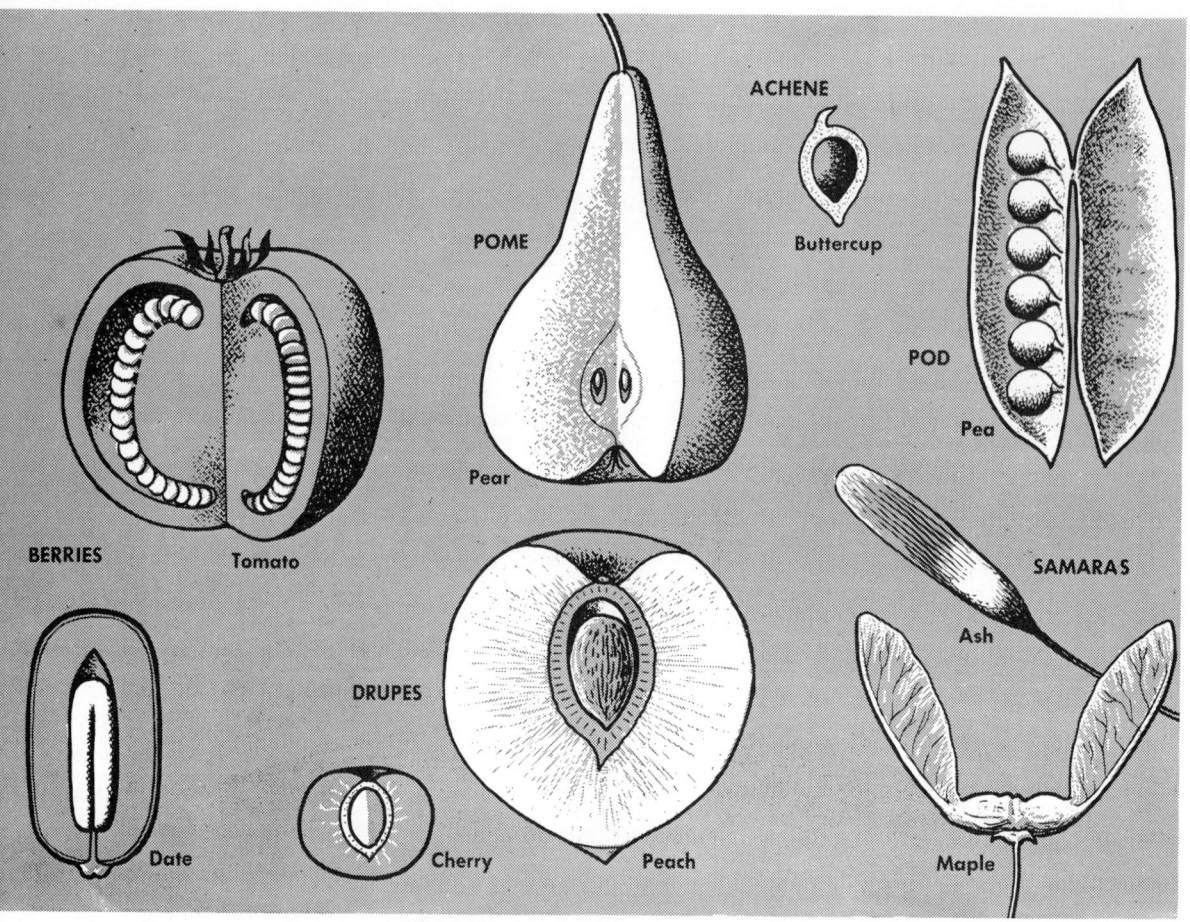

The art above summarizes the distinction between dry and fleshy fruits. In fleshy fruits, such as berries, drupes, and pomes, a fleshy structure surrounds the seeds. In dry fruits, such as pods, samaras, follicles, capsules, grains, and nuts, the seed is protected by a hard coat.

or ovules are developing into seeds. In the maturing seed, one or more protective seed coats form on the outside. Within these arises a miniature plant, called the *embryo,* and a mass of tissue, called *endosperm,* which serves to store food.

In some seeds, such as those of peas and beans, food storage in the endosperm is temporary. The stored food is soon transferred to the embryo, so that the mature seed has no endosperm at all. In other seeds, including those of corn, wheat, and castor beans, the embryo does not store food but obtains it as needed from the endosperm. Thus, a good part of the mature seed is made up of endosperm.

The embryo within the seed represents an entire small plant. It contains one or two seed leaves, called *cotyledons,* and a short *axis.* The upper part of this axis is called the *epicotyl;* the lower part, the *hypocotyl.* The lower end of the hypocotyl—the *radicle*—gives rise to the first root of the sprouting plant. Most or all of the shoot system of the new plant grows from the epicotyl.

If you examine a large seed, such as a lima bean, you can see the two large, fleshy cotyledons after you remove the seed coat. When the cotyledons are separated, the hypocotyl is revealed. In the seeds that contain endosperm when ma-

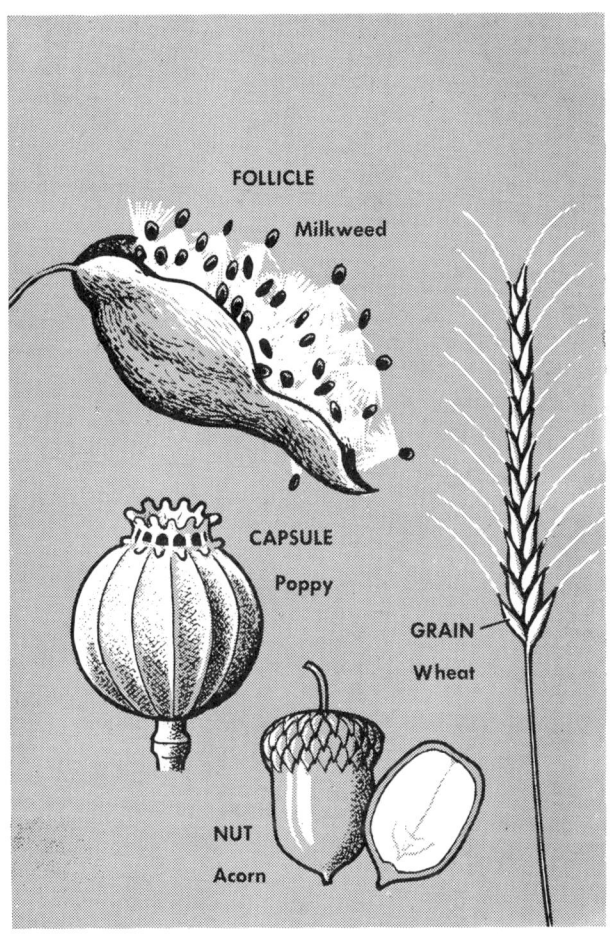

ture, such as the castor bean, the cotyledons are not fleshy.

The food stored in the endosperm of the seed or in the cotyledons is mostly in the form of carbohydrates (such as starch), fats and oils, and proteins. Carbohydrates are found in nearly all seeds. Proteins occur in large quantities in seeds such as peas and soybeans. The seeds of peanuts, coconuts, and cotton are particularly rich in fats and oils.

DISPERSAL OF FRUITS AND SEEDS; SEED GERMINATION

When fruits and seeds are mature, they are cast off from the plant on which they have developed. In plants with dehiscent fruits, it is the seeds that are scattered far and wide as the fruits split or burst open. If the fruits are indehiscent, the fruits, not the seeds, are scattered. In some cases certain plant structures or behavior patterns favor the dispersal of fruits and seeds. They may also be spread by wind, water, animals and man. If conditions are favorable, the seeds will germinate and will give rise to new plants. We discuss dispersal and germination in the articles "Seed and Fruit Dispersal" and "From Seed to Seedling."

PEOPLE USE FRUITS AND SEEDS

Fruits represent a considerable portion of the human diet the world over. Economically, the most important fruits of all are the cereal grains—wheat, corn, rice, oats, barley, and the like—which the average person would not consider fruits at all. Fleshy dessert-type fruits—apples, pears, plums, cherries, grapes, blueberries, dates, oranges, lemons, and grapefruit, to mention only a few—also form a valuable part of the diet. So do nut meats derived from plants such as hickory, pecan, and walnut trees.

Some well-known food items sold as vegetables—tomatoes, peas, and beans, among others—are really fruits, since they are derived, in part at least, from the ripened ovaries of flowers. Strictly speaking, we should reserve the word "vegetable" for edible parts of the plant not derived from the pistil. From this point of view, a potato and a Jerusalem artichoke are vege-

Left: the seed of the lima bean, showing one of the two cotyledons, or seed leaves, and the short axis, which is divided into two parts—the epicotyl and the hypocotyl. Food stored in the endosperm is soon transferred to the embryo; hence the mature seed has no endosperm. At the right is seen a castor bean, the seed of the castor-oil plant. In the castor bean a considerable portion of the mature seed is composed of endosperm.

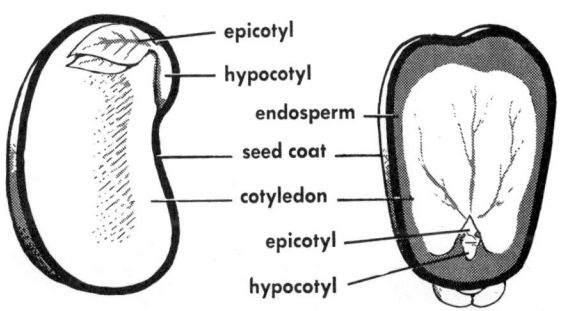

FRUITS AND SEEDS 117

tables, because they are tubers—that is, underground parts of the stem. Lettuce, cabbage, and spinach are vegetables because they are leaves of plants. This distinction between fruits and vegetables is a sound one from the botanist's viewpoint. Of course, it would not seem very convincing to a produce salesman.

The seeds of plants also serve us in many ways. We pointed out that cereal crops should really be considered as fruits, since they consist of grains. However, these grains are made mostly of seeds. The seeds of legumes such as beans and peas are important human foods. Edible oils and fats are extracted from the seeds of peanuts, soybeans, coconuts, and cotton.

Many oils derived from seeds are important in industry. For example, linseed oil, produced from flaxseeds, and tung oil, from the seeds of the tung tree, serve as drying oils in the manufacture of paints and other products. The tangled mass of hairs growing on the seed coats of the cotton plant yield cotton, one of the most important textile fibers. Certain popular beverages are derived from seeds. Coffee is brewed from the seeds of the coffee plant; cocoa from cacao seeds. Other seeds yield oils that have been useful in medicine.

Research also continues into new uses for seeds. The jojoba shrub, a plant that grows in desert regions of western North America, has recently been found to have seeds that yield a waxy oil that is very similar to the oil that is obtained from the sperm whale and widely used in the chemical industry. Scientists are looking into the possibility of using the plant's oil as a substitute for the whale oil, thereby reducing the commercial value of the endangered sperm whale.

Fruits are also, as explained in the text, grouped according to their development. A simple fruit develops from the ovary of a single pistil; an aggregate fruit develops from a number of pistils in the same flower; and a multiple fruit from the pistils of several flowers.

SEED AND FRUIT DISPERSAL

by Harry J. Fuller

The seeds or fruits of many plant species have certain special structures or patterns of behavior that favor their spread over wide areas. Various external forces or agencies, such as wind, water, and animals, may also contribute to this spread.

Seed dispersal may have important consequences. For one thing, it is likely to prevent the heightened competition for space, light, water, and soil nutrients that would result if seeds sprouted in the immediate vicinity of the parent plant. Dispersal may also help plant species to establish themselves in new areas. Hence, it may become an important factor in the survival of the species.

Seeds are the characteristic reproductive structures of seed plants. In some of these plants, called *gymnosperms* (they include pines, spruces, and firs), the seeds are produced upon the surfaces of scales. These are borne in clusters called *cones*. In other seed plants—the so-called *angiosperms,* or flowering plants—the seeds are produced within enclosing structures known as *fruits*. When a fruit is mature, it disintegrates or splits open, freeing the enclosed seeds. After one of these seeds sprouts, its embryo grows into a new plant.

In some species of plants, including the milkweed, poppy, catalpa, and pea, fruits remain attached to the parent plant. Seeds escape from the enclosing fruits and are then carried away by wind, water, or other agents. In other species, including the dandelion, elm, and beggar-ticks, whole fruits are dispersed from their parent plants, the seeds remaining within the fruits during the dispersal period. These fruits are usually small and are most often one-seeded.

The distances to which seeds and fruits may be dispersed vary greatly in different species of plants and with differing environmental conditions. Ordinarily, when seeds and fruits are spread by wind and water, they cover greater distances (often hundreds of kilometers) than when they are transported by animals or other agencies.

PLANT STRUCTURES THAT FAVOR DISPERSAL

In some plants, including members of the pink family, the fruit has a pore, serving as an opening, at its tip. The pore is surrounded by teeth that are sensitive to moisture in the atmosphere. These teeth curve backward—that is, outward—in dry air, thus allowing seeds to escape from the fruit through the opened pore. When the air is humid, the teeth bend upward and in-

The ripe seed vesicle of the Oriental poppy. Seeds sift out of the pores on these vesicles.

Hugh Spencer

ward, closing the pore and preventing the escape of the seeds. This is called a hygroscopic mechanism. It reacts to variations in the amount of moisture (*hygros,* in Greek) contained in the atmosphere. It is an effective arrangement for the spread of seeds. Since the seeds escape in dry weather, lack of moisture prevents them from becoming clumped together. Therefore, they are more widely scattered.

Other plants contain a sifting mechanism. Fruits such as the Oriental poppy and the bluebell *(Campanula)* contain small pores, corresponding to the holes of a salt shaker. As the seeds mature, they sift out of the fruits through these pores. The sifting process is speeded up as winds move the fruits to and fro.

In some species, the fruits suddenly rupture, often with explosive force. As the fruit splits, the enclosed seeds are flung or squeezed out, sometimes for distances of several meters. The splitting movements are caused by tissue strains and tensions that develop as a result of unequal growth and unequal drying of fruit tissues as fruits mature. Splitting mechanisms of this type are found in the fruits of legumes such as sweet pea and soybean, and of various other plants, such as oxalis, geranium, and touch-me-not.

Frequently, contact with some solid object will set off an explosive movement before the fruit is completely matured. If one touches the fruit of a common oxalis, sheep sorrel, or touch-me-not with a pencil or toothpick, the fruit will suddenly burst along its seams, flinging the seeds out. This will also happen if an insect of sufficient size and weight comes in contact with the fruit.

Sometimes a fruit will explode because of the increase of internal water pressure. This is a rather rare type of mechanism. It is best known in the squirting cucumber *(Ecballium),* whose fruit bursts at maturity, squirting out a portion of the pulp and the enclosed seeds.

WIND HELPS SPREAD PLANTS

Wind plays an extremely important part in spreading the seeds or fruits of many

Milkweed parachute seeds are blown away from the open pods. A single seed is shown in the drawing at left.

plant species. The effectiveness of the wind as a transporting agency is due to the minute size and weight of the seeds of orchids, heather, tobacco, and various other plants. An orchid seed, for example, may weigh as little as three hundredths of a milligram. It

Left: the pine's winged seeds drop from the cone cells when they mature. Middle: maple fruits behave like windborne gliders because of their flattened wings. Right: the winged fruit of the ash tree resembles that of the maple.

Lynwood Chace, Audubon

Feathery globes are formed by the dandelion. The seeds are attached to parachutes. They are blown over wide distances until they settle and germinate. A single seed is shown at left.

can be carried great distances by air currents.

Wind transportation is sometimes favored by specialized structures that give buoyancy to seeds and fruits in air currents. The flattened wings found on various seeds (catalpa, trumpet creeper) or fruits (elm, maple, ash) cause them to behave like gliders borne by air currents. In a few plant species a broad flat wing is attached to a cluster of fruits. This wing, which is really a specialized leaf and not part of the seed or fruit, is found on the fruit clusters of the linden, or basswood, tree.

Left: the one-seeded fruit of the elm tree has a green membranous covering. Middle: fruit of the lotus plant, a member of the water-lily family. Right: the fruit of the false buckwheat is extremely buoyant, enabling it to float on water.

all line drawings, General Biological Supply House, Chicago

In certain cases tufts of soft bristles or hairs serve as parachutes and enable fruits or seeds to float, sometimes for considerable distances, in air currents. Such tufts occur on the seeds of some species (milkweed, cottonwood, willow, fireweed) and on the fruits of others (dandelion, goatsbeard).

Certain plants, such as Russian thistle and tumbleweed, become detached from their roots when their seeds reach maturity. As wind causes the plants to roll over and over along the ground, they scatter their seeds far and wide.

WATER CARRIES SEEDS

Water is another important agent of seed dispersal. The seeds or fruits of some plant species are buoyant because they contain air spaces or tissues that are light in weight and resistant to water. They often float for great distances and for considerable periods of time. The fruits and seeds of several palms that grow in the Orinoco River basin of Venezuela are sometimes washed up on the shores of Norway after a voyage of thousands of kilometers in the Gulf Stream. The extensive spread of coconut palms from their original home in the islands off the southeastern coasts of Asia is due, in part at least, to the ocean travels of the buoyant coconut—the fruit of the coco palm. The fibrous husk of this fruit

SEED AND FRUIT DISPERSAL

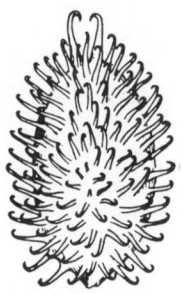  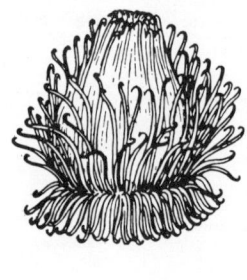 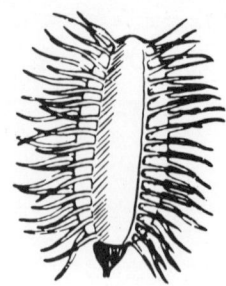

Left: cocklebur bur. The hooked ends help the seed stick to whatever has brushed against it. Second: the fruit of the sweet cicely splits to form two dry, one-seeded structures. Third: bur of the burdock plant. It is easily tangled in clothing and sheep's wool. Right: bur of the wild carrot, which is the cultivated carrot's ancestor.

makes it light, and its leathery skin prevents it from becoming waterlogged.

ANIMALS SCATTER SEEDS

Birds, mammals, and certain insects are often effective dispersal agents. Sometimes seeds or fruits are provided with specialized structures that adhere to the fur of animals as the animals brush against the parent plant. These clinging structures develop from the surface tissues. They include barbs, straight spines, and hooked spines, and are found in beggar-ticks, cockleburs, burdock, and other plants. When

Hugh Spencer

Left: close up of barbs on the seeds of the cocklebur. Right: A cluster of cocklebur seeds.

SEED AND FRUIT DISPERSAL

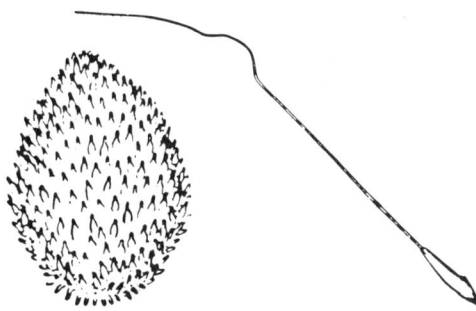

Left: the nutlet of the hound's tongue is covered with a number of barbed prickles. Right: the spikelet of the tall porcupine grass, which is found in the western part of the United States.

barbs or spines become entangled in a mammal's fur, they may reach and irritate the skin. The animal then scratches the seeds or fruits from its coat, thus aiding in dispersal.

The transportation of seeds or fruits on animal bodies does not always involve specialized structures. The coats of certain seeds become very sticky when moist. In this condition they will adhere to the hair of mammals or the feathers of birds. As they dry, they fall to the ground. Some seeds are imprisoned in the soil on which they have fallen as it cakes. Both soil and seeds may adhere to the feet of animals walking past. Seeds may be transported for a considerable distance in this way.

Birds and mammals often help scatter seeds by eating the fruits in which they are enclosed. These animals, especially the herbivorous varieties, eat many kinds of fruits, especially fleshy ones, such as berries and stone fruits, which become succulent and tasty as they mature. Many fruits apparently attract animals by their odors or by their red, orange, purple, blue, or yellow colors, which contrast sharply with the green color of leaves.

The seeds enclosed in the fruits may be digested and assimilated after they are swallowed by animals. In many plant species, however, the coats of some or all of the seeds are resistant to the digestive enzymes of animals' digestive tracts. They are scattered with the feces as the animals move about. Birds are especially effective in dispersing seeds in this way. Mammals also may drop large numbers of seeds in their feces. But they do not usually scatter seeds over such large areas as birds do, since their movements are ordinarily more restricted.

Birds also help spread the seeds of the mistletoe, which is parasitic on apple trees, poplars, maples, and other trees. The birds eat the white berries of the mistletoe. Later they wipe their bills on tree branches in order to get rid of the sticky seeds. The seeds remain attached to the branches until spring. Then they sprout. They thrust roots into the branches and produce a few pale green leaves.

Squirrels and other rodents eat many nuts. They also may bury many for future use and never return to claim them. In this fashion they help to plant such trees as oaks, walnuts, and hickories.

Some plant species are spread by ants, which derive food from oil-storing bodies

Hugh Spencer

Many fruits disperse their seeds by blasting them out of the fruiting body. They are propelled rather far. Some examples are: lupine fruits (left); witch-hazel shrubs (middle); squirting cucumber fruit.

Hugh Spencer

called elaiosomes, which are present on seeds, fruits, or structures attached to fruits. Such seeds may be carried by ants for considerable distances—sometimes as much as 60 meters.

PEOPLE ALSO DISPERSE SEEDS

Human beings are important agents in the dispersal of seeds. In some cases, we deliberately carry the seeds of useful plants—such as corn, wheat, sugar cane, coffee, and Pará rubber—to far-off lands. We have also transported the seeds of many ornamental plants, including tulips, hyacinths, lilacs, poinsettias, dahlias, and zinnias, far from their original habitats.

Some of the species introduced in this way may become naturalized. That is, they may ultimately propagate themselves without cultivation. They may come to be considered as weeds and, in some cases, as most unwelcome ones. The hemp plant (*Cannabis*) offers an example of this sort of transformation. It was introduced as a cultivated plant into the American colonies at an early period. Today it has become naturalized in many parts of the eastern and midwestern United States. It is an objectionable weed, since it yields marijuana. The soapwort, a native of Asia, was introduced into North America as a garden plant. It, too, has become naturalized as a troublesome weed in certain areas.

Like other mammals, people unwittingly transport seeds and fruits that are equipped with surface barbs and spines. These become fastened to the clothing of hunters, woodmen, and hikers walking through fields and woodland. The seeds or fruits of needle grass, beggar-ticks, tick trefoil, sandbur, and other plants are often transported in this way. Certain seeds have been carried considerable distances by "hitchhiking" on automobiles, trucks, and planes.

Sometimes the seeds of undesirable plants become mixed with the seeds of a cultivated variety that people seek to introduce into a new environment. As a result, the undesirable plants are able to win a foothold in the new environment, perhaps thousands of kilometers from their place of origin.

For an example, in the 1870s some South Dakota farmers decided to plant flax in their fields and obtained seed from abroad. Unfortunately the flax seed contained an admixture of Russian thistle seed, which the shipper had not removed. In just a few years the Russian thistle had established itself in South Dakota and Missouri. A decade or so later it had spread all the way to the Pacific coast.

WILL THE SEEDS THRIVE?

The effectiveness of a dispersal method varies greatly in different species. It may also vary enormously from time to time within the same species, particularly as environmental conditions change. If a new environment to which a plant has been transported is unfavorable, the plant will not thrive.

Venezuelan palm seeds may float across the Atlantic Ocean to the shores of Norway, but they will not develop into palm trees in Norway. The temperature and other environmental conditions in that northern country prevent germination of the seeds.

The seeds of many fruits are dispersed by animals. The ones shown above are scattered by birds. From left to right: fruit of poison ivy; fruit of sumac, typically found in thick cluster; fruit of hawthorne, known as haws; tiny fruits of chokeberry; and fruit of wild rose shrub.

In some cases, when the seeds of a plant species are dispersed into a new area, the temperature and other features of the environment may be favorable enough so that the seeds will sprout. However, other plant species already growing in that area may be such effective competitors in the struggle for existence that the seedlings of the newly introduced species never grow into mature plants.

Obviously, therefore, the dispersal of seeds and fruits does not necessarily result in the spread of plant species. It is effective in some cases, futile in others.

Squirrels and chipmunks, among others, hoard nuts and seeds. If the seeds are stored below ground and then forgotten, a new plant may germinate.

Lynwood Chace, Audubon

SEED AND FRUIT DISPERSAL

FROM SEED TO SEEDLING

by Harry J. Fuller

The next time you eat an orange or some grapes, save the seeds. Plant them in some soil. Keep the container moist and watch what happens. Soon, tiny green shoots will break through the soil's surface. With enough light and moisture, the grape plants will grow into vines. The tiny orange plants will develop attractive, shiny leaves. In a tropical or a subtropical environment, they will eventually develop into full-grown and fruit-bearing trees.

Not all the seeds that you plant will sprout, or germinate. This is true, too, in nature. There are many reasons why a seed may not sprout. The environment may be unfavorable. It may be attacked by disease. Or it may be eaten by an animal. If this happens, the reproductive cycle is broken, as far as that particular seed is concerned. It will not be able to develop any further into an adult plant.

But if all conditions are favorable, a transformation takes place within the seed. A tiny root pushes down into the soil, and a tiny shoot makes its way above the surface. The seed has germinated and has given rise to a seedling. If all goes well, the seedling will grow into a mature plant. This will bear flowers and the cycle will begin anew.

PARTS OF A SEED

A typical seed consists of three parts: an embryo, endosperm (a mass of food-storage tissue), and an envelope called the seed coat.

Embryo. The embryo is the part that will grow into a plant. It is somewhat cylindrical. Its lower end, called the *hypocotyl,* forms the first root of the seedling. The upper end of the embryo, the *epicotyl,* is the stem-forming portion. The embryo also bears one or more side growths, called *cotyledons,* or embryonic leaves. They usually serve for the digestion or storage of food, or for both digestion and storage.

In flowering plants, the number of cotyledons is almost always one or two. The number varies in cone-bearing plants; there may be as many as eight. Cotyledons may be thin or leaflike, as in the embryo of the castor bean. Or they may be thick and fleshy, as in the embryos of the peanut, garden bean, and pea.

In a few plant species, seeds may lack embryos because the reproductive processes of the species are abnormal. There is no embryo, for example, in the seed of the banana plant. Therefore it cannot germinate. Such plants must be reproduced by cuttings or other vegetative means.

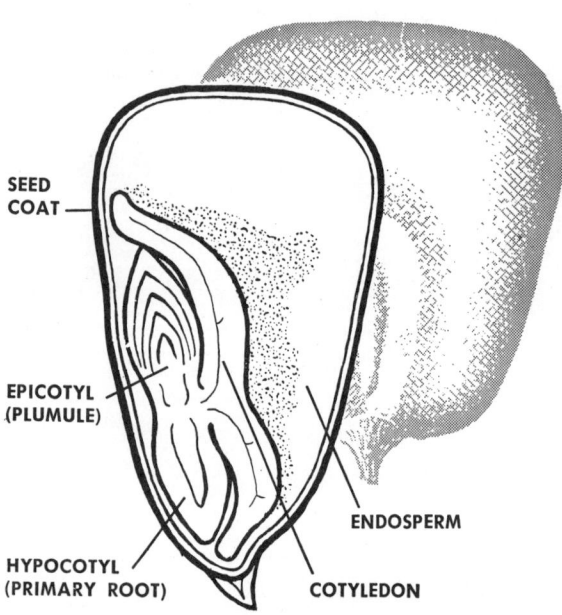

A typical seed consists of three parts: the embryo, endosperm, and the seed coat. The embryo is made up of the epicotyl, or stem-forming portion; the hypocotyl, or root-forming portion; and one or more cotyledons, or embryonic leaves.

Endosperm. Food-storage tissue is present in all seeds at some time in their development. In the seeds of certain plants, such as the bean and the peanut, the endosperm tissue is digested and absorbed by the cotyledons before the seeds leave the parent plant. The mature seeds, therefore, have no endosperm tissue.

In other plant species, including the corn and castor bean, the endosperm tissue of the seed is not digested and absorbed by the cotyledons until after the seeds have begun to germinate. Hence the endosperm is present in the mature seeds.

Seed coat. The seed coat is the surface layer of the seed. It is a protective structure. Sometimes, it is thin and papery, as in peanut and pea seeds. In other species, such as the Brazil nut, the protective coating of the seed is thick and hard.

The seed coat guards the embryo and endosperm against drying out and mechanical injury, as well as against the attacks of insects, bacteria, and fungi. A thick seed coat also serves as insulation; it provides effective protection for the embryo against extremes of temperature.

Seeds differ in size and structure. Few persons have difficulty in distinguishing among the seeds of green beans, peas, peanuts, radish, corn, and other familiar plants. There are noticeable differences in the shape and size of cotyledons, epicotyls, and hypocotyls, and in the thickness, texture, and color of seed coats.

A PERIOD OF DORMANCY

The seeds of most plant species germinate when the temperature is favorable, when there is an adequate supply of free oxygen, and when soil moisture is readily available. Yet in many species, the seeds are incapable of sprouting when they fall to the ground, no matter how favorable the environmental conditions may be. The seeds will not germinate until they have passed through a period of rest, called *dormancy*. This is true particularly of plants in temperate zones and in tropical regions with prolonged dry seasons. However,

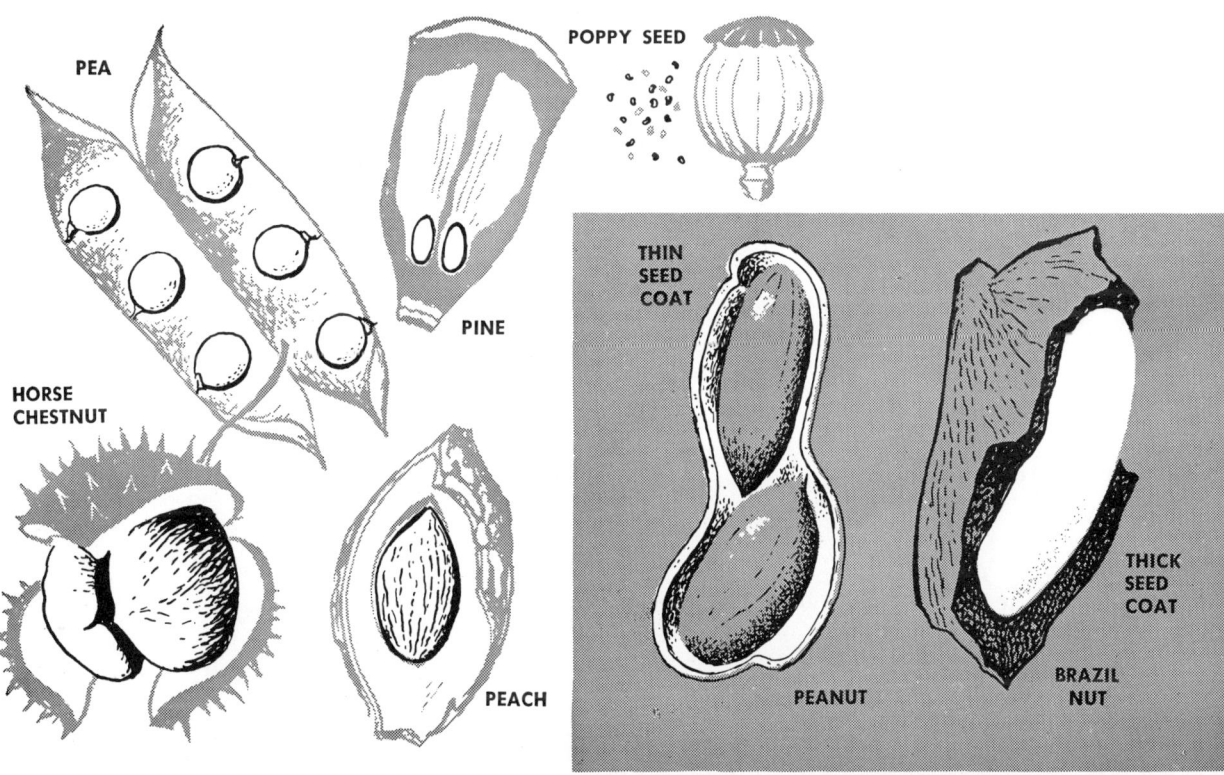

FROM SEED TO SEEDLING

dormancy does not result from any conditions of the environment. Rather, it is due to the peculiar structure or life processes of the seeds themselves.

Dormancy may be due to several factors. In certain seeds, such as those of sweet clover, water cannot penetrate the seed coat at first. Germination cannot take place until the seeds have been properly prepared by nature or people. Seed coats may be cracked by alternate freezing and thawing, as the temperature rises and falls. They may be eroded by the destructive action of soil bacteria and fungi. Some seed coats are subjected to the mechanical action of threshing machines or are scraped by files in the process known as *scarification*. Once a seed coat has been weakened in any of these ways, water can enter it and the seed can sprout.

In some instances, seed coats are impervious to oxygen. They will not germinate unless they are cracked or eroded in the ways just mentioned. Other seed plants may admit water and oxygen to the interior portions of seeds, but they are so hard that they form a barrier to the growing embryo. The first root and stem will not be able to emerge until the seed coats have been cracked or scarified.

The embryos of certain seeds may be immature at the time that the seeds are discharged from the parent fruit. Before germination can take place, the seeds will have to pass through a period of further development in the soil. The embryo and its parts will be fully mature at the end of this period.

The chemical condition of the embryo or endosperm at the time of seed release may be such that germination cannot take place. During the dormant period, certain chemical changes in the embryo or endosperm will occur. As a result, the seed will be able to sprout.

The seed coat may contain chemical inhibitors—substances that interfere with or retard a chemical reaction—which will prevent germination. These inhibitors must be disposed of during the rest period. They may be leached out of the seed coat by soil

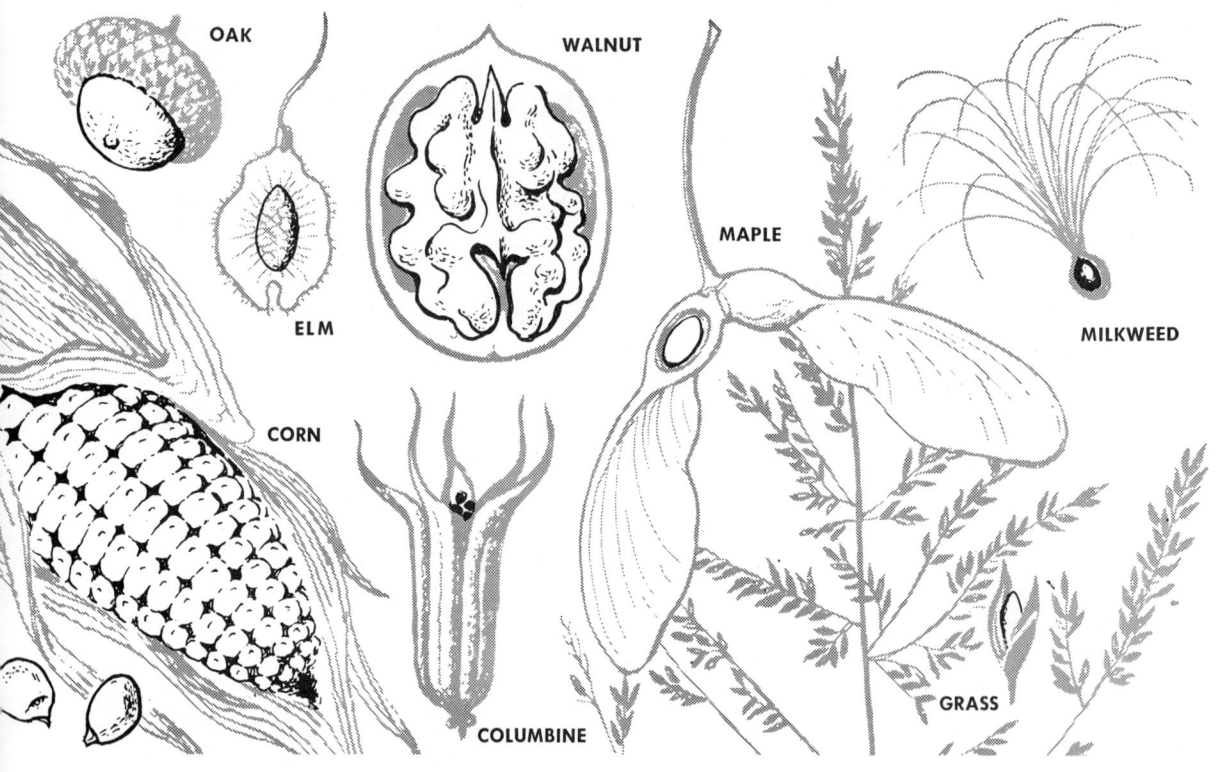

moisture. They may break down as they react with various chemical agents in the soil. As soon as the inhibitors have been removed, germination takes place.

Some botanists hold that the dormancy of seeds enables certain plant species to survive during periods when the environmental conditions are harmful. For example, if the seeds of plants in the temperate zones were to sprout as soon as they became mature, they might be killed by the low temperatures of late autumn and winter. The seeds are dormant during these seasons; their germination is delayed until the following spring. By that time there will no longer be danger of injury by low temperature. The dormancy of the seeds of plants in regions with prolonged dry seasons may also prove beneficial. It may prevent the germination of seeds at the beginning of the dry season, when drought conditions might make it hard for the seedlings to survive.

If a period of rest results from unfavorable conditions of the environment, it is known as *quiescence*. A seed may have passed through its period of dormancy. Yet if there is insufficient moisture or if the soil temperature is not suitable, the seed will not germinate. It will remain quiescent until conditions become more favorable.

READY TO GERMINATE

Once the period of dormancy has passed, seeds will sprout under the conditions we have already mentioned: suitable temperature and adequate supplies of oxygen and water. Contrary to popular belief, soil is not necessary for germination in most cases. Ordinarily, seeds will sprout in moist sand, sawdust, vermiculite, paper, or peat moss if the temperature is favorable and if there is enough oxygen and moisture. If seeds most commonly germinate in soil, it is simply because soil occurs so abundantly on the surface of the earth.

Germination may take place even if the soil does not supply nitrogen, phosphorus, potash, and other essential chemical nutrients. This is possible because seeds ordinarily contain enough food to nourish the growing embryos through the seedling

Top: the seeds of the well known Brazil rubber tree remain alive and capable of germination for only a few weeks after they have been cast off from the fruits. Bottom: the seeds of the Egyptian lotus remain viable, or capable of germination, for many years. The lotus is a member of the water-lily family.

FROM SEED TO SEEDLING 129

stage. Seeds will germinate, for example, in pure quartz sand supplied with distilled water. However, as the food stored within the cotyledons and endosperm is exhausted, the seedlings must receive nitrogen, phosphorus, and other chemical nutrients from the medium in which they are growing or they will die.

The most favorable temperature for the germination of most seeds lies between 15° Celsius and 38° Celsius. As the thermometer reading goes below 15°, the rate of germination decreases. No seeds at all will sprout when the temperature approaches the freezing point of water (0° Celsius). The seeds of many tropical plants fail to germinate when it is as cold as 10° Celsius.

As the temperature rises above 38° Celsius, the seeds of most plants show a declining germination rate. The point above which sprouting ceases altogether varies in different species. In most cases, however, a seed will not germinate if the temperature is above 60° Celsius.

Free oxygen is necessary for respiration, which releases the energy required for growth. That is why the absence of oxygen will retard or prevent germination in seeds. The seeds of some species of water plants, such as water lilies, can sprout when the oxygen concentration is very low. Most land-plant seeds, however, require abundant oxygen for normal germination.

If the soil becomes waterlogged, following very heavy rains, water may accumulate to such an extent that all oxygen is forced out of the soil spaces. Under such conditions, germination will not take place. The seeds in the waterlogged soil may be attacked by anaerobic bacteria, which flourish in the absence of oxygen. These seeds may die as a result. Or, if they germinate later, their respiration may develop abnormally.

Seeds must have just the right amount of water to germinate. If there is too much, as we have seen, it will drive out oxygen from cell spaces and the seed will not sprout. If, on the other hand, there is too little water, seeds will not be able to digest the food in their endosperm tissues; the epicotyls and hypocotyls will not grow. The seeds of different plants vary somewhat in their water requirements.

THE VIABLE PERIOD

Seeds remain viable—that is, alive and capable of germination—for a period that varies enormously, depending on the species and the environmental conditions. In many orchids and in the Brazil rubber tree, for example, seeds remain viable for only a few weeks after they are discharged from their parent fruits. If they do not germinate within this brief period, they will die.

The seeds of other species may be via-

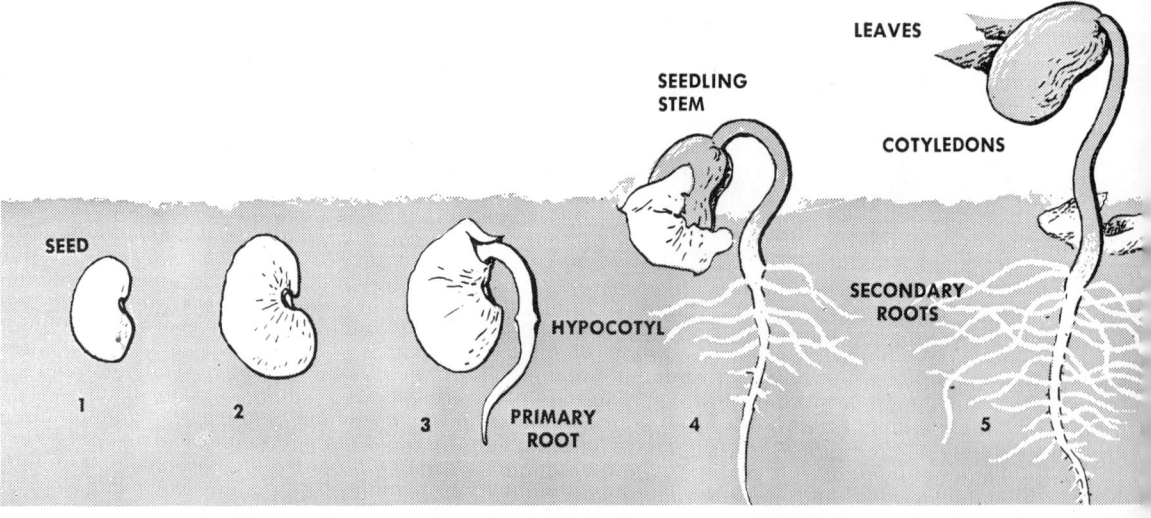

130 FROM SEED TO SEEDLING

ble for a century or longer. According to some accounts, grains of wheat and barley placed in ancient Egyptian tombs 4,000 years ago have kept their viability. Reputable plant scientists have never been able to verify these reports, and so they must be discounted. As far as botanists are aware, the record for viability is held by an oriental water lily species. Its seeds have remained capable of germination for 800 years.

The average period of viability ranges from 3 to 20 years. The percentage of viable seeds for a given species decreases sharply as the seeds age. Therefore, in selecting seeds of crop plants for planting in fields and gardens, it is important to use seeds that are not more than two or three years old to ensure a high percentage of germination. Storage conditions influence the length of the viable period of many seeds. Generally speaking, seeds keep their viability longer when they are stored under dry, cool conditions.

Botanists do not know much about the factors involved in the loss of viability. Apparently it is due to certain chemical and physical changes occurring in the cells of the embryo.

FROM EMBRYO TO SEEDLING

When conditions are favorable for germination, the seed begins to absorb water. As it does so, it swells, and the seed coat is ruptured. Water enters the cells of embryo and endosperm tissue. The moisture content of these cells rises from the pre-germination level of 10 per cent to 90 per cent or more. As the protoplasm of the cell absorbs moisture, various life processes are stirred into action. The plant-growth hormone, indoleacetic acid, which controls and promotes the various stages of growth, begins to function. It regulates the development of the hypocotyl and epicotyl.

Foods stored in the endosperm and the cotyledons are now digested. The rate of

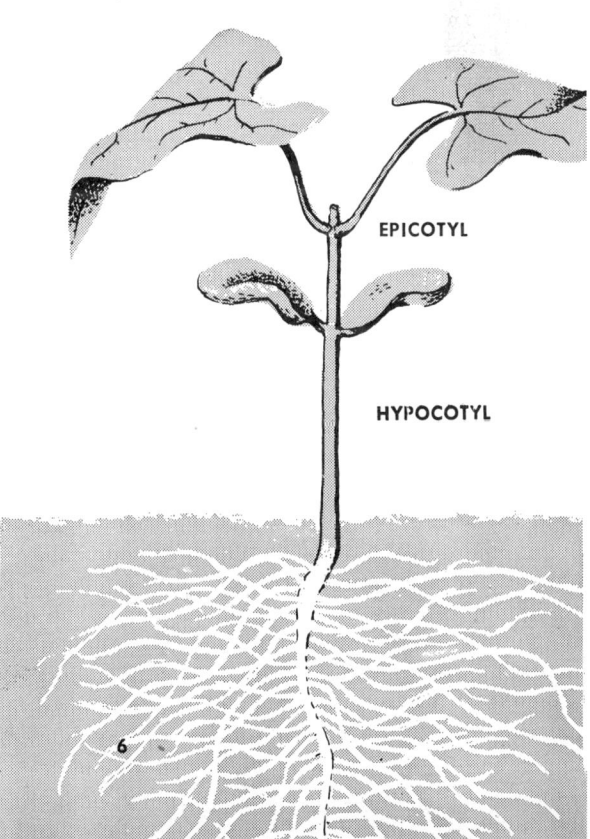

The germination of a garden bean. The bean (1) does not begin to germinate until the temperature is high enough, there is an adequate supply of oxygen, and soil moisture is readily available. When conditions become favorable at last, the seed begins to absorb water and it swells (2). The hypocotyl emerges through a split in the seed coat (3). It grows at both ends. The lower end forms the primary root. The upper end grows above the surface of the soil, carrying the cotyledons and epicotyl with it. The primary root then begins to put forth secondary roots (4). As time goes on, the secondary roots become plentiful. The stem grows higher, and leaves arise from the epicotyl. The growing plant uses the food stores contained in the cotyledons (5). As the stores are used up, the cotyledons shrink. Later, they will drop off. The stem grows higher. New leaves develop. By now, the seed has become a full-fledged seedling (6).

respiration increases rapidly. It results in the release of chemical energy, which is essential in cell division, protoplasm production, and other growth processes.

As digestion, respiration, and the assimilation of foods into living protoplasm occur, cells at the tips of the epicotyl and hypocotyl begin to divide, forming new cells. These cells begin to enlarge as they absorb water, and new protoplasm is formed.

The tip of the hypocotyl now emerges through a split in the seed coat—a split brought about by the absorption of water. The growing hypocotyl tip becomes the first, or primary, root of the seedling. It may reach a length of two centimeters or more before the growing tip of the epicotyl at the opposite end of the embryo emerges from the seed coat. This arrangement is highly advantageous. It ensures firm anchorage of the seedling in the soil. The water and the nutrient elements that the root draws from the soil will be necessary for the later growth of the seedling. After the primary root has become established, the epicotyl tip begins to grow upward; it forms the beginning of the seedling stem.

In the germinating seeds of some species, such as pea and corn, the hypocotyl remains in the soil and the epicotyl alone forms the seedling stem. In other species, including the garden bean and the castor bean, the hypocotyl grows at both ends. The lower end forms the primary root, as we have seen. The upper end grows above the surface of the soil, carrying the cotyledons and epicotyl with it. In the seedlings of such species, the lower portion of the stem develops from the upper part of the hypocotyl; the upper part of the stem is derived, not from the hypocotyl, but from the epicotyl portion of the seed.

The growing primary root soon begins to put forth root branches, or secondary roots. These may later branch to produce still more roots. As the stem develops, it begins to produce leaves and, at a later stage in seedling development, it produces stem branches.

We observed that, in certain cases, the cotyledons are lifted above the surface of the soil by the growth of the upper part of the hypocotyl. These cotyledons may persist for several days or even weeks. As the food stores they contain are used up, they begin to shrink. Finally, the cotyledons drop off of the seedling.

In some plants, the cotyledons contain chlorophyll and manufacture food. However, since cotyledons are small and short-lived, the total quantity of food they make is probably not very important to a plant. As for the cotyledons, such as those of peas, that never rise above the surface of the soil, they obviously cannot manufacture food, since exposure to sunlight is necessary for photosynthesis. They may contain chlorophyll, however.

The rate of seed germination varies greatly in seeds of different plant species. Under especially favorable conditions, radish seeds may put forth primary roots within 36 hours. Corn grains usually require 72 to 96 hours. It may take seeds of other species weeks or even months to germinate. These differences are due to various factors, such as seed-coat thickness and structure. Variations in rates of germination also involve variations in the plants' rates of digestion, respiration, cell division, and cell enlargement.

ENEMY ATTACKS

Insects and parasitic bacteria and fungi inhabiting the soil may attack seeds and retard or actually prevent their germination. Certain fungi groups are particularly dangerous to seeds at the germination stage. They attack seedling stems at the soil line, weakening them so that they fall over and cause the death of stem tissues. These fungi attacks bring about what plant pathologists (specialists in plant diseases) call "damping-off" diseases.

To prevent these and other diseases involving seedlings, seeds are often treated with a fungicidal, or "fungus-killing," powder. The powder adheres to the seed coat and destroys any fungus spores coming in contact with the coat after the seeds have been planted. Various liquid sprays are also used for this purpose and are quite effective as well.

PLANT ADAPTATION

Like a human being or an animal, a plant must adapt to its environment in order to survive. Survival usually means struggle. It may be hard to think of a plant as "struggling," yet that is often the case.

A plant, first, must compete with other plants for light, air, soil, and water. Second, it must contend with people and animals, who may eat it or make the environment unfit for it to live in. Third, the physical environment itself may be hostile or inadequate to a plant's needs. Is there enough light, water, air, and soil? Is it warm or cool enough for a given plant species?

NEEDED FOR SURVIVAL

Light is vital to a plant, especially a green one. Green plants manufacture foodstuffs in sunlight—a process known as photosynthesis. Light also affects the rate of water absorption and flower formation.

Temperature must also be taken into account. If it falls below 1° Celsius or rises above 43° Celsius, serious injury or death results for most vegetation. Temperature also directly influences the formation of seeds and the production of blossoms. It also helps determine the geographic distribution of plants.

Plants cannot survive without water. Many need immense quantities for their development. Therefore, the humidity of the atmosphere, the amount of precipitation, and the presence of streams, lakes, and soil moisture are important to plants.

Most plants need plenty of air. The atmosphere contains gases that are essential to their life: oxygen, carbon dioxide, water vapor, and nitrogen.

Not only the air itself, but its movement—the wind—affects plants greatly. Wind spreads pollen, spores, and certain seeds, enabling plants to grow in many places. But a powerful wind may distort or even kill vegetation, particularly trees. It may blow away needed soil. Also, wind increases the rate of evaporation and so may create a water shortage for plants.

Courtesy of The American Museum of Natural History

A tropical rain forest. Here, plants that have adapted to crowded conditions are successful.

Soil is extremely important to land plants. A plant's development depends in large part on the nature of the soil: its wetness, acidity, minerals, and the amount of oxygen it contains.

A plant is also influenced by other living things with which it competes. It has to withstand parasites, hungry birds, and grazing or gnawing mammals. Yet a plant often needs animals—for example, to spread its pollen or scatter its seeds.

GETTING LIGHT AND AIR

The stems and leaves of trees and shrubs show us some of the adaptations that enable a plant to get the maximum amount of light and air.

In such trees as beech, elm, oak, apple, and chestnut, there may be shoots that are

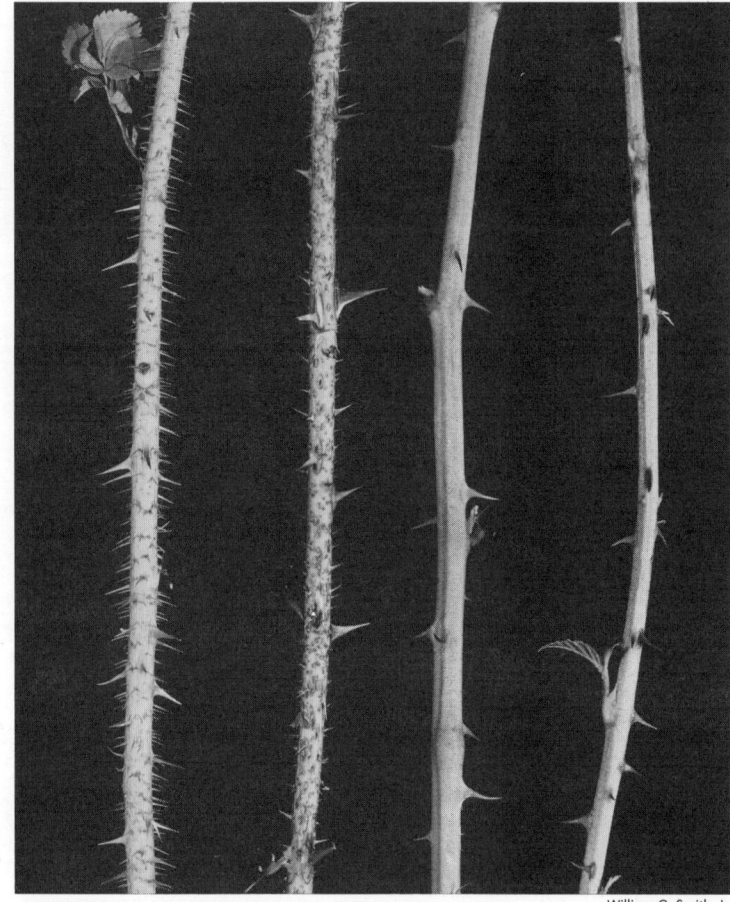

Wild-rose (left) and bramble (right) stems show adaptations both for defense and climbing.

primarily vertical while others are primarily horizontal. To secure a maximum amount of sunlight, the leaves on the vertical shoots are arranged in spirals so that any given leaf does not shade the leaf next below it on the shoot. Though the buds on a horizontal shoot of these trees may be arranged spirally, the leaves commonly arrange themselves alternately in a common plane on either side of the shoot.

In many trees that have their leaves placed opposite to each other, each pair occupies the space between the pair immediately below. You can see this by looking down on a shoot of the horse chestnut. Similar very perfect arrangements will be noted in many climbers. The arrangement of the leaves in the horse chestnut, when examined at the end of a shoot and vertically, has been termed a leaf mosaic, because of the variegated pattern that the leaves of this tree present. Another common device for a similar purpose is the fringelike dividing of the leaf. The carrot plant has this type of leaf. Such a leaf obtains good sunlight but does not shade to a great extent the leaves below it. Moreover, it is in less danger from strong winds.

WHERE LIGHT IS LITTLE

There are numerous plants that live in environments where light never penetrates. Many of these plants are fungi. They have no chlorophyll, but subsist on organic matter, produced by animals or other plants, that may be found wherever the fungi flourish. Rains washing into the underground nooks and crannies may be one of the agents responsible for getting organic matter into these places.

In caves and underground mines, or in pits and wells, where there is a certain amount of light penetration—even if very little—we find that the plants are principally green. Not only is this so, but the green coloring of vegetation in caves is often luxuriantly brilliant. In fact, it may appear even more vivid than that of a plant in the open air. This is true of liverworts, many mosses, and some of the ferns that flourish abundantly in these situations. In such cases the chlorophyll granules are developed in a very special way. The light falling on the plant cells is concentrated on the chlorophyll granules themselves, which thus receive a sufficient supply for food-making.

Plants also receive a minimum of light in the depths of the sea or at the bottom of lakes and pools, since sunlight is weak or absent there. The light under water diminishes in proportion to the depth. It is also greatly influenced by the amount of sediment in suspension and consequently varies before and after storms and at different distances from the mouths of rivers and streams.

Complete darkness reigns at depths greater than 200 meters below the surface of clear sea water. At a depth of 170 meters, the illumination is about equal to that at the surface of the water during moonlight. Under such conditions, plants possessing chlorophyll cannot manufacture the sugars ordinarily formed by green plants in sunlight. This explains why plants possessing green coloring matter are rarely seen at depths of more than 60 meters. As a matter of fact, by far the greatest part of the vegetation of the sea is found within about 30 meters of the surface.

In the process of photosynthesis, the leaves of plants do not absorb all the colors of the spectrum to the same degree. The red, orange, and yellow rays of light that reach the leaves are absorbed much more effectively than the rest. In deep marine water, vegetation receives more blue to green light than it does red to yellow. This is another reason why conditions are particularly unfavorable for plant growth at considerable depths below the surface of the sea.

Cornelia Clarke

U. S. Forest Service

The leaves of the carrot (top) and the beech (bottom) show adaptations for different life styles.

PLANT ADAPTATION

A striking adaptation is found in red algae. These plants contain, in addition to chlorophyll, pigments known as carotenoids and phycoerythrins. These pigments are present in considerable quantities and give the red algae their characteristic color. They also absorb blue to green colors of light more than does cholorophyll. It is suspected that after they absorb light, they transmit the energy from this light to chlorophyll, for photosynthesis.

In tropical forests the tree growth is often so profuse that sunlight may never reach the ground. In these hot and humid climates there are a great many *epiphytes* — plants — such as Spanish moss, orchids, and staghorn fern — that grow on other plants.

The epiphytes grow on upper limbs of trees and on poles. Their roots are suspended in air and they absorb moisture and gases from the humid atmosphere. The cells of epiphytes are not directly parasitic, for they contain chlorophyll and they manufacture their own food. However, they may injure trees by their sheer weight.

Top: a coniferous forest. The evergreen needles cover the trees all year, and the amount of sun reaching the ground is limited. Note that the new trees are growing in regions where light penetrates to the ground. Bottom: a beech-maple forest. These trees have adaptations for moderate temperatures and rainfall. Note the well-developed ground cover.

The liana is a vine with a woody stem, which encircles the trunk of a tree as it climbs to the highest branches. In this way, though the liana is rooted in the soil, it can reach the sunlight.

WHERE LIGHT IS STRONG

If we turn our attention to the flora found on the rocky slopes just above the level of the sea, but still exposed to spray in bad weather, we might expect to find—considering we are in the presence of direct sunshine—that the plant life would show certain adaptations to the intense light. These special features control the rate of photosynthesis, which is directly affected by the intensity of the sunlight falling on the leaf.

The foliage of plants living in strong sunlight is often covered with either a waxy coat or tiny hairs.

Photophiles is the name applied to plants whose growth is favored by strong sunlight. These include such types as sunflowers, rosinweeds, coneflowers, asters, milkweeds, spurges, and legumes. Many grasses such as buffalo grass, big bluestem, and Indian grass are also photophiles.

Plants that are better adapted to shady areas are known as *photophobes*. They have certain features that tend to enhance the absorption of sunlight by the plant. Frequently these plants have thin, broad leaves with stomata, or small openings, on both surfaces. The leaf surfaces are dull, and they lack any waxy covering or hairs. The photophobes grow best in light that is from about 15 to 25 per cent as intense as full sunlight. Such plants as mosses, liverworts, lady's-slippers, wood sorrel, hydrophyllums, figworts, and geraniums thrive in the shade of the forest.

LENGTH OF DAY

In addition to light intensity, the duration of light plays a role in the distribution and reproduction of plants. The summer days are long in the temperate climates of the northern and southern hemispheres. Here the so-called long-day plants flourish, for they will flower only after many days with long hours of sunshine. These species include beets, lettuce, clover, radish, henbane, and cereals.

Short-day plants occur in regions of the world where the day rarely exceeds 13 hours. Plants such as coleus, aster, dahlia, cosmos, and chrysanthemum bloom several times during the year.

Certain other plants, such as dandelion, tomato, buckwheat, cotton, and sunflower, grow regardless of the duration of sunlight. They will flower in any region of the world, providing other conditions are favorable to their growth.

ADAPTING TO WIND AND COLD

At high mountain elevations and at exposed positions along sea cliffs, trees will show marked distortions because of the strong prevalent winds. Often the trees are bent and twisted. The foliage may grow only on the leeward side of the tree. In general, only pine trees, scrub oaks, and some shrubs are able to withstand such conditions.

Winds also produce certain physiological effects, since they increase the rate of transpiration. Plants must balance the water that is evaporated from the leaf surfaces with water absorbed through the roots. If transpiration exceeds absorption, the plant will die. The narrow surface of the pine needle loses water less rapidly than does the broad surface of other plant leaves. That is why pines often predominate in mountain areas where the wind is strong.

Winds may influence the distribution of plants, for they affect the nature of the soil. High winds cause shifting of the soil, thereby preventing the growth of many plants with shallow root systems. Only perennials with deeply branched roots and special features can grow successfully in this type of light, wind-shifted soil. Perhaps an exception to this rule is the sagebrush. This plant has a shallow root system, which weakens as the plant is about to produce seeds. The winds may then uproot the sagebrush, which will roll along, shedding its seeds as it goes.

Plants exposed to cold or dry conditions may be found on high mountain peaks or in polar regions. They may take the form

The abundance of vegetation in this delta is due to high moisture and very rich soils. Plants have adapted to tidal fluctuations, wave actions, and temperature changes.

of cushions, mats, or rosettes. An example of such a plant is the lichen. Actually lichens are composed of two plants, an alga and a fungus. The green alga furnishes food for the fungus and the fungus provides shelter for the alga. Without the algal components that give the lichen its green color, the plant would not survive.

Lichens are widespread and hardy. They can cover rocks, trees, and boards that are exposed to sun and wind. The peculiar relationship of the two components of the plant makes lichens relatively independent of their environment.

WATER PLANTS

Plants such as water lilies and sedges that live in an aquatic region are known as *hydrophytes*. This name comes from the two Greek words meaning "water plants."

The soil at the bottom of a stream or pond is, in general, muddy and poorly ventilated. Hydrophytes frequently show many adaptations that aid in the passage and storage of air within the plant tissues. The stems of the water lily are usually hollow and connect directly with the stomatal openings of the leaf.

Some aquatic plants, such as the water crowfoot, have two different types of leaves. The submerged leaves are narrow and delicate. The leaves exposed to the outer air are broad, thus presenting the maximum surface area to the rays of the sun. Flowers may be present.

The water crowfoot and the water lily are attached to the bottom of the pond by roots. Certain other aquatic plants float freely, both at the surface and under the water. Duckweed floats on the surface. Roots arise from its broad, floating leaf structure. Hornwort, on the other hand, is a rootless, completely submerged plant that floats freely underwater. In this way it can move easily to areas where light, temperature, and air conditions are more favorable.

Many of the lower plants thrive in moist environments. Ferns, liverworts, and mosses have mobile reproductive cells that require a fluid medium for transportation.

Still another interesting adaptation of aquatic plants is their carnivorous habits. Insects abound in humid regions. Plants such as sundew and Venus's-flytrap catch these insects in their leaves. The leaves of the bladderwort, a completely submerged aquatic plant, are modified to form tiny trapdoors. As aquatic insects brush past the bladder-shaped leaves, the trap-doors open. The insects cannot escape and are eventually digested.

The cypress trees of the bayou regions have a special adaptation for transporting air to the submerged plant tissues. Root projections called *pneumatophores,* or "knees," grow above the surface of the water. They transmit air from the atmosphere to the submerged roots.

Generally, trees found in aquatic areas have shallow root systems. Swamp trees such as the banyan and mangrove often appear to be propped up. This is because roots grow from the lower branches down into the soil. The mangrove trees often catch debris in their exposed roots. In this way they build up the soil, and other plants can grow in it.

FROM LAKE TO FOREST

An interesting phenomenon of plant life is *succession.* As the environment changes, plants with different adaptations tend to replace the older ones. The amount of rainfall, the incidence of forest fires, the temperature, the elevation of the land—all influence the type of vegetation that will occur in an area.

Near a lake, pond, or stream where the water movement is slow, the first plants to appear are the underwater types. Their growth is limited by the amount of light penetrating into the water. As the bottom of the pond is built up and the water becomes shallower, rooted, floating-leaved species appear. Types with large leaves, such as water lilies, will exclude the previous underwater forms by shading the surface. This is an excellent example of natural selection.

As the water becomes still shallower, emerging plant types will predominate. These plants have roots in the mud and stems that extend above the surface of the water. They include rushes, reeds, cattails, and sedges. This type of close growth can hold organic sediment. It makes possible the formation of dirt beds that will be capable of supporting plants requiring a more substantial root system.

The plants that are better adapted to the new conditions will dominate the area. Thus, as the bayou fills in, the sedges will be replaced by myrtle and alder shrubs. Eventually these are succeeded by a swamp forest, which consists of ash, cedar, and tamarack trees. In freshwater bogs the sedge mats will be replaced by moorwort and laurel shrubs. The swamp forest that will rise here will consist mainly of tamarack and spruce trees.

Both the bog forest and the bayou forest will eventually be replaced by a so-called *climax formation,* which may be a forest consisting of firs, birch, and spruce. Various types of climax formations exist. They all represent plant communities best suited to certain stable climatic conditions. A few familiar examples of these formations are the tundra, the coniferous forest, the woodland, the desert sagebrush, the grassland, and the tropical rain forest.

ADAPTATIONS OF LOWER PLANTS

The competition for access to air and

Swamps contain variable amounts of water during the year and the plants must be able to withstand dry periods as well as wet periods.

Arthur Ambler, Audubon/PR

light does not cease in the sea. Many of the brown algae show interesting adaptations for buoyancy. The giant kelp, *Macrocystis,* has a holdfast at its base that secures it to a rock. The "branches" (stipes) of the plant have many air bladders. Each bladder supports a "leaf," or blade. *Laminaria,* or devil's apron, has a rootlike holdfast but no air bladders. Depending on the species, the yellow, rubbery blade is from 2 to 30 centimeters wide. Since it is broad, it floats easily.

The free-floating sargassum has a bladder at the base of each blade. This marine alga is transported by ocean currents from many coasts to the Sargasso Sea, which extends from the West Indies to the Azores.

Rockweed, or *Fucus,* is commonly seen along many seacoasts. This plant is short and is attached to rocks by means of a holdfast. The blade contains air bladders.

None of the brown algae have any structure resembling the xylem of higher plants. This tissue, which functions as a water-conducting and supporting element, is not needed by brown algae. Buoyed up by the sea, they obtain all the water they need directly from the ocean surrounding them. Some kelps have growth rings and sieve tube cells resembling those of phloem (food-conducting tissue).

Since the algae are in an aquatic environment, they may reproduce by shedding their gametes (reproductive cells) into the water. The gametes are mobile and propel themselves by means of whiplike flagella.

A great variety of adaptations exists among other lower plants. Microscopic living things such as bacteria, blue-green algae, and some fungi exist in every possible environment on earth.

The sulfur-purple bacteria live in sulfurous hot springs. They thrive in an atmosphere completely lacking in oxygen. In addition, they carry on photosynthesis without the aid of ordinary chlorophyll. They use the hydrogen-sulfide gas present in the mud as a source of energy, much as higher plants use water to make sugars.

Other bacteria, such as *Bacillus meg-*

Bladder wrack, an alga, covers a rock along the seashore. Algae live in water, but must be able to survive periodic exposure to the air.

N.Y. Botanical Garden

PLANT ADAPTATION

atheri, can form spores when placed in an unfavorable environment. The spores can withstand many hardships: temperatures as high as 190° Celsius; periods of drought lasting several decades; the action of disinfecting chemicals. They will sprout and produce the parent body when conditions become more favorable.

Pneumococcus is a parasitic bacterium that causes pneumonia. When it enters the human body, it secretes a jelly-like capsule about itself. The person's natural defenses (antibodies and phagocytes) are thus unable to attack the pneumococcus. Some bacteria possess flagella, which they use to propel themselves away from danger. *Vibrio comma* is an example.

Among the simplest plants are the blue-green algae. Their chlorophyll is not found in plastids, or bundles, but is scattered throughout the entire cell. They will frequently attach themselves to rocks or plants. Some types secrete jelly and float freely in the water.

The fungi and some bacteria lack chlorophyll. They cannot manufacture their own food. For this reason many of them are parasitic. The slime molds, when examined under a hand lens, appear to be long, entwined threads. This network, called a mycelium, is the body of the fungus. As it spreads over the host it absorbs nutrients through the threadlike structures.

In general all fungi are delicate; their cell walls are thin and very susceptible to drying. These plants never grow in a dry environment.

TOO WET OR TOO DRY

It is necessary for a plant to provide some means for promoting transpiration in cases where there is excessive moisture, and of checking transpiration in conditions of extreme dryness.

To bring about effective transpiration, the plant is provided with a large number of cells whose surfaces come in contact with the outer air. These cells are so constructed that they can "exhale" water. In these plants, too, there is always a considerable development of green, spongy tissue. Air can penetrate readily through the leaves.

Grant Heilman

The water lily lives in ponds. Its broad leaves give off a great deal of moisture to the air—an adaptation to an aquatic habitat.

An increase in the amount of transpiration that takes place is brought about by an increase of leaf surface. In some water plants whose stems and stalks are in the water and whose leaves float on the surface, the area of the leaf is a very large one; the leaves of the giant water lily, *Victoria regia,* may be two meters across. The whole surface of a pond may be covered with floating leaves, arranged in such a manner that the entire upper surface is exposed to sunlight. The under surface is colored a violet tint by a pigment called anthocyanin. According to certain physiologists, this pigment converts some of the sun's light rays into heat rays and as a consequence causes the leaves of the plant to be warmed.

For the sake of contrast, let us consider plants that have to sustain life under conditions of extreme dryness and which consequently are called *xerophytes*—a

Some plants have adapted to life in an arid environment. Palms (right) hold moisture well. An oasis (far right) can support a large amount of desert vegetation.

name derived from two Greek words meaning "dry plants."

Xerophytes grow in extremely dry soils in regions that are rainless during long periods. No doubt the first plants in the world were aquatic; we must therefore consider the xerophytes as highly evolved types, showing remakable adaptations to their environment.

One group of xerophytes, the *ephemerals,* live and reproduce only during the short wet season (a few weeks) and then die, leaving drought-resisting seeds that will sprout in the next rainy spell. Another group survives on little water, growing scantily or as dwarfed forms. Still a third group, such as many lichens, algae, mosses, and desert bushes, can literally withstand being dried out. They seem dead until moisture returns, when they revive and reproduce. The last, and most impressive, group is that of the water-conserving xerophytes, such as cacti, aloes, stonecrops, and Australian bottle trees. These store water in swollen tissues; because of this "juicy" condition, such plants are referred to as *succulents.*

The succulents, in order to save their water supply, must transpire slowly. The epidermis, or outer skin, is very thick. Even if it dries up, it has the power of revival when moisture comes. It is not only in the leaves, however, that drought-withstanding plants show special adaptations. The other body parts show them as well.

The stems of many xerophytes are peculiarly formed, being immensely thick in proportion to their length, at least when compared with ordinary land plants. These stems can hold a large quantity of water. In certain cases xerophytic plants consist of practically nothing but stems. This is true, for example, in cacti. Obviously, this reduces the transpiration surface to a minimum.

Even when a xerophyte has well-developed leaves, we find that they are so arranged as to be exposed to heat and light as little as possible. Certain xerophytes have no leaves at all. The whole stem is green and performs the work that is

FAO photo by G. Tortoli

Arabian American Oil Company

allotted to the leaves of most other plants. The cactus offers a particularly good example of this type of development. The leaves of this plant have been transformed into formidable spines. The process of photosynthesis must therefore be carried on by the stem, which is fully equal to this task.

ADAPTATIONS OF CACTI

The cactus runs counter to our ideas of what the parts of a plant should be like. We generally think of any large plant as having a somewhat brownish stem and a certain number of green leaves. The opposite is true of the cactus.

The stem is green and usually bears no leaves that could be recognized as such. Instead, it has brownish or grayish spines, which are, in many species, really modified leaves. The function of these modified leaves is to protect the green stem from injury—a most important task, indeed, since the stem is the producer of the organic substances that we normally would expect the leaf to manufacture.

Many of the protective spines grow to great length. They are so strong as to make extremely formidable weapons of defense against animals that might otherwise be tempted to feed on the stems, which are frequently the only green foodstuffs in the region.

The spines differ greatly in shape and in character, and it occasionally happens that the same species of cactus carries upon it spines of different kinds and sizes. The spines are borne on enlarged, cushionlike areoles. In the prickly pear, these areoles are spirally arranged. They bear numerous very fine, barbed bristles, called glochidia, which are often more troublesome than the spines.

Where spines are not present on cacti, other protective devices are provided. The outer covering of the plant may be exceedingly tough; poisonous substances may be present, as in *Lophophora;* the plants may be protectively colored and quite inconspicuous.

It is interesting to note that certain cac-

PLANT ADAPTATION

Courtesy of the American Museum of Natural History

Jack Dermid, Audubon/PR

The cactus (left) stores water in its thick stem. The very reduced leaves—the thorns—prevent water loss. The areoles of the prickly pear (below) carry a great number of fine, barbed bristles that are called glochidia. They are often more troublesome than the spines.

tuslike adaptations are also found in various African euphorbias and stapelias, which are not related to the cacti but which grow under quite similar conditions.

The stems of cacti offer about as great a variety as the spines. They may be wooded, branched, and treelike. More often, however, they are elongated, fleshy, and green. The formations of the latter description may be globular, unbranched, branched near the ground as in the case of the organpipe cactus, or branched higher up, as in some of the prickly pears and the Christmas cactus. The stems may be flattened, or they may be ridged and grooved. The flowers of cacti are large, showy, and brilliantly colored. In the night-blooming cereus (*Hylocereus undatus*), the blossoms may be as much as 30 centimeters in length.

All in all, cacti present a most striking appearance in their countless variations. For this reason, as well as for their hardiness, they are widely cultivated in ornamental gardens and in homes. In fact, a number of rare, handsome species of cactus are so sought after in their natural environment that they are in danger of extinction and have to be protected.

Bruce Roberts, Rapho/PR

VEGETABLES

What do the leaf of a spinach plant, the stem of an asparagus, the root of a turnip, the floret of a broccoli plant, and the fruit of a cucumber plant have in common? They are all vegetables. To botanists, all plants are vegetables. But in common use the term "vegetable" refers to a plant or plant part that is eaten during the main part of a meal, before dessert.

WHAT VEGETABLES ARE

In some plants—beets and carrots, for example—the edible part is the root; in some, such as asparagus, the stem; in others, including spinach and lettuce, the leaves; and in still others, such as the onion, the bulb. In yet another group—broccoli and other cauliflowerlike plants—it is the flowers that are eaten as vegetables. And finally there are a large number of plants in which the fruit is known and eaten as a vegetable. Included in this group are green beans, lima beans, sweet corn, cucumbers, tomatoes, and squash. The part of the potato plant that is eaten is the tuber—a much-enlarged, fleshy part of the stem that grows underground. Some authorities consider tree crops such as avocado and bread-fruit to be vegetables. The culinary herbs and edible mushrooms are also usually considered vegetables.

Vegetables are often cooked before being eaten, but not always. Tomatoes, carrots, and celery, for example, may be eaten raw or cooked, and cucumbers and lettuce are usually eaten raw. Vegetables do not have to be milled or otherwise processed before being eaten. To the home vegetable gardener this means that his produce can go quickly from the vegetable patch to the dining table—with a brief stop for rinsing off and for cooking, if necessary.

WHY WE NEED VEGETABLES

Vegetables are a good-tasting and attractive-looking part of our daily meals. In addition, they are an important source of vitamins and minerals, and they provide bulk, which aids in digestion.

All vegetables contain vitamin C. The

best sources of this vitamin are potatoes, sweet potatoes, tomatoes, cauliflower, cabbage, brussels sprouts, peppers, turnips, and broccoli. Raw vegetables contain a better supply of vitamin C than cooked ones do, because some of the vitamin is lost during cooking. Fresh vegetables—even when cooked—contain more vitamin C than frozen or canned ones.

The B vitamins, with the exception of vitamin B_{12}, are present in significant amounts in vegetables. Dried beans and peas contain thiamine, or vitamin B_1. Dark green leafy vegetables are good sources of folic acid and of riboflavin, or vitamin B_2. Another B vitamin, niacin, is present in asparagus, collards, sweet corn, mushrooms, peas, and lima beans.

Vitamin E is present in all green vegetables, and vitamin K is supplied by green leafy vegetables, tomatoes, and cauliflower.

Vegetables are also a source of minerals. During the growing process they extract certain trace minerals from the soil and then in turn pass these minerals on to the consumer. The most important minerals contained in vegetables are calcium, phosphorus, iron, and potassium. The potassium content is highest in the yellow vegetables, whereas calcium is present in collards, mustard greens, kale, dandelion greens, and broccoli.

VEGETABLES AND CLIMATE

Climate is the most important factor in growing vegetables. Indeed, only those regions of the world that have at least three to four months a year of warm, sunny days and moderate rainfall can support vegetable production.

Vegetables are frequently grouped into cool-season and warm-season crops. The cool-season vegetables are grown where the temperature averages between 15.5° and 18° Celsius, and the warm-season vegetables are grown where the average temperature is 18° Celsius or higher.

Cool-season crops include lettuce, spinach, cabbage, cauliflower, broccoli, asparagus, artichokes, peas, carrots, white potatoes, onions, celery, turnips, and radishes. In all of the cool-season vegetables with the exception of peas, the part of the plant that is eaten is the vegetative part. Vegetative parts are those sections that have to do with the plant's growth functions rather than with its reproductive functions. Roots, stems, and leaves are vegetative parts.

In all warm-season crops, except the sweet potato, it is the reproductive part of the plant—the fruit or seed—that is eaten. Warm-season crops include snap beans, lima beans, sweet corn, tomatoes, cucumbers, peppers, squash, eggplant, okra, and sweet potatoes.

When cool-season crops are stored, they should be kept at or near 0° Celsius. Warm-season crops should be stored at slightly above 10° Celsius.

VEGETABLE GROWING

Before seeds are planted, the farmer must prepare the soil. The first step in preparing the soil is *plowing*. The farmer drives the plow back and forth across the fields, turning up the soil at the surface.

The next step is *harrowing*. In this process the farmer uses an implement called a harrow, which is equipped with

Not all vegetables come from the farm. Small home gardens can produce vegetables like these tomatoes. Home gardens are becoming increasingly popular.

UPI

spikes or disks designed to break up the large chunks of soil left behind by the plow. The first harrowing is usually followed by a second one that makes the soil powdery.

The soil is now ready for the planting of seeds, also known as *seeding*. On commercial farms it is done with a special machine called a seeder, which is attached to a tractor. The seeds are planted in even rows at whatever depth is suited to the particular crop.

Fertilizer is added to the soil at the time of plowing, harrowing, or seeding, depending on the crop. Fertilizer enriches the soil with chemicals and makes it yield a larger, healthier crop.

When the crop begins to sprout, the farmer continues *tilling,* or cultivating the soil around the plants. This process prevents weeds from growing and helps keep moisture in the soil. The moisture is retained because constant tilling creates a dust mulch, or protective covering, that blocks evaporation.

When *harvest* time comes, the farmer may hire hand laborers known as pickers to harvest certain crops. It takes special skill to pick lettuce, for example, which has very tender leaves, and to harvest pea and bean plants, on which the pods ripen at different times. Mechanical harvesters are continually being improved, but they are still more often used to harvest crops destined for canning than for those that are sent fresh to market.

THE HOME VEGETABLE GARDEN

But what about a simple backyard vegetable garden? Or, for that matter, even a tiny window-box garden in the city? You do not need a great deal of space to raise vegetables. More and more people, wanting to be more nearly self-sufficient, and concerned about rising prices and the use of chemical pesticides in commercial vegetable growing, have begun to realize that their small backyards can be used for vegetable growing.

A good first step is to consult your state department of agriculture to learn what crops grow best in your area, and what the usual planting and harvesting

Jenny Tesar

A wide variety of vegetables—available in many markets—give color and interest to a meal while providing important vitamins and minerals.

dates are. Then decide whether you are going to make your garden an "organic" one. The term "organic" refers to the use of fertilizers and pest killers made from natural substances.

If you opt for an organic garden, you must prepare a compost heap even before starting the garden. Compost is decaying natural material that is used as fertilizer. First lay down a layer of kitchen garbage, which should contain eggshells, orange rinds, coffee grounds, and the like, but no greasy meat scraps. A layer of refuse—such things as vacuum-cleaner sweepings, sawdust, ashes, hair clippings, and bits of paper—can also be used. A layer of manure can be added, but manure is not available to most home gardeners and is not essential. Over this place a layer of grass clippings, and, at the very top, a layer of old leaves. The natural decay process prevents the growth of harmful bacteria and will eventually furnish a natural fertilizer.

Now carefully plan your garden. Choose a sunny place, marking off an

Different parts of some plants are eaten as vegetables. In the rhubarb it is a part of the leaf.

area at least 3 by 3.5 meters. Test the soil and see that it has a moist, crumbly consistency. Then it is time to go to work.

First turn over the soil with a shovel or a garden fork. This is the equivalent of plowing on a commercial farm. If you find any roots or stones, dig them out. If the turning of the soil has left big chunks of soil, break them up with the garden fork or the back of a rake. This procedure is the equivalent of harrowing. At this point spread some of the well-decayed compost, or some commercial fertilizer, over the garden. Mix it in well with the soil, raking back and forth until the soil is smooth.

Now you are ready to plant — seeds, following the instructions on the seed pack-

After carefully preparing the soil, a home gardener plants seeds, taking care to plant the correct number at the right soil depth and with an adequate amount of spacing.

ets, or any small, immature plants you may have bought or raised indoors from seed. In a few weeks, given proper weather conditions, most plants will start to sprout.

As growing progresses, the new plants require much care. They must be thinned out, as directed on the seed packets. Small fences should be put up for pea plants, poles for beans, and stakes for tomatoes. If no rain falls for a while, the garden should be watered thoroughly one evening a week, or more often, depending on weather and soil conditions.

The garden should also be kept free of weeds. Weeding by hand, using a simple hoe to get at underground stems and roots, is sufficient.

Weeds can be kept down and moisture held in the soil by covering the earth between the plants with a thick layer of straw, old hay, or sawdust. This covering, called a mulch, also keeps the soil temperature steady.

Insects frequently pose a problem. An organic gardener will want to remove many of them and their larvas by hand. A dusting of ashes and ground limestone can be used to deter many kinds of beetles. Tin cans with tops and bottoms removed can be pushed into the earth to turn away cutworms. Some organic gardeners make and use a natural insect repellant. They blend an onion, several cloves of garlic, and a piece of hot pepper in two or three cups of water. Excess water is then drained off and the residue applied to the plants with a paintbrush or a sprayer. Chemical insect repellants are also available.

You will be able to harvest and eat your crops all summer long, but by fall only a few plants will still be growing. After you pick the last of these, don't simply forget your garden patch. Bed it down for the winter. First pull up all remaining stalks and add them to the compost heap. Then turn the soil over lightly and cover the whole garden with a thick layer of compost, hay, or leaves. Last, lay a few evergreen branches over the garden to keep the topping from blowing away. In that way you will have a head start on next year's garden.

CACTUS PLANTS

by Lyman Benson

A tall, thick stalk. Something that looks like a mass of organ pipes. A spine-covered barrel. A bizarre-looking assemblage of prickly arms extending from a short stalk. Are these unusual looking inhabitants of desert regions somehow mistakes of nature? No. They are plants specially adapted for life under harsh, dry conditions. They are different types of cactus plants.

Cacti are unusual plants. They often have weird shapes. They have unusually thick, fleshy stems and are commonly leafless. They generally stand inconspicuously against the desert soil. In spring and summer, though, they blossom forth with some of the largest and most colorful flowers known in the plant world. All these unusual features of cactus plants together make them able to survive and multiply in their desert environment.

WATER NOT HARMFUL

Although cacti are not necessarily desert plants, most are found there or in adjacent dry regions. Their spread beyond the dry areas is limited mainly by competition from those other plants that do well where there is more water. Contrary to common belief, water does not injure cacti, unless it stands around the roots and makes the soil soggy. Nearly all species require good drainage, though a few cacti occur even in areas such as the Florida Everglades where the water table is close to the surface of the soil.

Each group of plants tends to live where it has an advantage. The special body form of a cactus adapts it to areas of high light intensity and low moisture or high evaporation rate. On local sandy areas or rocky outcrops, where competing species are at a disadvantage, cacti can occur in a variety of regions, including forests. Some species even grow in mountainous areas at elevations of more than 3,000 meters.

A LARGE FAMILY

The cactus family, Cactaceae, is very large, made up of perhaps 2,000 or more species. The plant's natural habitat is almost entirely limited to the New World, and it may have been completely so during pre-Columbian times. This indicates that the family developed during the last 135,000,000 years, since the continents of the Old and New World are believed to have separated about that time.

The giant saguaro cactus provides homes for desert birds and other animals.

Porterfield-Chickering/PR

Left: *Mammillaria elongata*, the golden star cactus. Right: *Astrophytum myriostigma*, bishop's cap cactus. Some are in bloom.

both photos, Derek Fell

In the New World, cacti can be found from northern British Columbia, Canada, eastward to Ontario, Canada, and southward through most of the states of the United States, through Mexico and Central America, and on through South America.

The few cacti in the Old World belong to a highly specialized genus, *Rhipsalis*. All or nearly all these species are identical to species occurring in Central and South America. The presence of *Rhipsalis* in the Old World may be due to relatively recent natural long-distance dispersal of the seeds by birds, or of whole plants attached to drifting tree trunks. More likely, though, the spread of these cacti can be attributed to dispersal by sea captains and explorers who took the plants to gardens in Europe and elsewhere soon after the first vessels traveled to the Americas.

In some of the warmer parts of Europe, Asia, Africa, and Australia, several prickly pear cactus species have gone wild, and a few species have become pests. The climate of many parts of the Old World is suitable for cacti and, lacking their normal competitors, some introduced species of prickly pears have tended to overrun large tracts of land.

NO LEAVES

Deserts are dry places. Little rain falls there. What does fall evaporates very quickly. For a plant to grow in such an environment, it must be able to retain water. It cannot survive if its moisture evaporates into the air. Evaporation occurs from the surface, so the more surface area a plant has, the greater the area from which water can evaporate. Thin structures such as leaves present a far greater proportion of surface to the surrounding atmosphere than do thicker, larger structures.

The following illustrates the relationship between surface area and volume: a cube one centimeter on a side has six surfaces, each of which exposes one square centimeter to the outside for a total area of six square centimeters. The cube's volume is one cubic centimeter. The ratio of surface to volume is thus 6:1.

A cube two centimeters on a side has four square centimeters on each surface— 24 square centimeters altogether. The volume is eight cubic centimeters. Thus the ratio is 3:1.

A cube three centimeters on a side has an area of nine square centimeters on each of its six surfaces—a total surface area of 54 square centimeters. The volume is 27 cubic centimeters—a ratio of only 2:1.

Therefore, to decrease surface area and thus evaporation rate, thin structures such as leaves must be eliminated or reduced in size. Cacti have done this admirably: they have no true leaves.

Cactus plants evolved from plants that once had leaves, but in most species the leaves have evolved into thin, hard spines that leave very little surface open to the atmosphere. The few primitive cacti species that retain ancestral leaves grow in warm, moist areas in the tropics.

SPINES

The spines of a cactus grow from enlarged cushionlike structures called *areoles*. An areole is really a bud growing from the side of the stem. Usually the spines are very sharp—and barbed, in many species. Spines serve not only to conserve water but also to keep animals away.

Most animals are deterred by the spines, but a few eat the plants anyway. One day I saw an old bull chewing the joints of a teddy bear cactus, or cholla. The joints were covered by long, stout, strongly barbed spines. A mass of joints clung to almost every part of the bull's face, but he kept on eating. Apparently the spines become encased in a mass of mucus in the bull's intestinal tract and do not harm him. I also remember one day during a very dry summer in New Mexico seeing wood rats working very hard to find an opening between the spines of a prickly pear cactus so that they could eat and drink from the cactus stem.

Spines serve in a special way to help conserve water. Downward-directed spines serve as drip-tips. If there is a light rain or mist, water accumulates on the spines that turn downward and forms drops. Whenever a drop becomes large enough, it falls to the ground, wetting the soil and being absorbed by the roots. This is particularly important when the plant is young and its roots are concentrated around the base of the plant. As a cactus grows, the root system grows. The spines serve as water catchers, directing water to the roots.

WIDE SPREADING ROOTS

As a cactus plant grows, its root system grows into an ever-widening circle. Unlike many other flowering plants, the mature cactus rarely has a large taproot. Instead, it has a system of numerous small roots, most of which grow one to ten centimeters beneath the surface. The roots spread widely in all directions, forming a circle many times larger than the circle bounding the area of the plant above ground. In some cacti, the roots spread 15 to 20 meters in all directions from the stem.

The roots remain receptive to water even through long drought periods. The cut stem of a cactus or a stem with the roots removed will sprout new roots—even if it is lying on a shelf or table. If these roots are placed in water they, like the ones in the soil, immediately absorb the moisture. Thus the cactus is ready at all times to blot up water, even from a light rain that may

Left: Coville's barrel cactus, *Ferocactus covillei*, shown in full bloom. Right: rainbow cactus, *Echinocereus rigidissimus*.

George Gerster, Rapho/PR   H. F. Flanders/PR

CACTUS PLANTS

penetrate only four or five centimeters below the surface. This gives the plant a great advantage over its deeper-rooted competitors, which can absorb quantities of water only when there are heavier rains. If the rainfall is greater, the cacti do better. But if the rains are light, they can still do fairly well — or at least get by.

PHOTOSYNTHESIS IN THE STEM

Plants normally carry out photosynthesis — the manufacture of foods — in the leaves. In cacti, the stem takes over this function. The raw materials for photosynthesis are chlorophyll, carbon dioxide, water, and light. Although a cactus stem has a relatively limited surface exposed to light, it is able to form adequate amounts of sugars. This is possible partly because of a special process occurring in the plants.

Stomata, or pores, on the cactus stem open only at night, when the air is cooler and evaporation is comparatively low. Carbon dioxide from the air enters the stem. It is combined into organic acids and stored until daylight. When light is present, the carbon dioxide is released, and it combines with water in the presence of chlorophyll in the stem to produce food.

THE EXQUISITE FLOWERS

For most of the year, the cactus plant clings inconspicuously to the desert soil. Then, in spring or summer, it briefly blossoms. Its flowers are among the most colorful and beautiful in the plant world.

The California barrel cactus, for example, has yellow, bell-shaped flowers, some four centimeters in diameter. The flowers form a circle near the top of the plant's stem.

The ladyfinger cactus, a plant that grows close to the ground, has pink or reddish-purple flowers that are some eight or nine centimeters in diameter. The flowers bloom for several days, opening at noon and closing by sundown.

The creamy white flowers of the giant saguaro, eight or more centimeters long and four centimeters wide, crown the plant's tall, ribbed stems. Inside the petals of each is a central pistil, surrounded by some 3,000 stamens. The flowers bloom for only one night and then mature into edible fruits.

Left: *Brachicereus,* a cactus that grows on the Galápagos Islands. Right: the teddy bear cactus, a cholla whose name is *Opuntia bigelowii.*

The fruits of cacti also vary in size, shape, and color. The fruit of the leafy lemon vine is a yellow berry. The fruit of the prickly pear is pear-shaped, juicy, and edible. It has been a part of the diet of Mexicans and Indians for centuries and can be made into candies, preserves, and syrups. The hedgehog cactus gets its name from the fact that the fruits are covered with spines. It is also known as the strawberry cactus because its bright red fruits look like strawberries — and can be eaten the same way once the spines are rubbed off.

ENDANGERED SPECIES

Many species of cacti are rare — some because they occur only in limited areas, some because they have been extensively collected, others because their habitats were destroyed by housing developments, farms, and so on.

In fact, the cactus family may be the most endangered of all the major groups of plants. For example, about 72 species of the 268 kinds of native cacti in the United States alone are so rare or so restricted in their occurrence as to easily become extinct.

The best means of saving endangered species is preservation of the habitat, or community, in which they live. This can be done by national and local organizations. Conservation laws that prohibit the sale of plants collected in the field have been, or can be, passed by governments.

It is also possible to remove the seeds from cactus plants and cultivate them. This has little effect on the native population, since most of the seeds are surplus. Normally, only a very small number will germinate naturally, and a still smaller number of the new individuals will become established as seedlings and grow into mature plants.

GROWING CACTI IN YOUR HOME

Their unusual and varied appearance, the formidable covering of spines, and the beautiful flowers, together with ease of cultivation, have made cacti favorite house and greenhouse plants. And, where the weather is warm enough, people can grow these plants outdoors.

A prickly pear cactus, *Opuntia oricola,* in full bloom. Animals feed on these cacti.

The danger to cacti is not, as many believe, from watering. Rather, it is from poor drainage. Nearly all species require good drainage. It is important to prevent water from standing around the roots and making the soil soggy.

If the plants are grown indoors, the soil should be relatively sandy so that the pots will drain quickly after watering. If they are to be grown outdoors, the soil should not include unusual amounts of sand, because it will dry faster outdoors.

Pots used outdoors should be larger than those used indoors. For example, 13-centimeter pots are about right for most small or average size cacti outdoors. Indoors, where evaporation is relatively slow, 8- to 10-centimeter pots should be used, because they will drain more quickly.

If you have a cutting from a large cactus, leave it in a dry place for ten days or two weeks. A layer of callus tissue will form just under the cut surface. This tissue will prevent or reduce the entry of organisms that cause rot after planting. After callusing over, the cutting may be placed in the ground or in a pot with a potting mixture.

A number of fine books provide greater details on the cultivation of cacti. In addition, there are more than 20 major cactus and succulent societies in the world, some of which have many branch organizations. Interest runs high, and those bitten by the "cactus bug" rarely recover.

Houseplants are either flowering or foliage. Above is the exotic hybrid cattleya, an orchid.

HOUSEPLANTS

It's a living thing that is kept in a house. It's not a dog or a cat or some other kind of animal pet. But it gives its owners pleasure as they care for it, and it brings color and brightness to the room that it's in. What is it?

It's a houseplant—a philodendron, a corn plant, a Chinese evergreen, or a rubber plant. It's an African violet, a spider plant, a grape ivy, or a chrysanthemum. If you've decided to paint your thumb green, these are the plants to get—they're the easy ones to grow.

WHERE AND WHAT TO BUY

Whether you live in a dark basement apartment or in a sun-bathed house, there are many plants that will add life and color to your rooms. In a large city, the best place to buy plants is from a shop that sells houseplants exclusively. They can help you in selecting the right plant for your apartment or house. They can also give you good advice on how to care for the plant. A very good shop will not sell you a plant that has a poor chance of surviving in your apartment.

Your next best source is a florist shop—"next best" because their main interest is flowers rather than houseplants.

If you live in a small town or a suburb, chances are there's a nursery close by. You're in luck—here you'll get the highest quality plants and at the lowest prices.

When you buy, check the plant out carefully. Avoid scrawny-looking plants and those that have yellowing or splotchy leaves. Get one with full foliage. If it's a flowering plant, it should have buds. You do want to see some flowers soon.

Be sure to look at the undersides of the leaves for insects or insect damage. Brown tips on the fronds of palms, though, are not a sign of insects; they are caused by low humidity.

If you feel that you don't have a green thumb, it's hard to go wrong with a corn plant, *Dracaena massangeana*. In addition to *D. massangeana*, there are two other easy-to-grow *Dracaenas*. One is *D. marginata*; the other is *D. warnecki*. None of the three resembles the others. They do not need much light and thrive with minimum care. You water the *massangeana* once a week, the *marginata* twice.

FOUR LIGHT EXPOSURES

The lighting conditions of your apartment or house are the most important consideration in selecting the right plant. The four different exposures—northern, southern, eastern, and western—are not defined by horticulturists as you might think, by the direction a window faces. Rather, each is described in terms of available sunlight.

An eastern exposure, considered the best for houseplants, receives three to five hours of morning sunlight. Next best is a western exposure—a window receiving three to five hours of afternoon sunlight. Third best is a northern exposure—a window that receives no sunlight. The least desirable, according to many horticulturists, is a southern exposure—a window that receives at least six to eight hours of sunlight each day.

And what about that dark, damp basement apartment? You'll have to use an artificial light. For a group of very small plants, an ordinary 100-watt light bulb will do. Your best bet, though, will be to get a sealed plant light. You can find one at almost any hardware or lighting-fixtures store.

A good one to use is the 150-watt size, placed about 2 to 3½ meters from the plants. You can also use a 100-watt bulb, placed 1 to 2 meters from the plants, or a 75-watt bulb, placed ½ to 1 meter from the plants.

Some houseplants do well in low light. These include the philodendron, rubber plant, Boston fern, sanseveria, grape ivy, maranta (prayer plant), bamboo palm, and Kentia palm. Some plants need medium light—about 1 to 2 hours of sunlight. If that's the maximum light available in your home, get an African violet, azalea, dieffenbachia, avocado, spider plant, asparagus fern, gloxinia, or begonia.

Some plants—the flowering plants—require strong light—about 3 to 5 hours of it. They will grow, some may even thrive, without hours of light. But they probably will not bloom.

A houseplant is not damaged by lack of light, however, or by too much light. They are harmed more by improper watering.

HOW MUCH WATER?

Nurserymen often group houseplants into three watering classes:

both photos, Derek Fell

Top: cacti and succulents make a stunning window box. Cascading over the side is a serpent cactus. Bottom: foliage plants on display: croton, red dracaena, dieffenbachia, and coccoloba. In line with the door frame is *Dracaena massangeana*.

HOUSEPLANTS 155

1. Wet family. You should water these plants when the surface of the soil is dry.

2. Moist family. Water when the soil is dry from 2½ centimeters to 7½–10 centimeters into the pot.

The white calla lily (top) should have direct sunlight except at midday, when its light should be indirect. Soil of the African violet (center) should always be a little dry. The *Aechmea chantini* (bottom) should have plenty of light if you wish it to flower.

(top) Derek Fell; (middle) George Kalmbacher, Brooklyn Botanic Gardens; (bottom) Graf photo

3. Dry family. The soil should be dry completely, to the bottom of the pot, before watering.

This doesn't seem to help much, for how do you tell when the soil is dry to a depth of 10 centimeters, or to the bottom of the pot? One way is to use a wooden spoon or tongue depressor to probe into the pot of soil. Another way is to add just enough water so that the soil is wet to a depth of 2½ centimeters. Then see how many days it takes the soil to dry out. If it takes two days, then a 12½ centimeter pot will take ten days to dry out.

Of course, you can always use the droop test. Wait till the plant wilts. Count the number of days since the last watering. This establishes the maximum time between waterings.

There are strong and differing opinions on watering houseplants. Some people say water them heavily once a week—in the bathtub. Others insist that each plant should be watered according to its particular needs.

To water hanging plants, it is best to take them down first. Some people water hanging plants by putting ice cubes on top of the soil, while the plant is still hanging. This is decidedly poor practice, for nearly all houseplants are tropical plants and, if anything, should be nurtured with water on the warm side instead of with nearly freezing water.

You treat a cactus differently. The soil of all cacti should dry out thoroughly before rewatering. A cactus can easily do without water during the winter months. Cacti usually bloom during the summer only if they have "dried out" for four months during the winter.

It's best to water a plant from the top of the pot rather than from the bottom, by pouring water into its saucer. Two exceptions to this rule are baby tears and African violets.

What do you do about watering when you go on vacation? That depends. If you're away one week or less during the summer you can forget about watering altogether. If you're going away for a longer time it will be best to have a friend care for

all photos, Derek Fell

The Boston fern (left) requires low light, moist soil, and high humidity. In the center is a *Philodendron selloum*, also called the saddle-leaved philodendron. It requires medium light soil that is barely moist. A *Dracaena warnecki* (right) will grow well in low light.

the plants. Or, you can buy a wick-type device to insert into the soil. The other end goes into a container of water.

Another suggestion: water the plant thoroughly. Then pull an air-tight plastic bag over the pot and fasten the bag around the stem, just above the soil. The plant will not need water for three to four weeks.

MOISTURE IN THE AIR

High humidity will make almost any houseplant, except a cactus, thrive. A good way to increase humidity is to fill a large tray with gravel to a depth of 5 centimeters. Add only enough water so that the water line is just below the gravel top. Then place your plants on top of the gravel.

It's a good idea to spray the foliage of your plants at least once a day. The real value of doing this is not so much that you increase the humidity—the spray usually dries up after about 15 minutes—but that you keep the leaves free of dust.

If your apartment or house is heated during the winter, you'll need a humidifier. For a reasonably large room, get the nine-liter size. The humidifier can bring the humidity up to the 65–75 per cent range, which is the level that houseplants prefer.

To see what the humidity is, you should buy a hygrometer. Your best bet is a combination thermometer/hygrometer, which measures both temperature and humidity.

Lighting, watering, humidity—these are the basic factors in plant care. Every indoor gardener, though, needs to know more—about things such as fertilizers, pest control, potting, and propagation.

FOOD FOR GROWTH

When do you fertilize a houseplant? Which fertilizer do you use?

About once every two weeks for flowering plants and once a month for foliage plants is fine. Use any indoor houseplant fertilizer. It usually comes in liquid, powder, or pill form. It's a good idea to use a weaker dosage than that prescribed by the manufacturer. Too strong a dose may "burn" the roots, and even kill the plant.

Any of the major brand-name fertilizers will be satisfactory. Regardless of which one you buy, mix the fertilizer with water before applying it.

A word of caution about dried cow manure, now available commercially in many shops. It is an extremely strong fertilizer and should be used only sparingly for

HOUSEPLANTS

houseplants. (It is best used for growing vegetables.)

MEALY BUGS, MITES, AND SCALES

Even though you inspect your plant at the time you purchase it, you should isolate the plant as soon as you bring it home. Its foliage should not touch that of any other plant. If after three weeks no disease appears, the plant is safe.

The four most common pests of houseplants are the aphid, mealy bug, red spider mite, and scale. The aphid is a little larger than a flea and is usually green, but may also be gray, black, or red. It will always be found on the plant's new growth. Aphids can be controlled by washing the affected leaves and stems with liquid soap (not detergent). Don't let the soap get on the soil, and be sure to rinse the soap off the leaves with clear water.

The mealy bug attacks almost every kind of plant. It is very easy to spot because it covers itself with a small white cottonlike ball. Mealy bugs can be controlled by swabbing them with a Q-stick dipped into alcohol. Please: touch only the little white balls and not the plant.

The red spider mite is so tiny that you probably won't find it without a magnifying glass. Tan in color, it builds a small web that gives it away. If the leaves of a plant begin to yellow with green dots left on the leaves, you have red spider. The green dots are water cells, which the red spider does not touch.

Red spider is probably the worst pest of houseplants. It does not like water. Therefore, the best way to control it is with a forceful spray of water against the affected areas. If this doesn't work, an insecticide must be used.

Scale, of which there are three kinds, may occur on the stems and leaves of houseplants. Soft scale has a tan color; oyster scale is white and elongated; and pin scale is dark-colored and shaped like a turtle shell.

Like mealy bugs, scale is controlled by washing it with liquid soap. It may be necessary to wash it two or three times in order to dislodge the bugs.

The Zebra plant (left) is easy to care for and will grow in full or low sunlight. The Paphiopedilum orchid (center) is also easy to grow. It needs an hour of sunlight. Gloxinias (right) need a great deal of indirect light and high humidity.

Derek Fell; (left) George Kalmbacher, Brooklyn Botanic Gardens; (right) Derek Fell

(left) Derek Fell; (middle) Derek Fell; (right) Graf photo

Begonias (left) are the most common of all houseplants. The cineraria (center) needs lots of sunshine and water and a cool room. Flowering and foliage houseplants (right) make a lovely window display, if you don't have a cat.

A word about insecticides: If you must use one, use a safe one. An aerosol insecticide should contain pyrethrum or rotenone. These are natural compounds derived from chrysanthemums. Mums, in fact, are good to have around in the fall to serve as protection for other plants, because insects do not like their odor.

CLAY OR PLASTIC?

Next to watering, indoor gardeners argue most about pots. As far as the plant is concerned, it couldn't care less whether you use a clay pot or a plastic one. As far as you are concerned, the main difference will show up in the length of the watering cycle. The palm you watered once a week in a clay pot will need to be watered only once every 2½ weeks when transplanted to a plastic pot.

It is not true that a clay pot enables a plant's roots "to breathe," as clay enthusiasts often claim. Air cannot pass through clay. Moisture inside the pot, though, does pass through the pot and evaporates on the outside of the pot. That's why soil in a clay pot dries out faster than soil that is placed in a plastic pot.

Left: the Easter cactus *(Rhipsalidopsis gaertneri)* produces bright red drooping flowers in April and May and sometimes in September. Right: cyclamens need abundant sunshine and a cool room.

both, Graf photo

HOUSEPLANTS   159

all photos this column, Graf photo

Terrariums (top left and right) require little attention. Desert and jungle scenes can be created in miniature. Terrariums can be made in a variety of containers.

George A. Elbert

RE-POTTING YOUR PLANTS

It's time to re-pot when the roots of a plant appear outside the pot. You should also re-pot a plant every two years whether or not it needs transplanting, in order to change the soil.

Use sterilized potting soil, which is available at shops that sell plants. You can also buy "enriched" soil as well as special soils—African violet mix and cactus mix, for example.

Before re-potting, you have to do three things:

1. Water the plant. You don't want the soil to be too crumbly.

Crotons (next to top) require 3 to 5 hours of sunlight daily to produce their striking foliage. The Opuntia cactus (middle) needs lots of sunshine and little water. The dieffenbachia (bottom) grows well in low light but it is poisonous to pets.

2. Get a new pot that is slightly larger than the old one. If the new pot is too large, all the plant's energy will go into unnecessary root growth.

3. Place a piece of broken pottery or a bottle cap over the drainage hole of the new pot. This keeps the hole from being plugged up by the soil.

To transplant, slide the palm of your hand over the top of the pot so that the stem of the plant goes between the second and third fingers. Turn the plant upside down, whack the brim of the pot against the top of a table, and pull the old pot off. Put the plant in the center of its new pot, and with you free hand pour the new soil around the ball of the old. Tamp the soil down, water the plant, and you're finished.

NEW PLANTS FROM CUTTINGS

Most houseplants are not grown from seeds. They are propagated by stem or leaf cuttings. All vines, except the philodendron, can be propagated by stem cuttings. To take a cutting from one of your plants, use a razor blade to cut straight across the stem. The cutting should have at least two leaves. Dip the cutting into a rooting hormone and plant it in a soil medium.

The philodendron is propagated by a section cutting. You need a leaf, a section of stem, and a node that contains several air roots on the cutting.

The African violet, rex begonia, and Peperomia are propagated by leaf cuttings. These are grown in the same manner as stem cuttings.

THE PLEASURE OF PLANTS

Well, that's the gist of it. This article has described the basic elements involved in the care of houseplants—how and when to water them, how to provide proper lighting for their individual needs, what fertilizers to use, what to do about plant pests, and how to pot and re-pot your plants.

As you get more and more deeply into this fascinating activity, you will learn more and more "tricks of the trade"—and probably develop a few of your own, as you find out what procedures work best for a particular kind of plant. You may want to experi-

both, Riki H. Kondo

Is the *Dracaena* at the top trying to tell us something? A closer look at the pot (bottom) tells all: the plant needs a bigger pot.

ment with developing complex and lovely indoor gardens, if you have the room for them. Terrariums offer the pleasure of creating landscapes in miniature, from jungles to deserts.

Whatever you choose to do with houseplants, they will add immense joy to your living quarters. Just remember the basics: water, fertilizer, and light.

HOUSEPLANTS

Grant Heilman

TREES

by Thomas Gordon Lawrence

Walks and rides along city streets or secondary roads will give you more pleasure if you can identify the trees you see. It would probably take years of study to know all the species that live in your country. Yet you can easily acquaint yourself with the main groups of trees.

Some trees have distinctive features that identify them at first glance. If you see a tree with acorns, it must be an oak, since all oaks bear acorns and no other trees do. No other tree has seeds like those of the maple. The tulip tree's leaves are unique. In winter, its candlelike clusters of seeds proclaim its identity.

Some trees are harder to identify. You may be sure, for example, that a certain tree is either a spruce or a fir. But you will have to inspect it carefully to make sure which one it is.

Fruits, flowers, and seeds often tell you what a tree is, but only when they are actually on the tree or have recently fallen. Leaves are your best guide during most of the year—throughout the year in the case of evergreens. Buds, leaf scars (scars left on the twig when a leaf falls), bark, and the way the trunk and branches grow all help to identify a tree. Sometimes you can make sure what your tree is by feeling the leaves or twigs, by smelling them, or even by tasting them. A hand lens or a reading glass will help when you inspect leaves, twigs, and buds.

IDENTIFYING MARKS

Leaf structure. Pines and their kin have leaves like needles or scales. Almost all other trees have flat leaves, usually very thin. We refer to them as broad-leaved, meaning that their leaves are broader than the needles of pine, spruce, fir, and other needle-leaved trees. Some broad-leaved trees, such as willows, have long leaves less than 13 millimeters wide.

Evergreens, including most of the pine family, and a few broad-leaved trees keep their leaves all winter. The name *deciduous trees* is given to those whose leaves die each fall.

Most leaves, like those of oaks and elms, have a large vein, called the *midrib*, that runs the length of the leaf, with the other veins starting from the midrib. Botanists call leaves of this type *pinnate*, or feather-veined, since a feather also has a central shaft. Other leaves, like those of the maple, sycamore, and sweet gum, have three or more large veins starting out from one point at the base of the leaf. Such leaves are called *palmate*, or palm-veined, because the main veins extend from the same general area, somewhat as the fingers do from the palm of the hand.

Oaks, elms, and most other trees have *simple leaves*. A simple leaf has only one leaf blade, though this blade may be lobed or irregular in shape. Most simple leaves consist of a blade and a leafstalk. In some cases, however, the blades may grow directly from the stem without any leafstalk. A bud always forms in the angle between the leafstalk and the stem.

Leaves composed of a number of leaflets are called *compound leaves*. The leaves themselves may be quite large, as in the case of the honey locust, but the leaflets of which each leaf consists are usually small. In certain trees the leaflets are compound in their turn. We say of such a leaf that it is twice-compound.

Compound leaves may be feather-compound or palmate-compound. Trees with palmate-compound leaves are the horse chestnut and its relative, the buckeye. Hickory, walnut, ailanthus, locust, sumac, and ash have feather-compound leaves.

Arrangements of leaves. Leaves are

Trees can often be identified by their leaves. Below are some of the more important leaf types.

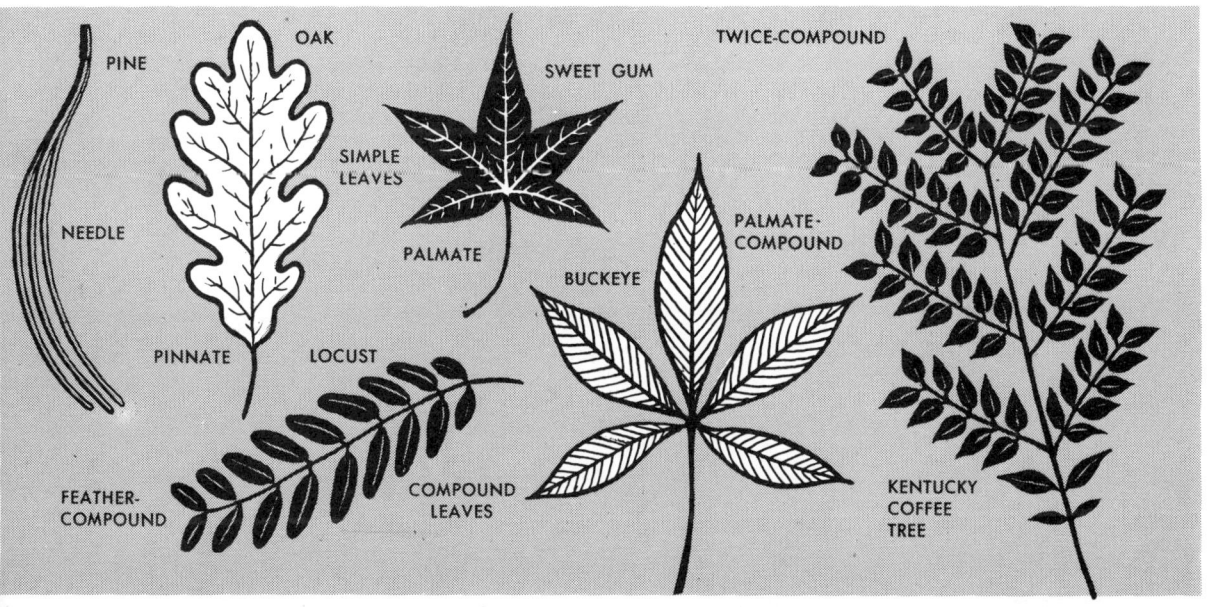

TREES 163

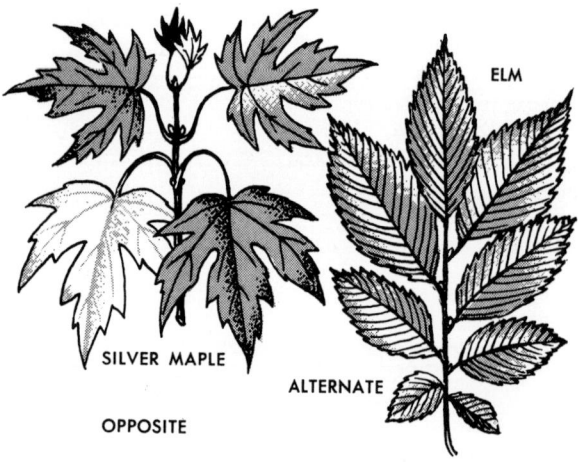

Opposite and alternate arrangement of leaves on stems.

either opposite or alternate in arrangement. *Opposite* leaves grow out in pairs from the stem. *Alternate leaves* grow out along the stem one after the other. Maple, horse chestnut, buckeye, ash, and catalpa are large, broad-leaved trees with opposite leaves. Look at a low branch or a seedling of a maple or an ash. You will see that two leaves always grow from the same point on the stem. Dogwood is a small tree with opposite leaves. Catalpa usually has three leaves growing from one node, as we call the point where leaves grow from the stem.

Buds and leaf scars. When a leaf falls, it leaves a scar typical of the tree it belonged to. You can recognize many trees in winter by leaf scars and buds. We call buds that live through the winter "winter buds." Both leaf scars and buds tell whether a tree has opposite or alternate leaves.

Bark. You can recognize some trees by their bark and be badly fooled by others. Dogwood, beech, shagbark hickory, and sycamore can usually be identified at a distance by means of their bark. Yet sometimes what you take for tulip tree bark may belong to an ash, what you think is a young beech may be a large hornbeam, and you may mistake a cherry birch for a cherry tree.

Bark may change a great deal as a tree grows. Once you get to know the trees, you will enjoy these contrasts. You may see all the variations from youth to old age on a large red maple tree—the bright red young twigs, the smooth, light gray younger branches, and the trunk that gets darker and rougher as it grows older. One reason the sycamore always looks both bright and beautiful is that it sheds large pieces of its old dark bark so that its pale green or almost white underbark gleams in the woods or on city streets.

Flowers. Magnolia, tulip tree, locust, and fruit trees blossom. Their large, bright flowers bring bees and butterflies hastening to the feast of nectar or pollen.

All trees bear flowers of one kind or another, from the conelike arrangements of

The three left-hand drawings show buds; the three right-hand drawings show distinctive types of bark.

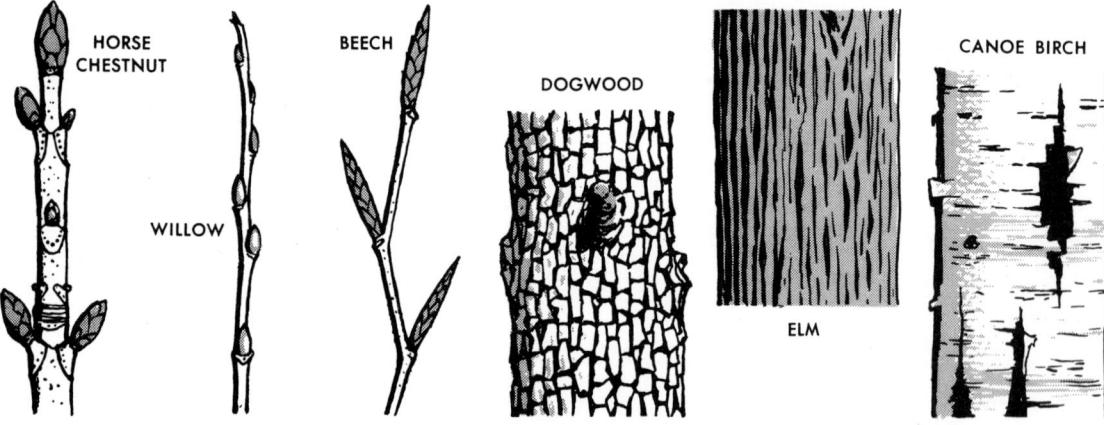

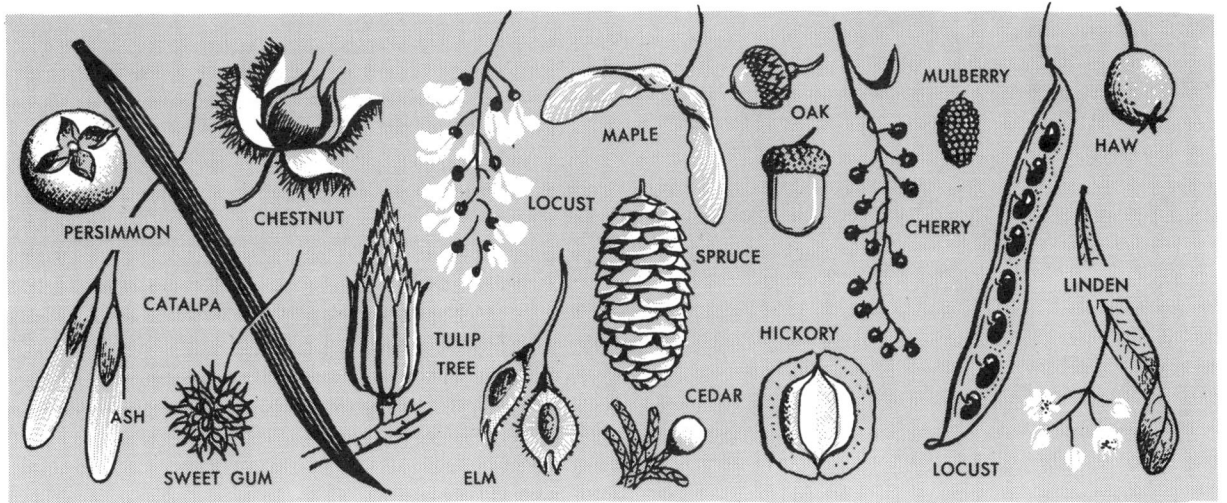

Fruits often provide positive identifying marks for trees. Above are some familiar fruits.

the pine family to the locust blossoms that look like clusters of sweet peas. Most tree flowers are small. Since they generally appear early in the growing season and are borne high up among the branches, many people never see them. This is a pity, for they are usually beautiful and always interesting.

Fruits and seeds. Flowers last a short time, but fruits and seeds can often be found long after they have dropped from the tree. Some pine trees cling to their cones—with the seeds inside—until the twigs grow around the cones and the live seeds are actually buried in the wood.

In many cases, the fruits and seeds are the most positive identification marks the tree can offer. If you find apples on a tree, you know it is an apple tree. Only oaks bear acorns; only catalpas display catalpa pods. A number of other trees have fruits, seeds, or seed clusters that you could not possibly mistake for anything else. Among these are pine, maple, cherry, magnolia, flowering dogwood, mulberry, and eucalyptus trees.

You would not expect to recognize a person merely by his nose, eyes, or hands. When you examine a tree, you will need to consider several things. If you cannot positively identify a tree by its bark, examine the leaves. If there are flowers, fruits, or seeds on the tree, your problem will be that much easier. Look at a number of leaves, buds, or leaf scars. Leaves at some distance back from the tip of the stem are usually most typical.

In the pages that follow, we discuss some of the more important and interesting trees.

CONIFERS: PINES AND THEIR RELATIVES

The conifers are an ancient group of plants. Forests of conifers covered the moister parts of the continents when the early dinosaurs were lumbering and leaping about the earth. This plant group includes the tallest, largest, and longest-living of all organisms—the redwoods of the western United States are believed to live as long as 2,000 years. Most conifers have their seeds in woody cones, although in a few, such as the junipers, the cones have been modified to form juicy berries. All conifers are evergreens, except larches and the bald cypress of southern North America. Some broad-leafed trees, such as the birch, alder, magnolia, and tulip tree, also bear their seeds in cones.

Pines. Pines are among the most widely distributed and most important of all trees. Some of the finest lumber comes from pine trees. They are also the source of pulpwood, pitch, tar, turpentine, and other products. More than 80 species make up the genus *Pinus*. Most are natives of the Northern Hemisphere, though people have introduced them to parts of the Southern Hemisphere. Species have also been transported from one northern continent to another. Pines live in a variety of habitats, but prefer well-drained soils that are stony and poor in nutrients.

If a tree has evergreen needles for leaves, and if the leaves come in bundles or clusters wrapped together by a dry sheath at the base, you can be sure it is a pine. One species, the single-leaf pinyon, as its name indicates, has leaves that grow singly. All the other pines have from two to five leaves in a cluster.

Close to 40 species of pines are native to North America. The eastern white pine is a beautiful and economically valuable tree that is the tallest tree in eastern North America, reaching heights of 60 meters or more. It is also an important lumber tree in Europe, where it was first introduced at the beginning of the eighteenth century. The only thing white about it is the smooth wood; its bark is rather dark.

The sugar pine of western North America grows to heights of well over 60 meters. When it is wounded, it forms masses of sugar. It has enormous cones, up to 50 centimeters long. The yellow, or ponderosa, pine is the most valuable timber tree in the Rockies. It, too, grows to well over 60 meters in height. Unlike the eastern white and sugar pines, which have five needles in a cluster, the ponderosa has three long needles in a cluster. Its broken shoots smell something like oranges. The pitch pine is found in the eastern United States. It averages 12 to 20 meters in height. Needles are in threes. Its lumber is of low quality but makes, because of its high pitch content, excellent firewood and charcoal. The slash pine is the most important pine of Central America. It may reach heights of 30 meters. Needles are usually in threes.

The Scotch pine is very widely distributed throughout Europe and Siberia. It is a valuable source of lumber, turpentine, and other products. Usually well under 30 meters in height, it has dark bluish-green needles in clusters of two. The bark is often reddish-orange. The Aleppo, or Jerusalem, pine is a very attractive native of the Mediterranean region. Ever since people established civilizations in this area, the Aleppo has been an important source of turpentine, tar, and rosin.

Japanese artists have made the Japanese black pine familiar to people everywhere. It is an exceptionally picturesque tree, with crooked, horizontal branches and rough, deeply fissured bark that is dark gray or black in color. Growing to heights of 30 meters or more, it is an important timber tree. Another interesting species is the lacebark pine of northwestern China. It often has more than one trunk. Unlike other pines, its bark peels off in large pieces, exposing the pale underbark. The short, light green needles are in threes. This tree is often planted around temples and in burial grounds.

Yews. These evergreens are native to the Northern Hemisphere. They are common ornamental plants. The English and Japanese yews are the two most important tree species in the genus. The English yew is native to Europe, western Asia, and northern Africa. It grows to 18 meters in height. The trunk is thick. The reddish, flaky bark of a young tree gradually becomes deeply furrowed as the tree ages. The fruit is a fleshy red cup that is edible. But the seed in its center is poisonous. The English yew may live 1,000 years or more.

Spruces. There are approximately 50 species of spruces. Important sources of timber, they are natives of the Northern Hemisphere and are found primarily in temperate and subarctic zones. They are most abundant in mountainous areas. In appearance, spruces resemble firs, and often live together. Enormous forests of these trees cover millions of hectares in Canada, growing so close together that you would find it difficult to force your way through the younger forests. The needles of spruces and firs are shorter, thicker, and stiffer than those of pines. Each needle grows out separately from the twig, with no sheath. The trunks are very straight; circles of branches grow straight out from the main trunk.

Two identifying marks help tell a spruce from a fir. Look at small twigs that have lost all their leaves. If the leafless twigs are rough with projecting stumps, the tree is a spruce. If the twigs are smooth, it is a fir. Spruces have cones that hang down from the twigs, while the cones of true firs stand upright. (The Douglas fir, which is not a true fir, has cones like those of the spruces.)

Of the seven species native to North America, the Colorado blue spruce is one of the best known.

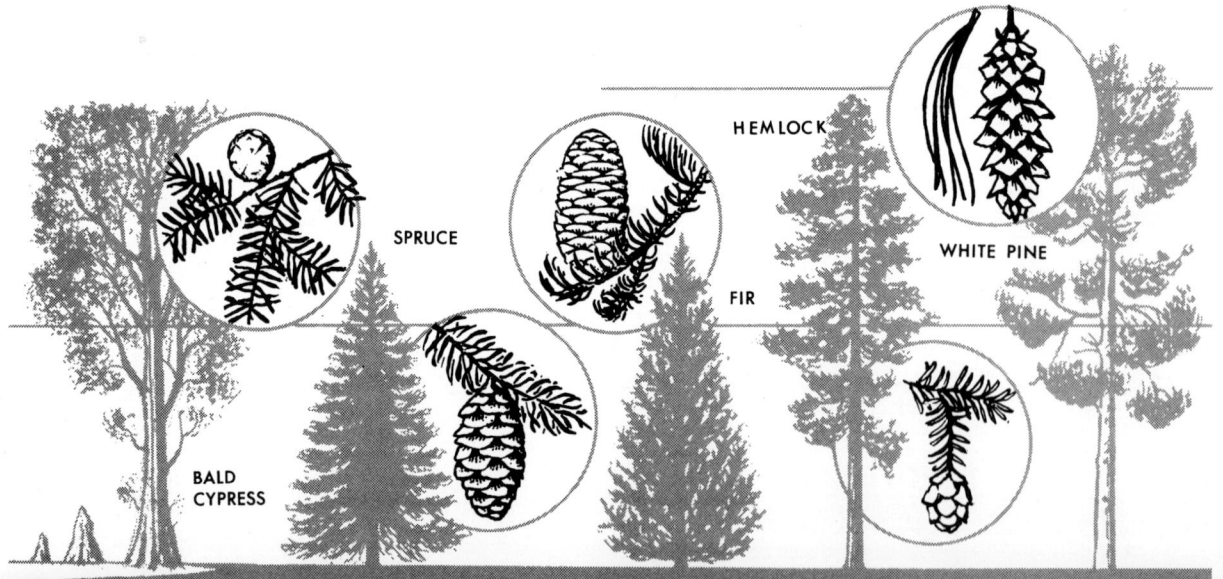

It may be 45 meters tall. The adult foliage is a dull bluish-green, a feature that has made this a popular ornamental, both in its native land and in other countries to which it has been introduced.

The Norway spruce is the best-known European species. It is a major timber tree of Europe and North America. The wood is not particularly strong but is ideal for making boat oars and masts, violins, and baskets. A cone-shaped or pyramidlike tree, it may reach heights of well over 45 meters. The needles are dark green.

The majority of spruce species are natives of Asia. The tigertail spruce of Japan grows to 40 meters. Its distinctive leaves look like flat sickles, and spread from all sides of a twig. The Yeddo spruce is widely distributed in eastern Asia. It has flat, dark green leaves that are bluish or silvery on the undersides. The tree is shaped like a pyramid, reaching heights of 45 meters and a trunk diameter of over 2 meters.

Firs. Firs, as mentioned, may be confused with spruces, though it is easy to distinguish between them. Natives of the Northern Hemisphere, firs are valuable timber trees. They require a lot of moisture and are particularly intolerant of air pollution. These are attractive trees and often are used in gardens or as Christmas trees.

About a dozen firs are native to North America. The tallest, the giant fir of the western United States and Canada, reaches heights of 90 meters. The trunk is tapered, narrowing as it rises. The branches are downswept and bear leaves that are green above and silvery on the underside. The widely distributed balsam fir is grown as a Christmas tree and for its resin, which is called "oil of balsam." The narrow needles are green above, silvery below, and have a lovely scent. The tree may be 18 to 21 meters tall.

The Douglas fir belongs to a different genus than the other firs. It is found as far south as northern Mexico, but reaches its greatest size in the north-

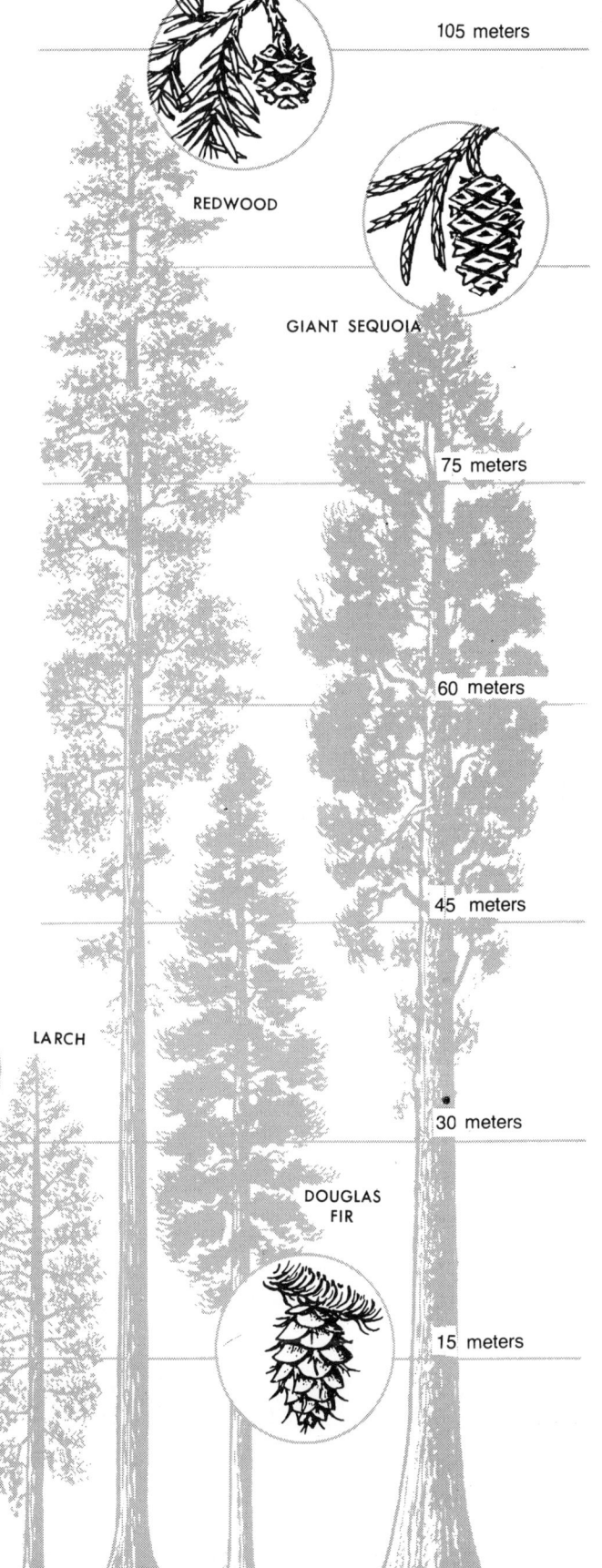

western United States and in British Columbia, where it may be close to 90 meters tall and have a trunk 3½ to 4½ meters in diameter. On young trees the bark is smooth and thin. On old trees, it may be 30 centimeters thick and is broken into large, scaly plates. Douglas firs are very important lumber trees.

The silver fir of Europe may be 45 meters tall and have a trunk 2 meters in diameter. The leaves are all arranged in two comblike rows on the branches. Its lumber is used for building interiors and for musical instruments.

Most firs are natives of Asia, though they have been introduced to other continents. The most widely distributed is the Siberian fir, which covers huge areas of the Soviet Union. The Himalayan and West Himalayan firs also form extensive forests, growing at altitudes up to 4,000 meters.

Hemlocks. Hemlocks are trees of North America and Asia. You could mistake a hemlock for a spruce or fir, except that the hemlock does not grow stiffly erect. It seems much more flexible. It has rough twigs and cones hanging down, like a spruce, but its needles and cones are smaller. The leaves have leafstalks, and there are two white lines on the underside of each leaf. The western hemlock grows in moist coastal regions from Alaska to California. It is found in huge forests together with Sitka spruces. The hemlocks grow under the spruces, but the shade cast by the hemlocks is so deep that no other trees can grow below them.

Larches. Larches are attractive trees with tapering trunks and a cone-shaped appearance. Unlike most conifers, they lose their leaves every fall. The needles are a beautiful light green and feel much softer than other conifer needles. Larch needles grow in thick clusters on the old branches and trunk, but singly on young green twigs.

Larches are native to cooler parts of the Northern Hemisphere. The widely distributed American larch is usually called by its Indian name, tamarack. When Hiawatha built a birch-bark canoe, he cried "Give me of your roots, O tamarack" because he—and other Indians—used the fine, tough roots of the larch for cord to sew canoes together.

A widely used ornamental species is the Japanese larch, which may be 30 meters tall, though at high elevations the twisted branches are close to the ground and the plant is shrublike in form. The European larch also reaches heights of 30 meters. It is an important source of lumber and turpentine. The wood is very durable and thus especially good for heavy outdoor construction.

Redwoods and sequoias. These natives of California are the world's biggest trees. The redwood is the tallest, sometimes reaching a height of 106 meters. One redwood is over 110 meters tall. The trunk is straight and graceful. The bark is reddish-brown and shows deep furrows. When the tree is young, its branches sweep down to the ground. In older trees, the lower part of the trunk has no branches. The branches higher up are rather irregularly spaced and stand out stiffly from the trunk.

The giant sequoia is the bulkiest tree in the world. The General Sherman tree, a giant sequoia in California's Sequoia National Park, has a diameter of 9.7 meters at the base. Its weight has been estimated at about 6,000 metric tons or more. One branch of the tree, growing out at right angles to the trunk high above the ground, is 39.5 meters long and 2 meters thick. The leaves of giant sequoias are scale-like and sharp-pointed. The cones are egg-shaped and remain on the tree for years.

Cedars. The four species of true cedars, genus *Cedrus*, are natives of the Mediterranean region and Asia. The best-known species is the cedar of Lebanon, which is frequently mentioned in the Bible. Ancient kings, anxious to use the fragrant wood in their palaces, overcut the vast forests of these majestic trees, so that today the tree is found only in scattered areas. The cedar of Lebanon reaches heights of up to 38 meters. The dark green needles radiate from the tips of short, spurlike branches.

Junipers. These are Northern-Hemisphere trees, though one species grows naturally south of the equator, in the mountains of eastern Africa. The fruits look like berries and are an important food for birds. The leaves are either scales or needles or a mixture of the two, depending on the species. The best-known American species is the so-called red cedar of eastern Canada and the United States. Sometimes a shrub, sometimes a tree, it may reach heights of 30 meters and live to be 300 years old. The fragrant wood repels moths and was once used to line clothes closets. The common juniper is native to Europe, where it grows as a tree. In North America, it generally grows as a shrub. The berries of this species are used to flavor gin.

Cypresses. Cypresses are natives of the Northern Hemisphere and are generally found in mild climates. Most are graceful trees, with fragrant, somewhat feathery foliage. The cones are very small and spherical in shape. A well-known species is the Mediterranean cypress, found throughout southern Europe and western Asia. The gates of St. Peter's church in Rome were made from the durable wood of this tree. The Monterey cypress is found on the central California coast. Its horizontal branches and broad, spreading crown make a picturesque sight above the rocks and sand that edge the Pacific Ocean.

Bald cypress. This tree is only distantly related to the true cypresses. A native of Mexico and the southeastern United States, it thrives in swamps. It sheds its delicate, light green needles every fall. The tree may be more than 30 meters tall, with a trunk 3½ meters in diameter. Around the lower part of the trunk one often sees "knees"—unusual roots that grow upward from under water. These odd structures both help to support the tree and to provide air for submerged parts.

BROAD-LEAFED TREES

Cycads. Some 150,000,000 years ago, these plants were very common on earth. Today, there are fewer than 90 species, all subtropical or tropical in habitat. Only three species grow taller than 6 meters. In North

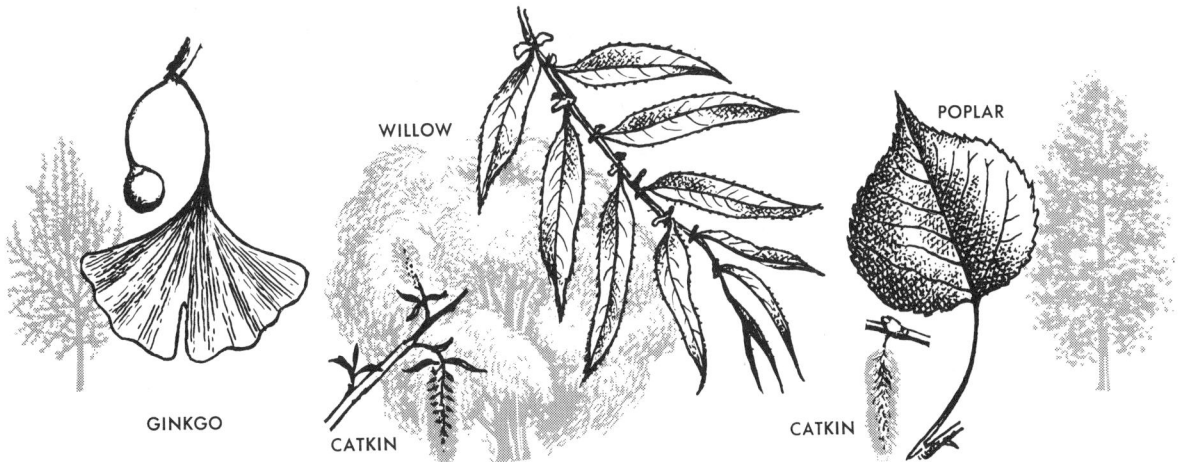

America, cycads are often called sago palms, but these plants are not related to palms. However, they do have stout, columnlike trunks that generally are not branched. The dark green leaves form a spreading crown atop the trunk. Cycads produce large cones—one reportedly weighed over 41 kilograms. Some species are grown as ornamentals in conservatories.

Ginkgos. The ginkgo, or maidenhair tree, is a widely cultivated and very ancient species. Apparently, it died out long ago in its native Chinese forests, and was saved from extinction because it was planted in temple gardens. It thrives in crowded cities, where gasoline fumes kill most other trees. The thick, fleshy leaves look like the leaflets of the maidenhair fern. They have no midrib and are fan-shaped. The outer rind of the seeds has a peculiar odor.

Monkey puzzle. This native of South America is a popular ornamental tree in many places. It may reach heights of more than 45 meters. The broad, stiff leaves are dark green. They overlap one another and closely cover the brittle, snakelike branches. The cones, which may approach a human head in size, contain as many as 180 seeds. These are called piñones and are very tasty.

Monkey puzzles, ginkgos, and cycads are more closely related to the conifers than to other broad-leafed trees.

Palms. Palms are easily recognized. The columnar trunks are generally unbranched. At the top, they bear a crown of large, stiff leaves. Palms are most common in tropical and subtropical areas. They are very valuable economically, producing lumber, foods, oils, and other substances used by people.

The betel palm is found from India through the Pacific islands to Australia. It reaches heights of 30 meters but has a trunk only 15 centimeters in diameter. Its seeds are chewed as a stimulant.

The coconut palm, found throughout the tropics, lives along the edge of the sea. It may be 30 meters tall, with dark green leaves 6 meters long. The fruit takes 10 to 12 months to mature. The meat of the large seed—the coconut—is eaten by people everywhere, either fresh or dried. The dried coconut meat is called copra and is a very important product. From it is made coconut oil, which is used in making soaps, margarines, and other products. The wood of the coconut palm is used for building, the leaves are dried and woven to make baskets, and the flowers yield a sugar.

The stout trunk of the date palm reaches heights of 30 meters. The feathery grayish-green leaves may be 6 meters in length. This species is native to North Africa and the Middle East, but is widely cultivated throughout the tropical world. It is able to live in dry climates, and is the palm found in desert oases. Individual clusters of fruit may weigh more than 13 kilograms. Although grown primarily for their fruits, date palms have hundreds of other uses. For example, sugar can be made from the sap, and the seeds can be roasted and used as a coffee substitute.

Raffia palms have leaves up to 21 meters long—the largest leaves of all plants. They are the source of raffia fiber. Natives of Africa, most raffia palms have stout trunks that may be as tall as 12 meters. The hanging flower clusters are very large—up to 3.5 meters long.

Casuarinas. Most of the 45 species in this family are Australian. Because of their foliage, they are sometimes called Australian pines. The leaves are tiny scales around the nodes of needlelike branchlets. The branchlets carry out photosynthesis for the plant. They are shed annually. Casuarinas are often planted along avenues, as windbreaks, or as ornamentals.

Poplars. Trees of the genus *Populus* have toothed leaves that are somewhat triangular in shape. The leaf stalks are exceptionally long. The bark is usually smooth when young, furrowed when older. The fruits develop in long, caterpillarlike clusters called *catkins*. These trees are native to the Northern Hemisphere and widely distributed, often forming extensive forests. The soft wood is used in construction and boxes. Some species are used as ornamentals. Poplars are brittle and thus easily damaged in storms.

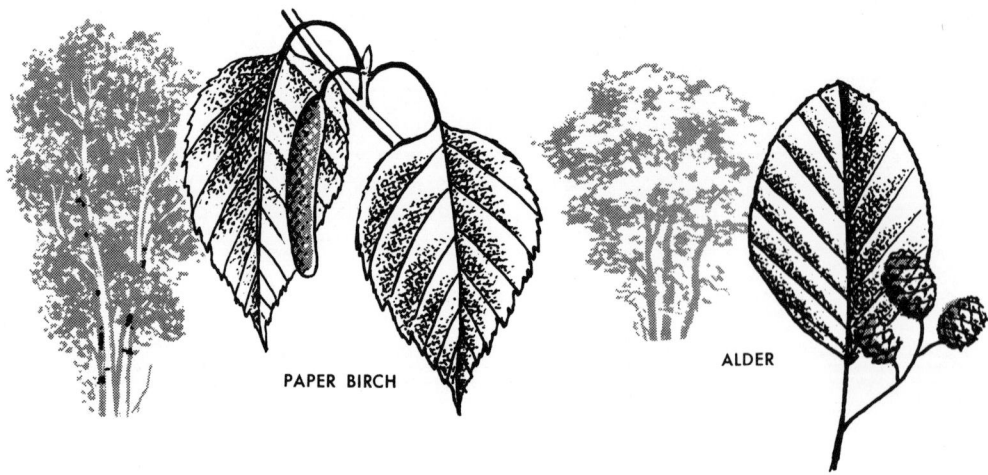

Well-known North American species include the cottonwoods and aspens. They have flat leaf stalks; the leaves tremble in even the slightest breeze. The Lombardy poplar, a native of Europe, has branches that grow almost straight up, so that the tree looks like a green exclamation point.

Willows. Willows range from shrubs like the pussy willow to the crack willow, sometimes more than 27 meters tall. Willows brighten the landscape in winter because many have olive green, red, or bright yellow twigs. The willow bud has just one scale, unlike most trees, which either have two or more bud scales or have buds unprotected by scales. Most willows have long, narrow leaves edged with small teeth. Bring cut stems of pussy willow into the house in spring and let them stand in vases of water until the catkins have opened and the large number of tiny flowers in each catkin have bloomed. Soon leaves will start to grow and white roots will sprout from the stems. When the wind breaks off branchlets from willow trees, the branchlets often grow into trees if they land in a favorable spot.

Willows are natives of all continents except Australia. Most prefer moist habitats, and are of great value in preventing erosion along rivers and around ponds. One very well-known species is the weeping willow, whose long, drooping branchlets wave with every breeze. A native of China, it is now common in many countries.

Birches. Birches look somewhat like poplars. However, the small seeds have no cottony covering like those of poplars and willows. Birch catkins reach a length of 2 to 5 centimeters by the time winter sets in, so that they are conspicuous at the ends of twigs all winter, long before their flowers open in spring.

Birches are natives of temperate and cold regions of the Northern Hemisphere. Many are favorite landscape trees because of their beautiful bark and their elegant appearance. The wood is used for furniture, interior construction, and plywood. In parts of Europe, it is used for roofing. The Indians of North America used birch wood for their canoes and wigwams.

Alders. Alders are small trees or shrubs with flowers, fruits, and seeds similar to those of birch. Both birch and alder bear their seeds in conelike structures, but the alder has woody cones that last for a long time. The 35 species are mostly natives of the Northern Hemisphere. The major North American species are the red alder and the white alder of the northwest. The black alder forms extensive forests in Central Europe and is found south into Africa and east to Siberia.

Hornbeams and hop hornbeams. These also belong to the birches. Trees of the Northern Hemisphere, they occur in woods under large trees. The New World hornbeam dangles clusters of tiny nuts, each attached to a small green leaf shaped like an arrowhead. It is also called blue beech because it has smooth blue-gray bark like a beech. The American hop hornbeam has rough bark. Its fruits are like little bags, each with one seed. They are in conelike clusters and look like the fruits of the hop vine. Both of these trees are often called ironwood because of their tough, hard wood.

Elms. Elms are trees of the Northern Hemisphere. Valued for their beauty and their lumber, elms are unfortunately subject to two widespread diseases that have greatly reduced their numbers: Dutch elm disease and phloem necrosis disease. In the elm tree, the main stem breaks up into a dozen or more arching trunks. You can always recognize an elm by the uneven base and the strongly marked vines of the leaf. Elm leaves are rough to the touch and have teeth along their margins. One of the world's most graceful trees is the American elm. Tall, with arching branches, it makes an excellent shade and street tree.

Hackberry. This North American tree is closely related to the elms. Once you see its purple-black berries and its gray bark ridged and sprinkled with hard, warty growths, you will remember it.

White stinkwood. In the same genus as the hackberry, this tree is widespread in Africa from Ethiopia south. It is a common ornamental and street tree. The

ELM

SWEET GUM

trunk is grayish-white. The newly cut lumber has an unpleasant odor — hence the name.

Fig trees. Approximately 800 species make up the genus *Ficus*. Natives of tropical and subtropical lands, these trees depend on gall wasps for pollination. Fig trees have fleshy fruits, many of which are eaten by people. The banyan tree is a fig tree native to India. As the banyan grows, roots from its spreading branches reach down to the soil and develop into secondary trunks. The tree keeps expanding. One measured over 600 meters in circumference — 20,000 people could stand beneath it.

Sweet gums. These trees of the Americas and Asia produce a fragrant resin used in medicines, incense, soaps, perfumes, and other products. The American species ranges from the northeastern United States through Central America and into Venezuela. The spherical, woody fruits swing from long stems and are covered with hornlike projections that split to release dry, winged seeds. With its five or seven pointed, triangular lobes, the sweet gum leaf is more starlike than any other leaf. The leaves turn brilliant red or orange in fall.

Rubber trees. These are native to India and Malaysia, where they reach heights of over 30 meters. They were the original source of India rubber, but are not, however, the source of natural rubber collected today. Many people in temperate lands raise rubber trees as houseplants.

Plane trees. These trees have leaves that look a bit like maple leaves, but they are alternate. Also, unlike other tree leaves, the leaf stalk fits tightly over a bud as its base. The bright appearance of plane trees comes from the mottled bark, which continually falls in patches to reveal the light green underbark. Summer and winter sunlight and even moonlight glisten on the dappled trunk and branches.

The North American planes are called sycamores and, in some places, buttonwoods, on account of their fruits. These are long-stemmed globes, each with hundreds of hard little seeds, surrounded by fuzzy hairs. The best of all trees for city parks is the London plane, which can stand city smoke and fumes unusually well. The London plane is actually a hybrid between the American sycamore and the Oriental plane.

Lindens. The American basswood, the English lime, and the European linden are all members of the same genus, *Tilia*. There are some 80 species in this Northern Hemisphere genus. They are sometimes called bee trees because bees eagerly seek out their blossoms in late spring. The leaves are usually heart-shaped. Clusters of fragrant cream-colored flowers are borne at the end of a slender stalk. The stalk springs from the center of a flat green stem that looks like a leaf. When the fruits mature, the whole arrangement breaks off and sails through the air.

Baobabs. These unusual-looking trees are native to Africa and northern Australia, where they inhabit hot, arid grasslands. The trunks are very thick, the branches are rather short, and foliage is sparse. The Australian baobab, or bottle tree, reaches a maximum height of about 12 meters, but its trunk may have a circumference of 18 meters. Its fruits and seeds are edible, and a white gum given off by the tree can be used to make pleasant-tasting drinks.

Kapok. This massive native of the American tropics is cultivated extensively in parts of Asia and Africa. Its fruits are 7½- to 15-centimeter-long pods that contain masses of brown, gray, or white fibers. The fibers, called kapok, are lightweight, waterproof, and very elastic. They are used as stuffing for sleeping bags, life preservers, pillows, and so on. The closely related silk-cotton trees also produce such fibers.

Balsa. This tree produces the lightest-weight lumber of all. The wood is used for rafts, model airplanes, and life preservers. The tree is native to the West Indies and the American tropics. It reaches heights of 24 meters. The leaves are heart-shaped and have reddish stalks.

Mangroves. These tropical trees grow to heights of 30 meters. The branches are high above the

PLANE TREE

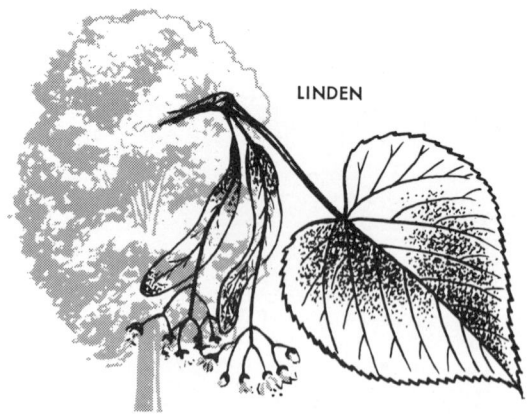
LINDEN

ground. From them dangle aerial roots that anchor into the swamp or coastal ground these trees favor. This helps build up the land. The mangrove's heavy seeds start to grow while still on the tree. Eventually they drop, root-side down, into the water.

Walnut and hickory trees. These are closely related and bear edible nuts. Leaves are alternate and compound. The walnuts have more leaflets on each leaf (from 11 to 23) than the hickories and the husk of a walnut does not split open. If you are in doubt, split a twig with a knife. If the soft pith is divided into little plates with open spaces between them, the tree is a walnut. If the pith is solid, it is a hickory.

Hickories, with one exception, have 5 to 11 leaflets on each leaf. The pecan tree is a hickory and the largest of the genus. Pecan trees in the Mississippi Valley are sometimes 48 meters tall.

Oaks. Oaks are strong, long-lived trees with tough wood. They all bear nuts (fruits) called acorns. These offer an infallible means of identification, since no other nuts resemble acorns. The leaves are alternate; in most species, they are lobed. The leaves usually are borne in clusters near the ends of branches. There are some 450 species of oaks, native to the Americas, Europe, northern Africa, and subtropical Asia. Oaks are among of the world's most important trees.

The New World oaks fall into two groups: the white oaks and the black oaks. White oaks usually have leaves with wavy or curved lobes. The leaves never have sharp lobes with bristle-pointed teeth at the tips of the lobes, as do the leaves of the black oak group. The white oak group includes the white oak itself, the evergreen, or live, oak, and all European oaks.

The black oak group includes the black oak, red oak, scarlet oak, pin oak, and other species. The leaves vary greatly in shape, even within the species.

The Asian oaks differ from other oaks in that the scales of the acorn cups are connected in concentric rings rather than being separate. Among the common and economically valuable species are the Mongolian, Daimyo, and Oriental white oaks.

Chestnuts. These are known by their long leaves with prominent midrib and the pointed teeth at the tips of the veins. The 12 species are natives of northern temperate zones. Years ago, one of the commonest trees in the eastern United States was the American chestnut, with its beautiful long leaves and its delicious nuts in their spiny burrs. Then came the chestnut blight, a deadly fungus. Few diseases ever killed off their victims so completely. All American chestnut trees of any size were killed. Year after year new stems grow up from the old roots, but the fungus kills these before they are more than several centimeters thick. The species is now almost extinct. However, Chinese and Japanese species are resistant to the chestnut blight and have been introduced into North America.

Beeches. These are natives of northern temperate regions. Many are grown as ornamentals. They are stately, beautiful trees. If you find large trees with light gray, very smooth bark, even on the older parts of the trunk, they are doubtless beech trees. Look for the very long, sharp-pointed buds. The leaves have veins arranged like those of the chestnut, but they are shorter and more nearly oval and the teeth at the margin are not so sharp. Beech nuts are triangular, small, sweet, and good to eat. They are enclosed in a spiny burr, but sometimes fall out. The pale dead leaves of the beech often cling to the trees all winter.

Horsechestnuts. These natives of central Europe are favorite landscape trees in many lands. This is one of the easiest trees to recognize. It has opposite leaves, each usually having seven leaflets arranged like the fingers of your hand—that is, palmately. It has big buds, covered with sticky resin, at the end of its branchlets. A large horsechestnut in spring is a beautiful sight, with hundreds of clusters of big flowers. Later, dry, spiny fruits develop. These split to free the big, smooth, bitter nuts. The American buckeyes (so called because their seeds suggest the eyes of a deer) are closely related to the horsechestnut. Their fruits are either smooth or else decidedly less spiny than those of the horsechestnut.

Magnolias. The magnolia tree bears beautiful

flowers. Inside the large, flashy petals are numerous stamens and pistils, arranged on a fleshy cone. After the petals fall, this cone grows and bears red, fleshy seeds. Magnolias also have large leaves. The cultivated magnolias, whose flowers bloom before the leaves appear, come from China or Japan.

Tulip trees. These are in the magnolia family. One species is native to China. The other, native to the eastern United States, is the tallest broad-leafed tree in North America, reaching heights of up to 60 meters. Its yellow and orange blossoms look like tulips. In summer, it displays glossy, lobed leaves that look as if someone had cut off their tips.

Camphor and cinnamon trees. These are lovely ornamentals native to Asia. The wood of the camphor is fragrant and strong, and repels insects. The dried bark of young cinnamon tree shoots is the spice we use for flavoring.

Sassafras. There are three species, one native to China, one to Taiwan, and one to eastern North America. A sassafras tree has three kinds of leaves, all on the same tree. Some are oval, some have two lobes and look like mittens, and some have three curved lobes. The roots of the American species provide oil of sassafras, which is used as flavoring in teas and tonics. Birds eat the purple fruits.

Fruit trees. Trees of the genus *Prunus* include plums, cherries, apricots, peaches, and almonds. Pears (genus *Pyrus*), apples *(Malus)*, and quinces *(Cydonia)* are closely related—all members of the rose family. Wild trees generally have small fruits that are popular with birds and small animals. Some, especially the larger cherry trees, are highly valued for their wood. The cultivated varieties are derived from wild species native to different parts of the northern temperate zone.

Acacias. There are some 750 species of these tropical and subtropical plants. Most live in dry habitats. They have thorns and showy flowers with numerous stamens. Many species are important sources of lumber, dyes, gum arabic, and other products. Acacias are divided into two groups. One group, which includes the African acacias and the cassie of the Americas, has feathery leaves composed of many leaflets. The second group, to which the koa of Hawaii and many Australian species belong, has undivided "leaves" that are actually flat, expanded leaf stalks.

Rosewoods. These are tropical and subtropical trees with pinnate leaves that consist of an uneven number of leaflets. The flowers look like those of pea plants and develop pod fruits. Most species have exceptionally beautiful wood that is used to make everything from furniture to salad bowls to umbrella handles.

Royal poinciana. This native of Madagascar is a colorful ornamental commonly grown in many warm lands. Up to 15 meters tall, it has feathery, pinnate leaves. The flowers are brilliant red or yellow and brighten any garden or roadside. They produce pods that may be close to 40 centimeters long. Eventually, the pods turn black, and remain on the tree long after the leaves have fallen.

Honey-locusts. These trees also have pinnate leaves. The flowers are greenish, the pods may be 45 centimeters long, and spines usually grow from the trunk and branches. The wood is durable. Except for one South American species, all are natives of the Northern Hemisphere.

Lignum vitae. This tree produces a very hard wood used to make such things as bowling balls and propeller shafts for steamships. A native of the West Indies, Central America, and northern South America, it reaches heights of 9 meters. Each leaf has 4 to 6 leaflets. Related species also produce durable wood of commercial importance.

Citrus trees. These natives of China and southeast Asia have been cultivated elsewhere for many centuries—the citron was introduced to Mediterranean countries around 300 B.C. by Alexander the Great's armies. Citrus plants have fragrant flowers and usually have undivided leaves. Commercially important species include citron, grapefruit, lime, sweet and sour oranges, and lemon.

CHESTNUT

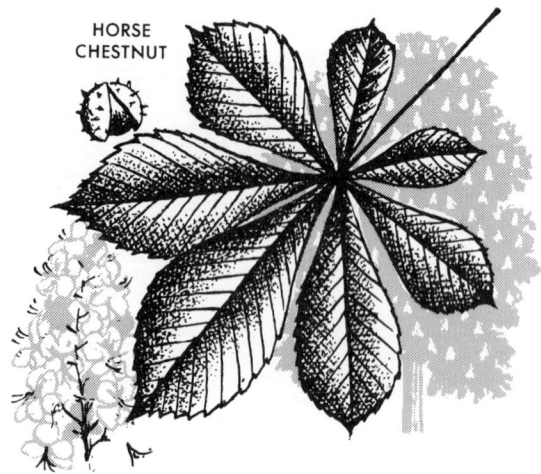

HORSE CHESTNUT

Mahogany trees. These natives of the American tropics produce beautiful wood used for making furniture. The best wood is probably that of the West Indian mahogany. This tree grows to heights of 18 meters. The pinnate leaves may be more than 17 centimeters long. It is an attractive tree and is often grown as an ornamental.

Tung oil and candlenut trees. These are Asian species. Oils from them are used in varnishes, paints, and other products. The tung oil tree has heart-shaped leaves and attractive flowers that are white with red and yellow markings. The candlenut tree's leaves may be lobed or similar to those of the tung oil tree. Its flowers are small and yellow.

Rubber tree. Rubber comes from *Hevea braziliensis,* a tropical South American tree that is now cultivated extensively, particularly in the Far East. This tree is the source of almost all natural rubber. When the trunk is cut, a juice called latex flows from the incision. This is collected in a very delicate, methodical operation. More than 15 liters of latex can be collected from a mature tree in a year. These trees have long-stalked, compound leaves that are purplish-bronze but turn red, orange, or brown when old. The whitish flowers are small and have a lovely scent.

Cashew and pistachio trees. These are noted for their nuts. The cashew tree is native to the Caribbean area, the pistachio to the Mediterranean region and Asia. The cashew is an evergreen; the pistachio is deciduous. Both have small flowers—those of the pistachio lack petals. In the same family is the mango, a native of India cultivated for its juicy, delicious fruit.

Sumacs. These are subtropical and temperate, with small flowers in clusters and small berrylike fruits. Some species have a deservedly bad reputation, for they are very poisonous. They cause severe skin inflammation, including blisters, when touched. Inhaling the smoke of a burning poisonous sumac can be even more dangerous. One highly poisonous species is the lacquer tree of China and Japan, which is commercially exploited as a source of a varnish called Chinese or Japanese lacquer.

Ebony. Ebony wood comes primarily from *Diospyros ebenum,* a tree of India and Sri Lanka. This tree may be 18 meters tall and have a trunk 1.8 meters in diameter. In the same genus is the persimmon tree of the United States and the Japanese date-plum, which is widely cultivated for its fleshy fruit.

Mulberry. This tree has alternate, palmate, toothed leaves, which may be heart-shaped or lobed. Pluck a leaf or cut a twig and notice the milky juice. The berries of these trees are good to eat. The most valuable mulberry tree is the white mulberry of Asia, whose leaves are fed to silkworms. No country can produce natural silk unless it has the proper mulberry trees.

Hollies. These are a varied group. Some are evergreens, some deciduous. Most have smooth bark and toothed leaves. The species famed for their scalloped spine-tipped leaves and red berries are the English holly and the American holly. Both are pyramid-shaped trees. The American holly is usually shorter and has duller leaves, and its berries do not form dense clusters as do those of the English holly.

Ailanthus. These natives of Asia and Australia have compound leaves. In autumn, the leaflets fall first. Eventually the midrib to which they were attached will fall from the branch. The best-known species is the tree-of-heaven, which is widely cultivated and seems to thrive in polluted urban environments. Its long leaves have 15 to 31 smooth-edged leaflets. At the base of each leaflet is a scent gland. The scent of the male flower is unpleasant. The male and female flowers are borne on separate trees. The grayish-brown bark is smooth in younger trees, cracked in older trees.

Maples. These trees have opposite leaves that are deeply lobed and palmately veined. No other tree has fruits like theirs. They are called samaras and consist of two seeds joined at the base, each with a wing. Maples are natives of the Northern Hemisphere. Many are grown as ornamentals and some are harvested for their wood. The best-known species of North America is the sugar maple. Its leaf is the

national emblem of Canada. This tree yields maple sugar and syrup. The wood is also highly prized. The sugar maple has thick foliage and beautiful orange, yellow, and scarlet colors in autumn.

The Japanese maple, especially varieties with red leaves, is widely cultivated as an ornamental. It grows to heights of about 7½ meters. Unlike American and Asian maples, European species do not turn brilliant colors in autumn. The Norway maple gives dense shade and is the only maple that yields a milky juice when you break a leaf or stem. The sycamore maple grows well in coastal land and in open upland areas. Its smooth, white lumber is excellent for carving.

Eucalyptus. The 500 species of this Australian group of trees range from plants 2 meters tall to the Australian mountain-ash, which may be well over 100 meters tall. They are by far the most common and most important trees of Australia. Some species have also become common in other lands to which they have been introduced. The trees are slender. The leaves often are a bluish or whitish green. They vary in shape with the age of the tree. Young leaves usually grow horizontally from the branch. Adult leaves hang. Most species are evergreens. The fruits are woody.

Some eucalyptus species are valuable sources of lumber. The wood is hard and very durable. Oils produced by the leaves are used in medicines, perfumes, and other products. The nectar of eucalyptus flowers is the primary source of food for the bees of Australia.

Dogwood. These trees have smooth-edged opposite leaves, rough, checkered bark, and clusters of berrylike fruits that often remain most of the winter. Examine a flowering dogwood in spring and you will see that the four or more large "petals" are really modified leaves. They surround a cluster of tiny flowers, each with petals of its own. The dogwood never grows to large size, but thrives under forest trees.

Balata rubber. This comes mainly from *Mimusops bidentata,* a tree of the West Indies and Central and South America that may be 30 meters tall. The thick leaves are alternate. The flowers are small and bell-shaped. The juice, or latex, of these trees is used for shoe soles, machine belts, and other products.

Ashes. These trees of northern temperate zones have leather-veined, opposite, usually compound leaves. The flowers are in dense catkinlike clusters. Ash seeds are winged but are easily distinguished from winged seeds of other trees. The wood of the tree is tough and elastic and is used to make baseball bats and other sporting goods.

Olive trees. Natives of the Mediterranean area, these are widely cultivated for their fruits. The tree grows to 9 meters in height. The dark leathery leaves are silvery underneath. The fragrant flowers are small and greenish. Olive trees may live more than 1,000 years.

Strychnine. This comes from seeds of *Strychnos nuxvomica,* a 15-meter-tall native of tropical Asia. The greenish flowers are pleasantly scented. The fruits look like small oranges. Closely related is the curare poison nut vine of Central and South America.

Catalpas. These are natives of North America, the West Indies, and Asia. The leaves are large and heart-shaped or lobed. The flowers are bell-shaped and produce long, beanlike seed pods. The wood of some species is durable but coarse. It is used for telephone poles, fences, boatbuilding, and construction, depending on the species and the location.

Teak trees. These are native to India and neighboring lands. Their wood is very durable and hard and is used for many products. The trees may reach heights of 30 meters. The leaves are opposite and large—often almost a meter long.

Coffee trees. These natives of the Old World tropics are now extensively cultivated elsewhere. The most important species, the Arabian coffee tree, may reach heights of 9 meters. The leaves are long and glossy. The white flowers have a lovely scent and produce fleshy fruits that gradually become purple. These contain the seeds, or coffee beans.

Cinchona trees. These tropical American plants are the source of quinine, a drug once important in the treatment of malaria. These medium-sized evergreens have large leaves and yellowish or pinkish flowers. The quinine is produced from the bark.

ANIMAL LIFE

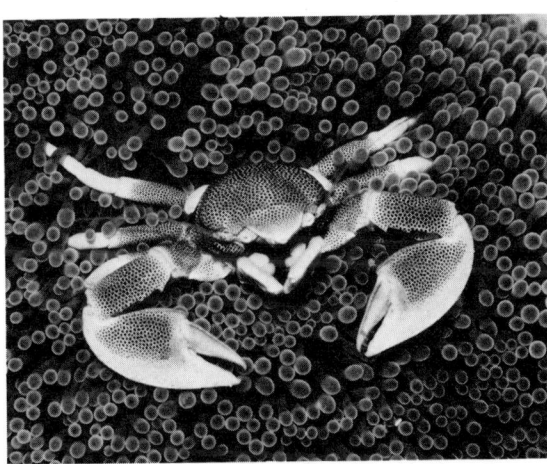

both photos, Jen and Des Bartlett/PR

Animals in their extremely varied forms, habitats, and life adaptations are found throughout the world in all different kinds of environments. Opposite page: a leopard, an agile climber, roams over mountainous terrain. Above: a crab sits on a sea anemone.

178-182	The Life of Animals
183-189	The Animal Kingdom
190-197	The Protozoans
198-204	Sponges and Jellyfish
205-208	Starfish and other Echinoderms
209-215	Flatworms and Roundworms
216-226	Mollusks
227-230	The Earthworm and Its Kin
231-238	Lobsters, Crabs, and their Kin
239-250	Spiders and their Kin
251-265	Insects
266-276	Ants, Bees, Wasps, and Termites
277-287	Beetles
288-295	Butterflies and Moths
296-299	The Primitive Chordates
300-315	Fishes
316-326	Sharks and their Kin
327-336	Amphibians
337-345	Lizards
346-357	Snakes
358-360	Turtles
361-365	Crocodiles and their Relatives
366-378	The Life of Birds
379-382	Flightless Birds
383-390	Birds of Prey
391-396	Wild Fowl
397-399	Pigeons and Doves
400-411	Other Non-Perching Birds
412-426	Water and Shore Birds
427-444	Perching Birds
445-452	Bird Observation
453-460	Animal Behavior
461-474	Animal Migration
475-489	Fossils

THE LIFE OF ANIMALS

The many animals that inhabit the earth show an almost endless variety in size, shape, internal and external structure, and habits. Well over one million species are now recognized and new ones are being constantly discovered.

Animals range in size from the microscopic protozoans to the mighty blue whale, which may reach a length of 30 meters and a weight of over 80 metric tons. Some animals, such as the deer fly and the cheetah, can move with great speed. Some, such as the sloth and the snail, move slowly. Some, including the flowerlike sea anemone and the coral, become permanently fixed to one place. Certain animals, like the weasel among vertebrates and the praying mantis among insects, are predators, eating other animals. Others, like deer, cattle, and Japanese beetles, live on vegetation. Still others—hyenas, jackals, and vultures, for example—are scavengers, feasting on carcasses.

ANIMAL STRUCTURE AND FUNCTION

The vastly different forms of animal life are all made up largely of protoplasm, a substance in which the physical and chemical processes of life take place. Protoplasm is a very complicated mixture of various materials and is contained in one or more basic units called *cells*. Generally speaking, animal cells are microscopic in size, but there are exceptions. The nerve cells of large animals (including man) may be almost a meter in length. The yolk of a bird egg is a single cell. It may be over 8 centimeters in diameter in the case of the ostrich, the largest of all birds.

The animal cell is bounded by a very thin membrane, called the plasma membrane. Outside this, there may be a flexible envelope—the pellicle. The protoplasm within the plasma membrane always contains a nucleus, or core, which may be at the center or at one end of the cell. Some cells have two or more nuclei and some mature cells lose their nuclei. Within the nucleus is a fine network, the chromatin. It carries the units, called genes, that determine the heredity of the animal. In the protoplasm outside the nucleus of the cell, there are various bodies that play an important part in life processes.

ANIMAL TISSUES

Certain microscopic animals, such as the ameba and the paramecium, consist of only one cell. Most animals, however, are multicellular, or many-celled. In the more highly organized varieties, there are millions upon millions of cells, grouped together to form *tissues*. In each tissue, the cells are more or less alike and perform the same functions. There are epithelial tissues that cover the outside of the body and various inner parts. Connective tissues bind together other tissues and help support the body. Muscular tissues bring about movement as they contract or lengthen. Nervous tissues transmit stimuli.

ORGANS

Tissues in turn are combined in larger units, known as *organs,* which perform certain definite tasks. The organ called the heart serves as a pump and keeps the blood circulating. The brain acts as a nerve center, receiving messages from the outside world and sending messages to different parts of the body. The lungs serve to draw in and exhale air in the process of respiration. The liver filters the blood, secretes bile and stores the sugar called glucose.

ORGAN SYSTEMS

Finally, organs are joined in still more complex units, called *systems.* In most multicellular animals, we find a skeletal system, for the support and protection of the body; a muscular system, capable of bringing about motion; a circulatory system, for the transportation of food and waste materials; a respiratory system, for the process of breathing; a digestive system, for preparing

Animals vary greatly in size, shape, internal structure, and behavior. Right: a sea anemone—a somewhat atypical animal in that it adheres to a rock or the sea bottom and does not move freely. Left: just the head region of one of the largest of animals—a rhinoceros. These two widely different animals share one characteristic: both must obtain their food from outside sources. The sea anemone closes its tentacles on food that falls on it. The rhinoceros actively goes after food, sometimes uprooting a tree to feed on the leaves.

food so that it can be absorbed by the body; an excretory system, for disposing of wastes; a nervous system, for regulating the body's internal activities and permitting the animal to adjust to its environment; and a reproductive system, for bringing new individuals into being.

FOOD AND DIGESTION

All these systems can function only if the animal can obtain food, which will be transformed in its body in such a way as to release energy. Most plants can manufacture their own food by the process known as photosynthesis, but animals cannot do this. Animals obtain essential food elements by eating plants or plant-eating animals. They obtain their food in various ways. The ameba enfolds microscopic animals or plants in temporary extensions of its body called pseudopodia, or "false feet". When the tentacles of the hydra touch its prey, certain structures they contain pierce the body of the victim and inject a paralyzing fluid. The tentacles then draw the prey to the hydra's mouth. To obtain its food, the earthworm swallows earth, digests any minute plant or animal life contained in it, and ejects the rest. The insect called the aphid sucks plant juices. The praying mantis nibbles at the soft parts of insects and discards the rest. Many whales draw in huge gulps of water and strain out of the water the plankton (tiny animals and plants) it contains. Certain mammals, such as the horse, deer, and sheep, are vegetarians. Others, including the lion, puma, and weasel, eat animal prey.

Much of the food that animals eat is in the form of complex substances—proteins, carbohydrates, and fats. These must be transformed in the body to simpler compounds that the cells can absorb—a process called *digestion*. Often food has to be broken down by mechanical means before chemical digestion can take place. This preliminary process is performed by the teeth of various animals. In birds, food is ground in the gizzard against particles of sand that have been swallowed. The churning action of the stomach also helps break down food.

Such mechanical processing is not required in certain cases, as when the ameba devours its prey or when the aphid sucks plant juices. However, in every case, chemical changes must take place in complex food substances before they can be absorbed by the cells. Such changes are brought about by digestive enzymes. As a result, proteins are reduced to amino acids; fats, to fatty acids; carbohydrates, to simple sugars.

THE LIFE OF ANIMALS

RESPIRATION

When food has been suitably prepared by the process of digestion, it must be burned in the cells so that energy may be provided for the animal's needs. Burning, or combustion, takes place when oxygen, derived from the air or water that surrounds animals, combines chemically with food, releasing energy and yielding the gaseous waste product carbon dioxide. The carbon dioxide is then discharged to the surrounding air or water. The process of drawing in oxygen and giving off carbon dioxide is called breathing, or *respiration*.

Animals breathe in many different ways. In the water-dwelling one-celled ameba, oxygen enters the body through the cell membrane from the surrounding water, and carbon dioxide passes out through the same membrane. The earthworm breathes through the moist outer layer of its skin. In insects, oxygen is drawn in and carbon dioxide expelled through fine tubes leading from the surface of the body to the tissues. There are special structures for respiration in the higher animals—the gills in water animals; the lungs in land animals. Oxygen is transferred from the gills or lungs to the blood and is carried in the blood stream to the cells. Carbon dioxide is carried from the cells in the blood to the gills or lungs, and is released from the body.

A certain amount of the energy derived from the combustion of food in the cells is given off in the form of heat. The rest is used by the animal in carrying on its various activities. Energy is required, for example, in building up complex proteins and other compounds from simple substances in the cell and in bringing about growth in cells and tissues. Energy is also needed for the many types of movement in which the animal engages.

LOCOMOTION

The type of movement called *locomotion*, or going about from one place to another, generally distinguishes animals from plants. There are many different types of locomotion. The ameba moves as its substance streams into temporary extensions of its body—the pseudopodia previously mentioned. The flagellates beat the water with whiplike structures called flagella. Some fishes, such as eels, make their way through the water by wiggling movements of the body. Others advance by moving their caudal fins, or tails, from side to side. The squid draws water into its body and then forces it out through a flexible tube located near its arms. The resulting jet moves the animal backward. Certain animals—birds, bats, insects—can fly by moving their wings; flying snakes can glide as they flatten their bodies. Land animals with legs that raise the body from the ground are often capable of very rapid locomotion.

REPRODUCTION

A vital activity of animals is reproduction, or the forming of new individuals. Some of the lower animals, including the one-celled paramecium, reproduce by dividing into two halves. The tunicate develops a bud, which gradually grows into a new animal as large as the old. Certain flatworms break up into two or three parts, and each part becomes a new individual.

Most animals, however, reproduce their kind through sexual reproduction. Male sex cells (sperm), produced by males,

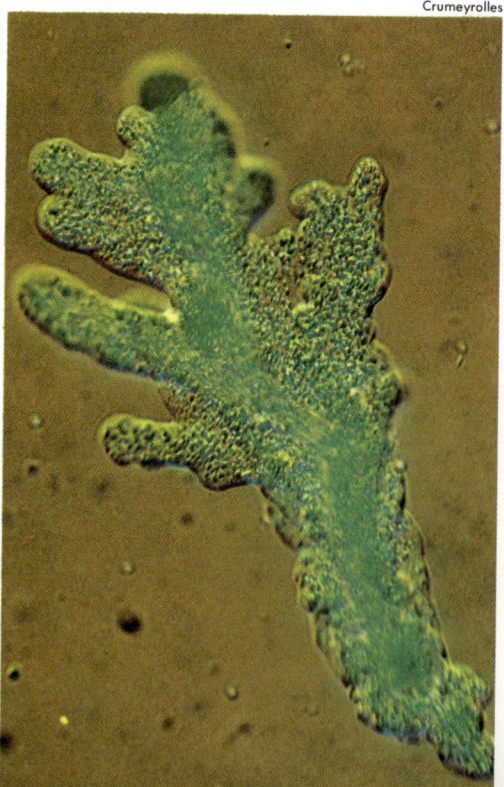
Crumeyrolles

and female sex cells (eggs), produced by females are united in the process of fertilization. (In some cases, as in snails, both male and female sex cells develop in the same individual.) The fertilized egg then grows into a new individual. In many cases—butterflies, salmon, and birds, for example—the fertilized egg develops outside the body of the female parent; in others (practically all mammals), the developing egg remains inside the body of the female and is separated from it only at the moment of birth.

Whatever the method of reproduction, traits are passed on from one generation to another through the operation of heredity. The ameba gives rise to another ameba; the snail, to another snail; the lion, to another lion.

PLACE IN THE ENVIRONMENT

The mode of life of animals is influenced by the environment in which they live. There are many factors in the environment; they include sunlight, temperature, air pressure (in the case of land animals), and water pressure (in the case of water animals). Animals are also affected by their relationships with other animals as well as with plants. Many of these relationships have to do with the obtaining of food. Thus, in this eternal quest, living things, plants and animals alike, are bound together in food chains. For example, plants manufacture food in sunlight; insects devour plants; songbirds eat insects; hawks prey on songbirds; when hawks die, beetles and other insects feed on their carcasses, and bacteria transform various compounds in these carcasses into nutrients that plants can absorb. Disease may also be considered as one phase of the search for food. For diseases are caused by living organisms that feed, often as parasites, upon other organisms.

THE STUDY OF ANIMALS

The study of the different aspects of animal life is the province of the science called *zoology*. The word comes from the Greek *zoios*: "animal" and *logos*: "science". Like other major sciences, zoology is divided into a number of different fields. *Anatomy* is the study of form and structure. *Histology* deals with tissues; *cytology*, with cells. The living processes of animals form the subject matter of *physiology*. *Nutrition* considers how animals take in and utilize food substances. *Embryology* examines the formation and development of a new individual from the fertilized egg. *Genetics* deals with heredity and variation. It shows how characteristics are passed on from one individual to another or from one population to another. *Parasitology* examines the animals that live in or on other animals. *Pathology* is the science of the causes,

Animals move in many different ways. Opposite page: an ameba moves as its body flows into temporary extensions, called pseudopodia. Bottom left: a cuttlefish swims by forcing water through a special structure in its body. Bottom right: a gazelle, an animal capable of running very rapidly.

De Sazo/Rapho   Varin/Jacana

Many animals are very important economically. Above, a herd of sheep raised for their meat and for their wool.

symptoms, and nature of diease. The relations between animals and their environment is the concern of *ecology*. *Natural history* considers the life of animals in their natural surroundings. *Paleontology* takes up fossil animals, which lived in past ages and the remains or imprints of which are preserved in rocks. The study of the origin of animal life and the gradual differentiation of types is the task of *evolutionary science*. *Taxonomy* deals with the classification of animals. Animal behavior is studied in the science of *ethology*.

In addition to the above sciences, which consider animals in general, certain subdivisions of zoology have to do with particular groups of animals. Thus *vertebrate zoology* is concerned with animals with backbones; *microbiology*, with one-celled animals; *entomology*, with insects; *ichthyology*, with fishes; *herpetology*, with amphibians and reptiles; *ornithology*, with birds; and *mammalogy*, with mammals.

ANIMALS AND PEOPLE

From the earliest days, animals have served as a source of food and clothing for man. Also, animals such as the horse, the donkey, the mule, the ox, the elephant, and the camel are still important as work animals in large areas of the world.

Animals such as dogs, cats, guinea pigs, rats, and mice have been indispensable as test animals in laboratory experiments in various fields of science. Without them, we would not have sulfa drugs, antibiotics, intricate operations that save lives, vaccines, and so on. Great advances in physiology and psychology have been due in large part to the use of test animals.

Animals serve man in various other ways. Various insects, such as bees and butterflies, aid in the production of fruit by pollinating the blossoms of plants. The secretions of certain animals are valuable. These secretions include beeswax, derived from the bee, and shellac, produced by the lac insect. The carmine-red pigment of the cochineal serves for cosmetics, medicines, and dyes. Cantharidin, processed from the blood of the Spanish fly, is a medicine.

Not all animals are useful to man. Some predators are considered pests. Various insects and rodents feed on crops, on different parts of valuable trees, or on man's supply of stored foods, and they exact a high toll. Certain snakes, spiders, scorpions, and insects are dangerous because of the poison in their fangs or stings. Animal parasites, such as ticks and various kinds of protozoans, worms, and insects, cause disease in man and his livestock. Other animals, including the *Anopheles* mosquito and the tsetse fly, are carriers of disease; they harbor harmful organisms which are transmitted to man.

Sponges, left, are members of the phylum Porifera—the pore-bearing animals, the simplest of the many-celled animals. The jellyfish, above, is a coelenterate. These multicellular animals can prey actively on other animals. Their close relatives are corals and hydrozoans.

THE ANIMAL KINGDOM

Living things are generally grouped as plants or animals and classified in the Plant Kingdom or the Animal Kingdom. Animals, in general, share certain characteristics—the ability to move from one place to another and the need to obtain food from external sources. Certain one-celled organisms share the characteristics of plants and animals. These organisms are difficult to classify, and some biologists place them in a third kingdom—that of the Protista. In general, however, a classification into two Kingdom is convenient and is what we have followed in *The New Book of Popular Science*.

The subdivisions of the Animal Kingdom are given below. Important groups of animals are discussed in separate articles.

PROTOZOA PHYLUM—"First Animals"
Single-celled animals (some plantlike). Generally microscopic; some extinct forms larger. Individual or colonial; free, attached, or parasitic. Take in food or make it. Reproduce by fission, budding, spores, or sexually. Inhabit land, sea, and air.

Mastigophora Class Whip-bearers	Cell firm, with one or more flagella for movement. Some are colonial. EUGLENA, VOLVOX.
Sarcodina Class Fleshy animals	Cell with no definite shape and sending out pseudopods for movement. Some with shells. AMOEBA, GLOBIGERINA.
Sporozoa Class Spore animals	Cell firm, often with thick wall; generally no structure for movement; parasitic. GREGARINA.
Ciliata Class Animals having cilia	Cell firm with cilia or cirri to provide movement. Some reproduce sexually. Also called Ciliophora. PARAMECIUM, STENTOR.

Jacana (Kernets)

A. Dowry/PR

Suctoria Class
 Sucking animals
Cell firm, with cilia in young; adult attached and has hollow tentacles; some parasitic. PODOPHRYA.

MESOZOA PHYLUM—"Middle Animals"

Small, wormlike; bilateral symmetry; body solid. Reproduce by fission or sexually. Marine; parasitic on other marine life. DICYEMA.

PORIFERA PHYLUM—"Pore-bearing Animals"

The sponges, most primitive of many-celled animals. Saclike; walls, riddled with pores and canals, surround central cavity. Flagellated cells circulate water; digestion within cells, not in a digestive tract. Attached, colonial, or single; reproduction sexual or asexual. Marine and freshwater.

Calcarea Class
 Lime animals
Sponges with internal skelton of lime spicules; marine. Also called Calcispongiae. SCYPHA.

Hexactinellida Class
 Animals with six-pointed spicules
Sponges with inner skeleton of siliceous spicules; marine. Also called Hyalospongiae. HYALONEMA, GLASS SPONGES.

Demospongiae Class
 Common sponges
Sponges with spicules of silica or spongin or with no internal skeleton at all. Marine and freshwater. COMMON BATH SPONGE.

COELENTERATA PHYLUM—"Animals with hollow intestine"

Polyps and jellyfish. Body with radial symmetry; digestive tract present; few tissues and organs; tentacles and stinging capsules. Attached or free, often colonial. Reproduction sexual or asexual (fission or budding), often with alternation of generations, with a sexual free or attached medusa stage and an asexual attached polyp stage. May prey actively on other animals. Marine and freshwater.

Hydrozoa Class
 Water animals
Simple polyps and medusae; alternation of generations. HYDRA, OBELIA.

Scyphozoa Class
 Cup animals
Often large jellyfish; alternation of generations, but polyp stage reduced. AURELIA.

Anthozoa Class
 Flower animals
Attached polyp; no medusa stage. Some have hard external shell or skeleton. Often colonial. Marine. CORAL, SEA ANEMONE, SEA FAN.

CTENOPHORA PHYLUM—"Comb-bearers"

The comb jellyfish. Body spherical to cylindrical, with bilateral or radial symmetry; stomach and internal tubes present. Moves by means of eight platelike bands of fused cilia; often has pair of tentacles; not colonial. Reproduces sexually; sexes not separate. Marine; feeds actively on other marine animals.

Tentaculata Class
 Animals with tentacles
Body with two or more tentacles. VENUS'S-GIRDLE.

THE ANIMAL KINGDOM

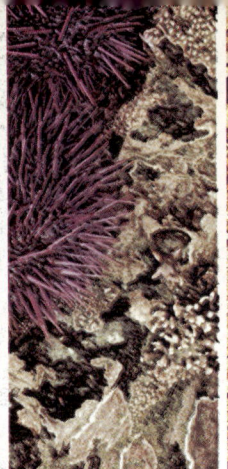

Opposite page, far left: a garden-variety worm—the earthworm. Earthworms are annelids. Middle: sea urchins. These colorful, spiny animals are echinoderms. Near left: a familiar crustacean—the lobster.

Lauros—Atlas—Photo

Nuda Class
    Naked animals — Body lacks tentacles. BEROE.

PLATYHELMINTHES PHYLUM — "Flatworms"

Unsegmented, flattened worms with no distinct head or legs; body with bilateral symmetry; paired nerve centers; no circulatory system. Reproduce asexually (by budding) or sexually; sexes not separate. Often parasitic.

Turbellaria Class
    Wrigglers — Small; move by cilia; not parasitic. PLANARIA.

Trematoda Class
    Hole animals — Young are free; adults are parasitic and are attached to host. LIVER FLUKE

Cestoidea Class
    Girdle animals — Long, narrow flatworms; hooks or suckers for attachment; no digestive tract; all parasitic. TAPEWORM.

NEMERTINEA PHYLUM — "Nemerteans" (from Neumertes, a sea nymph in Greek mythology)

Flattened worms not divided into sections, or segments; body with bilateral symmetry; circulatory or other organ systems; brain in "head." Move by means of cilia. Reproduce sexually or asexually; sexes separate. Have larval stage. Marine, not parasitic. Considered a class by some authorities. LINEUS.

ENTOPROCTA PHYLUM — "Animals with interior opening"

Small, attached mosslike forms. Body with bilateral symmetry; various organ systems; U-shaped digestive tract; tentacles with cilia surrounding mouth and waste opening. Colonial or single; reproduce by eggs; sexes separate; have larval stage. Marine and freshwater. Sometimes classed under the phylum Bryozoa. URNATELLA.

ASCHELMINTHES PHYLUM — "Sac worms"

Small, slender, often microscopic forms; free or attached. Body with bilateral symmetry; digestive tract and other organs. Reproduce by eggs; sexes separate; no larval stage. Marine and freshwater.

Rotifera Class
    Wheel-bearers — Tapering body; disc with cilia at front end for moving and feeding. Most are free-living; some are parasitic. ROTIFER.

Gastrotricha Class
    Hairy-stomach animals — Wormlike; provided with cilia; no jaws; mouth bristly. CHAETONOTUS.

Kinorhyncha Class
    Jaw-moving animals — Spiny, cylinder-shaped worms; circled by rings. ECHINODERA.

Nematoda Class — Round, tapered, thin worms; not segmented; may be parasitic. Sometimes

THE ANIMAL KINGDOM

Thread animals classed under the phylum called Nemathelminthes. ROUNDWORM, HOOKWORM.

Nematomorpha Class
Thread-shaped animals Long, very thin worms; larvae infest insects. HAIR SNAKE, HORSEHAIR WORM.

ACANTHOCEPHALA PHYLUM—"Spine-headed animals"
Elongated worms with spines at front end; body with bilateral symmetry; no digestive tract; circular body muscles; degenerate. Reproduce by eggs; sexes separate. Parasitic. Classed by some authorities under phylum Nemathelminthes. ECHINORHYNCHUS.

BRYOZOA PHYLUM—"Moss animals"
Minute, attached; form mosslike colonies. Each individual in a small hardened cup. Body with bilateral symmetry. U-shaped digestive tract and other organs; tentacles around mouth. Reproduce asexually (budding) or sexually; sexes not separate; larval stage. Marine and fresh-water. HORNERA, LICHENOPORA.

PHOEONIDEA PHYLUM—"Phoronians" (from Phoronis, a name given to Io, a maiden in Greek mythology)
Attached, unsegmented tubular worms with crown of tentacles; body with bilateral symmetry; organs well developed, U-shaped digestive tract. Colonial. Reproduce by eggs; sexes not separate; larval stage. Marine. PHORONIS.

BRACHIOPODA PHYLUM—"Arm-footed animals"
Attached shellfish; bivalve shell with bilateral symmetry; tentacular arms inside shell; organs well developed. Reproduce by eggs; sexes separate; larval stage. Not colonial; each individual attached to sea bottom by stalk. Marine. Much more common as a fossil. LAMP SHELL.

ECHINODERMATA PHYLUM—"Spiny-skinned animals"
Body with radial symmetry; usually five-armed; spaces between the arms may be filled in. Internal shell-like case of lime plates or spicules. Specialized organs; tubes for circulating water; tube feet for locomotion or breathing. Not colonial; attached or free. Reproduce by eggs; sexes separate; larval stage. Marine

Asteroidea Class
Starlike animals Body with five or more flexible arms, or rays. Active; prey on other animals. STARFISH.

Centipedes are multi-legged arthropods that live mostly on land.

Echinoidea Class
    Spiny animals
Round to flattened body with no projecting arms; smooth to spiny. Free-living. SEA URCHIN, SAND DOLLAR.

Ophiuroidea Class
    Snake-tailed animals
Small body with five long, jointed arms. Freeliving. BRITTLE STAR.

Holothurioidea Class
    Water polyps
Long, soft body, ten or more branched tentacles around mouth. Free or burrowing. SEA CUCUMBER.

Crinoidea Class
    Lilylike animals
Body usually attached to sea bottom by stalk; five to ten branched arms extending from body shell. SEA LILY, FEATHER STAR.

CHAETOGNATHA PHYLUM—"Bristle-jawed animals"

Small, elongated worms with finlike appendages; bristles around mouth; body with bilateral symmetry; nervous system and brain; no developed circulatory, waste and breathing systems. Reproduce by eggs; sexes not separate. Carnivorous: Marine. ARROW WORM.

MOLLUSCA PHYLUM—"Soft-bodied animals"

The familiar shellfish. Body unsegmented, with bilateral symmetry; mantle around body. Often with inner or outer shell, single, bivalve or compound; organs well developed; nervous system. Free or attached; not colonial. Reproduce by eggs; sexes usually separate; larval stage. Inhabit land and water.

Amphineura Class
    Nerve-circled animals
Simple, primitive forms, with convex, compound shell composed of many plates or spines. Creeping animals; marine. CHITON.

Pelecypoda Class
    Hatchet-footed animals
Bivalve shell; no head; massive footlike appendage; usually attached to bottom. Marine and fresh-water. OYSTER, CLAM, SCALLOP.

Scaphopoda Class
    Boat-footed animals
Elongated forms with tubular shell open at both ends; no head; foot and tentacles. Burrowing; marine. TOOTH SHELL.

Gastropoda Class
    Belly-footed animals
Single shell, usually coiled, or no shell. Inner organs twisted around; head, jaws and tentacles; creep on large solelike foot; some swim by fins. Found in water and on land; some breathe air. SNAIL, SLUG, CONCH.

Cephalopoda Class
    Head-footed animals
Actively swimming forms with or without inner or outer shell; straight or coiled. Many tentacles; head and eyes. Spurt out water in jets. All marine. May be tremendous in size; include largest invertebrates. CUTTLEFISH, OCTOPUS, SQUID, NAUTILUS.

ANNELIDA PHYLUM—"Ringed animals"

Worms with long, segmented bodies; bilateral symmetry. Various organs developed; brain and nerves; closed circulatory system. Often have bristles or other paired appendages for locomotion. Reproduction sexual or asexual; sexes separate or not separate; often have a larval stage. Found in seas, streams, soil. Some species parasitic.

Archiannelida Class
    Old ringed animals
Small; primitive; no appendages. Marine. DINOPHILUS.

Polychaeta Class
    Many-bristled animals
Bristles for movement; head tentacles; some forms free-swimming. Marine. SEA MOUSE, SANDWORM.

Oligochaeta Class
    Few-bristled animals
Few bristles or appendages; found in water and soil. EARTHWORM.

Hirudinea Class
    Leeches
Flattened body, with sucker at each end; usually no other appendages. Parasitic; found in streams. LEECH.

THE ANIMAL KINGDOM

SIPUNCULOIDEA PHYLUM — "Tubelike animals"

Small, nonsegmented worms; bilateral symmetry; body can be contracted; various organs present; tentacles. Reproduce by eggs; sexes separate; larval stage. Marine. Sometimes classified under phylum Annelida. PEANUT WORM.

PRIAPULOIDEA PHYLUM — "Priapuslike animals"

Thick worms, with no internal segmentation; bilateral symmetry; front end enlarged and provided with spines. Various organs, but many disappear in adult stage. Reproduce by eggs; sexes separate. Burrow in mud; marine. Sometimes classed under phylum Annelida. PRIAPULUS.

ECHIUROIDEA PHYLUM — "Adder-tailed animals"

Thick, bottle-shaped worms; adult body not segmented; show bilateral symmetry. Various organs present; body bristles; long snout or proboscis. Reproduce by eggs; sexes distinct, but male may be parasitic in female. Burrowing; marine. Sometimes classified under phylum Annelida. URECHIS.

ARTHROPODA PHYLUM — "Joint-footed animals"

Body with bilateral symmetry; usually divided into distinct segments, which may be absent, inconspicuous or fused together. Numerous paired appendages; nerve cord and brain; eyes simple to compound or may be absent. Antennae may be present. Body and limbs often covered with hardened outer skeleton or case, which is periodically shed to allow for growth. Usually reproduce by eggs; sexes generally separate; larval stage often occurs.

Crustacea Class Shell animals	Head and thorax united; antennae; jaws and other appendages, some branched; shell hard. Mostly marine and fresh-water. CRAB, LOBSTER.
Arachnoidea, or Arachnida Class Arachne animals	Eight legs and other appendages; often supplied with poison gland; bite or suck prey or host. Live mostly on land. SPIDER, SCORPION, TICK.
Xiphosuida Class Sword-tailed animals	Head and thorax fused; twenty-four appendages; unsegmented belly with twelve legs; long, pointed tail; broad horseshoe-shaped shell. Marine. Also known by class name of Merostomata. Sometimes classed under the Arachnoidea. KING (HORSESHOE) CRAB.
Chilopoda Class Lip-legged animals	Head and long, segmented body, antennae; many legs; live mostly on land. CENTIPEDE.
Diplopoda Class Double-footed animals	Animals with short head and thorax and many legs. Live mostly on land. Sometimes classed under Chilopoda. MILLIPEDE.
Onychophora Class Claw-bearing animals	Long, wormlike, with no distinct head; many legs; live on damp soil. PERIPATUS.
Pycnogonida Class Thick-kneed animals	Body small; segments may be fused no jaws; head appendages; eight to twelve long legs. Mostly marine. Sometimes classed under Arachnoidea. SEA SPIDER.
Pentastomida Class Animals with five openings	Small, wormlike, unsegmented; head and thorax fused; no appendages; no hard covering; internal organs absent or reduced; mouth provided with hooks. Parasitic on vertebrates. Also classed under Arachnoidea. LINGUATULA.
Tardigrada Class Slow-moving animals	Tiny, sluglike, unsegmented; eight clawed feet; no blood vessels or breathing organs. Marine and fresh-water. Sometimes, classed under Arachnoidea. WATER BEAR.

Symphyla Class
    Grouped-together animals

Long, colorless, wormlike; no eyes; antennae; jaws; twenty-four legs. Inhabit damp ground in gardens and feed on vegetables. Also classed under Chilopoda. GARDEN CENTIPEDE.

Pauropoda Class
    Small-footed animals

Small, segmented, wormlike; no eyes, antennae; eighteen to twenty legs. Inhabit damp soil under rocks and logs. Sometimes classed under Chilopoda. PAUROPUS.

Insecta Class
    Cut-in animals

Most successful and advanced of the invertebrates. Have head, thorax and abdomen; six legs and usually wings attached to thorax; some wingless varieties. Antennae; jaws and mouth varied; eyes simple to compound or absent. Lay eggs; found everywhere.

CHORDATA PHYLUM — "Animals with notochord"

Body with bilateral symmetry; minute to huge in size. Nerve cord with supporting notochord along back, enclosed with bone in higher forms; brain in most; organ systems may be well developed, particularly nerve system; appendages may occur. Reproduction usually sexual, asexual in lower forms; sexes separate or combined; larval stage may be present. Found almost everywhere.

Hemichordata Subphylum
    Animals with half a notochord

Wormlike forms with short notochord at front end of body; gills. Marine. TONGUE (ACORN) WORM, CEPHALODISCUS.

Tunicata Subphylum
    Tunic animals

Adult often with outer covering; free or attached; sometimes colonial; few chordate characters in adult but present in larva, which has notochord in tail. Reproduce sexually or by budding. Marine. Also called Urochordata. APPENDICULARIA, ASCIDIAN, SALPHA.

Cephalochordata Subphylum
    Head-notochord animals

Fishlike forms with complete notochord and nerve cord; gill slits; no true head; fins. Marine. LANCELET.

Vertebrata Subphylum
    Animals with backbone

Bone segments around nerve cord (spinal column); notochord absent in higher forms; internal bone skeleton; head, brain, senses well developed. Breathe by gills or lungs. Reproduction mostly sexual; lay eggs or bear live young.

Agnatha Class
    Jawless animals

Fishlike forms with no jaws and no scales; skeleton cartilaginous; spinal column very simple; toothed mouth, round and suckerlike. Mostly parasitic. Extinct armored forms were first vertebrates. Lamprey, Hagfish.

Chondrichthyes Class
    Soft-boned fish

Jawed fish with cartilaginous skeleton; scales; fins paired. Marine. Shark, Ray.

Osteichthyes Class
    Hard-boned fish

True jawed fish with bony skeleton; scales; fins paired; gills; air bladder.

Amphibia Class
    Land-and-water animals

Mostly four-legged, adapted to life in water and on land; some limbless; lay eggs in fresh water. Larva has gills; adult with lungs or gills.

Reptilia Class
    Creeping animals

Completely adapted to land living, though some have returned to water; skin naked or scaly; four limbs or legless; skeleton bony; heart incompletely four-chambered; lungs. Lay eggs.

Aves Class
    Birds

Blood at constant temperature (warm-blooded); heart four-chambered; lungs; forelimbs serve as wings; some flightless; walk on hind legs; jaws developed into beak, usually toothless; skin with feathers. Lay eggs.

Mammalia Class
    Animals with breasts

Generally four-limbed body covered with hair; blood temperature constant; four-chambered heart; lungs; skull bones fused; brain highly developed. Most bear young alive and suckle them with milk from breasts. Sexes separate.

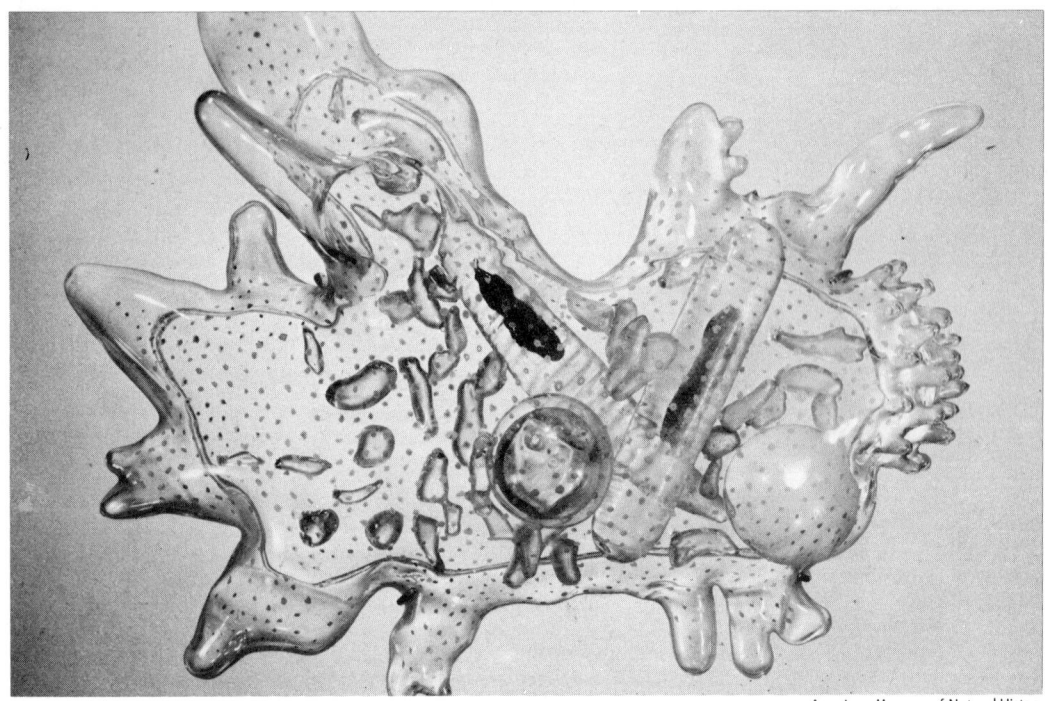

American Museum of Natural History

Glass model of *Amoeba proteus*. This protozoan, which is common in fresh water, resembles a mass of living jelly. It moves about by extending fingerlike projections of its substance and flowing into them.

THE PROTOZOANS

When a drop or two of pond water is viewed through a microscope, a fascinating world in miniature is revealed—a world populated by a host of minute organisms. Many of them are tiny creatures called protozoans, which range in size from several microns to a few millimeters. (A micron is equal to 1/1,000 of a millimeter.)

Protozoans have often been described as single-celled animals, with a simple organization. This definition is not quite exact.

For one thing, a certain number of protozoan species are plantlike in one important respect: they possess pigments by means of which they can manufacture foods through the process of photosynthesis. These species are often classified as algae. Yet they are closely related to other protozoan species that lack pigments, cannot manufacture food, and must obtain it after the fashion of animals.

Perhaps it would be best to consider protozoans, together with the algae, as connecting links between the plant and animal kingdoms.

Is the protozoan single-celled? It is true that most protozoans consist of a single unit of protoplasm containing a nucleus and surrounded by a cellular membrane or wall of some kind. But some forms, as we shall see, have a great many nuclei. Others consist of colonies made up of a number of individuals.

The protozoan can hardly be called a simple organism. This becomes evident when we compare it with the specialized cells in the bodies of multicellular, or many-celled, animals. In the latter, each type of specialized cell forms a different kind of tissue. The tissues make up the different organs by means of which the animals move about, catch and chew and digest food, circulate body fluids, breathe, eliminate wastes, reproduce, and so on. In the protozoan, all these activities occur within the tiny blob of protoplasm of which its body consists. Some of these processes take place in special structures called *organelles.* The specialized cells of multicellular animals have no independent life of their own. The protozoan is completely self-sufficient.

It is because they are comparatively complex that protozoans can adapt themselves to a wide range of environments. They flourish in fresh or stagnant permanent ponds, in semipermanent rain-water ponds, in marshes, and in streams. Many inhabit mud, moist soil, brine pools, and hot springs where temperatures range from 35° to 65° Celsius. Protozoans are abundant in the seas, from the open surface waters to the bottom muck. They are found even in snow drifts. Many species live as parasites in the body cavities, tissues, and cells of animals and plants.

There are at least 15,000 to 20,000 protozoan species. They have not yet been adequately classified. However, we can divide protozoans into five major groups, or classes: flagellates, amebalike protozoans, spore-producing protozoans, ciliates, and suctorians.

Algae and fungi are quite closely related to some of these forms. The multicellular animals probably evolved from the flagellates or ciliates.

THE FLAGELLATES

The flagellates form the class Mastigophora. They possess one to several long filaments called *flagella* (singular: flagellum). These are usually attached to the front end of the body.

A flagellum may be used for swimming or for creating water currents that bring in food. It may also serve as a sensitive organelle for exploring the environment. In swimming, the flagellum ordinarily makes first a sidewise or backward beat and then a relaxed recovery stroke to the forward position again. This causes the organism to move forward, often in a more or less spiral path.

The body of the flagellate usually assumes a definite shape and is covered by a firm *pellicle,* or skin. Some species may be encased in a shell, a cover of plates, or some other kind of armor. Often a flagellate will develop *pseudopods,* or "false feet," which are formed by a flowing of the organism's protoplasm. Such a flagellate moves as its protoplasm streams into the newly formed pseudopods. These "false feet" are only temporary extensions of the protoplasm.

Flagellates are divided into plantlike and animallike forms. The plantlike forms possess *chromatophores,* structures that contain the green pigment chlorophyll. In many species this green color is masked to some extent by red, brown, or yellow pigments. The plantlike flagellates manufacture their own food—carbohydrates and protein—from carbon dioxide, water, salts, and the energy derived from sunlight.

The flagellates readily form cysts. That is, they become rounded and secrete a more or less impermeable membrane over the body surface. In this stage the organism remains alive even though the cyst may be thoroughly dried and blown about by the wind. Many species of protozoans are widely distributed over the world because they are transported as cysts by the wind or on the feet of wading birds.

Besides forming protective cysts, many plantlike flagellates enter upon what is called the *palmella stage.* In this, the body becomes round and the flagella are

A typical dinoflagellate. The body wall is grooved by an encircling girdle and several long flagella extend out from the main part of the body.

Eric V. Grave/PR

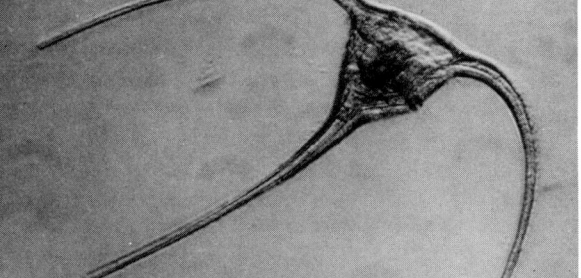

THE PROTOZOANS   191

Marine Research Laboratory, Florida Dept. of Natural Resources

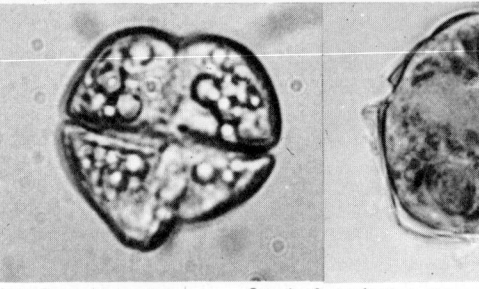

Marine Research Laboratory, Florida Dept. of Natural Resources

Cape Ann Society for Marine Sciences, Inc., Gloucester, Mass.

Some species of dinoflagellates, such as the two shown above, *Gymnodinium breve* (left) and *Gonyaulax tamarensis* (right) produce poisons that discolor the water, producing a "red tide" that can kill enormous numbers of fish. Such a red tide occurred in Florida's Tampa Bay.

lost. The organisms then grow and reproduce by fission (splitting the body in two), forming extensive green scums on ponds and other bodies of water.

Plantlike flagellates. The plantlike flagellate *Chromulina* is quite simple. This animal, commonly found in fresh water, is naked—that is, it has neither pellicle, nor shell, nor other such covering. It is spherical or oval in shape. It has only one flagellum and one or two yellow-brown, bandlike chromatophores. Sometimes *Chromulina* occurs in such numbers that the water is colored a golden brown. This flagellate is apparently closely related to certain brown algae and to the diatoms.

Cryptomonas is a somewhat more complex organism, having two flagella, a single yellowish to brown-green chromatophore, and a distinct gullet. Its body is flattened, oval in shape, and surrounded by a firm pellicle.

The dinoflagellates are unique among plantlike flagellates in that the body wall is grooved by an encircling furrow, or girdle, and a longitudinal furrow. Each of these furrows contains a flagellum. Usually there are two pink cavities called *vacuoles* within the body. Canals lead from them to the outside. The vacuoles serve to draw in nutritious liquids and possibly solid particles. The chromatophores, when present, range from yellow-brown to blue-green. The body surface may be naked, or covered with a thin cellulose wall, or armored with a thick cellulose wall that is divided into plates. Most dinoflagellates are marine organisms. They form part of the plankton—the passively floating or weakly swimming plant and animal life of a body of water. Many dinoflagellates produce poisons that kill enormous numbers of fish.

The plantlike flagellate *Chlamydomonas* belongs to a freshwater group in which the chlorophyll, when present, is not masked by other pigments. These organisms, therefore, are usually grass-green in color. *Chlamydomonas* is small and oval-shaped and has two flagella. A red form of *Chlamydomonas* is found in melting snow, giving the snow a red color.

Eudorina is an odd-looking flagellate. It is made up of 32 individual organisms arranged, in loosely packed form, near the surface of a jellylike sphere. *Eudorina* is closely related to *Volvox*, which is generally considered to be an alga.

Euglena is another freshwater green flagellate that is sometimes classified as a plant. There are many species of *Euglena*; they are usually spindle-, cigar-, or oval-shaped. One flagellum arises from the wall of a flask-shaped gullet at the front end of the body. In the presence of light, *Euglena* uses its chorophyll to manufacture carbohydrate. In darkness, or if it loses its chlorophyll, *Euglena* must absorb organic acids or other organic sources of carbon in order to produce its carbohydrate.

Peranema is very much like *Euglena* but it is colorless and has a mouth opening. A pair of rodlike structures extend alongside the gullet and serve to support the mouth while food is being drawn in. This rod apparatus may also be used for piercing *Peranema's* prey.

Animallike flagellates. None of the animallike flagellates have chromatophores and, therefore, they cannot manufacture their own food. Some forms, especially the parasitic ones, absorb dissolved food materials from the medium in which they live. Most species, however, feed on microorganisms or nutritious particles in the water. This food material is digested in food vacuoles that form within the flagellate's protoplasm. In general, the body shape of the animallike flagellates is rather plastic, with no cellulose wall.

Among the most important flagellates, as far as people are concerned, are several species of the genus *Trypanosoma*. *Trypanosoma gambiense* is the cause of sleeping sickness. It has a slender, curving body that tapers at both ends. The single flagellum is attached throughout the length of the body by a thin layer of pellicle. This forms a membrane that undulates when a wave passes down the flagellum. Trypanosomes, as members of the genus *Trypanosoma* are called, are sucked up with the blood when a tsetse fly bites an infected host. The trypanosomes undergo development first in the gut and then in the salivary glands of the fly. Finally, when the fly bites another victim, the flagellates are introduced into the blood stream of the new host. They invade the lymph nodes and sometimes the cerebrospinal fluid (a fluid in the brain and spinal cord), causing sleeping sickness.

Trypanosoma cruzi is the cause of Chagas' disease, or South American trypanosomiasis, which affects the muscles, heart, and nervous system. Other trypanosomes are found in the blood of fish, amphibians, reptiles, birds, and mammals.

Bodo, a somewhat different flagellate but apparently related to *Trypanosoma*, is a small, oval-shaped organism that inhabits stagnant fresh water. It has two flagella, one of which is trailed in swimming.

In fresh water we find two strange kinds of flagellates, *Codosiga* and *Protospongia*. These organisms have an oval-shaped body surmounted by a collar which encircles the base of the single flagellum. The collar, a membrane made of protoplasm, is a device for obtaining food. Food particles or bacteria adhere to the collar and slowly pass down it to the body proper. In *Codosiga*, a number of these transparent collar cells, as they are called, cluster at the end of a simple or branching stalk. *Protospongia* is a colony of from 6 to 60 organisms embedded irregularly in a gelatinous mass. The collar cells occur at the surface of the mass. The organisms on the inside are collarless. The only other animals with collar cells are the sponges. It may be, therefore, that the sponges evolved from organisms similar to *Protospongia*.

Various flagellate species belonging to the genus *Trichomonas* inhabit the intestines of vertebrates. They feed mostly on

Euglena are common one-celled organisms commonly found in freshwater ponds, often forming a green scum on the water's surface. They have plant characteristics: chloroplasts and the ability to carry on photosynthesis, as well as animal characteristics: a long whiplike flagellum used for movement.

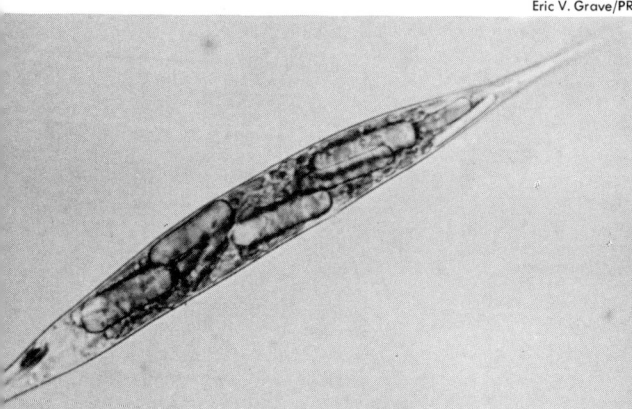
Eric V. Grave/PR

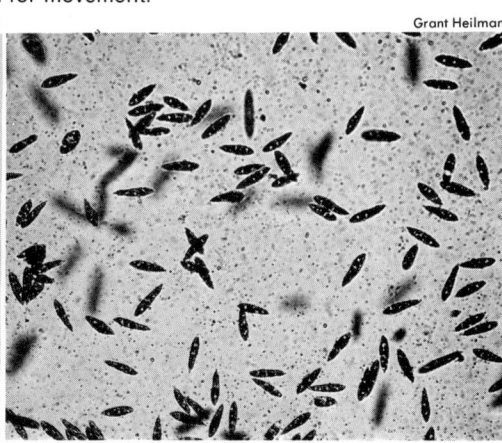
Grant Heilman

bacteria and yeasts found in this environment. In human beings, different forms of *Trichomonas* are found in the mouth, colon, and vagina. A typical *Trichomonas* is small, oval in shape, and has four free flagella; a fifth flagellum is attached to an undulating membrane.

The large animallike flagellate *Trichonympha* lives in the gut of termites. Here it digests wood fragments swallowed by the termite and makes some of the products of this digestion available to the insect. If the termite loses its flagellates, it will die of starvation, even though it continues to swallow large quantities of wood. *Trichonympha* is a bell-shaped organism covered with a great number of flagella. It is one of the most complex of the flagellates. Closely related forms inhabit the alimentary canal of cockroaches and woodroaches.

AMEBALIKE PROTOZOANS

The amebalike protozoans, or Sarcodina, float or creep about in their liquid environment. A thin membrane surrounds the protoplasm of the body, allowing for the formation of pseudopods. These "false feet" are used both for movement and for capturing food. Some species are plastic, naked organisms. Others develop internal or external skeletal structures that protect the body and give it some rigidity. The Sarcodina live almost entirely on small organisms such as other protozoans, tiny multicellular animals, and algae.

Perhaps the best-known of the Sarcodina is *Amoeba,* a freshwater form. It is quite large as protozoans go, ranging up to three-fifths of a millimeter in size. *Amoeba* puts forth one to several fingerlike pseudopods. The protoplasm of the body contains a nucleus, numerous food vacuoles, granules, crystals, and a *contractile vacuole.* The contractile vacuole is a sort of water pump. Since *Amoeba* lives in fresh water, water diffuses into its body from the external environment. The contractile vacuole pumps this excess water out, thus preventing *Amoeba* from swelling unduly and perhaps bursting. Contractile vacuoles are found in all freshwater protozoans.

The giant ameba, *Chaos,* has several hundred small nuclei. This animal may grow to be as much as five millimeters across.

Various amebas are found in people. *Entamoeba histolytica* occurs in the human large intestine and is responsible for the disease known as amebic dysentery. This ameba secretes a substance that dissolves the intestinal lining. The ameba then enters the connective tissue and muscular layers, where it feeds on blood cells and tissue-cell fragments. It may invade the liver, where it

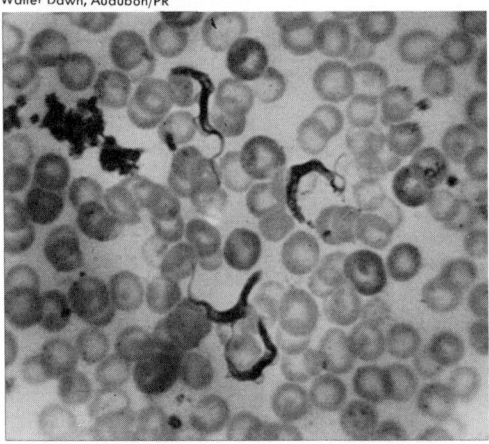

Trypanosoma gambiense, the slender, curving flagellate that causes sleeping sickness. Magnified 2,500 times in this photo.
Walter Dawn, Audubon/PR

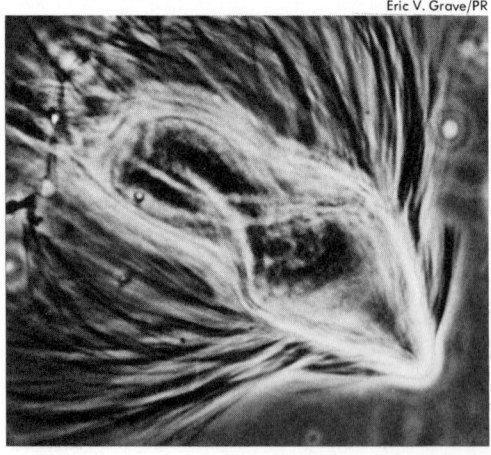

Trichonympha flagellates live in the guts of termites, digesting the wood fragments swallowed by the termites as they feed.
Eric V. Grave/PR

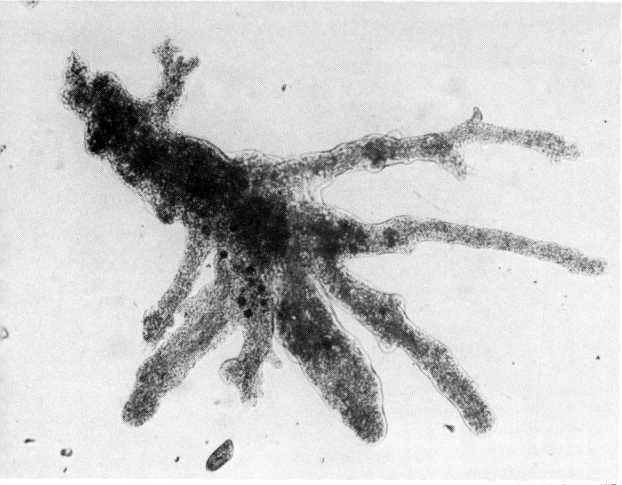
The giant ameba *(Chaos chaos)*. Two small paramecia are swimming from the left toward the ameba and may be quickly captured by the ameba.

does great damage by causing abscesses. Another *Entamoeba* species seems to contribute to pyorrhea, one of the most serious disorders of the teeth and gums.

Arcella and *Difflugia* are amebas that house themselves in single-chambered shells. *Arcella* secretes a transparent to yellowish-brown shell that is made of tiny prisms, fitted together. The shell is formed of siliceous, or silica-containing, substances. It is domelike above and concave below, with a central opening through which the pseudopods extend. *Difflugia* fashions a globular or flask-shaped shell out of sand grains, which it "glues" together. Both of these shelled amebas occur in fresh water and moist soils.

The *foraminiferans* are shelled amebas that live almost exclusively in the sea. The shells usually contain many chambers, for the young organism starts life with one chamber and adds new chambers as it grows. Many foraminiferans secrete siliceous or calcareous (containing calcium carbonate) shells. Others use foreign materials that are picked up by the pseudopods and pulled into the body, where they are held together in the shell wall by a secreted cement. The chambers of the shell may be connected like a string of beads; they may be coiled in a flat or conical spiral like a snail's shell; they may be in two or three alternating rows like a braid; or they may be arranged irregularly. The shell is just inside the body of the foraminiferan, so that a layer of protoplasm covers it. The pseudopods form a meshwork, which traps the small organisms the foraminiferan eats. Among the best-known foraminiferans are those belonging to the genus *Globigerina*. Their shells form a thick deposit, called globigerina ooze, on the ocean bottom.

Other Sarcodina groups are the *heliozoans* and *radiolarians*. The heliozoans are freshwater organisms with spherical bodies that are almost frothlike in appearance. Numerous stiff, long, and thin pseudopods radiate from the body. The radiolarians are floating marine animals. They differ from heliozoans in that the body is divided into an inner central capsule and an outer layer, both of protoplasm. They also have radiating and stiff pseudopods. The radiolarians secrete skeletons of silicon or strontium sulfate, which take the form of radiating spines or latticed networks. The skeletons assume the form of spheres, helmets, disks, bells, and various other shapes. They form a mud deposit known as radiolarian ooze on the ocean floor.

SPORE-FORMING PROTOZOANS

These make up the class Sporozoa. They are all parasitic protozoans with a very complicated life cycle. Spores are produced at some stage or other of the cycle. The Sporozoa spore is a cell that is usually surrounded by a resistant membrane; it is called a *sporozoite*.

In general, sporozoans are incapable of locomotion, though the young sometimes move by means of pseudopods. These protozoans absorb from their hosts such food materials as dissolved protoplasm, body fluids, or tissue fluids.

The most famous sporozoan is *Plasmodium*, the malarial parasite. Naked sporozoites, exceedingly small and spindle-shaped, are inoculated into a person's bloodstream by the bite of an infected female *Anopheles* mosquito. Eventually the parasites enter the red blood cells. *Plasmodium* also causes malaria in birds, reptiles, frogs, monkeys, apes, bats, squirrels, buffalo, and antelopes.

Babesia bigemina, whose sporozoites are inoculated by the bite of the tick, is responsible for Texas cattle fever. *Nosema bombycis* brings about pebrine, the fatal disease of silkworms. *Eimeria* infects chickens, causing coccidiosis. Sporozoan parasites, as a matter of fact, are found in almost any animal you can think of.

THE CILIATES

The ciliates, or Ciliophora, are so called because they are provided with hair-like processes called *cilia*, which serve for locomotion.

The cilia are shorter and much more numerous than flagella. They are arranged on the ciliate's body in diagonal or horizontal rows. Their action suggests that of the oars in a multi-oar racing hull. The cilia push backward for the power stroke. In the recovery stroke they return to the forward position. Since at any one time some of the

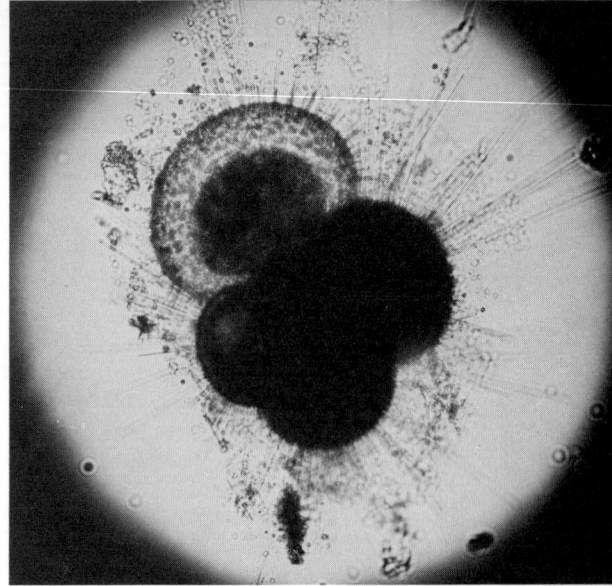
Walter Dawn, Audubon/PR

Globigerina, a foraminiferan, or shelled ameba. Their shells form a thick ooze on the ocean bottom.

Paramecium (left) is the well-known slipper-shaped ciliated protozoan commonly found in fresh water. *Paramecia* reproduce sexually and asexually—sexually by conjugation (below) with two individuals exchanging hereditary material before each divides; asexually (bottom) by fission.

both photos, Hugh Spencer, Audubon/PR

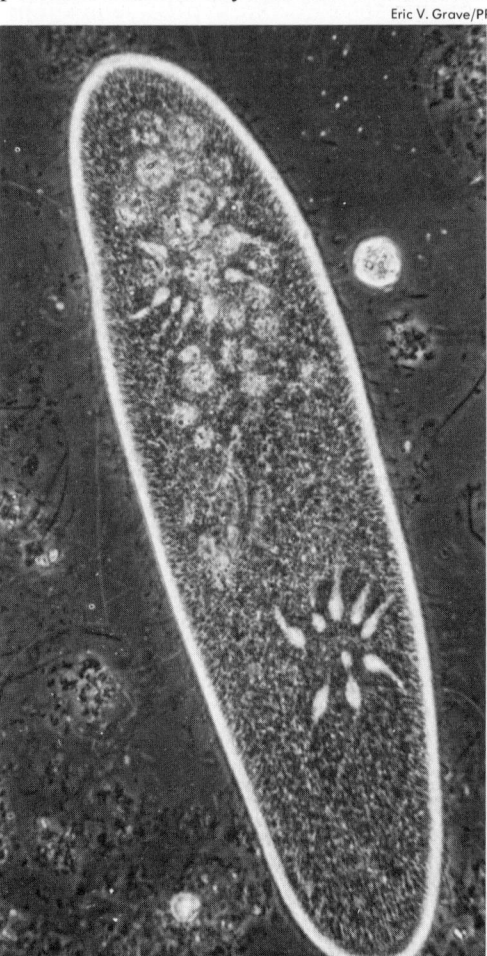
Eric V. Grave/PR

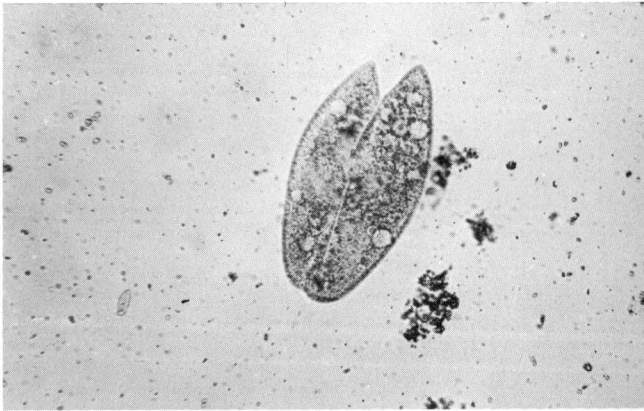

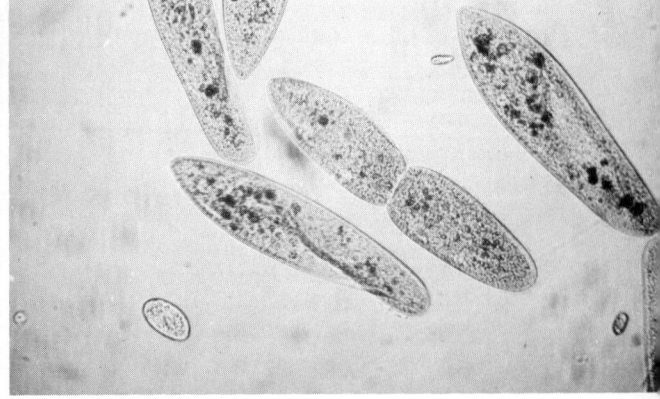

196   THE PROTOZOANS

cilia are engaged in the power stroke and some in the recovery stroke, the result is a continuous flow of power, and the animal moves ahead smoothly.

Though the body of a ciliate may twist or turn to some extent, or lengthen or shorten, it still retains a more or less permanent shape. The ciliates feed on dead organic matter or on various live microorganisms.

Opalina is a very much flattened ciliate possessing many nuclei of the same size. It is found in the intestine of frogs. The rest of the ciliates possess two kinds of nuclei: small ones, or micronuclei, and large ones, or macronuclei. The macronucleus seems to control many of the metabolic activities of the animal; the micronucleus is concerned with the reproductive process. Most ciliates possess a mouth. All of the freshwater species have contractile vacuoles.

Paramecium is a well-known slipper-shaped ciliate found in a fresh water. It has a uniform covering of cilia and a long groove leading into the mouth. Some species of *Paramecium* may reach a length of a third of a millimeter. *Paramecium* often falls victim to one of its relatives, *Didinium.* This is a barrel-shaped organism, at the front end of which there is a projecting cone, with the mouth at the tip.

In one group of ciliates—order Peritrichida—we find a disk-shaped head end, with two or more rows of cilia surrounding the mouth. The beat of the cilia creates water currents that bring food into the mouth. There are few, if any, cilia on the rest of the body. Most species of this group are fixed to some attachment point by means of a stalk. *Vorticella,* a bell-shaped organism, is a common ciliate of this type.

Blepharisma is a typical representative of the ciliate order Spirotricha. Its elongated-oval body is covered with cilia for swimming. Rows of fused cilia surround the mouth and help push food toward it. *Spirostomum,* a common freshwater relative of *Blepharisma,* has an elongated, cylinder-shaped body from one to three millimeters in length; it is a giant among protozoans. *Stylonychia,* another freshwater form, has an oval body, with a flat bottom surface and a convex top surface. This ciliate has cir-

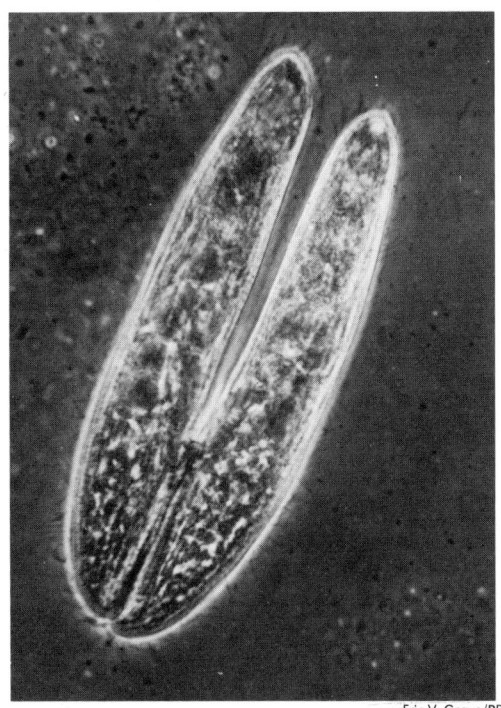

Blepharisma conjugating. Conjugation is a primitive form of sexual reproduction that does not involve true sex cells.

ri—large, stiff bristles composed of fused tufts of cilia; they are moved like legs in walking.

Epidinium, a related form, is found in the digestive tracts of cattle and reindeer. It has special structures for moving about, feeding, swallowing, digesting, excreting, and contracting, and nuclei for maintaining metabolism and controlling reproduction. There is an anus and skeletal plates for maintaining the body's form.

THE SUCTORIANS

The suctorians comprise the class Suctoria. They are common in fresh and salt water. The young are ciliated and free-swimming. Adults have no cilia and are attached by stalks to inanimate objects, plants, or small animals. They lack a mouth. They have tentacles with which they seize their victims—small ciliates—whose protoplasm they suck. Suctorians may be spherical, conical, or branched in shape. Some are parasitic.

SPONGES AND COELENTERATES

by Francis J. Ryan and Elizabeth J. Ryan

Orange and blue cyaneas: in centuries gone by, they may have been responsible for some of the sea-serpent stories told by frightened sailors. Their bodies are more than two meters in diameter. Long tentacles trail behind the body. But cyaneas aren't sea serpents. They are giant lion's mane jellyfish. However, sailors—and swimmers—are right to be afraid, for the sting of a cyanea can be very dangerous to people.

Fortunately, relatively few people come in contact with cyaneas. But if you have swum in the ocean or walked along its coast, you may have seen other, smaller jellyfish and their relatives, the anemones. You may also have seen sponges, though it's probable that you did not recognize them, since most living sponges do not look at all like the bath sponges sold in stores.

Jellyfish, anemones, and sponges are among the simplest of the multicellular, or many-celled, animals. Like all multicellular animals, different cells in each organism have different functions. Some are responsible for getting food, others for protection. Some specialize in movement, others in reproduction.

SPONGES

At first glance sponges resemble odd gelatinous plants. They are fan-shaped or dome-shaped, formed like vases, bowls, goblets, or trumpets, branched like trees, or flattened out in lichen fashion. They are animals, however, although very unusual ones. Basically, the sponge is a hollow tube, attached at one end to a support and open at the upper end. Some sponges exist in colonies of many individuals united to each other at their bases.

Sponges are many-celled animals lacking specialized organs and incapable of movement. A layer of flattened, protective cells covering the body's surface is perforated by tiny pores. This feature is responsible for the phylum name, Porifera, which means "hole-bearing".

The pores open to canals that run through the jellylike substance, or *mesenchyme,* of the body. The canals, in turn, open into a large central cavity. In the more complex sponges, these canals lead to spherical chambers and pass from them into the cavity.

Lining the chambers or central cavity are *collar cells,* or *choanocytes.* Each has a collar of protoplasm, which encircles the base of a whiplike structure known as a *flagellum.* As the flagellum undulates, it creates a current of water. The waving of many flagella causes water to enter through

Not all sponges are the bathtub variety. This one is a basket sponge, a member of the class of glass sponges.

Chuck Nicklin, Woodfin Camp

the sponge's pores, circulate in the canals and chambers, and flow into the central cavity. Microscopic plants and animals and organic debris are brought in with the water. The food particles are drawn to the collar cells, where they are engulfed. They are then digested or passed on to cells that creep about like amebas in the mesenchyme. Incoming water currents also bring oxygen to the cells.

After water circulates in the central cavity, it passes out through a large opening, the *osculum*. Carbon dioxide and other wastes discharged by the cells are eliminated in the escaping water.

A skeletal framework supports the soft mass of the sponge. This prevents the canals from collapsing and allows considerable growth. (Some sponges may be almost two meters high.) Special mesenchyme cells secrete this skeleton, which is made up of needles, called *spicules,* or of protein fibers, known as *spongin*.

Spicules may be either straight or curved and are often pronged. They may be sharply pointed, knobbed, or frayed at the ends. Often they project beyond the body, making the sponge appear bristly. The spicules, together with the unpleasant secretions and odors produced by the sponge, protect it from enemies.

Sponges probably evolved from an aggregate of individual protozoans, perhaps the choanoflagellates, which are much like the sponge's collar cells. The sponge has no sense receptors, no nervous system. However, it can close its pores and osculum and contract its body cells when harmful substances are in the water. A unique animal indeed, the sponge is set apart from all other many-celled creatures by its simple structure and the somewhat specialized but un-coordinated cells of its body.

REPRODUCTION

Sponges may reproduce by sexual means. Special mesenchyme cells increase in size and become egg cells. Others divide into sperm cells, which are discharged into the water. The egg remains in the mesenchyme, where it is fertilized by a sperm from another sponge.

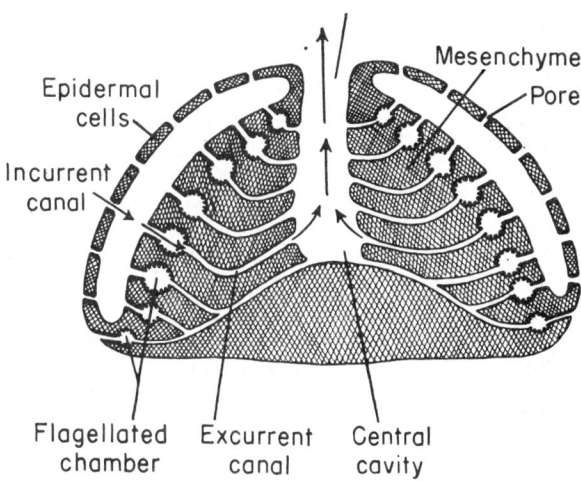

Cross section of a typical freshwater sponge, showing incurrent and excurrent canals, pores, and osculum.

The fertilized egg becomes a flagellated larva (a larva with flagella), which escapes through the osculum and swims away. Soon the tiny larva attaches to a support and begins to grow as a young sponge.

Diagram of a typical colonial sponge, which develops tubular branches. The branches may grow to a height of two meters.

all diagrams, Gen. Biol. Supply House, Inc.

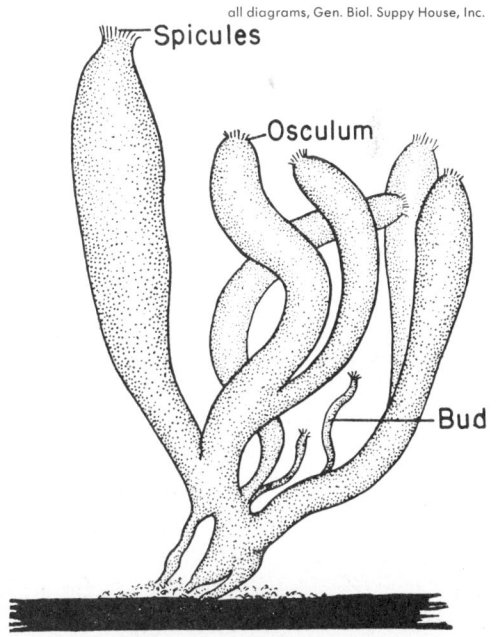

SPONGES AND COELENTERATES

Reproduction may also be by *budding*. Cells grow out from the body and develop into miniature sponges. Depending on the species, these either remain attached as members of a branching colony or drop off from the parent to lead an independent life.

Some sponges form internal buds called *gemmules*. These are small masses of food-enriched mesenchyme cells, protected by a resistant coat and often strengthened with spicules. Gemmules are commonly produced by freshwater sponges—those dull-colored or greenish irregular blobs that grow on submerged leaves and watersoaked logs in the clean water of lakes, ponds, and streams. When the parent sponge dies, the gemmules withstand freezing and drying. They grow into adults when conditions for sponge life again become favorable.

TYPES OF SPONGES

Sponges are grouped in three classes, depending on the type of skeleton they possess. Members of the class Calcispongiae secrete spicules of calcium carbonate. Members of the class Hyalospongiae have siliceous spicules—that is, spicules made of silica, which is familiar to us as quartz or sand. They are known as glass sponges. Deep-sea glass sponges form a skeletal network suggesting spun glass.

The third class, Demospongiae, have either siliceous spicules, a skeleton of spongin, or a framework made of both spicules and spongin. Some have no skeleton at all. The Demospongiae are the most common sponges. They include the freshwater sponges and the boring sponge, which protects itself by etching its way into rock and mollusk shells. The sponge that we use in our homes—the bath sponge—is not the whole animal but only its framework of spongin, which is elastic, chemically inert, and similar to silk and horn.

One genus of sponge, *Suberites,* grows on empty snail shells that house hermit crabs. The sponge absorbs the shell. Thereafter it serves as a covering for the crab as the crustacean moves about. This arrangement is mutually beneficial. The crab gains protection; the sponge is transported from one place to another by the crab and thus comes in contact with new sources of food.

Most sponges are veritable "apartment houses" for a host of animals. Marine worms, pistol crabs, shrimps, and slender fishes find a haven in the canals and chambers; barnacles attach themselves to the surface. They all get food from the water passing through the sponge's body.

COELENTERATES

The phylum Coelenterata includes the hydroids, jellyfish, and sea anemones. The phylum name comes from Greek words meaning "hollow intestine."

Like sponges, coelenterates are essentially tubular animals with a central cavity. Here food is digested as well as circulated; therefore, it is called the gastrovascular (digestive and circulating) cavity.

Food enters by the mouth at the upper end of the body and is broken down by enzymes secreted by gland cells. The nutrients then diffuse into the body. Often food is only reduced to particles, which are engulfed and fully digested, as with the sponges, by certain cells lining the cavity. The beating of flagella gives rise to water

Kitchen-Kinne, Audubon/PR

Right: several types of coral in a tropical environment. A colony of sea whips is in the foreground, a brain coral behind them. Opposite page, left: a branch of a tropical stony coral, showing expanded polyps and a contracted polyp.

currents, which bring in food particles and oxygen. Countercurrents carry wastes out through the mouth.

The outer surface of the body consists of a layer of tightly packed protective cells. Interspersed in this layer are sensory cells, sensitive to touch and chemical substances, and specialized cells called thread capsules, or *nematocysts*. Each nematocyst contains a fluid under pressure and a spirally coiled, hollow thread. When the capsule is stimulated by touch, and possibly by chemicals as well, the thread is forcibly ejected. Some of these threads pierce the coelenterate's prey and then inject a benumbing poison into it. Other threads either stick to the prey or wrap around its appendages. Nematocysts are very abundant on the coelenterate's tentacles, which grow as a crown around the mouth. Once prey is paralyzed and held fast by the threads, the tentacles enfold it and draw it into the mouth.

Between the surface cells and those lining the gastrovascular cavity is a layer of supporting jellylike substance. Mesenchyme cells are found in this layer. These are unspecialized cells that form nematocysts and sex cells. Numerous nerve cells lie below the outer covering and join together to form an extensive nerve net.

The animal just described—a sort of living tube crowned with tentacles—is known as a *polyp*. The different types of coelenterates are merely variants of this polyp form. They are grouped into three classes: Hydrozoa, the hydroids and freshwater hydras; Scyphozoa, the jellyfishes; and Anthozoa, the sea anemones and corals.

THE HYDROIDS

Hydroids consist of hundreds of tiny polyps united by a stalk to form a branching colony. The gastrovascular cavity of each polyp joins with that of the stalk so that there is a cavity common to the entire colony. Often the stalk and polyps are held erect and protected by a horny sheath. Hydroids feed on minute worms and various small crustaceans.

A hydroid colony resembles a fern or other "feathery" plant. Such colonies are commonly found attached to wharf pilings, rocks, and kelp. The colony reproduces by giving off buds. Some buds form a mouth and a circlet of tentacles at the tip to become what are known as feeding polyps. Others develop into reproductive polyps; they have neither mouth nor tentacles but produce tiny saucerlike appendages. When mature, these saucers break off and swim away. They are called *medusas*. They look like miniature jellyfish with tentacles hanging from the rim of the saucer.

The function of the medusas is to reproduce sexually. Sex cells, on the underside of the saucer, produce eggs or sperm, which are shed into the water where fertilization takes place. Each fertilized egg becomes a ciliated larva—one provided with hairlike processes called *cilia*. The larva

Below center: the reproductive form of some hydroids looks like a small jellyfish. Below right: a hydroid colony resembling a feathery plant with the polyps of the colony joined by a common stem.

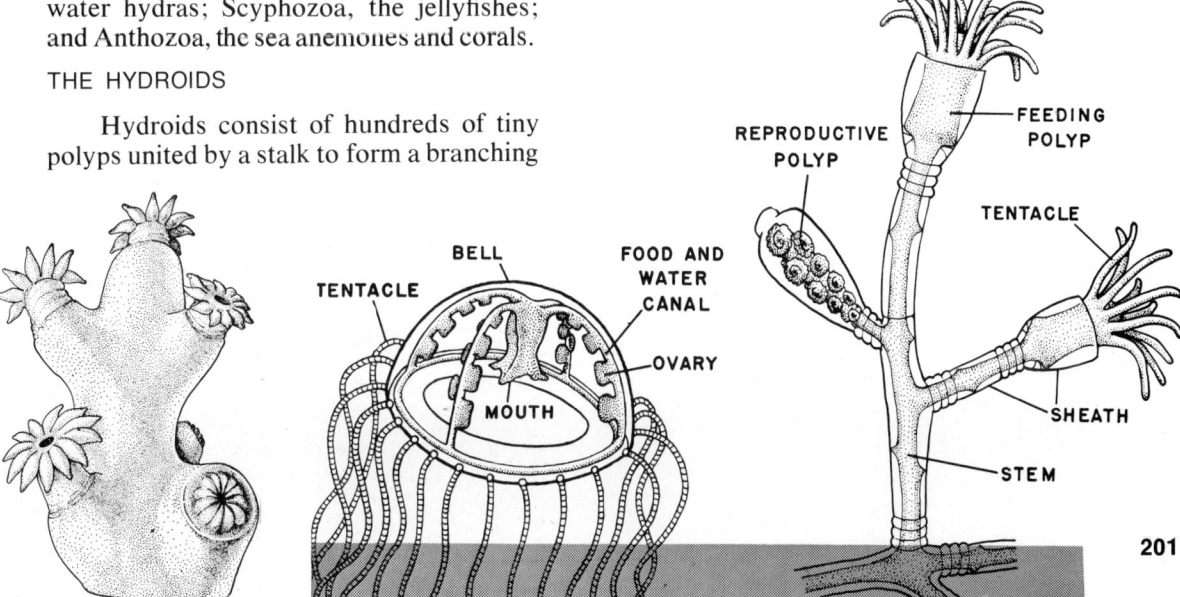

swims about for a time before attaching itself and forming a polyp. The polyp then buds and a new hydroid colony is produced. The process whereby the colony forms asexual buds that give rise to sexually reproducing organisms is called *alternation of generations*. This behavior, which is also found in plants, serves to spread the species into new localities.

Some species of Hydrozoa have insignificant polyps; the medusa is the conspicuous stage in their lives. Others flourish as hydroid colonies producing only attached, degenerate medusalike structures that shed eggs and sperm into the water.

The freshwater hydra, a minute individual polyp, does not go through a medusa stage. Instead, the fertilized egg remains affixed to the outside of the body until a heavy membrane forms around the embryo. Then it separates from the parent and later develops into a young hydra. The hydra also produces asexual buds that grow from the body, form tentacles, and finally pinch off from the parent.

Hydras prefer the clean waters of lakes and ponds, where they feed on tiny worms, insect larvae, young fish, and microscopic crustaceans. They glide along the bottom and on submerged plant stems by creeping movements of the cells at the base of the body. They also move by somersaulting. First, they bend over and attach their tentacles to a support while releasing the base. Then they swing the base over and attach it, freeing the tentacles, and so on.

The coelenterate known as the Portuguese man-of-war is related to the hydroids and hydras. It is a complex colonial animal supplied with a crested, gas-filled float, from which hang feeding polyps, clusters of attached medusas, and long, trailing tentacles armed with stinging thread capsules.

THE JELLYFISHES

The typical jellyfish has a bell-shaped, gelatinous body. Under the central part of the bell is a short process bearing the mouth. The corners of the mouth are pulled out into grooved oral arms. These carry nematocysts that paralyze and entangle small aquatic animals. The prey is swept along the ciliated grooves of the arms, through the mouth, and into the spacious gastrovascular cavity, which has branched radial canals going to the margin of the bell. Numerous tentacles fringe the edge of the bell. Sense organs that are sensitive to light, chemicals, and the directions of movement are also located on the bell's margin. The jellyfish weakly swims by rhythmically contracting its bell.

Ovaries or testes, as the case may be, lie on the floor of the gastrovascular cavity. Sperm are released in the water and fertilize the eggs in the cavity of another jellyfish. Each fertilized egg then lodges in a fold of an oral arm, where it develops into a ciliated larva. This escapes and grows into an inconspicuous polyp. Eventually the polyp develops a number of horizontal constrictions, so that it comes to look like a pile of saucers. The saucers break away as medusas and develop into adult jellyfish.

SEA ANEMONES AND RELATIVES

The stout-bodied sea anemone is a noncolonial polyp that attaches to rocks or shells and rarely changes its place. Tenta-

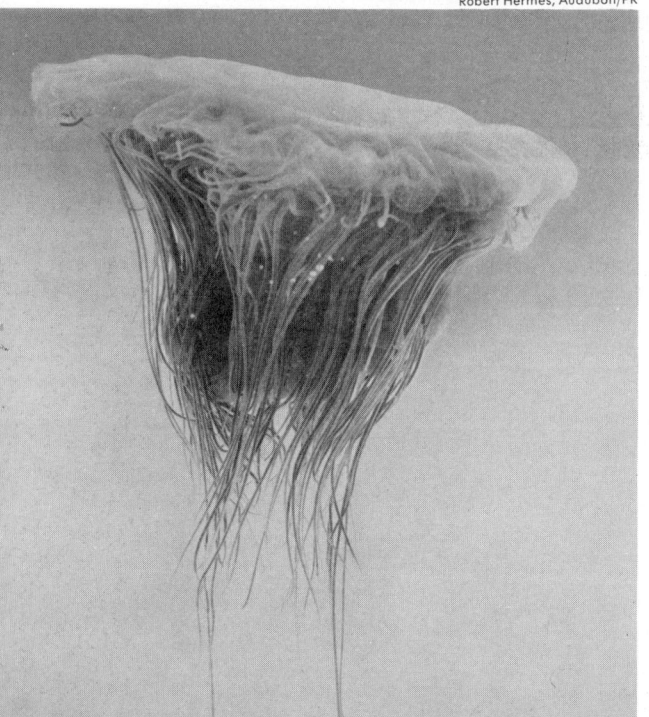

A lion's mane jellyfish. It is very common in temperate coastal waters, and its sting is very painful.

Robert Hermes, Audubon/PR

cles rim the upper part of the animal and surround the mouth, which leads to a gullet. Below this is the gastrovascular cavity, which is divided by partitions. These increase the digestive capacity of the animal so that large prey, such as crabs and fish, can be consumed.

The sea anemone can reproduce itself by dividing its body in half longitudinally. Sometimes, too, as the animal slides along on its slimy basal disk, or foot, fragments of its body are left behind. Small anemones are regenerated from these pieces. Eggs and sperms are produced on the partitions of the gastrovascular cavity and are released through the mouth. Ciliated larvae develop from the fertilized eggs and form single anemones. There are no medusas in the anemone's life cycle.

The stony corals are colonial animals similar to anemones. They remain attached to one spot and secrete cups of calcium carbonate into which they can retract. Stony corals are found in deep, cold water, but it is only in the tropical and subtropical seas that they contribute to the building of reefs.

The anemones and stony corals have various kin. These include the organ-pipe corals, which live in calcareous tubes joined together by platforms, and the precious, or red, corals, which are stiffened by calcareous spicules and are used in making jewelry. The related sea whips, sea fans, and sea plumes are branching colonies of polyps supported by a flexible horny material.

EVOLUTION OF COELENTERATES

The coelenterates display a somewhat higher organization than the sponges, particularly in the development of a true digestive cavity and in the elaboration of special sensory cells, a nerve net, and muscle fibers. External stimuli affect the sensory cells, and impulses are conducted by way of the nerve net to the muscle fibers. Longitudinal muscles cause the polyp to shorten; circular muscles cause it to lengthen. Muscle fibers circling the mouth can close it off when harmful substances are in the water or when a falling tide leaves the animal high

N.Y. Zoological Society

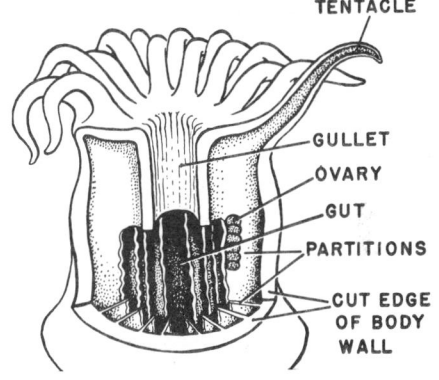

A sea anemone. Its mouth is surrounded by tentacles. The sea anemone's gastrovascular cavity is divided by partitions. The anemone is anchored to the bottom by a so-called pedal, or basal, disk.

and dry. The various muscle fibers, coordinated by the nerve net, also allow polyps to bend in one direction or another and move their tentacles. The muscles of the medusa provide contractions of the bell for swimming.

It has been commonly held that coelenterates evolved from colonial protozoans, much as did the sponges. But a revolutionary theory that has gained considerable support assumes that the coelenterates evolved from a primitive flatworm. If this theory is true, the sea anemones with no medusa life-cycle stage are the most primitive of coelenterates, and the jellyfish and hydroids, with their life cycles of alternating polyps and medusas, are a secondary development.

A sun starfish, one of the largest and most beautiful of the starfishes. It is found in shallow waters throughout much of the world.

STARFISH AND OTHER ECHINODERMS

by Francis J. Ryan and Elizabeth J. Ryan

A starfish isn't a fish. A sand dollar isn't a form of money. And a sea cucumber isn't a vegetable. All of these misleadingly named animals are echinoderms, or "spiny-skinned" animals. The phylum name, Echinodermata, comes from the words *echinos,* meaning "hedgehog," and *derma,* meaning "skin."

In one very obvious way, the animals in this phylum are different from other animals. Most animals, including human beings, have *bilateral symmetry.* Their bodies are dividable into more-or-less identical right and left halves. But echinoderms have *radial symmetry.* Their bodies are built on a circular, or radial, plan. In the center of the body is the mouth. From here, arms or other structures extend outward, rather like the spokes of a wheel.

The echinoderms have no near kinship with any of the other invertebrates. They are highly interesting to biologists, nonetheless, because of their larval forms. These show close affinities with the larvae of the protochordates, the animals whose ancestors gave rise to backboned animals.

A group of more typical—that is, five-armed—starfish. Starfish range in size from about one centimeter to one meter.

PARTS OF THE BODY

The skin of a typical echinoderm covers an internal skeleton of calcareous ossicles, or small bones, which give a more or less rigid structure. Projecting outward from the ossicles are numerous calcareous spines.

Inside the skeleton is the large body cavity, or *coelom*, in which lie the internal organs. The coelom contains a lymphlike fluid that bathes the organs. Amebalike cells creep about in it removing wastes and carrying nutrients to all parts of the animal's body.

There is a complete digestive system leading from the mouth, on the underside of the body, to the anus. Digestive glands pour their secretions into the stomach. A unique arrangement called the *water vascular system* allows water to enter a seive plate on the body's surface and then circulate by way of another canal to a ring canal, which branches into radial canals. Each radial canal gives off many pairs of *tube feet*. When these are distended with water, they are used in locomotion and serve as respiratory surfaces.

Encircling the echinoderm's mouth is a *nerve ring*. Five branches radiate from this ring and tiny nerves go to the internal organs, the skin, and the tube feet. There are no well-developed sense organs, though the tube feet probably function as organs of touch.

The sex glands shed their products into the water, where fertilization occurs. Fertilized eggs give rise eventually to larvae that swim freely by means of ciliated bands. The larvae go through many stages of development before they begin to look like miniatures of their parents.

There are more than 5,000 known species of echinoderms. They live in either marine or brackish waters. They are grouped in five classes.

STARFISH

Starfish, or sea stars, make up the class

Russ Kinne/PR

George Whiteley/PR

A sand dollar. Like other echinoderms, it has radial symmetry. Five zones containing tube feet radiate from a central disk.

A basket star, a graceful-looking echinoderm with a flattened body and very thin arms that branch repeatedly.

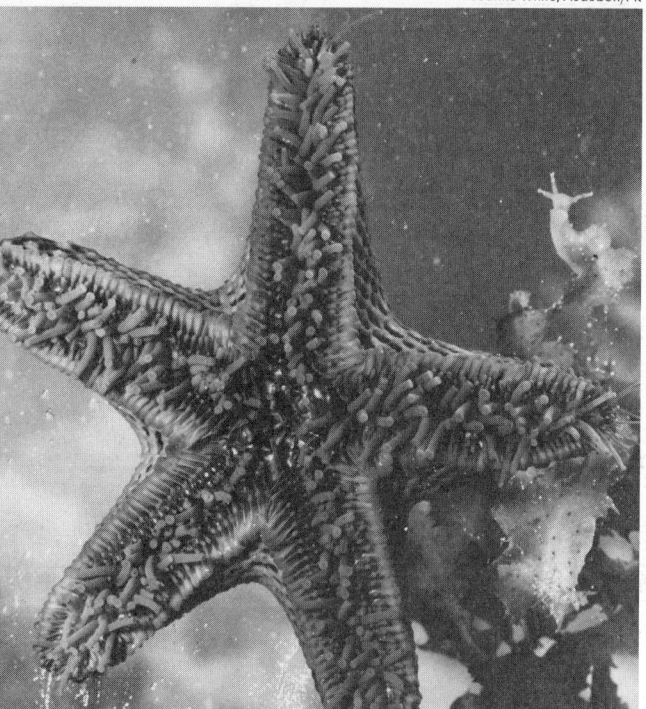

Top: starfish crawl along the bottom, feeding on mollusks, mainly clams, mussels and oysters, which they pull open by means of the suckers at the ends of their tube feet (bottom).

Asteroidea. They are perhaps the best known of the echinoderms. The body is a central disk from which radiate five or more arms. Between the blunt spines on the upper surface project skin gills, which are fingerlike extensions of the coelom; they are excretory and respiratory organs. Small pincers interspersed among the spines protect the skin gills and clear the surface of foreign matter.

On the underside of each arm is a groove from which protrude the slender tube feet. A light-sensitive eyespot and a short tentacle, which may be sensitive to chemicals, are located at the tip of each arm. Sea stars prey on tube worms, crustaceans, and mollusks, particularly clams and oysters.

Commercial clammers and oystermen have long despaired of the competition they have gotten from starfish. At one time they tried to get rid of the starfish by chopping them up into several pieces. Assuming that this killed the starfish, the people tossed the pieces back into the sea.

But the clammers and oystermen found this greatly increased their enemies, for starfish, like many other animals without backbones, have the ability to regenerate themselves. If a starfish is cut into a number of pieces, each piece that includes at least a tiny part of the central disk will grow into a new, complete animal.

BRITTLE STARS

Brittle stars and serpent stars (class Ophiuroidea) have a small, flattened body disk with five or more many-jointed arms. In some cases, as in the basket star, the

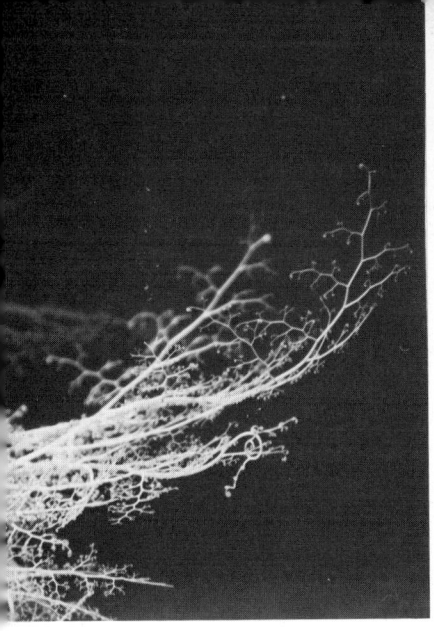

arms repeatedly branch. The stomach is saclike, and there is no anus.

These animals use their flexible arms to move jerkily about and to swim. The arms also catch worms, mollusks, and other animals and bring the prey to the mouth.

SEA URCHINS AND SAND DOLLARS

Sea urchins and sand dollars (class Echinoidea) are globular-shaped, oval-shaped, or flattened into thin disks. Five teeth surround the mouth. The skeleton forms a hard shell of flattened immovable ossicles. The spines are numerous, stiff, and movable. Five rows of tube feet radiate over the surface, converging at the upper and lower centers of the body.

Most sea urchins live near rocky shores, where they feed on algae, small marine animals, and decaying matter. Many people, particularly natives of Mediterranean countries, consider sea urchin eggs a delicacy.

Sand dollars usually live buried in the sand. They have very short spines, giving the animals a furry appearance.

Closely related to sand dollars are the heart urchins. They, too, burrow into the sea floor. They feed on microscopic plants and animals.

Right top: large sea urchin with very long, slender, tentaclelike tube feet seen projecting between rigid spines. Right middle: another sea urchin—this one with shorter tube feet. Right bottom: underside of a sea urchin showing centrally located mouth opening.

Starfish, sea cucumbers, and some other echinoderms are able to regenerate lost body parts. Photo above shows a starfish regenerating three of its five arms.

SEA CUCUMBERS

Sea cucumbers (class Holothurioidea) are elongated animals with a mouth surrounded by tentacles at one end and the anus at the other. The skin is either leathery and muscular or delicate and transparent and possesses only microscopic calcareous ossicles. Sea cucumbers move by muscular contractions or by using the tube feet, which extend in five rows the length of the body. They eat organic material taken from mud or small animals entangled by the tentacles at one end of the sea cucumber's body.

Sea cucumbers have an unusual protective device. If disturbed by an enemy, the sea cucumber may eject, or disgorge, its internal organs. The predator eats these, forgetting about the sea cucumber itself. Later, the lost organs can be completely regenerated.

The Chinese consider sea cucumbers a delicacy. The animals are dried, then used in making soups.

SEA LILIES

Sea lilies and feather stars (class Crinoidea) are flowerlike, brilliantly colored animals having flexible branching arms. Many are attached, mouth upward, to the sea bottom by means of a horny stalk. Others have no stalk and swim freely by using the arms. Microscopic plants and animals, caught by the arms and swept to the mouth, form the food.

Sea lilies and feather stars inhabit all ocean depths from just below the low-tide line to more than 3,500 meters below the surface.

Sea cucumbers are found in seas throughout the world, usually lying on the bottom. They move in a caterpillarlike manner by muscular contraction of their body and by using rows of tube feet.

FLATWORMS AND ROUNDWORMS

by F. L. Fitzpatrick

The word "worms" is popularly applied to a great variety of long, slender, and limbless animals: earthworms, tapeworms, hookworms, shipworms (which are mollusks), blindworms (which are lizards), and the larvae of various insects.

To a zoologist the word "worms" has a much more restricted meaning. It is applied particularly to three groups of animals: the flatworms, or platyhelminths; the roundworms, or nemathelminths; and the segmented worms, or annelids. The animals belonging to these groups are found in almost all parts of the world. Some of them burrow in the earth. Others crawl along the ground. Still others swim in the water. A considerable number are parasites on other animals or on plants. Some worms are considered to be benefactors of mankind. Others, including tapeworms and liver flukes, rank high among the dangerous pests of people and domestic animals.

FLATWORMS

The flatworms make up the phylum Platyhelminthes. This name comes from the Greek words *platys,* meaning "flat," and *helmins,* meaning "worm." These are the simplest of the worms. There are about 10,000 known species. In some of them, the outer layer of the body is provided with vibrating, hairlike structures called *cilia.* In others, the outer layer is smooth or spiny.

TAPEWORMS

Perhaps the best-known flatworms are the tapeworms, or cestodes. Many species are harmful parasites of people and domestic animals.

The body of a tapeworm generally consists of a rounded head, or *scolex,* and a number of flattened segments, or *proglottids.* The head bears hooks or suckers (or both), with which the worm attaches itself to the host—usually to the lining of the intestine. New segments are formed next to the head. As the older segments keep on growing, the largest of all, naturally, are to be found at the tail end of the body. A fully grown tapeworm looks like a long, narrow ribbon. It may reach a length of 9 or 10 meters.

The head is the anchor that keeps all the segments within the body of the host. Each segment functions more or less as a self-contained unit. There is no digestive system; each segment absorbs, through the body wall, digested food from the digestive cavity of the host. As a segment gets older and larger, it becomes filled with eggs. When it is mature at last, it breaks off from the rest of the body and passes to the exterior with the wastes of the host.

The tapeworm is a flatworm. Many tapeworms are harmful parasites of man and of domestic animals.

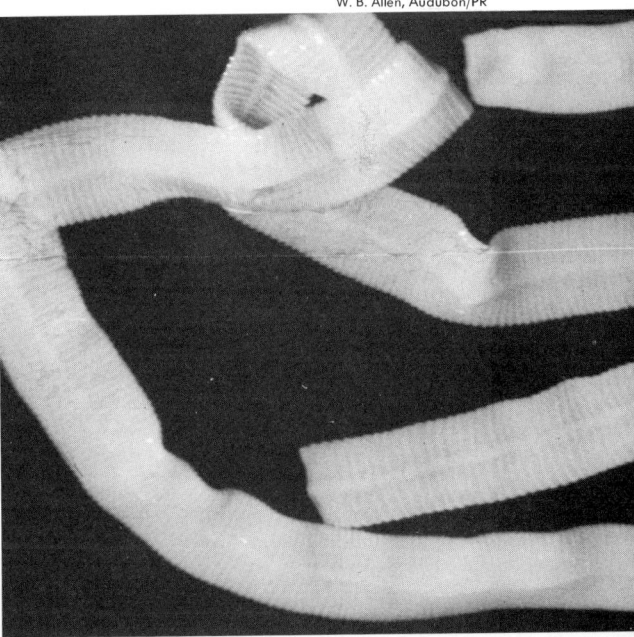

W. B. Allen, Audubon/PR

A common tapeworm that attacks human beings is the so-called beef tapeworm, *Taenia saginata*. This animal must live in the bodies of two different hosts to complete its life cycle. The adult beef tapeworm is found in the human intestine, where it sometimes reaches a length of more than 9 meters. When the tail-end segments, filled with eggs, break off, they pass to the exterior with the wastes of the human host.

If the wastes are not disposed of in a sanitary manner, the eggs may be deposited on grass that is eventually swallowed by cattle. Boring larvae are then freed from the eggs. They migrate into the muscle tissues of the host and form cysts. If a person eats raw or partly cooked meat from one of these infected cattle, the cyst around the young worm will dissolve in the stomach of the human host. The worm will then take up its place of abode in the intestine.

The presence of a beef tapeworm in the intestine is not likely to prove fatal or even dangerous, because the animal cannot fill human tissues with cysts. Its eggs must always pass out of the human body before they can hatch. The worm is an undesirable boarder, however, since it deprives the human host of a part of the nourishment he should obtain from his food. Fortunately, it can be eliminated from the body quite easily by means of simple drugs administered under a doctor's direction.

The pork tapeworm, *Taenia solium*, is far more dangerous to people. Normally the adult worm lives in the human intestine. Here it develops the usual mature, egg-filled segments, which break off and pass to the exterior from time to time. If any of the eggs are swallowed by pigs, the larvae that hatch from them will form cysts in pig muscle. When infected pork that is only partly cooked is eaten by a human being, the cysts dissolve and the adult worms become established in the human intestine. The pests can be disposed of, as in the case of beef tapeworms, by prescribed drugs.

Flukes in sheep liver (1) lay eggs (2) that become embryos (3). Embryos develop in a snail (4) into flukes, which escape (5-7) and encyst (8). Cysts are eaten by sheep and flukes enter the liver (9).

Gen. Biol. Supply House, Inc.

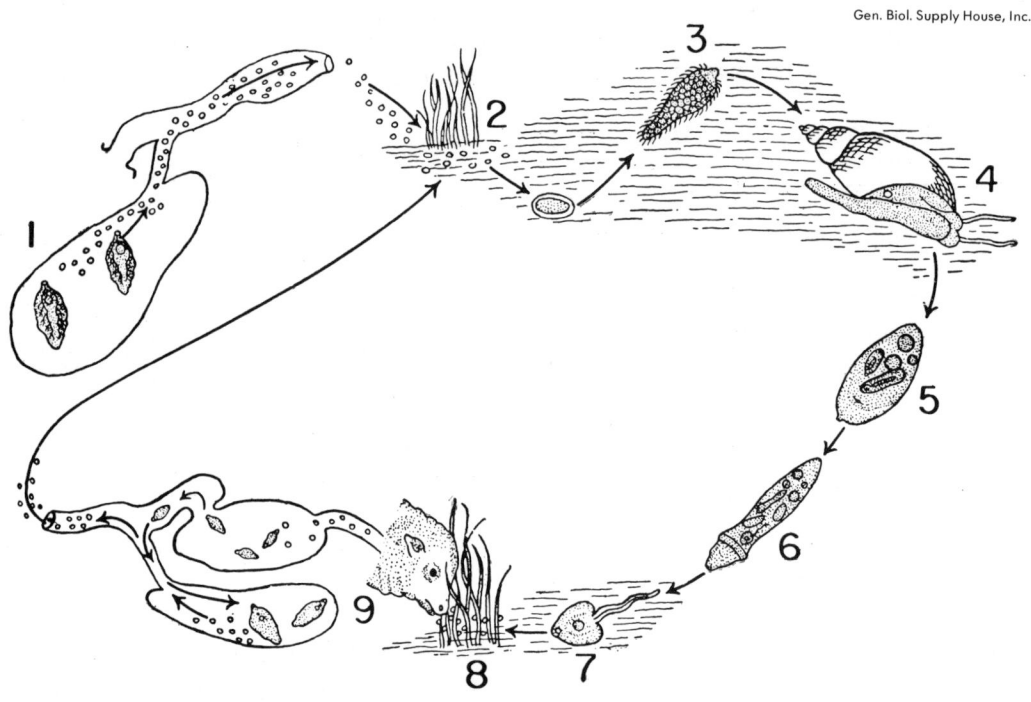

Unfortunately, some of the pork-tapeworm eggs may develop into young in the human body, before they can pass out to the exterior with the body wastes. When this happens, the young ultimately penetrate into muscle tissues, causing inflammation. The consequences are particularly serious when the encysted larvae lodge in vital areas of the human body, such as the eye and brain.

Dogs and rabbits act as hosts for many kinds of tapeworms. Most of the parasites form their cysts in fish, mice, rats, rabbits, sheep, cows, and pigs. One tapeworm, *Echinococcus granulosus,* that occurs as an adult in dogs forms cysts in human muscle, liver, lung, brain, and bone tissue. We become infected by eating unwashed vegetables that have been contaminated by dogs, by drinking contaminated water, or by kissing dogs to whose fur the eggs are clinging.

Domestic birds are also subject to tapeworm attacks. Adult worms develop in the bodies of birds when the birds eat infected earthworms or insects.

FLUKES

Another important flatworm group are the flukes, or trematodes. The best-known member of the group is the liver fluke, *Fasciola hepatica,* which lives as a parasite in the liver of sheep, cattle, pigs, and sometimes human beings. This fluke is widely distributed throughout the world. It has a short and flattened body, about 2 centimeters long. A sucker, containing the mouth, is located at one end. There is also a ventral sucker on the lower surface of the body; this is used for gripping. Unlike tapeworms, flukes possess a digestive tract; this has only one opening to the exterior—the mouth, which also serves as an anus.

The common liver fluke of sheep has a most amazing life cycle. About 200 adult flukes may be housed in a sheep's liver. These commonly produce numerous eggs. As many as 100,000,000 eggs may be formed within the body of a single sheep host. After the eggs have begun to develop, they pass through the bile duct of the liver into the intestine and then to the exterior with the wastes.

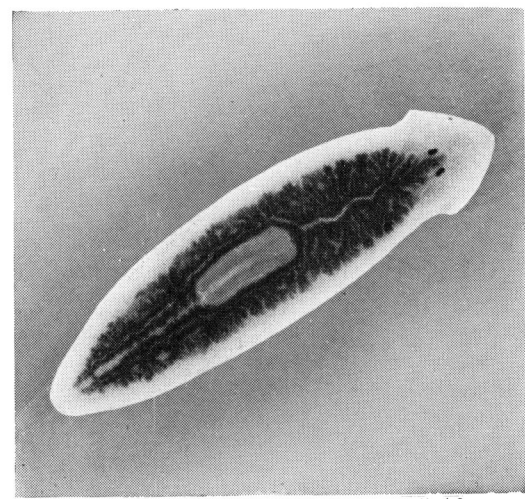
Hugh Spencer

Stained preparation of turbellarian *Planaria*. Eyespots are visible at the anterior, or front end, of the body.

The egg of a sheep liver fluke must get into water of the right temperature if it is to continue its development. When this happens, the egg gives rise to a tiny ciliated

A land *Planaria* found most frequently on tropical and subtropical forest floors.
Robert Mitchell, Tom Stack and Associates

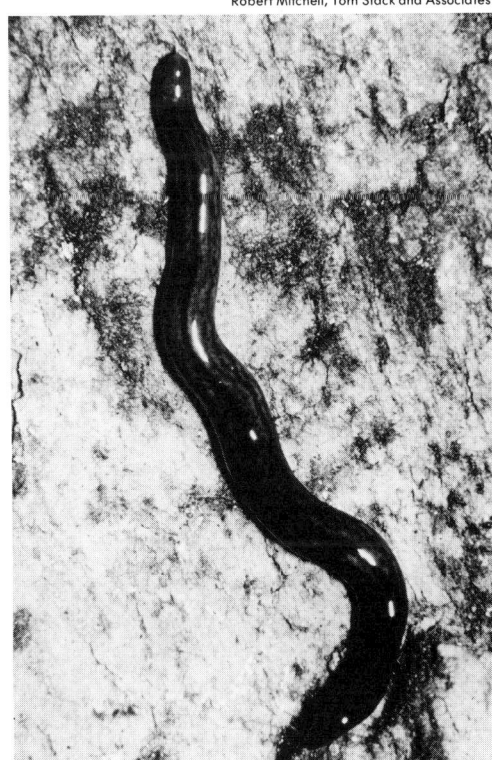

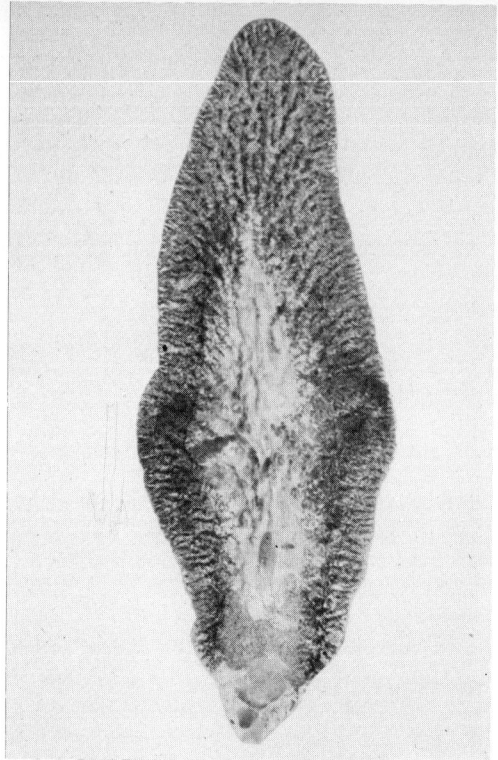

Liver fluke. This and other species attack sheep, cattle, pigs, and sometimes man. If untreated, they can inflict serious injury to their hosts.

embryo. The little creature will die unless it finds a certain type of snail to serve as its host. If this host is forthcoming, the embryo burrows into it. Within the snail great numbers of young flukes are produced by a complicated budding process. The young escape from the snail and swim about freely for a time. Then they climb upon blades of grass at the water's edge and form cysts about themselves. If a sheep swallows the cysts, the young flukes will gradually work their way into the liver.

Naturally, very few sheep liver flukes are likely to meet all the conditions set by such an exacting life cycle. If these worms continue to thrive and flourish, it is because of the enormous number of eggs they produce.

Sheep may be seriously injured, and sometimes killed, by liver flukes. If the victims are not too far gone, it is possible to cure them by means of drugs that kill the worms.

There are many other species of parasitic flukes. They live in such organs or structures as the intestines, lungs, and blood vessels. Various kinds of flukes attack people. There are several species of human liver flukes in Africa and the Far East.

TURBELLARIANS

The turbellarians, or true flatworms, are generally free-swimming animals that do not lead parasitic lives. They have short, flattened bodies and a digestive system with a single opening on the underside of the body. Their bodies are covered with cilia, which can be used to produce swimming movements.

These worms are found in the sea, in fresh water, and in some moist places on the land. Some feed on worms, insects, and tiny mollusks. Others eat microscopic organisms.

Turbellarians belonging to the genus *Planaria* are particularly interesting because they can regenerate, or regrow, lost parts of the body. If the head or tail end is removed, a new head or tail will develop. If the body is cut in two, the head end will grow a new tail and the tail end will develop a new head. *Planaria* are often studied in biology courses.

ROUNDWORMS

The roundworms, or threadworms, are also called nemathelminths—the equivalent to "thread worm" in Greek. There are about 12,000 species. A number of these are parasitic on people and animals. Some attack plants. The roundworms are slender creatures. They do not have cilia. Nor are they segmented.

TRICHINA WORMS

The most notorious of the roundworms, from the human standpoint, is the trichina worm, *Trichinella spiralis*. This animal is parasitic on human beings, pigs, house rats, and probably other animals. It is a tiny creature. Adult males are only a millimeter and a half in length while female adults are three to four millimeters long.

The presence of the trichina worm in the human body causes the disease known as *trichinosis*. People are infected when they eat raw or rare pork from pigs that had adult trichina worms in their intestines.

The female worms in the pig host bear tiny young, which penetrate the walls of the intestine, enter the blood vessels, and work their way into muscle tissues. Here each tiny worm embryo forms a cyst about itself. Life in the muscle tissues of the host represents a dead end for the trichina unless a person or some other carnivore eats the tissue in which it is located. If the host is not devoured, the cyst wall will begin to harden after about six months and the embryo within the cyst will eventually die.

If the infected pig is butchered and used as human food, the story may be very different. The meat looks perfectly wholesome, because the cysts are too small to be seen by the naked eye. Thorough cooking will kill the trichina embryo within the cyst. But if the meat is eaten raw or only partly cooked or smoked, there is acute danger of human infection.

By the time the meat containing the parasites has reached the person's intestine, the cyst walls have dissolved. The young worms soon grow to adult size. The human victim may suffer from digestive disturbances—including nausea, diarrhea, and abdominal pain—if enough of the worms are present. If the disease is recognized at this stage, the parasites may be driven out of the intestines through the use of drugs under a doctor's direction, and no lasting harm will be done.

If the worms are not molested, the females bear a new generation of young. These bore into blood vessels of the intestinal wall, are carried to all parts of the body, and form cysts in the muscle tissues. Muscle soreness and fever are typical symptoms of the disease at this stage.

Spiny-headed worms of the phylum Acanthocephala. These parasitic worms are closely related to roundworms.

Jack Dermid, Audubon/PR

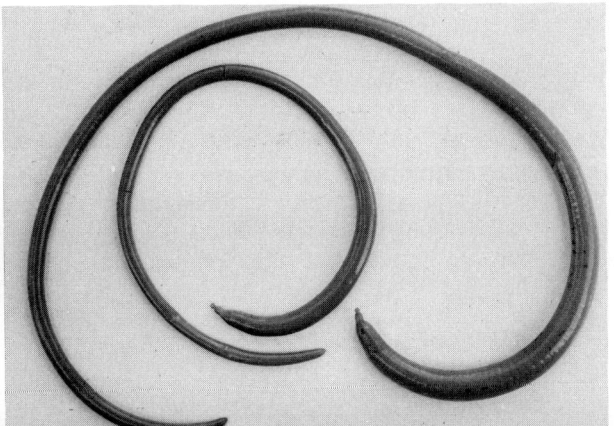

Breathing, swallowing, and chewing movements are likely to be painful for a period of three or four weeks. The consequences are particularly serious when the trichina larvae damage vital areas of the body, such as the heart and diaphragm.

Once the young trichina parasites have gotten into the muscle tissues, they cannot be disposed of by means of drugs. It is possible to kill adults remaining in the intestines, so that no more young will be produced. Otherwise little can be done. In the course of time calcareous matter is laid down in and around the cyst, which is ultimately transformed into a granule of lime in the muscle.

It has been estimated that there may be as many as 100,000,000 trichina cysts in some human hosts. In other cases, the number of cysts is comparatively small. A number of people have had trichinosis without being aware of the fact. There have not been enough worms in their bodies to cause serious damage.

HOOKWORMS

The hookworms are also members of the roundworm group. They are abundant in tropical and subtropical regions. The species known as the American hookworm, *Necator americanus,* is about a centimeter long—considerably larger that the trichina worm.

Adult American hookworms suck blood from the wall of the host's small intestine. The female produces large numbers of eggs, which pass to the exterior with the wastes. If these eggs are deposited upon warm and loose, moist soil, they develop into tiny larvae.

The larvae generally get into a human body through the soles of the feet. They bore through the skin, producing a sensation known variously as "ground itch," "dew itch," or "skin itch." Entering the blood vessels, the larvae start on an amazing journey through the body. First they are transported by the blood stream through the heart to the tiny blood vessels of the lung tissue. They bore their way out into the air spaces of the lungs and then move upward through the bronchial tubes and

FLATWORMS AND ROUNDWORMS

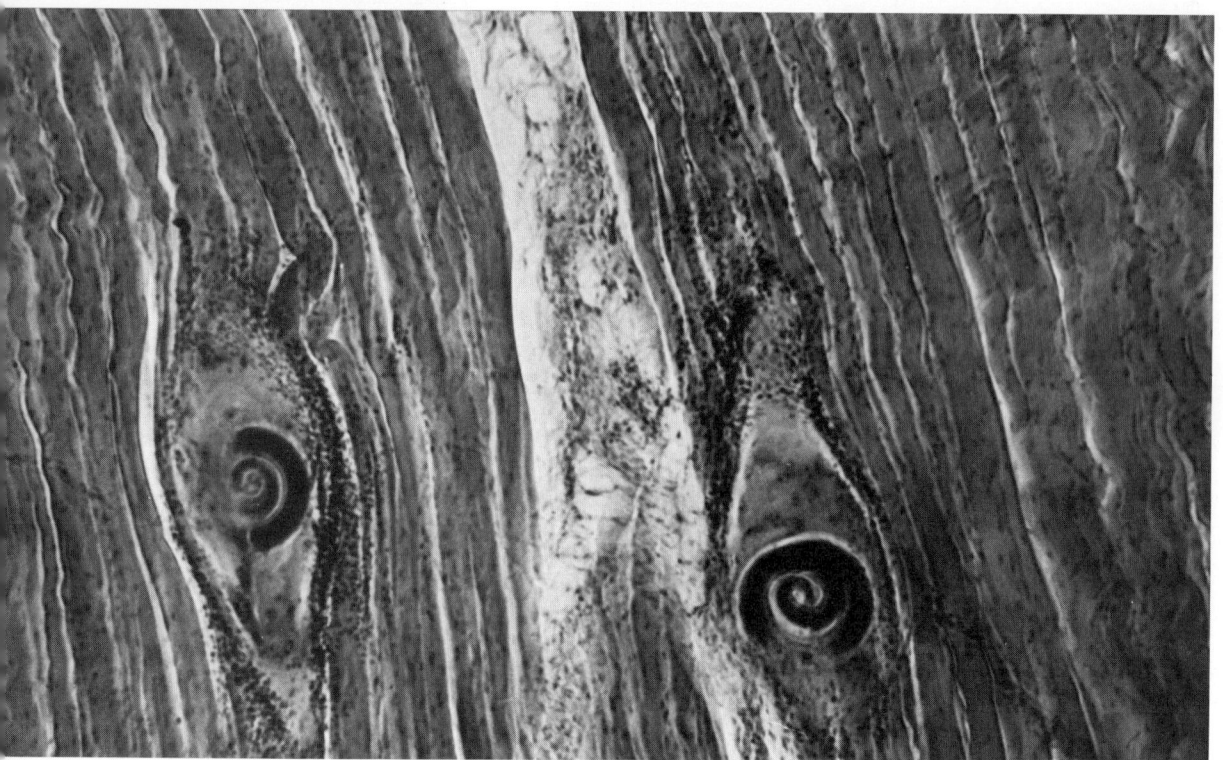
The trichina roundworm causes trichinosis. The worm embryos penetrate muscle fibers and form cysts as shown in the above photo.

windpipe to the back of the mouth cavity. They then pass down the esophagus and through the stomach, reaching the small intestine. Here the hookworms settle down to perhaps six or seven years of life.

Hookworm victims usually suffer from loss of blood and become thin and anemic; their skin takes on a waxy appearance. They often suffer from heartburn and constipation, and are easy prey to various diseases. Child victims are often retarded in their development.

The first line of defense against hookworms is to adopt the practice of wearing shoes. This will prevent most of the parasites from entering the body, even in localities where they are abundant. Another effective measure is to arrange for the sanitary disposal of human wastes containing hookworm eggs.

Adult worms may be driven out of the human intestine by certain drugs used under the guidance of a doctor. Of course, hookworms in the human body will die of old age in time, but six or seven years is a long time to wait.

THE FILARIAL WORM

Another parasitic nematode of tropical and subtropical regions is the filarial worm, *Wuchereria bancrofti,* which is transmitted from one human to another by various species of mosquitoes. The adult female worms measure between 7 and 10 centimeters, while the males are half that size. They usually occur in the lymphatic vessels and lymph nodes.

The females produce eggs that develop into embryos known as microfilariae, some of which reach the blood circulating near the host's skin. Here the embryos are taken up by a mosquito as it feeds on its human victim. Within the tissues of the mosquito, the young filarial worms develop into infective larvae. These enter a human host the next time the mosquito feeds and eventually reach the lymphatic system, where they mature.

The condition known as *filariasis* results from inflammation and from the obstruction of lymphatic channels by both the bodies of the mature worms and by scar tissue caused by the worms' presence. A further complication, in which the limbs and other regions of the body swell to enormous size, is known as *elephantiasis*.

OTHER PARASITIC ROUNDWORMS

The Guinea worm, *Dracunculus medinensis,* another parasitic roundworm of human beings, is found in Africa, parts of the Middle East, and India. People become infected by drinking water containing copepods (crustaceans) that carry the infective stage of the worm. After the infected copepods are swallowed, the worms mate and the female migrates through the tissues of its human host. About ten months later the fertile female, which may now reach a length of a meter or more, comes to lie just under the skin, through which it frees its young. To extract the worm, one end of the animal is rolled up on a stick. Each day the stick is given a few turns until the entire worm is drawn out.

Dogs, cats, poultry, cattle, sheep, horses, pigs, and goats are also attacked by parasitic roundworms. The heart worm, *Dirofilaria immitis,* lives as an adult in one chamber of a dog's heart or in the arteries that lead from the heart to the lungs. The tiny young of the heart worm are transmitted to dogs through the bites of mosquitoes.

The stomach worm, *Haemonchus contortus,* attacks both sheep and cattle. A fully grown female stomach worm is about three centimeters in length; a male is somewhat smaller. Eggs produced by the female pass to the exterior with the wastes and hatch within a very short time. The young pass through several stages of development. As larvae they crawl up on blades of grass, where they are likely to be swallowed by grazing sheep or cattle.

Various species of roundworms attack plants. The sugar-beet nematode, *Heterod-*

These sugar-beet roots have been infested with small white female nematodes. They suck out the root contents, inflicting serious damage.

era schachtii, is a good example. It is never more than 1.2 millimeters in length. Its small size, together with the fact that it lives in the roots of the sugar beet, make control a difficult problem. This worm has been present in European beet fields for many years. It appears to have been brought into the United States accidentally some time after 1900. The most effective control measure is to change to another crop when sugar-beet nematodes appear in a beet field.

The octopus, a cephalopod, is unusual in that it lacks the shell carried by most mollusks. It has eight tentacles that have sucker pads on the bottom. Above: octopus showing tentacles and pads. Right: closeup of sucker pads. Far right: octopus enveloping a crab with its arms.

MOLLUSKS

by F. L. Fitzpatrick

It would seem to be a far cry from the tiny snail in a home aquarium to a 15 meter-long giant squid; from the edible oyster, firmly attached throughout life to the same rock or shell, to the freely swimming scallop; from the vegetarian slug to the sinister-looking, carnivorous octopus. These animals vary greatly in size, appearance, and habits. Yet all belong to the phylum of the Mollusca, or mollusks. This is one of the largest groups in the animal kingdom. More than 70,000 species have already been described. Many mollusks live in the sea. Others are found in freshwater lakes, ponds, and streams. Some dwell on land.

The word "mollusk" comes from the Latin *molluscus,* meaning "soft." The name is apt enough, for all mollusks have soft bodies. In most cases the body is protected by a shell, made up largely of calcium carbonate. This shell is secreted by the body covering known as the *mantle.*

Most mollusks also have an unusual structure called the *foot,* which takes quite different forms in various species. In clams, for example, the foot is a muscular extension of the body and is used in plowing through mud and sand. In snails, it is flat and is used for creeping. In squids and octopuses, it is divided into arms, which serve to seize the animals' prey. Certain oysters have no foot.

The phylum Mollusca is divided into five classes. The class Cephalopoda includes the squid, octopus, cuttlefish, and nautilus. The Pelecypoda are represented by oysters, clams, scallops, mussels, and teredos. The Gastropoda include snails, slugs, limpets, abalones, and conches. The Scaphopoda are the tooth shells. And the Amphineura, the most primitive mollusk class, consists of the chitons. Many mollusks are economically important.

SQUIDS, OCTOPUSES, AND THEIR RELATIVES

Among the best-known mollusks are the striking creatures known as squids and octopuses. They belong to the class of the Cephalopoda, or cephalopods. The name means "head-feet" in Greek. These animals are so called because the foot, which is separated into a number of "arms," encircles the head.

The cephalopods differ from most other mollusks in one important respect: they generally do not develop shells. Instead, the mantle forms the outer part of the naked body. In some species, however, there is an inner skeleton.

All cephalopods dwell in the sea. They are provided with arms, often called *tentacles*, that have suckers or hooks or both. Almost all cephalopods secrete an inklike fluid, which is stored in a special sac. When they wish to escape a pursuer, the animals squirt the ink into the water, making it turbid and thus confusing the foe. Most cephalopods are capable of chameleonlike color changes. The skin contains cells called *chromatophores* ("color-bearers") that contain different pigments. When these cells become larger or smaller, the color of the skin changes rapidly. Because of such color changes, the animals generally blend effectively with the background.

SQUIDS

The expert cephalopod swimmer called the squid is a streamlined, spindle-shaped creature. It is sometimes called the *sea arrow* because of the way in which it darts through the water. The foot is divided into ten arms. Two of the arms are longer than the rest; these bear suckers and are used to seize and hold prey. The eyes have no lids but otherwise they look startlingly like human eyes.

The squid draws water through a central cavity of the body — the *mantle cavity* — and forces it out through a flexible tube, the *siphon*, when the mantle is contracted. The siphon is located just back of the arms. The jet of water that spurts through it serves to propel the animal swiftly backward. Ink is also discharged through this siphon.

The fins, which are two flaplike extensions of the mantle, are used chiefly for steering. They also serve to propel the squid slowly forward or backward.

One of the most familiar species is the common squid, *Loligo pealei*. It is found in Mediterranean and east Asian waters and along the eastern coast of North America. Some fishermen use it as bait. It also serves as human food, particularly in the Mediterranean area and the Far East.

The squid known as the flying squid,

MOLLUSKS 217

The streamlined shape of the squid makes it an excellent swimmer. Two of the ten arms have suckers.

Ommastrephes bartrami, has been compared to the flying fish. It often shoots out of the water, particularly when the weather is rough, and sometimes lands on the decks of ships.

The most formidable of the squids is the giant squid, *Architeuthis princeps*. This is the largest of all the invertebrates, or animals without backbones. The total length of the animal, including its body and arms, may be 15 meters or more. This large animal lives in the open sea, making its way far beneath the surface. Live giant squids have but infrequently been encountered at sea. However, they are sometimes cast up on beaches, especially along the shores of Newfoundland. According to some authorities, the giant squid may have given rise to the numerous legends of mighty serpents frequenting the depths of the sea and occasionally engulfing ships.

OCTOPUSES

Few dwellers of the deep have stirred people's imagination more than the octopuses (genus *Octopus*). Many tales have been told of these creatures lurking in crevices and amid rocks and suddenly darting out to attack some hapless wader or diver. Such bloodcurdling tales are totally false or at least grossly exaggerated. It is true that a large octopus with eight long, powerful arms, two large staring eyes, and a vicious-looking beak, would be a rather unpleasant customer to meet under water. Yet there is little evidence to show that even the larger octopuses attack humans.

The foot of the octopus is divided into eight arms. It is this feature that gives rise to the name, which means "eight feet" in Greek. The animal has a parrotlike beak with which it rends its prey. Octopuses range in arm-and-body length from 5 centimeters to 9 meters. The larger species, sometimes called *devilfish,* may attain a weight of 35 kilograms. The octopus can crawl along the sea bottom on its arms. Sometimes it swims about by sucking water into the body and then squirting it out.

Most octopuses are shy, retiring creatures. During the day they are generally hidden in crevices. At nightfall they steal out in search of prey. Stealthily, an octopus creeps up on some unsuspecting fish or crab. Once the powerful arms are entwined about the victim, there is no escape. The beaklike jaws quickly end the captive's struggles and the octopus feeds upon the prey. With the coming of dawn, the animal retreats to its lair. The octopus is itself the prey of certain animals, including conger eels, whales, and sharks.

Octopuses are eaten in coastal areas of

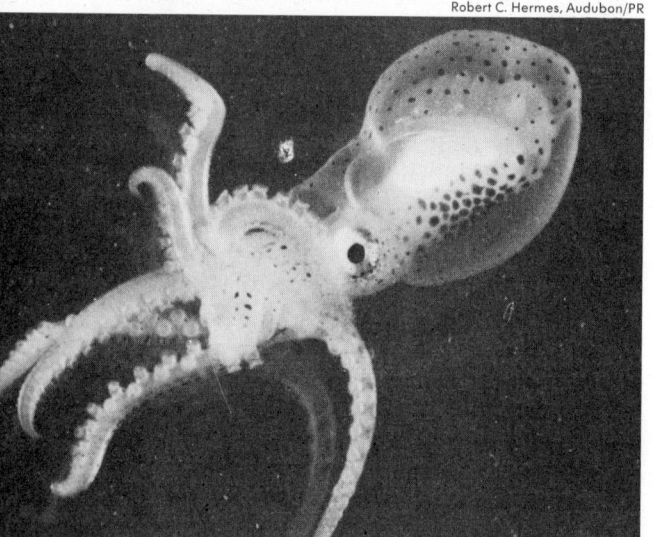

A newly hatched octopus. It is now only about one centimeter long, but it may grow to quite a large size.

Europe and North America. They are also eagerly sought after in various parts of the Far East and in the islands of the South Pacific.

CUTTLEFISHES AND NAUTILUSES

A well-known relative of the squids and octopuses is the common sepia, or cuttlefish, *Sepia officinalis*. This small creature, ranging from 15 to 25 centimeters in length, secretes a calcareous inner shell known as *cuttlebone*. This substance is used to supply lime to canaries and other caged birds. It also serves as a polishing agent. The pigment called sepia is prepared from the deep brown fluid that the cuttlefish ejects in order to conceal its retreat.

The pearly, or chambered, nautilus, *Nautilus pompilius,* found in the South Pacific and the Indian Ocean is a member of an ancient group of animals. Only a few species survive today.

The pearly nautilus shell is spirally coiled and divided into compartments, each of which represents a chamber in which the nautilus lived at some stage of its growth. Naturally, the animal is found in the outermost chamber. About 90 tentacles are set around the mouth. Though these tentacles have no suckers, they can cling tenaciously to solid objects. The head can be withdrawn into the shell. A hood at the back of the head partly closes the opening.

The female of the paper nautilus, or argonaut, *Argonauta argo,* secretes a spirally coiled and symmetrical white shell. This shell serves as an egg case; the argonaut can drop it at will. The female may reach a length of 20 centimeters. The male, however, is a little creature, only about 2 to 3 centimeters long. It never secretes a shell.

CLAMS AND OTHER BIVALVES

The clams, oysters, mussels, and teredos belong to the class Pelecypoda, or Lamellibranchia. They are also known as *bivalves,* because their shells are divided into two parts, known as *valves.* The inner surface of the shell is coated with a layer known as *nacre,* or *mother-of-pearl.* This layer is fine-grained. It may be white or it may show all the colors of the rainbow.

Russ Kinne, Audubon/PR

The cuttlefish, a small octopus-and-squid relative. The cuttlefish is the source of cuttlebone and of the pigment sepia.

The two valves are joined by one or two strong muscles, which can hold the shell tightly closed. It is these muscles that are cut when a mussel or clam is opened. Some bivalves, such as clams, have a well-developed foot, which can be extended beyond the shell to move the animal from place to place. True oysters, however, cannot move about as adults. They are firmly attached to solid objects on the bottom of the sea. Bivalves have no specially differentiated head.

Some bivalves have two tubes, or *siphons,* through which water is drawn in and forced out. The water that is sucked in contains the tiny organisms that serve as food: protozoans, eggs, larvae, the spores of algae, and minute plants called diatoms. Food is taken into the digestive canal by way of a mouth opening. Oxygen enters the blood through the two gills. Wastes are carried away as water is forced out of the excurrent siphon.

OYSTERS

The true, or edible, oysters (genus *Ostrea*), lead sedentary lives attached to an underwater object. The shell is quite asymmetrical. The valve that is fastened to a submerged object is large and quite thick.

The other one is smaller and thinner. The two parts of the shell are closed by a single muscle, popularly called the "heart," which extends from about the center of one valve through the animal's body to the other. True oysters occur in many parts of the world, particularly along the coasts of Europe, North America, and Japan.

When the first white settlers came to North America, they found that Indian tribes along the coast depended on oysters for a considerable part of their food. Evidently they had been eating these mollusks for generations, because large piles of oyster shells had collected around Indian towns and encampments. The first settlers and those who followed picked and dredged oysters from the shallow bays. For a long time it was thought that the supply would never be exhausted.

Increasing demand, however, led to overfishing in the late nineteenth century. It then became necessary to supplement the natural supply by planting barren bottoms with young oysters, thus starting new beds. Today a considerable portion of the oyster supply in North America comes from privately owned beds. Oysters are also raised in Japan and in various European countries, particularly France and Holland.

To raise oysters successfully, one must be familiar with the life cycle of these shellfish. The female of a typical species, such as *Ostrea virginica,* which is found along the eastern coast of North America, produces millions of eggs a year. They are discharged into the water, where many are fertilized by sperm cells ejected by male oysters. If the egg is fertilized, it develops into a tiny larva, which swims about freely at first. After a couple of days it begins to develop a shell. In a week it is entirely enclosed. Dropping to the bottom it becomes attached to a solid object such as a rock or shell. The *spat,* as the young oyster is now called, grows rapidly and in time becomes a mature oyster.

In spite of the vast numbers of eggs produced by female oysters, the oyster population is not constantly on the increase. For one thing, many of the eggs are not fertilized. Also, vast numbers of the little larvae are eaten by fishes during the period when they are swimming about in the water. Even after they drop to the bottom and become securely attached, they are by no means safe. They may be smothered by shifting sand and mud, or devoured by starfish, drumfish, or other natural foes. And then, once oysters reach the adult stage they are sacrificed by the millions to meet the demands of the market.

During the breeding season, oystermen locate places where the surface of the sea is covered with oyster larvae. They pave the bottom in such places with various hard materials, such as old bricks, tile, empty bivalve shells, brush, and discarded metal parts. After a time the spat drop to the bottom and become attached to the paving materials. These materials are then dredged up and planted in spots that have been selected as favorable for the development of oyster beds.

Oysters are often planted in moderately shallow water, where the bottom is of hard mud. In such a place there are likely to be marine plants, which will provide food for the microscopic organisms upon which oysters feed. Oystermen avoid places where there is shifting mud and sand, or where starfish or other natural enemies of oysters abound, or where the waters may be contaminated with sewage.

Oysters that are ready for market are collected in shallow waters by means of oyster tongs. These are like two long-tined rakes, hinged so as to open and close like shears. In deeper waters the oysters are taken by means of a dredge.

In France young oysters are removed to partially enclosed growing ponds, admitting the tides through sluices and floodgates. When fully grown, the oysters are fattened in small enclosed ponds called *claires.*

Japanese oyster farms are generally in shallow, brackish water. Each farm is enclosed by a bamboo fence or hedge. The spat are collected and held on bamboo stakes thrust into the bed. When the oysters are fully grown, the stakes are pulled out and the oysters harvested.

Cross-section of the chambered nautilus shell. This type of animal is very ancient. Only a few species of nautilus survive today.

Grant Heilman

CLAMS

Many of these bivalves are also eaten by people. One of the most sought after is the soft-shelled clam, *Mya arenaria*—so called because of its rather thin and fragile shell. It is found in Europe and along the Atlantic and Pacific coasts of North America. The soft-shelled clam is also called the long-necked clam because of the unusual length of its "neck." This consists of two tubular siphons that are joined together and covered with tough skin.

This clam burrows into mud or sand with its tongue-shaped foot to a depth of 7 to 10 centimeters. The "neck" extends just out of the sand at high tide, as the animal feeds. At low tide, holes in the mud or sand show where the clam is buried. It is then a simple matter for a clammer to walk along the beach and dig out quantities of the bivalves.

The hard-shelled clam, *Venus mercenaria*, differs in various respects from the soft-shelled variety. Its thick, solid shell is a rather dirty white in color and is marked with concentric rings. The inner part of the shell is whitish, turning to purple at the outer edges. In the Americas, this purple section was used by coastal Indians for the money known as wampum. The hard-shelled clam is also known as the quahog and as the littleneck clam, since its siphons are much shorter than those of the soft-shelled variety.

The hard-shelled clam is found in great numbers along North America's Atlantic coast. It dwells on sandy or muddy bottoms at depths ranging up to 15 meters. It makes its way through sand or mud with its large foot. Clammers usually go out in boats to fish for hard-shelled clams, gathering them with a rake or dredge. The clams are served raw on the half shell or are used for clam fries and chowders.

Perhaps the most remarkable member of the clam group is the giant clam, *Tridacna gigas*, found in the coral reefs of the Pacific. This is the largest of the bivalves. Its shell may be almost a meter long and may weigh over 200 kilograms. The edible portion may come to 9 kilograms or more. Giant clam shells have been used for baptismal fonts and babies' bathtubs.

SCALLOPS

The bivalves known as scallops are found in many parts of the world. Their range extends from shallow water to fairly deep water. The shell is fan-shaped and the valves are arched and rounded. There are two winglike projections at either end of the

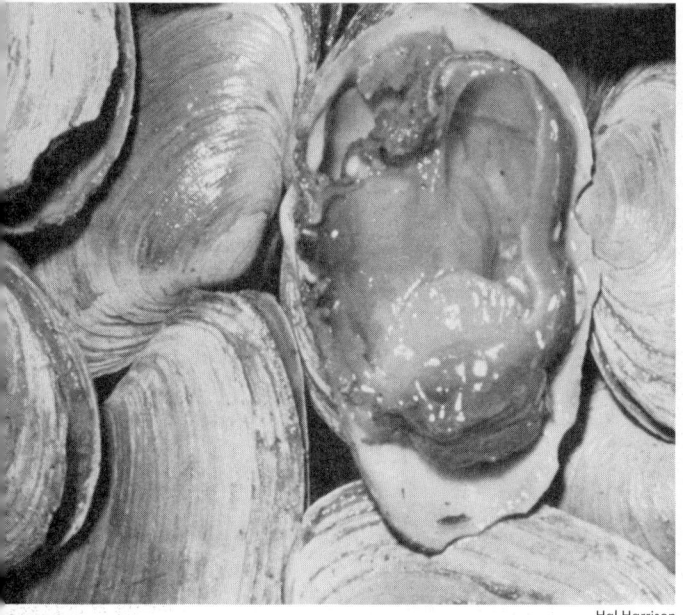

Clams are bivalves: they have shells divided into two parts. The photo shows a collection of clams with one open, showing the clam body.

hinge of the shell. About 20 ridges radiate from the hinge, increasing in width as they extend outward.

Scallops are good swimmers, especially when young. The jets of water they spout as they alternately open and close their shells propel them through the water in a series of jumps.

Scallops can swim by opening and closing their shells. The beadlike structures around the edge of the shells are eyes.

Several species of scallops are highly esteemed as food. Only a small portion of the body is eaten: the single large muscle that in life serves to hold the two valves of the shell together.

MUSSELS

The marine mussel has a wedge-shaped black or bluish shell. A bunch of threads, called a *byssus,* is secreted by a gland located immediately behind the foot. These threads harden when they come in contact with sea water and cause the animal to be firmly attached to solid objects such as rocks. The byssus can be discarded and a new one secreted. In this way the animal can move to new surroundings if unfavorable conditions arise.

The edible mussel known as *Mytilus edulis* is popular in various parts of Europe. It is served in restaurants of U.S. cities along the eastern coast. It abounds in Atlantic coastal waters and also in the Mediterranean.

The freshwater mussel is found mostly in rivers and lakes, where it burrows in mud. The pearly shells of fresh-water mussels are used in the manufacture of "pearl" buttons.

TEREDOS

The teredo, or shipworm, is a boring bivalve. It excavates burrows in wood that is under salt water. The two valves of the teredo are provided with fine ridges, suggesting the teeth of a file. Soon after it hatches from the egg, the teredo begins rasping with its double file at the wood of a pile, breakwater, or ship bottom. As the burrow that is formed is deepened, it is lined with a pearly coating. In time the teredo becomes a long, wormlike creature. The tapering body dwarfs the tiny valves, which are at the innermost part of the burrow. Siphons protrude from the opening of the burrow to draw in water and food and force out wastes. When the siphons are drawn in, the hole is closed by means of two plates attached to the rear end of the body.

Outwardly, a piece of timber attacked by teredos shows only a number of small holes. Inwardly, it may be honeycombed

The shell of a giant clam. This one is about one-half meter wide, but clams can grow to a width of nearly one meter.

Runk/Schoenberger/Grant Heilman

with teredo burrows, sometimes so close together that the wood between them is as thin as paper. In time the most solid timbers are so riddled that they collapse. Metal or concrete sheathing is used to protect timber from the attacks of teredos. Heavy impregnation with creosote has also proved effective.

SNAILS AND RELATED FORMS

Snails, slugs, limpets, abalones, and conches are included in the large class of mollusks known as gastropods. These animals have a foot and mantle cavity, like other mollusks. They have a well-developed head region and generally possess a spirally coiled shell that is all in one piece.

SNAILS

Snails are particularly widespread. Some dwell in the ocean, others in the fresh water of rivers, ponds, and lakes. There are numberless land snails, too. They abound in tropical jungles and are also found in damp places in temperate zones.

The snail's head bears the mouth opening and one or two pairs of tentacles. The eyes are set upon or at the base of the tentacles. The animal creeps upon its flat foot from place to place. Certain gland cells of the foot secrete mucus, which lubricates the path over which the snail crawls. This accounts for the slick trail that the animal leaves as it passes over a more or less flat surface. Both the head and foot of a snail can be withdrawn into the shell.

Freshwater snails and land snails were probably eaten by people long before recorded history. Today they are regarded as delicacies in many countries. The market supply comes largely from snails that are raised in captivity. The snail farms of southern France, Italy, and Spain are well known. About 10,000 snails can be kept in a pen 8 or 9 meters square. The animals are fed meal, vegetables, and bran.

In many areas snails are a pest because they feed voraciously on garden crops. The giant African snail, *Achatina fulica,* has become a particularly serious menace. This creature, a native of east Africa but now found in many other lands, is sometimes more than 15 centimeters long and as big around as a tennis ball. Its diet is varied, including garden plants, flower petals, decaying tissues, and manure. It is long-lived and fertile and can thrive under the most unfavorable conditions.

Whelks and *periwinkles* are marine snails that are commonly used as food by

Snails and slugs are gastropods. Top: a land snail commonly found in gardens. These animals eat a great deal of plant matter. Bottom: a slug. Slugs have no shell. They are plant eaters, usually feeding at night.

Europeans. The whelk is widely distributed in the North Atlantic. Besides serving as food, it is used as bait in cod fishing. Periwinkles are found in temperate and cold seas in many areas. They abound on rocks and in seaweed and feed on seaweed. The long tongue, or radula, of the periwinkle is a remarkable structure. It is provided with many rows of sharp, curved teeth.

The rasping radula of the snail known as the *oyster drill* is particularly well developed. This tiny creature, less than 2½ centimeters long, drills a hole through the shell of an oyster near the hinge and then sucks out the soft body of the victim through the hole. The oyster drill is one of the chief foes of oystermen.

SLUGS

Among the snails' kin are the curious animals called slugs. These mollusks, which range in length from 2 to 10 centimeters, have no external shell. Land slugs live in moist places; they are often found under stones and in holes in the ground. At night they emerge from their retreats to feed on vegetation. They sometimes invade vegetable gardens. Sea slugs crawl on rocks or seaweed in shallow water along the coasts of North America, Europe, and Asia. They also feed on plant matter.

LIMPETS

The gastropod known as the limpet has a rounded or oval shell and looks like a diminutive volcanic cone. Some limpets even have a small opening at the top of the shell, suggesting a crater. Limpets adhere so firmly to rocks near the low-water mark by means of the suckerlike foot that they withstand the beating of the surf. At high tide they move about in search of the algae on which they feed. After their feeding forays they again attach themselves to the rocks. Limpets are found in many parts of the world.

ABALONES

The shell of the abalone has a rather startling resemblance to a human ear; for that reason this gastropod is sometimes called an *ear shell*. The large shell is very ornamental, particularly after the rough outer surface has been polished. Abalones are found in the Far East and on the Atlantic and Pacific coasts of the New World. They live on rocks near the shore, feeding on seaweed. When disturbed, they cling with surprising strength to rocky surfaces. The flesh is often used in stews and chowders. Sometimes it is prepared in the form of a steak. In the Far East it is generally dried or smoked.

CONCHES

The conch is a large gastropod that is especially common along the coasts of the southern United States and the West Indies. The shell is sometimes 25 centimeters long and may weigh as much as 2½ kilograms. It has a small spire with a large lower whorl. The foot of the conch is provided with a clawlike appendage. The animal moves in a series of leaps, sometimes turning quickly to avoid capture. Conch shells are sometimes made into horns. They are also used for cameos and buttons. The flesh serves as food in certain areas, including the Bahamas and the Florida Keys.

TOOTH SHELLS AND CHITONS

The mollusk class of the scaphopods, or tooth shells, is a small one, numbering only about 200 species. In most species the long, curved, tapering, ivory-colored shell looks something like a boar's tusk. In some varieties, known as elephant-tusk shells, the shell is not curved. The tooth shell generally lives in fairly deep waters off the coasts, in many regions.

The chitons and their kin make up the class Amphineura. They are found all over the world, except in the far north and south. Chitons have a shell consisting of overlapping plates. They may be seen at low tide crawling about or clinging to seaweed or rocks. The larger chitons are edible; the flesh is generally called *sea beef*. Some Amphineura have no shells and look like grubs.

ECONOMIC IMPORTANCE OF MOLLUSKS

The mollusks are exceedingly useful to mankind. We have already considered their importance as a source of food, and we have mentioned some of the other ways in which they serve us.

Mollusk shells are particularly valuable. The beautiful shells of abalones, conches, and other varieties are commonly sold as souvenirs and used as ornaments. The mother-of-pearl inner layer of various mollusk shells is used for pearl inlays and knife handles and in hundreds of other ways. Tons of bivalve and other mollusk shells are ground up every year and used as material for surfacing roads. Because of their lime content, the ground-up shells are used as fertilizer and are also fed to domesticated birds, such as chickens.

Pearls. Pearls are undoubtedly the most glamorous products yielded by mol-

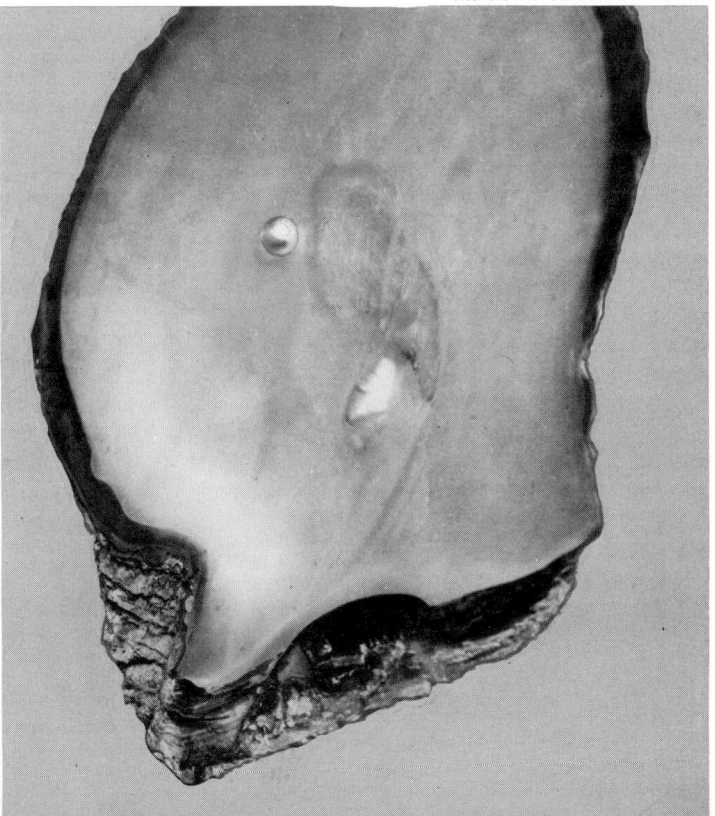

Australian News and Information Bureau

A blister on an oyster shell. Blisters are formed the same way as pearls. Nacre is secreted around an irritant that has managed to get inside the oyster shell.

lusks. Like mollusk shells, these precious gems represent the secretions of the mantle that envelops the body. They are made up of the same lustrous substance—mother-of-pearl, or nacre—of which the inner layer of the shell consists.

How a pearl is produced. Suppose that a foreign body, such as a grain of sand or a parasite, finds a lodging place between the mantle and the shell. It will serve as an irritant and will greatly stimulate secretion at that point. As the grain or parasite is slowly encased in a nacreous coating, it may be rolled about by slight contractions of the mantle. It thus remains free from the shell and takes on a rounded form. In time it will be enclosed in many layers of nacre and will be a full-fledged pearl.

By far the greatest number of pearls used in commerce come from the so-called pearl oysters, belonging to the genera *Avicula* and *Pinctada*. However, various other mollusks, including some edible oysters and some freshwater mussels, also yield pearls.

More or less perfectly formed spherical pearls are few and far between. In many cases such pearls originate when parasitic worms form cysts in the mantle layer. As the cysts are more or less rounded in form, they make ideal development centers.

A great many pearls, called baroque, are more or less irregular in form. They may be pear-shaped, dome-shaped, or rather flat. Baroque pearls are often made into pendants, brooches, and rings. Baroque pearls are not nearly so valuable as spherical pearls.

Pearls are of many different colors, including white, cream, rose, brown, blue, yellow, and green. The color depends on the oyster's diet, the temperature of the water, and various other factors. The favorite colors for gems are white, cream, rose, steel-blue, and black.

The matching of pearls has much to do with their commercial value. When large pearls of the same size and shape are matched to form a necklace, they bring a far higher price than if the individual pearls were sold separately. Necklaces made up of pearls that are matched in a perfectly graded series—smaller to larger to smaller—are also valuable.

Pearl fisheries are found in various parts of the world. The most valuable pearls come from the pearl oysters that grow in the warm waters of the Persian Gulf and the Red Sea and off the coasts of India, northern Australia, certain South Pacific islands, and Central America.

Pearl collecting is an uncertain business at best. In one instance a week's catch of 35,000 pearl oysters yielded only 21 pearls, of which only three were suitable for commercial use. In a number of cases the collecting of pearls is carried on in connection with the more dependable business of collecting mollusk shells and preparing them for market.

Cultured pearls. Humans have succeeded in "growing" pearls by deliberately inserting a foreign substance within the shell of an oyster or mussel. A pearl produced in this way is called a cultured pearl. It is very different in origin from the artificial pearl created by the chemist and used in much costume jewelry.

The Chinese Buddhists were pioneers in the production of cultured pearls. They inserted small plates bearing the image of Buddha between the shell and mantle of marine clams. These plates became coated in time with nacre. They were then sold as souvenirs and also as objects of religious veneration. This practice has continued to the present day.

The Japanese have succeeded in producing spherical cultured pearls on a commercial scale, and they have almost a monopoly of the cultured-pearl industry. A mother-of-pearl bead is carefully inserted in the mantle of a pearl oyster, which proceeds to cover the bead with a thin coating of nacre. The original bead inserted in the oyster makes up most of the finished product. The nacre coating is generally only about a millimeter thick. Of course, the longer the pearl is permitted to develop, the thicker the coating—and the thicker the coating of nacre, the more valuable the pearl.

THE EARTHWORM AND ITS KIN

By F. L. Fitzpatrick

Thrust a spade into soft, rich soil and you are sure to dig up some earthworms. Sometimes after a heavy spring rain you'll see them crawling across a wet sidewalk.

Earthworms are common in almost everyone's garden. The giant earthworm of Australia, shown here, is larger than the garden variety. It may grow to nearly four meters.

Australian News and Information Bureau

And if at night you look closely at a plot of grass, you may find worms partly out of their burrows, enjoying the cool, moist air. If you plan to go fishing the next day, this is an ideal time to gather worms to use as bait.

Earthworms are probably the best known of the segmented worms, or annelids. Together with some 7,000 other species, they make up the phylum Annelida. The name is derived from the Latin *anellus,* meaning "small ring." It refers to the segmented nature of the organisms' bodies.

Annelids are slender creatures. They have a complete digestive tract, with a mouth at one end and an anus at the other. Most species have *setae*—short, bristlelike hairs extending from the body wall and used in locomotion. The hairs also serve another purpose. It is because earthworms cling so tenaciously with their setae to the walls of their burrows that birds find it so hard to pull them out of the soil.

There are three main classes in this phylum: the oligochaetes, which include the earthworms; the leeches; and the polychaetes, or bristle worms.

EARTHWORMS AND THEIR RELATIVES

The oligochaetes, or "few-bristled worms," are among the most common members of the annelid group. They show a rather striking variety of colors: brown, purple, blue, green, and a nondescript pallid color.

The common earthworm, belonging to the genus *Lumbricus,* has long served as the classical example of the phylum. The body of this worm has more than 100 segments, which are very noticeable because of the grooves extending around the body. Each segment, except the first and the last,

The marine annelids may be found burrowing among the roots of eelgrass in the muddy bottoms of coastal waters.

bears setae, which are moved by muscles within the body wall.

An earthworm's body is covered by a thin, transparent membrane called the *cuticle,* which is secreted by the skin layer just beneath it. The cuticle is always kept moist by glandular secretions. It is through this membrane that an earthworm breathes. There are hundreds of different earthworm species distributed all over the earth except in localities that are very cold or very dry.

The earthworms are exceedingly valuable because of the way in which they turn up the soil. The English naturalist Charles Darwin called attention to the "plowing" activities of worms in his *Formation of Vegetable Mould Through the Action of Worms,* published in 1881. "The plough," he observed, "is one of the most ancient and most valuable of man's inventions, but long before he existed the land was, in fact, regularly plowed and still continues to be thus plowed by earthworms." Darwin pointed out that a certain field had once been covered with stones. These had entirely disappeared after some 30 years passed. They had been completely covered by the castings, or wastes, of earthworms.

Earthworms literally eat their way through the soil, obtaining nourishment from organic matter contained in it. They bring their wastes to the surface. It is in this way that they turn over the soil. It has been estimated that over 100,000 worms may be found in a single hectare. The earthworm population in black loam will bring a one-centimeter layer to the surface, on the average, in two years. The burrows of the worms make soil porous and enable rain water to penetrate within the earth.

During and following heavy rains, large numbers of earthworms are often seen crawling about on the surface of the ground. The rain water has flooded them out of their burrows. Fishermen looking for bait go out on rainy nights in order to collect some of the larger worms, which are called night crawlers.

Certain oligochaetes are very small, barely visible to the naked eye. Others, such as the giant earthworm of Australia, are imposing animals two meters or more in length. They are sometimes mistaken for snakes. There are many variations in length between these two extremes.

Earthworms serve as food for other animals, such as birds and mammals (including people). Some authorities think that they may do a certain amount of harm by spreading disease. If they have previously burrowed through the decaying bodies of diseased animals, the worms may transmit diseases to domestic animals that feed upon them. It is known that some earthworms contain the young of parasitic roundworms that live as adults in domestic poultry.

Certain popular beliefs about earthworms are either entirely erroneous or only partly true. For example, some people maintain that the worms turn into fireflies — an absurd notion, based perhaps on the fact that the firefly is a beetle that passes through a wormlike larval stage.

Another belief, which is only partly true, is that if an earthworm is cut in two, both parts will continue to live and finally will develop new segments to replace those that have been lost. It is true that the head part of an earthworm that has been cut in two often continues to live and may add, or regenerate, tail segments as time goes on. If the cut has been made too near the tail, however, the tail part never succeeds in redeveloping the important internal organs in the forward part of the body. It dies after a comparatively short time.

Earthworms reproduce hermaphroditically. That is, each worm has both male and female sex organs. When copulating, paired earthworms exchange sperm. The eggs develop in coccoons, taking about two or three months to develop. Worms may live more than ten years.

LEECHES

Leeches are also well-known members of the annelid group. They are usually somewhat flattened in form. The body is divided into segments, but, except in one species, no setae are developed.

Leeches have a sucker at the rear end of the body. Many species, too, have a sucker surrounding the mouth at the head end. Some species are found in the sea, some in fresh water, and some on land.

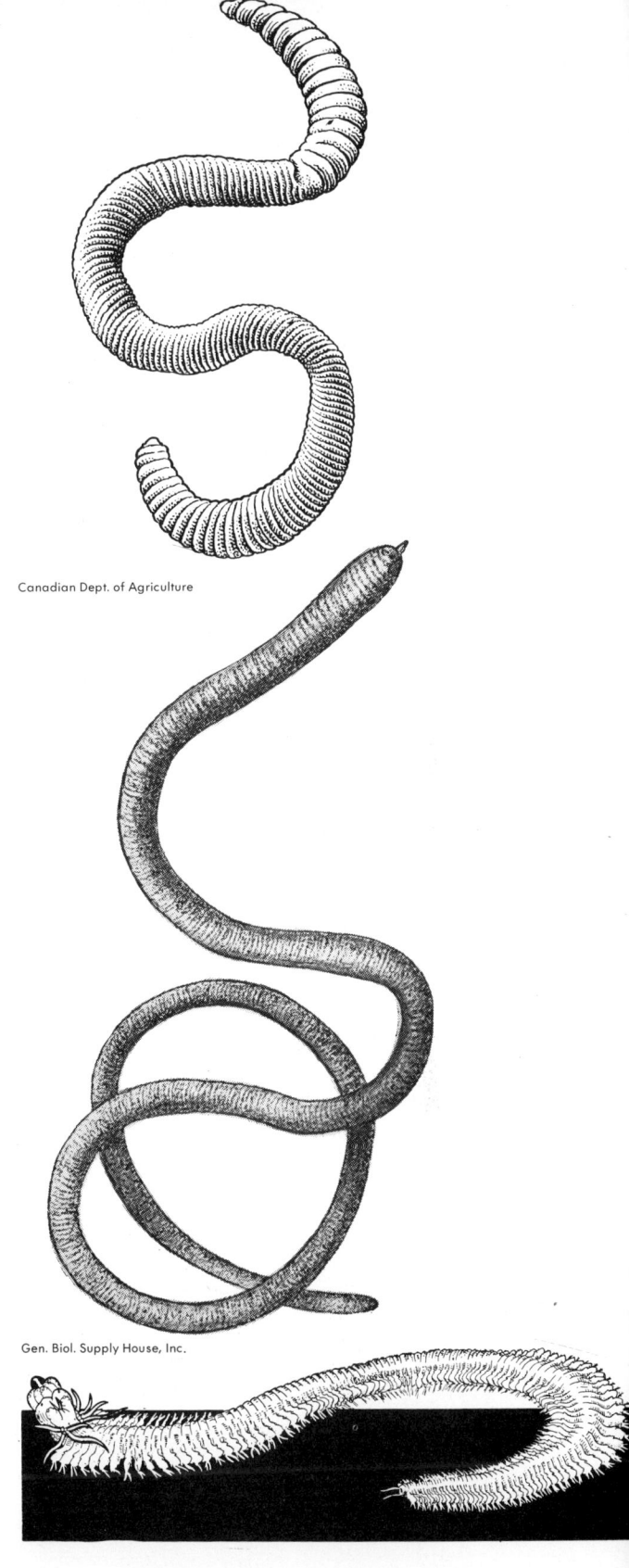

Canadian Dept. of Agriculture

Gen. Biol. Supply House, Inc.

Top: typical earthworm, showing segments. Middle: giant thorn-headed worm that infests pigs. Bottom: the clamworm. Unlike the earthworm, the clamworm has a head that bears sensory appendages. It lives near the low-tide mark of ocean beaches.

THE EARTHWORM AND ITS KIN 229

They are generally parasitic animals.

A parasitic leech attaches itself to the host by means of its suckers. Once this is done, the animal makes an incision in the host's skin and gorges itself with blood. Then it drops off. Some non-parasitic leeches capture and devour small forms of life that live in the water. Leeches also use their suckers to help them move. They make looping movements — like a measuring worm — over solid surfaces. They swim through the water with undulating movements of the body.

During medieval times and for several centuries thereafter, bloodsucking leeches were used by physicians to draw blood from patients. After it was discovered that bleeding had little or no curative value, the practice of using leeches to draw blood was largely, although not entirely, abandoned.

In temperate regions of the world, leeches annoy bathers. However, they are not particularly harmful to people. Sometimes, they get into the throat and nasal passages of certain forms of wildlife, including ducks, and strangle their victims.

Certain species, such as *Limnatis nilotica* in the Middle East and *Dinobdella ferox* in India, live in springs or wells. They enter the mouth or nasal passages of people or animals that drink from these springs or wells. Attaching themselves to the walls of the nasal passages, they obstruct respiration and cause hemorrhages.

Bloodsucking land leeches, found in great numbers in tropical rain forests, also enter the nostrils, which they block until they have had their fill of blood. It is a very serious matter if they enter the sinuses, because once the leeches have become swollen with blood they cannot make their way out again.

BRISTLE WORMS

Among the most interesting of the annelids are those living in the sea. Some of them, such as the clamworms belonging to the genus *Nereis,* are free-swimming animals that prey on other marine creatures. Certain marine forms have beautifully colored gills — large, plumelike projections on the head or sides of the body. The odd sea mouse, *Aphrodite,* has an oval body covered on top with a thick mat of long, silky, and iridescent hairlike setae.

Several of the marine annelid families are well-known tube builders. Some line the burrows in which they live with a thin, limelike secretion. Others build tubes in the sand. The parchment worm, *Chaetopterus,* constructs a parchmentlike U-shaped tube that is buried with only the openings jutting above the sandy mud. The appendages of the middle segments of this worm are united to form three pairs of circular fans that draw water in through one opening of the tube and force it out through the other. The numerous tiny organisms carried into the tube by this water current serve as food for the worm.

The palolo worm, *Eunice viridis,* found in the South Pacific, is famous for its breeding habits. During most of the year the worm lies coiled up in its burrow, generally in a coral reef. As times goes on, eggs or spermatozoa develop in the hind-end segments of the animal. In the last quarter of the October-November moon, the hind ends of the palolo worms are cast off. They make their way to the surface of the sea in vast swarms and the eggs and spermatozoa are then discharged into the water. The natives collect the worms by means of nets and use them as food. A related species, *Eunice fucata,* of the West Indies, breeds similarly in the third quarter of the June-July moon.

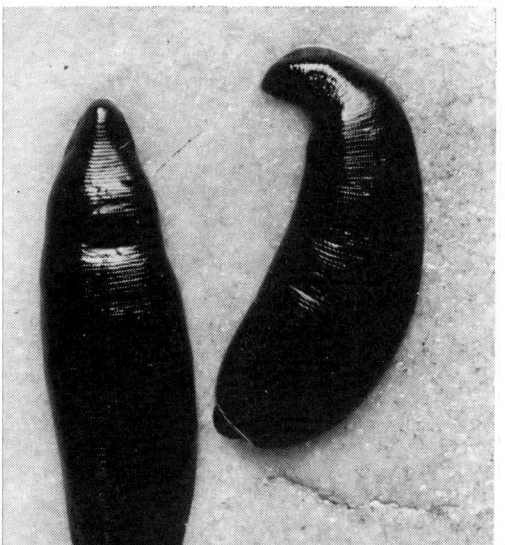

Leeches are annoying and sometimes dangerous worms, attaching themselves to their victims and sucking blood from them.

Cornelia Clark

LOBSTERS, CRABS, AND THEIR KIN

by F. L. Fitzpatrick

Broiled lobster from the waters off South Africa or New England. A salad made with tiny shrimp gathered off the coast of Mexico or Greenland. Hot, creamy soups made from the European crab *Maja squinado*. These are just a few tasty dishes that can be prepared using the many edible species of invertebrates called crustaceans.

Crustaceans derive their name from their tough, crustlike shell: the Latin word *crusta* means "crust" or "shell." In addition to lobsters, shrimps, and crabs, the class Crustacea includes crayfish, barnacles, water fleas, and a host of less well-known animals. Almost all live in water.

These animals belong to the largest group in the animal kingdom, the phylum Arthropoda, or "animals with jointed feet." This phylum contains well over 1,000,000 species, of which some 30,000 are crustaceans. In other articles we look at such arthropods as the spiders and their kin and the insects.

GENERAL CHARACTERISTICS

The body of a typical crustacean is composed of distinct segments. It is usually divided into two chief areas: the cephalothorax, made up of the head and thorax, and the abdomen. There is an exoskeleton, or outer skeleton, consisting of a flexible substance called *chitin*. In most crustaceans the chitin becomes impregnated with calcium carbonate ($CaCO_3$), most familiar to us as the principal component of limestone. This results in the animals' tough, crustlike shell.

The tough outer skeleton is shed at intervals so that the animal may continue to grow. The old exoskeleton splits along the back and the crustacean works its way out of its covering. It can then grow considerably before the new exoskeleton hardens. During this time the crustacean is in what is called the soft-shell stage. It generally retires to a secluded spot in order to avoid its enemies.

A crustacean has two sets of antennae, which are provided with sense organs. In very small representatives of the group the antennae are also swimming structures. In the water flea *Daphnia*, for example, they

A well-known crustacean—the American lobster. These lobsters may grow to over 50 centimeters.

American Museum of Natural History

serve as "oars." There is a pair of hard mandibles for chewing. Two pairs of maxillae and the maxillipeds test food and pass it into the mouth.

The eyes of many species are mounted on movable stalks. In some species the eyes are compound—that is, they consist of a number of units. The crayfish, for example, has two compound eyes, borne on short stalks; each of these eyes consists of about 2,500 units.

The number of legs, or appendages, which occur in pairs, differs according to the species. The appendages located on the thorax serve as walking legs. The first pair, however, may be greatly enlarged to form grasping structures, or *pincers*. The rather small appendages borne on the abdomens of some crustaceans are called *swimmerets*. The eggs of female lobsters and crayfish are attached to these structures.

In many cases appendages and even eyes may be regenerated, or grown again, if they are damaged or lost. The walking legs of crayfish, for example, have a breaking point near the base. If one of these legs is injured, the animal may snap it off, and in time a new growth will replace it.

The larger crustaceans have well-developed internal organs and systems for carrying on such processes as digestion, respiration, circulation, excretion, reproduction, and nervous responses. The smaller representatives of the group lack a definite respiratory system. They breathe through the entire surface of the body.

Most crustaceans live either in the sea or in fresh water. Some have invaded the deeper regions of the ocean. They may dwell one and one-half kilometers or more below the surface. A few species live on land.

The eggs of marine crustaceans generally develop into free-swimming larvae, which are often decidedly odd-looking creatures. They usually pass through several larval stages before they attain the adult form. Freshwater crustaceans commonly skip the larval stage. The young have the adult form when they hatch.

Many crustaceans, including lobsters and crabs, eat both animal and vegetable matter and often decaying substances. To a certain extent, therefore, these animals act as scavengers. The so-called fish lice are crustaceans that live as parasites on fishes.

LOBSTERS

A large crustacean whose meat is especially prized is the lobster, belonging to the family Homaridae. The larger members of the family may weigh 13 kilograms or more and may reach a body length of 60 centimeters, with pincers 35 to 40 centimeters long. However, such giants are not particularly common. In the United States most of the lobsters reaching the market are under one kilogram in weight.

Some lobsters dwell on the sea bottom in shallow water throughout the year. Others move into shallow water in the spring and migrate to greater depths as the water grows colder in the fall. Lobsters generally prefer rocky bottoms.

A female lobster produces thousands of eggs a season. The eggs are attached to the swimmerets on the lower side of the abdomen. When the larvae hatch, they make their way to the surface of the sea. There they swim or float about for a period of three or four weeks. During this period, the young lobsters are destroyed in vast numbers by surface-feeding fish. The survivors eventually drop down to the bottom, where they take on the appearance and habits of adults.

In nature, the percentage of young that attain maturity is relatively small. In the United States, many young lobsters reach the larval stage in the comparative security of salt-water pools or tanks at hatcheries. They are kept in these waters until they are past the vulnerable stage in the early part of their life cycle.

For many years the preferred lobster-catching device has been the trap called the lobster pot. It is a cratelike affair, made of wooden slats that are spaced far enough apart so that small lobsters, which would not be acceptable for the market, can crawl out and escape. The trap is provided with a funnel-shaped entrance that makes it easy for a lobster to enter but practically impossible to get out.

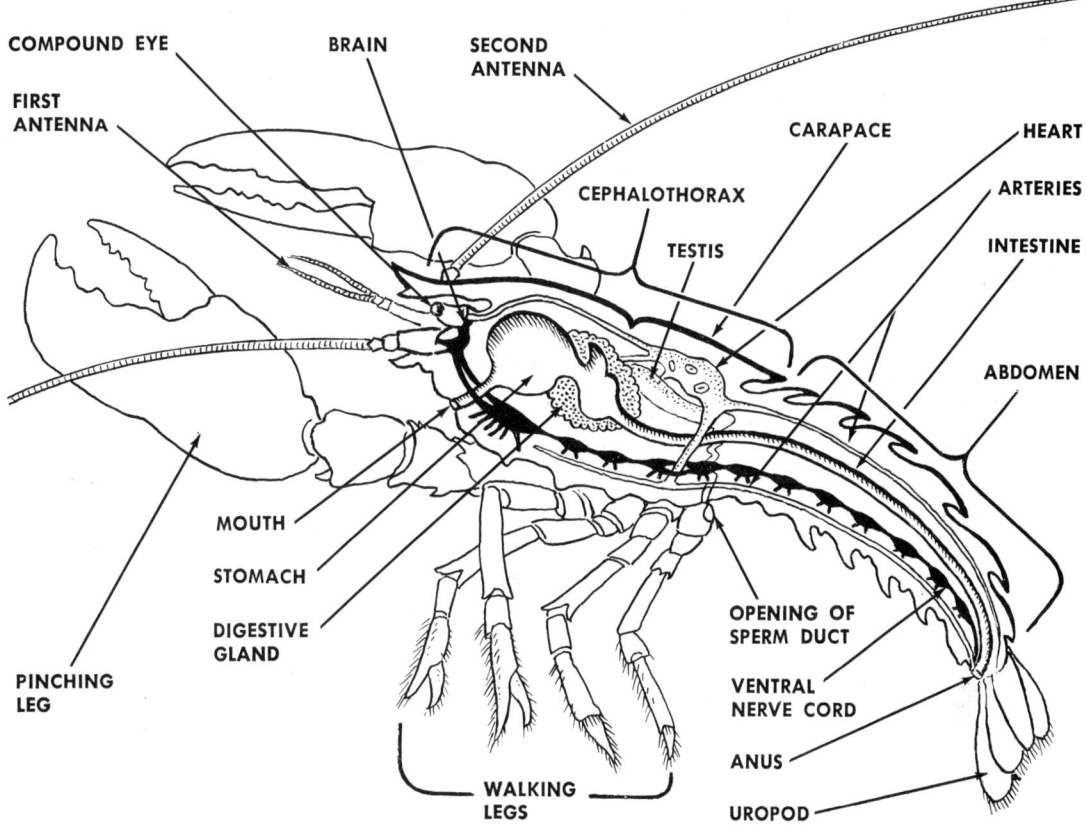

Diagram of the external and internal anatomy of male common lobster. Note the simple construction of the digestive, nervous, and circulatory systems.

The lobster pot is carefully baited, generally with decaying fish. Then it is weighted, so that it will sink to the bottom, and set out in shallow-water areas where lobsters are likely to be found. The position of the trap is marked by a buoy on the end of a rope attached to the trap. The person who sets the pots visits them at regular intervals to remove the catch through a hinged door and to replenish the bait.

The American lobster, *Homarus americanus,* is found northward from the Carolinas along the Atlantic coast of the United States and Canada. In the not too distant past, 1000,000,000 American lobsters were sent to market each year, and lobster canneries worked overtime along the New England coast. In those days lobsters could be bought very cheaply; they were not generally considered to be a luxury food. They are now much scarcer and the market price has soared accordingly.

Several varieties of lobsters thrive in European waters. The common lobster, *Homarus gammarus,* occurs off coastal areas from the Mediterranean to Norway, and is usually caught by means of lobster pots. A smaller species—the Norway lobster, *Nephrops norvegicus*—is an inhabitant of the same regions. It is found in deeper waters and is usually taken by trawling.

Lobsters without pincers occur in warm waters off various land areas, including Florida, the West Indies, South Africa, Australia, and New Zealand. These animals, which belong to the family Palinuridae, are armed with spines, with which they can stun their prey or hold off pursuers. The meat of these lobsters is also highly prized.

CRAYFISH

The freshwater crustaceans called crayfish make up the family Astacidae. They look much like diminutive lobsters. The compartively small varieties belonging to the genus *Cambarus* are a familiar sight in and about streams, lakes, and ponds in the eastern part of North America. A larger variety, *Astacus,* is found in western North America, and in Europe and Asia.

Crayfish normally live in shallow water. Here they avoid their natural enemies by hiding in aquatic vegetation or under stones. They sometimes come out onto the banks under cover of darkness. Ordinarily they crawl forward, but if they are disturbed they can shoot backward with considerable speed. They feed on a wide variety of plant and animal life, including other crayfish, tadpoles, small fish, snails, insects, and aquatic plants. They also serve as scavengers. They are themselves the prey of raccoons, opossums, bears, alligators, ducks, fish, and other animals.

In some relatively moist areas, crayfish may desert their ponds and streams and live on land. Here they dig burrows down into the soil until they strike the water that will enable them to keep their gills moist. From time to time they sally forth from their retreats to feed upon young vegetation, including crop plants such as cotton and corn. In irrigated areas, their burrowing activities sometimes weaken earth dams

The male fiddler crab has one claw that is much larger than the other.

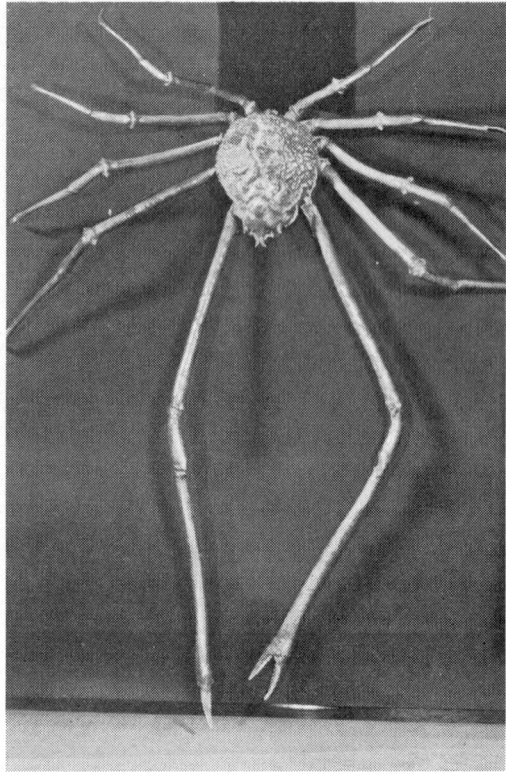

both photos, American Museum of Natural History

The Japanese spider crab is the largest of all crabs. It may grow to a length of 3 meters.

and cause water to be lost from irrigation ditches.

Crayfish are considered edible in many areas where they are abundant. The larger varieties belonging to the genus *Astacus* are preferred because of their size. Crayfish also provide food for many fishes, and fishermen often use them for bait.

SHRIMPS

Like crayfish, shrimps look somewhat like small lobsters, with unusually long antennae and fragile shells. The larger shrimps are often called *prawns*. Various species of these small crustaceans are found in the deep sea. The varieties we eat, however, occur in shallow waters along coastal areas and in freshwater streams. Vast numbers of shrimps are found along the Gulf and Pacific coasts of North America and in European waters. They are caught in nets and are sold fresh, dried, or

The hermit crab uses the shells of other animals as its home. When it outgrows its borrowed shell, it finds a bigger one. Often, it shares its home with anemones.

American Museum of Natural History

in cans. Only the abdomens are eaten by humans.

CRABS

The crabs are distinguished from such crustaceans as lobsters and crayfish by their small abdomen, which is folded up under the body. They have five pairs of legs. The legs of the first pair end in pincers and are larger than the rest. The last pair of walking legs are often flattened and are used as swimming structures. Crabs usually walk or crawl with a curious, sideways gait. They generally have a varied diet, which includes dead animals and plants.

There are a great many species of crabs in various parts of the world. Most of them live in the sea; some are found in fresh water; a few species live on land. Water-dwelling crabs breathe by gills set in cavities at the sides of the body. These cavities are larger in the true land-dwelling species and serve as lungs.

One of the most important food crabs is the blue crab, *Callinectes sapidus,* found along the Atlantic coast from New England to Brazil. The so-called hard-shelled crabs and soft-shelled crabs of commerce are really blue crabs at different stages of the molting process. The hard-shelled variety is a blue crab with a fully developed exoskeleton. A soft-shelled crab is the same animal after a molt has taken place and before the new covering has had time to harden. Hard-shelled crabs are caught in traps or nets of various kinds. Soft-shelled crabs commonly hide in the vegetation of shallow waters and may be collected with dip nets.

Various other crabs are used as food. Among these are the edible crab, *Cancer pagurus,* found off the coasts of Great Britain and continental Europe, and the Dungeness crab, *Cancer magister,* which occurs off the Pacific coast of North America.

Among the most interesting crustaceans are the small hermit crabs belonging to the families Paguridae and Parapaguridae. The abdomen of these crabs is soft and quite unprotected. To shield this vulnerable part of its body, the animal seeks out an empty marine-snail shell and backs into it. The hermit crab's pincers and legs protrude from the front end when the animal crawls about in search of its food; it can withdraw entirely within its shelter if danger threatens. When a hermit crab outgrows one shell, it abandons it and seeks a larger one.

In some cases sea anemones establish themselves on the shells inhabited by hermit crabs. As the anemones are well armed with stinging cells, their presence undoubtedly brings added security to the crabs. The sea anemones also profit from the relation-

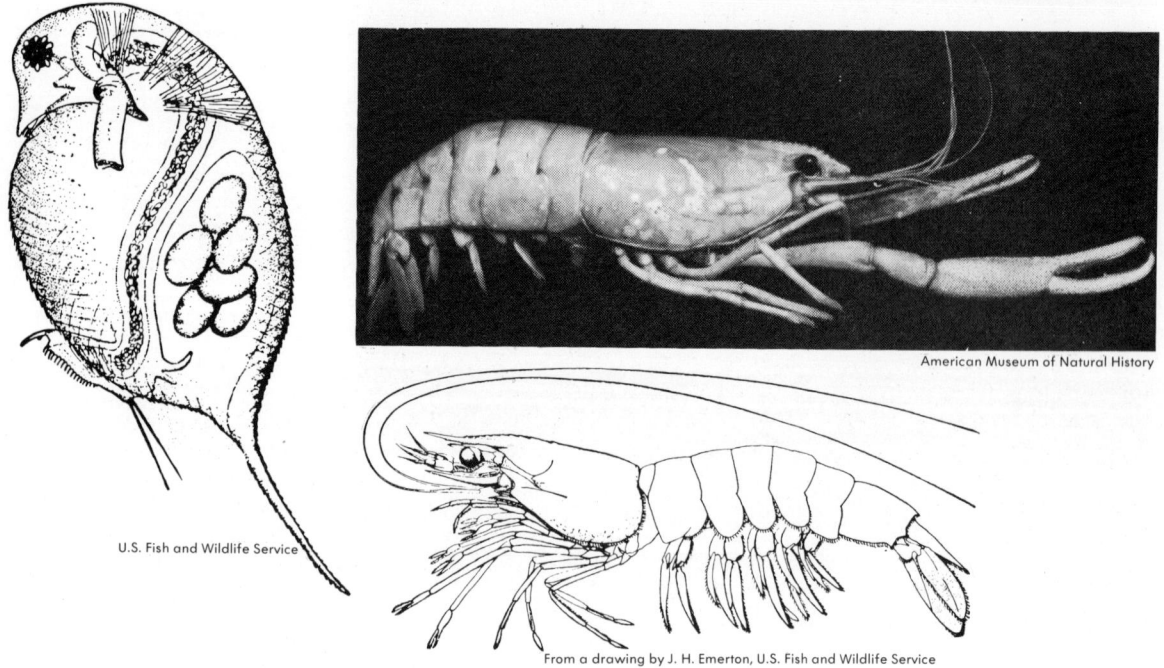

Top left: *Daphnia*, a water flea. Notice the eye spot at the head and the egg sac at the opposite end. Top right: shrimps resemble lobsters. Photo shows a freshwater shrimp from Panama. Bottom right: southern shrimp.

ship, since they are carried about from place to place. It has been reported that certain varieties of hermit crabs, upon changing from old snail shells to new ones, carefully transfer their sea anemone companions to the new shelters. The anemones do not appear to resist this action, though they generally sting other animals that try to uproot them.

A crustacean often found on beaches of temperate regions is the fiddler crab, belonging to the genus, *Uca*. One claw of the male—generally, but not always, the right claw—is much larger than the other. The fiddler crab lives in a burrow that it tunnels in soft, moist soil above the highwater mark. Generally it sits at the mouth of its burrow moving its large claw slowly, somewhat after the fashion of a fiddler manipulating a bow. This is the animal's way of attracting the attention of the female of the species.

The largest of all crabs is the giant spider crab, *Macrocheira kaempferi*, of Japan, which measures three meters or more from the tip of one of its fully extended pincers to the tip of the other. It is a rather inactive animal with a comparatively small, rounded body and long, ungainly legs. It commonly covers itself with seaweeds or sponges. Thus camouflaged, it stalks its prey deliberately until the victim is within reach of the slender claws.

Not all spider crabs are as long as the giant Japanese variety. A common species found on the Atlantic and Pacific coasts of North America, *Libinia emarginata*, measures only one-third meter or so across the extended pincers. Crawling about the sea bottom, it often makes its way into lobster pots and devours the bait.

One of the best-known land crabs is the robber crab, *Birgus latro*, found on the islands of the Indian and Pacific oceans. This large animal, often one-third meter or more in length, lives in a burrow at the foot of a coconut palm, lining its retreat with the fibers of coconut husks. Its pincers are very strong and heavy. It carefully strips the fiber from a coconut and then patiently hammers at the nut with its strong claws until the shell has been broken. It then extracts the meat of the nut.

The king, or horseshoe, crabs are a remnant of an ancient group, which is now almost extinct. They were at one time classified with the crustaceans but are now put in a separate class—the Merostomata. They are large arthropods measuring 30 centimeters or so across the shield, or carapace, which covers the top part of the body. When seen from above, the carapace has the outline of a horseshoe; this accounts for the name "horseshoe crab." The tail is long, pointed, and hard. The king crab lives in shallow water. It plows through sand or mud seeking its prey—worms and mollusks.

PILL BUGS AND GRIBBLES

The small land crustacean known as the pill bug (family Armadillididae) is often found far from ponds and streams. This little animal, which is about 15 millimeters long when fully grown, rolls itself up in a ball when disturbed. It commonly lives in cellars and under stones, boards, dead leaves, and other debris, in comparatively moist surroundings. Pill bugs feed upon plant materials, both living and decayed. They rarely become so numerous as to be a menace.

Some crustaceans, such as the gribbles, *Limnoria lignorum,* do not look much like their crustacean kin. The gribble, a small animal belonging to the isopod group, is a destructive creature. It looks like a small worm not more than four millimeters in length. It has two pairs of antennae and six pairs of legs.

Swarms of gribbles burrow their way into submerged timbers of all kinds, including the piling of wharves and the bottoms of ships. The boring activities of the gribbles are not entirely undesirable from the viewpoint of human beings, since they help break up floating wreckage that might otherwise be a menace to shipping.

BARNACLES AND CIRRIPEDS

Barnacles look like crustaceans only in the larval stage. The larva is a free-swimming creature. After swimming about for a time, the larva attaches itself to some solid object in the water and secretes a calcareous outer shell.

The adult of the variety known as the goose barnacle (family Lepadidae) is attached to the supporting object by a fleshy stalk. The body of the animal, enclosed in its shell, is at the other end of the stalk. The shell is made up of several plates and is roughly conical in shape. Six pairs of long appendages, fringed with hairs, can be thrust out from a slit in the shell. They strain minute organisms from the water and sweep them toward the animal's mouth inside the shell. Certain barnacles, called acorn shells, or acorn barnacles (family Balanidae), lack a stalk; the shell is attached directly to the support. An opening at the top of the shell can be closed by four movable plates, or valves.

The blue crab has bright blue legs, the last pair of which are paddle-shaped. The blue crab is important commercially. The popular menu item, a soft-shelled crab, is a blue crab that has just completed its molt, and lacks a hard shell.

N.Y. Zoological Society

When barnacles are firmly anchored in place, they look much more like mollusks such as clams or oysters than like crustaceans. As a matter of fact, they were classified as mollusks by zoologists until 1830. In that year, the naturalist J. V. Thompson showed that barnacles developed from free-swimming larvae that were typically crustacean. Since that time the animals have been grouped with the crustaceans.

Barnacles are very common in shallow sea waters, especially in the tropics and subtropics. The adults may often be seen attached to rock surfaces, seaweed, various marine animals, the piling of wharves, and the hulls of ships and boats. In most cases they affect us neither for good nor for evil. However, when they become attached in great numbers to the hulls of ships, they become exceedingly annoying. An accumulation of barnacles, seaweed, and other marine organisms may gradually build up on the underwater surface of the vessel. The ship's speed may be so reduced as a consequence that it will have to be drydocked, scraped, and repainted. Quick-drying, antifouling paints are often applied to ship bottoms to prevent the accumulation of barnacles and other growths. These paints consist of iron oxides to which poisonous materials, such as white arsenic, have been added. The poisonous substances gradually dissolve, and they discourage the growth of barnacles and other fouling organisms.

Among the close kin of the barnacles are cirripeds of the genus *Sacculina,* which are parasites of crabs. The saclike body, from which this crustacean derives its name, is attached under the abdomen of the crab by means of a short stalk. Rootlike processes extend from the stalk to all parts of the body, avoiding the vital organs. *Sacculina* causes sterility in both male and female victims. It has a particularly remarkable effect upon male crabs, causing them to take on female characteristics.

COPEPODS AND WATER FLEAS

The order Copepoda is composed of animals that are often microscopic in size. They make up a considerable part of the plankton—the vast floating population of the sea consisting of tiny crustaceans, protozoa, the eggs and young of various marine animals, and microscopic plants such as diatoms and other small algae. Since plankton forms an important part of the food supply of fish, including many varieties eaten by people, the copepods are even more important to us than are lobsters, crabs, and other comparatively large crustaceans that find their way to our tables.

The typical free-swimming copepods have a streamlined body ending in a forked tail. There are two pairs of antennae, of which one is much larger than the other. Both pairs are used in swimming. The four or five pairs of feet are also used for swimming. A number of species of copepods are parasitic, either in the larval or adult stage, on a number of marine and freshwater animals, including worms, fishes, and whales. In these cases, the various species display a full range of adaptations to their parasitic life. Some of them live within the body of their host, while others have modifications to attach themselves to the outside of the body. Sometimes the modifications are so great that the parasitic species bear no resemblance to free-living species of copepods. The bodies may be completely without segmentation or appendages and may look rather wormlike. Internal parasitic copepods are often without mouthparts, absorbing their food directly from the host.

The name "water fleas" is sometimes applied to certain copepods. The name is also used for a different group of minute crustaceans, belonging to the subclass Branchiopoda. Among the best-known of the branchiopod water fleas is *Daphnia,* which is found in vast numbers in ponds and ditches. Less than two and one-half millimeters in length, *Daphnia* has a fairly complicated system of internal organs. The feathered antennae are the principal swimming organs; they propel the animal through the water in a series of jerking movements. *Daphnia* feeds on microscopic organisms in the water. It is the prey of tadpoles, fish fry, and other small aquatic creatures.

SPIDERS AND THEIR KIN

by Willis J. Gertsch

The spider is commonly regarded as the very symbol of cold and malevolent cruelty. A fly or a mosquito entangled in a spider web would certainly consider this reputation well deserved. But humans have little reason to think harshly of spiders. To be sure, some of them plague us by laying down their dragline silk on walls and ceilings. Nor does anyone like funnel webs that spoil the beauty of evergreens. It is true, also, that a very few species of spiders have a venom that is dangerous to humans. On the other hand, spiders serve us (unwittingly, of course) by destroying a vast number of bothersome insects—roaches, mosquitoes, flies, and other unwelcome creatures. For certain peoples the spider is an article of food; it is cooked and eaten with great relish.

Spiders belong to the arthropods, or jointed-leg creatures. All members of this phylum have jointed legs and other appendages. All, too, have segmented bodies encased in stiff skeletons.

The spider is sometimes called an insect, but it is definitely not an insect. Insects have six legs; spiders have eight. Insects—and indeed most arthropods—have antennae, or feelers; spiders have none. Most insects have wings; there is no such thing as a winged spider.

The supreme accomplishment of spiders, perhaps, is the art of spinning silk made within their bodies. It is to this activity that they and their kin owe the class name Arachnida. The story has it that Arachne, a mythical princess of ancient Lydia in Asia Minor, was famed for her skill in spinning and weaving. She became so proud of her work that she rashly challenged Athena, the Greek goddess of handicrafts, to a test of skill. The goddess accepted the challenge, which proved to be disastrous to the Lydian princess. For Athena was enraged at the perfection of Arachne's handiwork and she tore it to pieces. In despair Arachne hanged herself. Thereupon Athena changed the rope into a cobweb and the maiden herself into a spider.

A garden spider, showing typical spider body with two regions separated by a narrow waist and light legs.

Dur Morton, Audubon/PR

GENERAL CHARACTERISTICS

If we examine a spider at close range, we find, first of all, that its body is divided by a narrow waist into two principal regions, the cephalothorax and the abdomen. "Cephalothorax" comes from the Greek words *kephalē,* meaning "head," and *thorax,* meaning "chest." As the name indicates, the cephalothorax is made up of the head and thorax (chest), which are fused together to form a single piece. The upper portion consists of a hardened and ordinarily rounded shield called the *carapace,* which bears the eyes at the front end. Most spiders have eight eyes. There are also six-eyed, four-eyed, and two-eyed spiders. A few that live in caves have completely lost the eyes or retain only traces of them. The carapace shows considerable variations in some species; it may be ornamented with humps and projections.

Behind the cephalothorax is the abdomen, ordinarily a saclike structure joined to the cephalothorax by a narrow waist. In nearly all spiders the abdomen is in one piece and is rather soft. The abdomen of some of the orb weavers, or geometric spiders, is brightly colored, hardened, and armed with curious spines and other outgrowths. Perhaps this armor serves as a protective device against birds. In some other spiders the abdomen is drawn out.

The underside of the abdomen bears the openings to the breathing organs: the *tracheae,* or air-conducting tubes, and the *book lungs,* which are peculiar breathing devices found only in arachnids. There are usually two book lungs. Each is a sac within the abdomen containing 15 to 20 shelves, suggesting the pages of a book; the blood circulates through these shelves.

All spiders are air-breathing. None of them are truly aquatic, though certain species living on the banks of streams and lakes run over the water and also dive. The water spider of Europe and Asia can remain submerged for long periods by car-

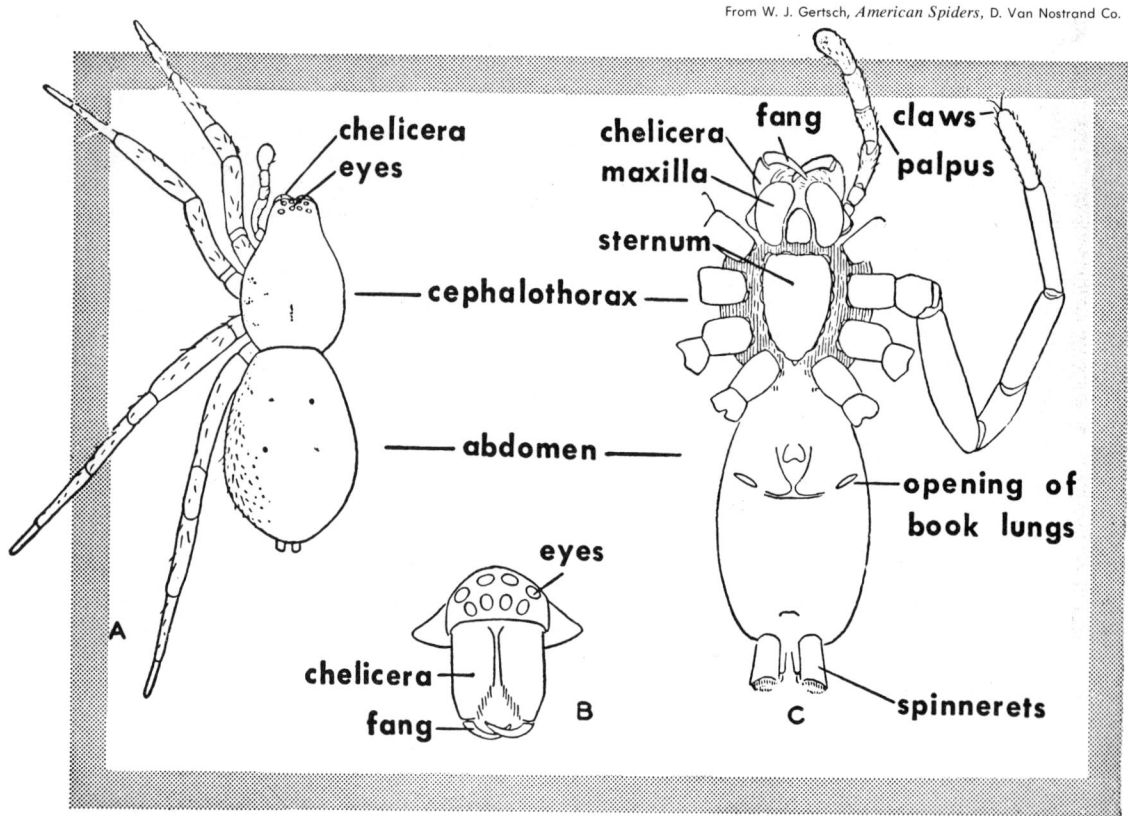

A diagram of the body structure of a spider. Left: the body as seen from above. Right: the body as seen from below. Note that the head and thorax are not separated, but are fused together to form one piece—the cephalothorax.

From W. J. Gertsch, *American Spiders,* D. Van Nostrand Co.

rying a bubble of air with it.

Directly beneath the spider's cephalothorax, at the front end, are two jaws, called *chelicerae*. Each has a stout, fixed segment and a sharp, movable fang, which is the part thrust into the prey. Connected with the jaws are a pair of poison glands located in the fixed segment of the jaws or in the head. These glands send their venom coursing through a duct opening near the end of the fang. The venom of most spiders has only slight effect on warm-blooded creatures. Only in a few instances can this venom cause severe bodily distress to humans.

All spiders are predaceous, or preying, animals. They eat insects almost exclusively, subduing them with venom. With the aid of their sharp jaws and the edges of their maxillae—another pair of appendages lying immediately behind the jaws—they cut and break the body of the prey. At the same time they bathe it with digestive liquids from glands located near the mouth. In this way the softer parts of the insect are predigested, and are then sucked into the spider's stomach by powerful muscles. The prey is rolled and chewed until only a little ball of indigestible matter remains, and this is finally cast aside. Spiders often take several hours to suck an ordinary fly completely dry.

The spider's eight legs are important, of course, as a means of getting from place to place. They are also useful to the spider because they are covered with extremely sensitive hairs and spines, which make it aware of its surroundings—fortunately for the spider, because it can neither hear nor smell. The spider has other sensory organs, called *palpi*. These are leglike appendages attached to the maxillae. Like the legs, they are covered with hairs and spines.

The male spider is rather short-lived; it dies after a brief period of intense activity. It is generally much smaller than the female and it runs considerable risk in approaching its formidable mate, which may be hungry and interested only in satisfying its appetite. Males are sometimes killed and eaten by their females under these circumstances. In many cases, however, males and females live together without this happening.

Leonard Lee Rue IV, Audubon/PR

Many spiders spin webs in the air to trap flying insects. Here a black and yellow garden spider waits on its orb-shaped web. It has spun a zig-zagging band of silk across the middle of the web.

Each spider begins its life as an egg, which is usually one of many encased in a silken sac, or *cocoon*. The largest spiders lay a great number of eggs, sometimes as many as 2,000 or 3,000 in a single mass. Smaller spiders produce fewer eggs; some lay only 6 to 12 and do not leave more than 2 or 3 in any one spot.

In general, the eggs in a sac produced in the fall hatch before winter. The *spiderlings* (young spiders) sometimes remain in the sac until spring. But in many cases they leave it and live under leaves during the winter. They are quite similar in appearance to the adult. Growth takes place through molting, a process of shedding the old skin at intervals as the creature grows. Most spiders molt from 6 to 12 times and cease molting after they become mature.

SILK SPINNING

The *spinnerets*, or spinning organs, of spiders are fingerlike appendages, usually located near the rear of the abdomen on the lower surface. There are generally six spinnerets, and from them may be spun several different kinds of silk. The silk is produced as a liquid in various glands within the abdomen. This hardens to form the familiar

both photos, Hugh Spencer, Audubon/PR

Spiders construct many different types of webs. Top: twin orb webs. Orb webs are the most beautiful and complicated of all webs. Threads of dry silk extend from the center like spokes of a wheel. Coiling threads of sticky silk connect the spokes. Bottom: a hammock web constructed between the branches of a shrub. The spider waits for an insect to fall into the web.

silken line as soon as it issues from the tiny spigots on the ends of the spinnerets. Some spiders produce a viscous, or sticky, line that is used to entangle insects.

The copious use of silk by spiders sets them apart. Although some of the other arachnids and many insect larvae spin silk, they make only occasional and limited use of this product. But wherever a spider goes, it always leaves behind a dragline, or guiding thread, attached at intervals to the surface on which it crawls. This practice suggests the paying out of a rope when a person enters a deep cave or climbs a steep cliff. It accounts for the many silk threads, or cobwebs, we see on plants and in buildings and particularly on the walls and ceilings of homes. A spider lines its retreat or burrow with silk; as we have mentioned, it encloses its eggs in a silken cocoon, often durably and beautifully made. Particularly interesting is the silken snare that certain spiders spin to capture their prey.

Spiders also use silk in the amazing activity called *ballooning*. On balmy days in the fall or spring, young spiders of many different species allow the breezes to pull out threads from their spinnerets. The tiny creatures are then raised aloft and wafted through the air. Tremendous distances can be covered in this way by spiderlings. The great 19th century English naturalist Charles Darwin recorded the arrival of ballooning spiders on the ship *Beagle* when it was some 95 kilometers from land.

At one time spider silk was used widely for lines, or markers, in the lenses of certain optical instruments. Silk that was to serve in this way was pulled from the spinnerets of the living spider and wound upon cards. Etched glass lines and platinum wire are now largely replacing spider silk.

TARANTULAS AND TRAP-DOOR SPIDERS

The more than 30,000 different kinds of spiders are divided into two major groups. One is made up of tarantulas and trap-door spiders. The other consists of the so-called true spiders.

The members of the former group are older and in many cases more primitive than the true spiders. They are also much fewer in number, being restricted largely to the tropics and subtropics. Their jaws are parallel with the long axis of the body, are set side by side, and move up and down. The spider lifts its body and jaws high up before it strikes. There is a swift downward thrust and the fangs pierce the prey, making two parallel punctures.

Some members of this group spin sheet and funnel webs similar to those of the true grass spiders. Certain species are hunters; they have thick foot pads that enable them to climb steep surfaces with ease as they pursue their prey. The members of other species are skilled engineers, digging deep tunnels in the soil and covering the openings with hinged trapdoors.

Tarantulas. Giants among spiders are the hairy creatures called tarantulas by Americans and Canadians, and bird-eating spiders by Europeans. They belong to the family Aviculariidae. The largest ones, which dwell in the steaming jungles of northern South America, attain a body length of nine centimeters and a leg span of over 25 centimeters. By comparison, the species found in the southwestern United States are pygmies, with bodies only about five centimeters long and legs that span only 15 to 18 centimeters. Tarantulas are longer-lived than most other invertebrates; some of them reach the age of 25 or even 30 years.

Most tarantulas live on the ground, making their homes in burrows that they line with heavy sheets of silk. They rarely go far from the mouths of these burrows but wait patiently for their prey to come close. Insects form a large part of their diet. Tarantulas also subdue and devour frogs, lizards, and small snakes. Some of the large tree-dwelling species occasionally feed on small birds; hence the name bird-eating spiders.

The venom of tarantulas is deadly to cold-blooded animals but ordinarily harmless to people. The great size and hairy body of this spider make it appear to be more dangerous—this is, to humans—than it really is.

Tarantulas have an odd protective device. They scrape the back of the abdomen with their hind legs when they are irritated—an act that sends the fine hairs flying in a small cloud. Some of these hairs may lodge in the nose and eyes of a pursuer. This must cause a most unpleasant stinging sensation and at least temporarily confuse the pursuer. The hairs of some species can also be irritating to human beings. Parasitic flies and wasps are the worst enemies of tarantulas. The great digger wasp of the genus *Pepsis* hunts out tarantulas, paralyzes them, and then lays its eggs upon their bodies.

Trap-door spiders. Nearly as well known as tarantulas, trap-door spiders are stout, moderately hairy, and famed for their skill in digging tunnels in the earth. Their digging instrument is a comb of stout spines on the margins of the jaws. After preparing their burrows, they waterproof the walls with saliva and earth, line them with silk, and then cap the entrance with a hinged trapdoor. Some of the more active species construct what is known as a wafer door—a flimsy trapdoor of silk that lies loosely over the burrow. In some cases, the makers of wafer doors have a secret side burrow within the main chamber into which they can retreat when menaced by enemies; this burrow is closed with a second trapdoor.

The most expert artisans among the trap-door spiders build a thick door of alternating layers of soil and silk. It is beveled to fit into the burrow opening much as a cork fits into a bottle and, appropriately enough, is called a cork door. It is heavy enough to close of its own weight; it can be held firmly from within.

TRUE SPIDERS

The true spiders far outnumber tarantulas and trap-door spiders and are the common forms occurring in fields and woods in temperate zones. The jaws of true spiders are set side by side, as in the case of tarantulas, but the fangs of true spiders are

A wasp inserting its sting into a tarantula. The wasp deposits its eggs on the tarantula's body. Despite its reputation, the venom of most species of tarantulas is harmless to people.

Lee Passmore

SPIDERS AND THEIR KIN 243

Life photographer Andreas Feininger, © Time, Inc.

N.Y. Zool. Soc.

Top: a cross section of the burrow of a trap-door spider. Bottom: a California trap-door spider, at the entrance to its burrow. Trap-door spiders are famous for their skill as efficient tunnel-builders.

turned inward and move toward each other as they pierce the body of the prey. Few true spiders live longer than one year. They may be conveniently grouped in two classes — the sedentary, or settled, web species and the more active wanderers, or vagabonds.

Sedentary webbuilders. These spiders rely greatly on silk throughout their lives.

Their sense of sight is poor. They make up for this deficiency by building large and efficient web-snares. The struggles of an insect in even the most remote recesses of the web are communicated to the spider, which has been patiently waiting for its meal to come to it.

Familiar to most of us are the tangled webs spun by the commonest of all house spiders, *Theridion tepidariorum,* which is as much at home in the tropics as in temperate regions. Males and females of this species live amicably together in the same web, thus disproving the popular belief that the female of the house spider always eats the male.

A rather distant relative of the common house spider is the funnel-web spider, which places its sheet web in the corner of a basement, rushing out to seize any insect that alights on the web.

Quite closely allied to the common house spider is one that has become notorious throughout the world for its venom. This is the black widow, or shoe-button, spider, *Latrodectus mactans.* It spins a small, tangled net in some dark crevice and hides away during the day. The glossy black abdomen of the mature female does indeed resemble an old-fashioned shoe button; the rather long legs are only slightly less black. Red spots and dashes adorn the upper surface of the abdomen, especially in the males and in the immature stages; but mature females usually lose all bright upper markings. On the underside is a blood-red marking resembling an hourglass. This is visible in living specimens because these spiders always hang back-downward in the web. The female sometimes destroys the male, whence the name "black widow"; but she is quite as likely to live on friendly terms with her mate.

The venom of the black widow and of its close relatives throughout the world is an extremely potent one, fatal to small mammals and frequently serious in the case of humans. Although few humans succumb to the bite, the symptoms are often most alarming and are accompanied by excruciating pain. Fortunately for us, the black widow spider is a rather timid and retiring

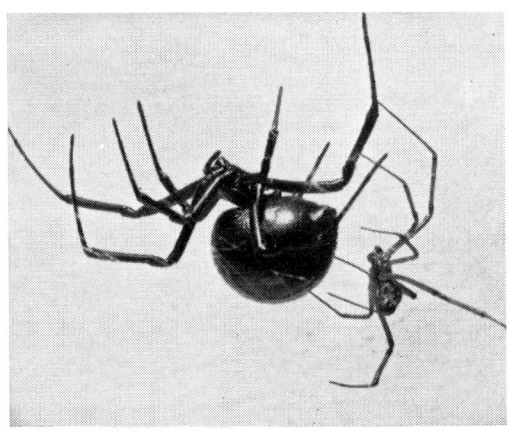 

both photos, Walker Van Riper, Denver Museum of Natural History

Left: the tiny male black widow spider is shown courting the female. The picture shows the characteristic profile of the female. Right: the female black widow devouring her mate.

creature. Furthermore, her striking markings make her easy to recognize.

The orb weavers, or geometric spiders, form a numerous group. They construct the familiar type of web known as the *orb web* — a roughly rectangular framework of silk from which extend threads that meet in the center, like the radii of a circle. A sticky spiral is laid down upon these radial threads. The web is usually placed at an angle; the spider hangs away from the sticky spiral lines and touches the others only with the claws of its feet.

The shimmering lightness and intricate symmetrical design of the orb web make it a thing of wonder and beauty. To the American Indians the orb web is a symbol of the heavens. The four corners of the framework point to the four cardinal points of the compass. The silky spiral represents, so it is believed, the mystery and the power of the Great Spirit.

Walker Van Riper, Denver Museum of Natural History

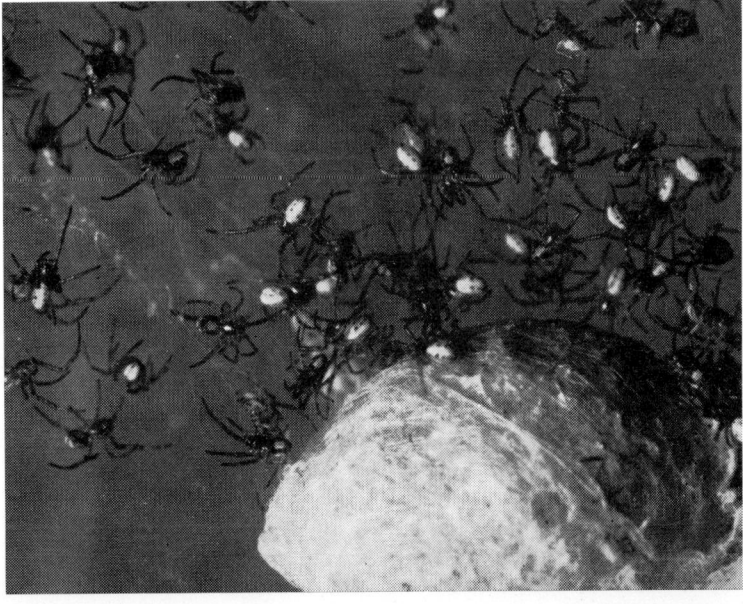

Swarms of tiny black widow spiderlings are emerging from their egg sac.

SPIDERS AND THEIR KIN

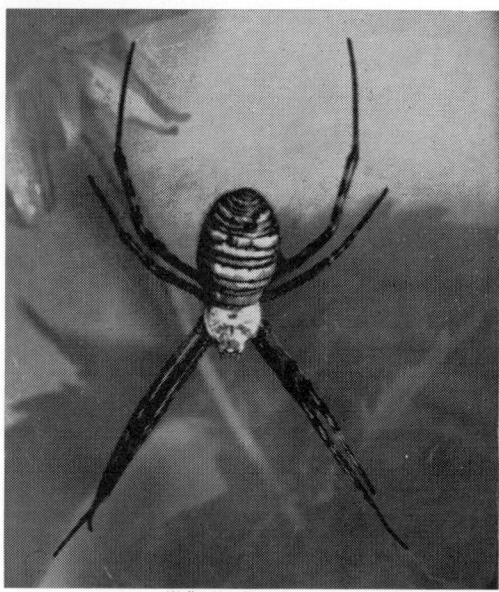

Walker Van Riper, Denver Museum of Natural History

The full-grown female orb-web spider, as seen from above. The design of the web that this spider weaves is beautifully symmetrical and intricate.

A spider guarding its large and flimsy egg case which may contain a large number of eggs and even some spiderlings, or young spiders.

Walker Van Riper, Denver Museum of Natural History

Vagabond spiders. Vagabond, or wandering, spiders place much less reliance on silk than do their sedentary relatives. They use silk chiefly to cover their eggs and line their retreats. They have fairly good eyesight and actively pursue prey.

The wolf spiders are swift hunters, living on the ground. They drag their egg sacs behind them. These look like white balls and are attached to the spinnerets. When the young hatch, they climb onto the back of the mother and stay there until they are able to provide for themselves.

The name "tarantula" is given by Europeans to the largest of all wolf spiders, *Lycosa tarentula.* Many legends are associated with this notorious creature, named after the city of Taranto in Italy. According to one of these legends, accepted as gospel truth in the Middle Ages, the bite of the Italian tarantula brought on a nervous ailment called tarantism. The victims of this malady were inclined to be melancholy and often were seized with an irresistible desire to dance. A sure cure for tarantism, it was said, was to whirl about in the rapid measures of the dance called the tarantella. Modern science scoffs at all this, for we know that the venom of the Italian tarantula, for all its ill repute, has little effect on people.

Vagabond spiders also include the fisher spiders, or pisaurids. Some of them run over the surface of ponds and streams and dive and stay under water for long periods. They are known to capture small fishes and frogs, which they drag to the edge of the pond or stream and suck dry. The females carry large egg sacs around in their jaws. Later, at the tip of a plant, they spin a substantial nursery web in which the young hatch. The mother stands guard over the young until they are old enough to look out for themselves.

The water spider, *Argyroneta aquatica,* of Europe is able to live in the fresh water of streams for weeks, enclosed in a little diving bell of silk. This spider carries air bubbles beneath the surface to its retreat and keeps a supply of air imprisoned in the silken chamber. Even the eggs are laid and

The female wolf spider is dragging her egg sac. She carries it about until her young are hatched.

Walker Van Riper, Denver Museum of Natural History

the family hatched out under water in the security of the nest.

The jumping spiders are the most highly developed of the vagabonds. Brightly colored and ornamented with tufts of hair, hanging scales, and curious spines, these strongly built creatures run about actively on the ground and on vegetation. They have excellent eyesight. While courting, the little males dance and pose before the females, apparently to show off their bright colors.

SCORPIONS AND HARVESTMEN

Other arachnids, closely related to spiders, are the scorpions and the harvestmen, or daddy longlegs. They differ quite markedly from the spiders in appearance.

Scorpions. A scorpion has formidable weapons. In front it is armed with a pair of powerful pincers. Behind, the abdomen narrows to form a long tail, ending with a curved sting. When the scorpion is ready for the kill, it seizes its prey — an insect or a spider — in its pincers. Then, bringing its tail over its back, it thrusts its sting in the victim and injects the latter with a dose of poison. The scorpion will fearlessly attack any animal that molests it, no matter what its size.

Gordon S. Smith, Audubon, PR

A female scorpion carrying its young on its back.

SPIDERS AND THEIR KIN 247

A scorpion with its sting-tipped tail raised over its back ready to be thrust into a victim.

Walker Van Riper, Denver Museum of Natural History

The sting of several scorpion species is very painful and dangerous to people, especially children. In the case of most species, however, the sting is painful but not fatal. There is a legend that scorpions sometimes commit suicide by stinging themselves, but the fact of the matter is that they are not affected by their own venom.

The female scorpion often devours the male after mating. It brings forth its young alive and carries them on its back until they are ready to get along by themselves. Scorpions generally do their hunting at night. When dawn comes, they hide in some convenient place of refuge—under a stone or a rotting log or in some person's shoe.

Daddy longlegs. An odd creature, the daddy longlegs has long, wispy legs and a small egg-shaped body fused into a single piece. It is a wandering scavenger. It has no burrows or nest but roves about the fields in search of its food, which consists of small insects, living and dead. Its chief protection against enemies is the nauseating odor of its stink glands. It is quite harmless to humans.

MITES

Mites are the smallest of the arachnids. Their bodies are egg-shaped or round. The head, thorax, and abdomen form a single unit. Many mites are reddish in color. They live in soil or debris or move freely over plants. Some of the most gaily colored species have taken to living in water and swim with the aid of long hairs on their legs.

The tiniest mites are wormlike and suck plant juices, causing galls, spots, and blemishes on some garden plants and trees. Other pygmies live in the air tubes of the honey bee and in the hair cavities of mammals, including humans. The free-living forms feed on tiny animals or eat decaying vegetable or animal matter. About half the

The familiar daddy longlegs. Note the long legs and the small body of this scorpion relative.

Walker Van Riper, Denver Museum of Natural History

mites are parasitic and live on the bodies of invertebrates or vertebrates during all or part of their lives.

Mites hatch from eggs as six-legged larvae. After a certain period of feeding, the larva changes into a nymph, an eight-legged form that is not yet mature. It undergoes one or more nymphal stages and then becomes an adult.

Among the most troublesome mites are the larvae of the harvest mites. These larvae, more commonly known as red bugs, or *chiggers,* abound in grassland areas. They attach themselves to the skin of people, causing violent itching and irritation. Their victims also include rats, rabbits, birds, and other animals. Certain species transmit the bacterialike organisms that cause scrub typhus, a disease that is often fatal to humans.

Ticks. Ticks are the largest of the mites. When a female is swollen with food, it may be more than two centimeters in length. The skin of a tick is leathery; its upper surface is provided with a tough plate, called the scutum. The beaklike mouth parts are forced into the skin of the tick's hosts. After a female tick has become swollen with food, she falls to the ground and lays her eggs, which usually number several thousand. The six-legged ticks that emerge from the eggs climb plants; as a cow or other host brushes by, the ticks attach themselves to it.

Some ticks use the same host throughout their lives. Others require two or three different kinds of hosts in order to complete their life cycle. Many ticks attack people and are a source of great annoyance because of their irritating bites. Among tick-borne diseases are Texas fever, which primarily affects cattle, and Rocky Mountain spotted fever, which is a very serious disease of mankind.

MILLIPEDES AND CENTIPEDES

These many-legged arthropods are rather distant relatives of the spiders. They belong to the Myriapoda, or myriapods. The name means "countless legs" and refers to the many pairs of jointed legs. The head is not fused with the thorax (as it is in

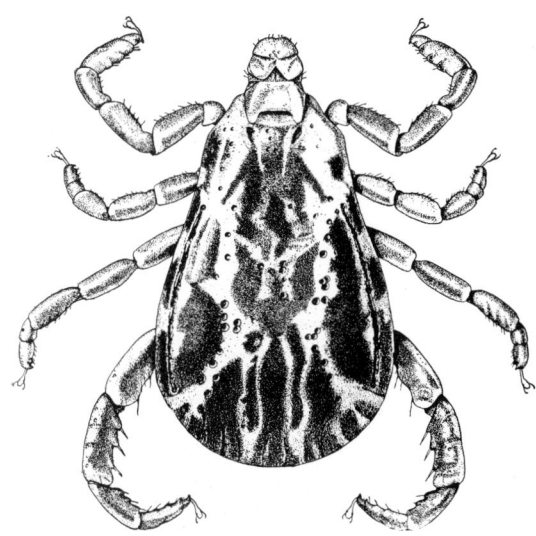

U.S.D.A. Bureau of Entomology and Plant Quarantine

The male American dog tick. Dog ticks, which are the largest of the mites, are arachnids. They feed on the blood of canines. Some ticks are transmitters of serious diseases of people and animals.

the case of the spider); it bears antennae and a pair of eyes. Below the head are jaws and accessory mouth parts similar to those found in insects.

Millipedes. The millipedes make up the class Diplopoda. They are very sluggish animals that live under stones, debris, and bark and in deep soil and humus—in fact, wherever it is sufficiently moist and dark. They feed for the most part on decaying vegetable matter. A few speices of millipedes are pests in gardens and greenhouses, where they often attack the bulbs and roots of plants.

The millipede body is divided into ringlike segments. Each segment bears two pairs of walking legs. For this reason the millipedes are called diplopods, which means "double-legs."

When millipedes hatch from eggs, they have three or four pairs of legs. They gradually add additional pairs as they molt. At maturity some of them have as few as nine pairs, while others have more than a hundred. Even this respectable total is a far cry from the "thousand legs" indicated by their name.

Some species of millipedes are very

pretty creatures, with chestnut bodies margined in yellow or pink. The large, brown members of the genus *Spirobulus* are common in North America and often exceed five centimeters in length. These millipedes give off a strong-smelling yellow secretion when menaced. Some millipede species curl up into a coil when menaced, to protect their softer underside. All millipedes are harmless to humans.

Centipedes. The centipedes ("hundred legs") are the wolves of the myriapod clan: they actively hunt small animals that live on or in the soil. They subdue their prey by poison from the first pair of legs, called the poison claws.

Unlike the millipedes, each body segment of the centipedes has only a single pair of legs. Freshly hatched centipedes have six or seven pairs of legs, or even more. Adults commonly have 15 pairs. Certain species that dwell deep within the soil have nearly 200 pairs, thus exceeding the quota of the millipedes.

Centipedes make up the class Chilopoda. Some tropical species are over 30 centimeters long and able to inflict a very severe bite. Smaller species, found in temperate regions, bite viciously. Their venom causes a disagreeable irritation but is not dangerous.

The house centipede, *Scutigera coleoptrata,* which crawls on the walls of damp cellars, differs markedly in appearance from most of its relatives. It has 15 pairs of very long legs, which drop off when they are touched. Most people find the house centipede a repulsive creature. Yet it should be hailed as an ally, for it feeds on insect pests.

Millipedes and centipedes are very distant relatives of arachnids. Top: a sluggish millipede from Ecuador. Bottom: a centipede. This multilegged animal hunts smaller animals that live in the soil.

both photos, Walker Van Riper, Denver Museum of Natural History

INSECTS

By John C. Pallister

The insects are the most numerous, the most varied, and the most widespread of all animals. The adult insect is a six-legged animal that generally has two pairs of wings.

Insects range in size from the 0.002 centimeters long fairy fly and the 0.025 centimeters long fungus beetle to the elephant beetle, which is as large as a man's fist, stick insects which may reach a length of 30 centimeters, and the huge Atlas moth, with a wing span of 25 centimeters.

Already, something like a million species of insects have been described, and new ones are being discovered every year. There are probably more than two million in all.

Insects live in almost every land area of the globe. They are found well up beyond the Arctic Circle, on high mountain peaks, in deep caves, and in forbidding deserts, as well as in the areas, such as the tropics, where they particularly thrive. They are by no means restricted to the land. Some adult insects and many larvae dwell in lakes, ponds, and streams. A few are found in or near brackish water and along the tidal limits. A very few live in the offshore waters of the ocean.

Certain adult insects have just one important function—to reproduce; they mate, deposit eggs, and die. A number of insects live only through the summer season. Others survive the winter. Certain social insects may live for a few years.

These widely distributed little creatures make up the class Insecta in the phylum Arthropoda. This phylum, made up of segmented animals with jointed legs, includes such familiar animals as spiders, ticks, mites, centipedes, milipedes, scorpions, shrimps, crabs, and lobsters.

INSECT BIOLOGY

Insects lack an internal skeleton. To protect their soft internal organs insects have evolved a hard outer skeleton known

Nature in Pictures—Ross E. Hutchins

Insects have evolved a great many habits to take advantage of the world's diverse habitats. Here, a weevil feeds on a cotton boll.

as an *exoskeleton*. The exoskeleton is made hard by a substance called sclerotin. The exoskeleton also contains chitin, proteins, and sometimes minerals—calcium salts, mostly. The exoskeleton is resistant to water, weak acids and bases, alcohols, and the digestive enzymes of most mammals.

Some insects—beetles, for example—have a very hard body covering. Aphids and some others, on the other hand, have a very fragile covering.

The body of the insect is divided into three parts: the *head, thorax,* and *abdomen*.

HEAD

The head is oval, globular, or simply roundish. It may be large or small in proportion to the body.

Antennae. Except for a few primitive varieties, all insects are equipped with one pair of antennae, which arise from the front of the head, usually near the eyes. They function as organs of touch. The antennae of some insects also act as organs of smell.

The antennae of mosquitoes, some flies, some butterflies, and some wasps are able to receive sound waves. Antennae show great diversity in size, shape, and structure. They may resemble threads, or strings of beads, or tiny combs, feathers, clubs, or hammers, to list just a few forms.

Eyes. Insect eyes are of two kinds, simple and compound. The compound eyes, often large and prominent, are composed of small hexagonal lenses called *ommatidia.* There are usually only a few lenses in the compound eye of an ant or other ground insect. There may be as many as 28,000 in the case of the dragonfly.

The simple eyes, called *ocelli,* are very small. In adult insects, they are usually two or three in number and are placed near the compound eyes. Probably they permit the insect to distinguish between light and darkness. The immature insect usually has only simple eyes. Most adults have compound eyes. Many adults have simple ones as well. A considerable number have no eyes.

Mouth parts. Some insects bite and chew their food, which may be pretty tough material, such as wood. Others pierce plant tissues and suck the juice. Others may suck the blood of animals.

An insect has three pairs of jaws. The *mandibles,* or true jaws, make up the first pair. In sucking insects, such as mosquitoes, the mandibles are fused together to form a needle-sharp piercing tube. Some male beetles and soldier ants have enormously enlarged mandibles that look and perform like claws. Certain butterflies have no mandibles at all. Below the mandibles there is a second pair of jaws, the *maxillae.* These hold food, carry it to the mouth, and chew it sideways instead of up and down. On the maxillae are a pair of feelers, called *palpi,* which may be quite long and conspicuous. The final pair of jaws is often fused together to form a lower lip called a *labium.* It too has palpi. Both sets of palpi are sensory organs, conveying sensations of taste and smell as well as of touch.

Many insects have a kind of short, fleshy tongue called a *hypopharynx.*

THORAX.

Behind the head is the thorax, which supports the legs and wings. It is composed

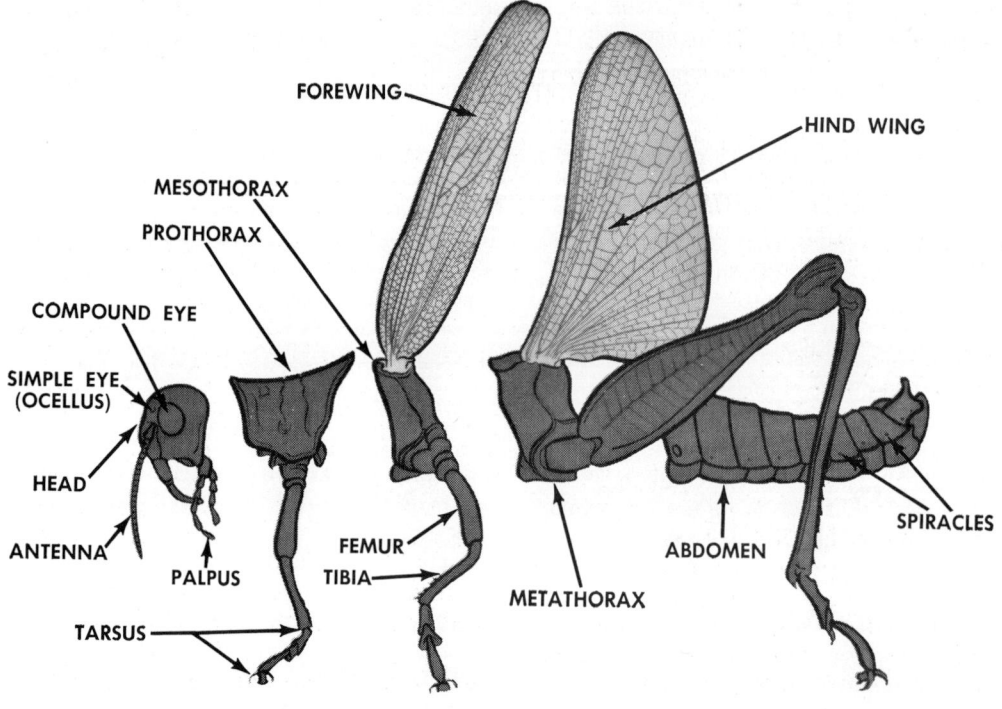

of three segments, each holding one pair of legs.

Legs. Almost all insects have six true legs, and none has more than six. Some immature insects have no legs. The immature insects called *caterpillars* have three pairs of true thoracic legs and from two to ten other pairs on the abdomen. Certain adult forms use their two front legs for digging; others, for seizing their prey; still others, for holding the female during mating. Insects that swim have paddlelike middle and hind legs. The hind legs of grasshoppers, fleas, and other jumping insects are greatly enlarged. Some insects have a comblike structure on their front legs, which is used for cleaning their antennae. Usually insects move their legs in sets of three: the first and third legs on one side move with the second leg on the other.

Wings. Certain insects are wingless throughout their entire lives. Others may have only one pair of wings. The majority, however, have two pairs, one pair attached to the middle segment of the thorax and the other to the last segment. In flies, the second pair of wings is represented by a pair of little stumps, called *halteres*. In effect, therefore, flies have only two wings. The forewings of beetles are stiffened and thickened into a shieldlike covering, called an *elytra*, for the delicate rear wings.

Insect wings are supported by a network of tubular veins. Wings may be smooth and transparent or covered lightly or heavily with fine or coarse hairs or with minute colored scales. The iridescent colors seen on some moth and butterfly wings are usually due to diffraction by these scales.

ABDOMEN

The abdomen is usually the largest part of an insect's body. It is composed of a series of nine to eleven ring-shaped segments. These may be closely jointed, making the abdomen stiff and rigid, or so loosely jointed that the insect is able to curve its abdomen up over its head. Commonly, the first segment is as broad or nearly as broad as the thorax to which it is attached. In wasps, ants, and other "wasp-waisted" insects, the first and often the second and third seg-

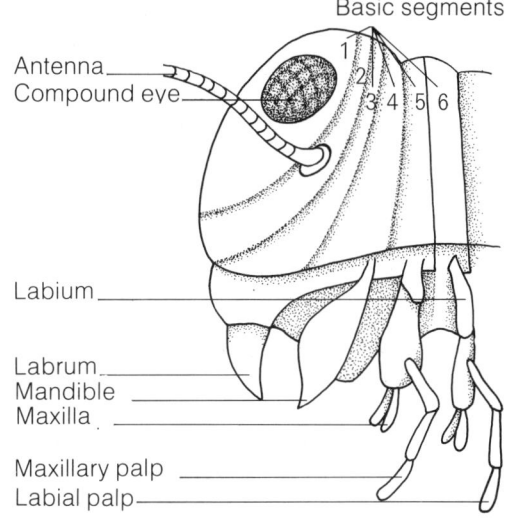

TYPICAL STRUCTURE OF THE INSECT HEAD

ments are extremely narrow, while the rest of the abdomen is globular.

Many insects carry prominent appendages at the end of the abdomen. These may be a pair of long bristles, or tails, looking like antennae and serving much the same purpose. In the case of earwigs, the appendages consist of a pair of strong pincers at the end of the abdomen. Some female insects have a long *ovipositor,* or egg-laying tube, projecting from the tip of the abdomen. Ants, wasps, and bees use the ovipositor as a stinger.

COLORATION

Many insects wear odd or beautiful decorations, at whose purpose we can only guess. They may have horns, spines, or ridges on the back or head. Hairs, stiff or silky, may cover them all over or in patches. On the wings of moths and butterflies the hairs are flat and scalelike.

Insect colors are often striking. Certain colors—browns, blacks, and yellows—are usually due to pigmentation, or coloring matter occuring in the exoskeleton. Metallic and iridescent colors are due to light diffraction from the surface structure. Green and magenta are due to both these factors.

PHYSIOLOGY

An insect has no lungs. It takes air through openings along its sides into a system of air tubes, called *tracheae,* which spread throughout the body and wings.

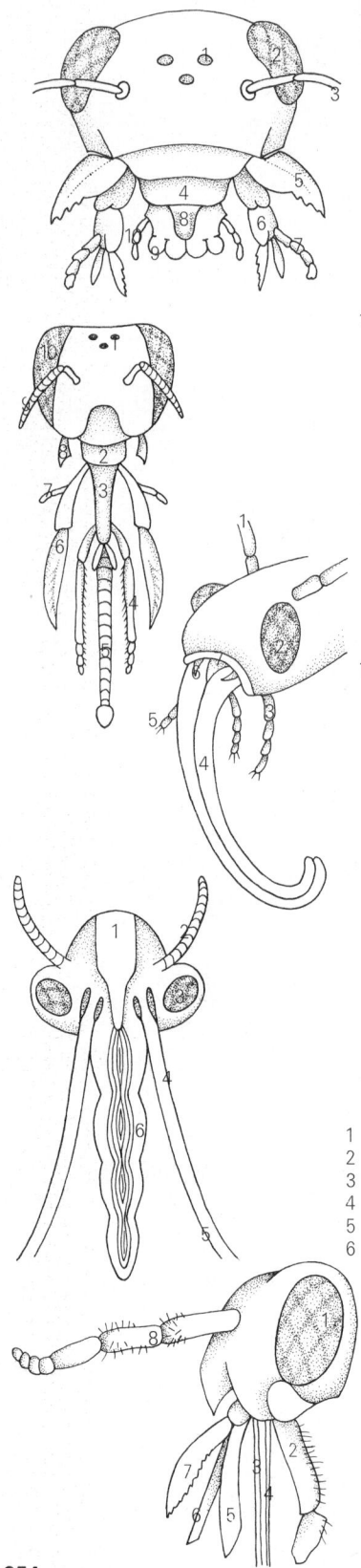

HEAD AND MOUTH PARTS OF A GRASSHOPPER, A CHEWING TYPE OF INSECT
1 Ocellus (simple eye)
2 Compound eye
3 Antenna
4 Labrum
5 Mandible
6 Maxilla
7 Maxillary palp
8 Hypopharynx
9 Labium
10 Labial palp

HEAD AND MOUTH PARTS OF A BEE, A LICKING-SUCKING TYPE OF INSECT
1 Ocellus
2 Labrum
3 Mentum
4 Labial palp
5 Glossa (tongue)
6 Maxilla
7 Maxillary palp
8 Mandible
9 Antenna
10 Compound eye

HEAD AND MOUTH PARTS OF THE BUTTERFLY, A SUCKING TYPE OF INSECT
1 Antenna
2 Compound eye
3 Labial palp
4 Proboscis

5 Maxillary palp
6 Mandible
7 Labrum

HEAD AND MOUTH PARTS OF AN HETEROPTERON, A PIERCING-SUCKING TYPE OF INSECT
1 Labrum
2 Antenna
3 Compound eye
4 Mandible ⎫ forming
5 Maxilla   ⎬ a stinging
6 Labium   ⎭ stylet

HEAD AND MOUTH PARTS OF A HORSEFLY, A PIERCING-SUCKING TYPE OF INSECT
1 Compound eye
2 Labium
3 Mandible
4 Hypopharynx
5 Maxilla
6 Labrum and epipharynx
7 Maxillary palp
8 Antenna

Aquatic insects breathe through *tracheal gills,* or they have chambers that they can fill with fresh air before they submerge.

A simple pulsating tube, open at both ends, serves to keep the blood in motion throughout the body. In some insects pulsating sacs in the knee joints drive the blood through the legs. The blood is usually colorless or greenish. Some aquatic insects and some larvae have red blood.

The digestive system varies with the food habits of insects. Some take only liquid food. In those which eat solids, the food passes from the gullet into a storage chamber, or *crop.* From there it goes into a *gizzard,* where it is ground fine, and thence into the gut for digestion.

Special glands account for some of the extraordinary activities of insects. Caterpillars and all other insect larvae that spin silk are equipped with silk glands. The froth of spittle bugs is whipped up from a secretion that comes from a gland near the tip of the abdomen. Small glands provide certain insects with poisonous or ill-smelling liquids or gases with which they confound their enemies. Honeybees have wax glands. The glands of aphids, lac insects, and most other members of the Homoptera secrete waxy substances. The wings of some male butterflies have scent glands that attract females.

Considering its size, an insect has an extremely large and complex nervous system. The nerve centers of the brain, in the head, control the eyes, the antennae, and the mouth parts. Double nerve cords lead from the brain back to other nerve centers, each of which controls an adjacent part or function.

The sense of touch of insects is centered chiefly in their antennae, but many species have abdominal *cerci,* or bristles, which also react to touch. Caterpillars and many other larvae and some adults are equipped with sensitive hairs and spines.

Organs of smell are often found on the antennae. They sometimes occur on the hind legs, feet, or some of the mouth parts. The insect can detect sounds through organs located in various parts of the body. These auditory organs are on the antennae of the true flies and of some butterflies and wasps. Katydids and crickets "hear" through so-called "ears" on their front legs. Grasshoppers hear through structures on each side of the abdomen, near the base. Many insects have organs or specialized structures for producing sound. These are never in the mouth or throat, but somewhere on the abdomen, legs, or wings.

REPRODUCTION

All insects develop from eggs, which vary greatly in size and shape. Some are microscopic. Others are spheres about three millimeters in diameter or slender, cone-shaped structures about six millimeters long.

In most species the female deposits her fertilized eggs singly or in groups or masses in or near the particular food that the young will eat—manure or the body of an insect or other animal. In some aphids and flies, the eggs hatch within the mother's body. Certain ants, bees, wasps, and aphids produce a generation of perfectly formed adults from unfertilized eggs. This process is called *parthenogenesis*. In aphids, a generation produced in this way usually alternates with a generation of insects from fertilized eggs.

The number of eggs laid by female insects ranges from a few to many thousands. In general, the solitary insects that provide good protection for their eggs deposit a comparatively small number. Those that drop their eggs in exposed places—on leaves, twigs, or in water, where the mortality will be great—may lay several hundred. Among the social bees, termites, ants, and wasps, where large colonies are necessary, the queen may deposit several thousand eggs in a single day. Her eggs and young are cared for and fed by a special caste of adults.

The young of the noncolonizing species are not attended in this way. As soon as they are hatched they are able to care for themselves. Since the eggs were deposited in or near the food supply, the larval insect starts at once to feed on its immediate environment. Where eggs have been laid in masses, the first to hatch are quite likely to devour the later arrivals.

GROWTH AND METAMORPHOSIS

In the process of growth, from egg to adult, most insects pass through a series of drastic changes, called *metamorphosis*. In the case of many insects, there are four stages in this development: (1) the egg stage, which ends when the young insect, called the *larva,* emerges from the egg; (2) the larval stage, in which the insect constantly feeds, sheds its exoskeleton several times, and becomes an inactive *pupa* after the final larval molt; (3) the pupal stage, a period of inactivity in which the body of the insect undergoes striking changes; (4) the adult stage. If the insects go through all four stages, the metamorphosis is said to be *complete*. It is called *incomplete* if the pupal stage is omitted.

Larval stage. Various special names are applied to the larvae of insects. The young of grasshoppers, mantids, crickets, cockroaches, cicadas, and others are often called *nymphs*. Young dragonflies, stone flies and May flies, which live under water, are known as *naiads*. The name *caterpillar* is given to the young of moths and butterflies. Other names applied to larvae are *maggots, grubs,* and *worms*.

In the case of many species, the larval stage is the longest period of the insect's life. The periodical cicada spends either thirteen or seventeen years as a nymph underground. Many May flies live from three to five years as naiads and only a few hours or a few days as adults. Some woodboring beetles are said to spend from twenty-five to thirty years as grubs within a tree trunk.

The larvae of the insects with complete metamorphosis bear no resemblance to the adults, as witness the caterpillars that become moths and butterflies, the maggots that become flies, the wireworms that grow to be click beetles. After undergoing a number of molts, the larvae are ready to enter upon the pupal stage. Those that have been dwelling underground or within a plant or animal will stay quietly in their quarters. Some of those that have been living in the open will surround themselves during this critical stage with a protective

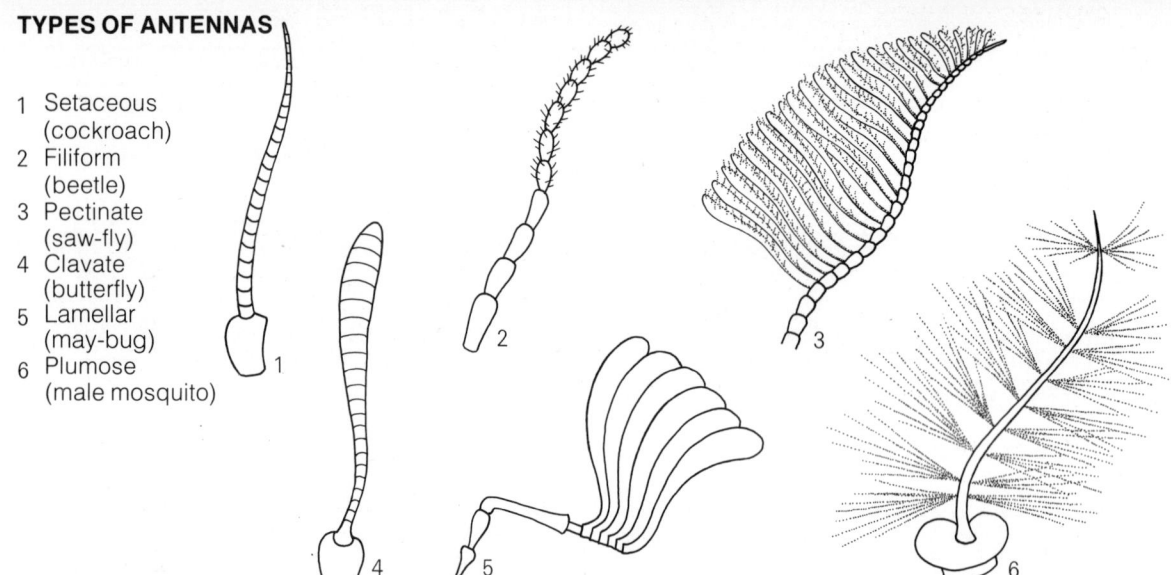

TYPES OF ANTENNAS

1. Setaceous (cockroach)
2. Filiform (beetle)
3. Pectinate (saw-fly)
4. Clavate (butterfly)
5. Lamellar (may-bug)
6. Plumose (male mosquito)

case. The cocoon of the silk-spinning moths is the best-known example. People use the cocoon material to make silk.

Pupal stage. The *pupa,* as the insect is called in the quiescent stage, undergoes a startling transformation into the adult form. The entire larval body is not transformed into an adult. Rather, the adult comes from a group of cells, or *imaginal disk,* situated on the larval body. The transformation may be completed within a few days, or it may take all winter. Finally, the adult insect breaks out of the pupal skin (and also the protective covering, if it had one) and dries its wings. The adult insect is called the *imago.* It will never change again, nor will it ever increase in size.

Miniature adults. The young of insects with incomplete metamorphosis, such as dragonflies, cicadas, and grasshoppers, resemble the adult form somewhat but have no wings. Wing pads appear only after the larvae have molted several times. Certain insects with incomplete metamorphosis, such as lice and camel crickets, have no wings as adults. When the dragonfly naiad has reached its full growth, it comes out of the water in which it has developed onto some convenient plant stalk. The cicada nymph pushes out of the ground and climbs a meter or so up a tree trunk. The grasshopper nymph stays where it is. In each case, the hard exoskeleton splits down the back and a brand new creature emerges, with small damp wings. Soon the wings expand and dry, and the insect flies off.

VARIATIONS WITHIN SPECIES

Even the individuals within a species may vary in color or size, sometimes enormously so. This phenomenon is called *polymorphism.* It occurs to a certain extent in most species. A mass of eggs of the tiger swallowtail butterfly may produce some females that are black and yellow like the males, and some that are dark gray and black. Spring, summer, and fall broods of butterflies often differ in color and size.

Social insects. Variation within species reaches its maximum development among the social insects. All termites, all ants, a few families of wasps, and a few families of bees live in highly organized communities, whose members are separated by physical differences into several types or *castes,* each with its own tasks. In general, each colony is founded by or around a *queen,* who is ordinarily much larger than the other members and who does nothing but lay eggs. Besides the queen, there are the true males, the sexless workers, the soldiers and, at times, the young queens. Generally, the ants have the most complicated social organization. They use several kinds of both workers and soldiers to maintain a colony. Bees and wasps get along with a simpler system of queen, males called drones, and workers. Insect colonies range in size from hornet nests with a few dozen individuals to tropical ant and termite "cities," which have populations in the millions.

AN INSECT'S WING

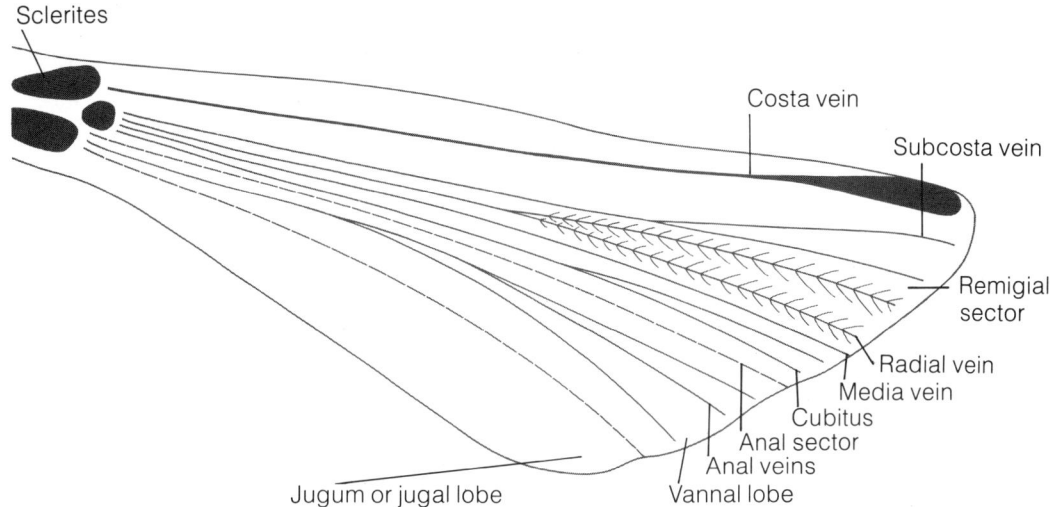

Solitary insects. The great majority of insects are solitary; that is, they do not cooperate in forming nests, feeding the young or fighting, even though they may collect in masses, as the ladybirds do, or travel in great hordes, as do the grasshoppers.

DEFENSES

All insects have many enemies, big and small. Man wages unceasing warfare upon the species that infect him with disease, or that attack his crops, animals, and goods, or that annoy him with their stings and bites. Many species are the prey of other insects or of birds, snakes, spiders, or fishes. Insects have developed a variety of defensive weapons against such enemies as these.

Stink glands, from which nauseating odors can be ejected, protect stinkbugs, bedbugs, many different beetles, and some caterpillars from attack. Other insects discharge malodorous or poisonous fluids. The grasshopper covers his would-be captor with a brownish liquid called "tobacco juice". Blister beetles, oil beetles, and others discourage attack with caustic secretions. The hairs or bristles on many caterpillars irritate the skin, either through mere contact or because they carry poison from glands at their base.

It would be hard to imagine a method of camouflage that has not been used by some insect species or other. Ground insects and those that do not fly too far afield are likely to be colored in monochromatic shades of brown, or green, or gray, matching their backgrounds, or in mottled patterns that blend with sunlight and shadow. Others, whose activities might make them more noticeable, often wear brilliant or startling colors. This is supposed to warn that the insect carries stink fluids or is otherwise unappetizing. Sometimes the conspicuous colorings of dangerous or ill-tasting insects are mimicked by more vulnerable species. The viceroy butterfly apes the inedible monarch butterfly. The harmless yellow and black syrphid fly resembles the wasp called the yellow jacket.

INSECTS AND MAN

Since man is almost everywhere and insects are almost everywhere on the land areas of the earth, and since both eat all kinds of food and use all kinds of materials to build their homes, it is obvious that their interests will often conflict. To be sure, there are surprisingly few insect species that are at any time dangerous to man's health or possessions. It has been estimated that less than one tenth of one per cent of all the insects in the world have any harmful relationship with man. But this relatively small percentage takes a heavy toll in human health, lives, and possessions.

EFFECT ON HUMAN HEALTH

Most of the insects that seriously affect man's health belong to the Diptera, or true flies. Bot flies, warble flies, and some other Diptera species are parasitic on man and other mammals, spending all their larval life within the body of the host. The common housefly spreads disease by walking first over germ-laden filth, and next over human food or hands. Typhoid fever, dysentery, and cholera have been transmitted in this way.

Insects that feed on human or animal blood often carry disease organisms in their salivary juices. One kind of mosquito transmits malaria; another, yellow fever. Still others are *vectors,* or carriers, of dengue, filariasis, and encephalitis. Tsetse flies carry the African sleeping sickness. Deer flies transmit tularemia. Fleas, lice, and bedbugs carry typhus, relapsing fever, and one form of bubonic plague.

Many insects irritate us without seriously disturbing our health. We are annoyed by the bites of mosquitoes, black flies, horseflies, and midges. The caterpillars of the brown-tail moth are covered with bristles that sting like nettles. The stings of bees, wasps, and ants are often painful and dangerous to people allergic to them.

EFFECT ON CROPS

Insects that are injurious to our agricultural crops, food products, clothing, and wooden buildings are more numerous than those that affect our health. They belong to many orders, particularly the Coleoptera (beetles), the Isoptera (termites), and the Lepidoptera (moths and butterflies). Some of the most damaging of these pests are of foreign origin. In their native lands they do little or no damage, usually because at home they have too many or too formidable enemies to attain a destructive population. Brought to other countries by accident, they are free to feed and multiply on whatever crop is attractive to them. In the United States, the Hessian fly, the European corn borer, the codling moth, the gypsy moth, the brown-tail moth, the Mediterranean fruit fly, and the Japanese beetle are prominent examples of insects that are relatively harmless in their native regions, but cause great damage elsewhere.

Besides damaging crops by feeding on them, insects often transmit plant diseases. They may carry the spores of a destructive fungus on their feet or mouth parts from one plant to another. Thus the engraver beetles carry the Dutch elm disease. They may harbor a plant virus through part of its development within their alimentary systems. Aphids, leafhoppers, and other plant-sucking insects transmit certain blights in this manner.

CONTROLLING INSECTS

Man spends a great deal of time and money trying to defend himself from insects, and with only partial success. To keep out immigrant insects of known destructiveness, governmental authorities establish quarantines; but their effectiveness is limited by the smallness and persistence of the incoming pests and the lawlessness, ignorance, and carelessness of many people. Chemical warfare upon insect pests is waged with insecticides. Man also carries on biological warfare against the pests. The breeding places of the offending insects are destroyed. Patients suffering from diseases such as malaria and yellow fever are screened from insects that could transmit the diseases to other human victims. Entomologists (scientists who specialize in the

A TYPICAL INSECT'S LEG (GROUND BEETLE)

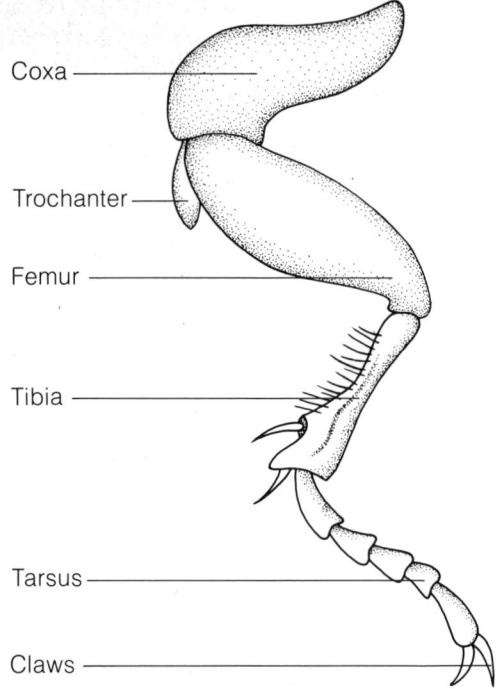

Coxa
Trochanter
Femur
Tibia
Tarsus
Claws

study of insects) seek out the insect pests' natural enemies and turn these enemies against the pests.

BENEFICIAL INSECTS

As we have pointed out, comparatively few species of insects are to be reckoned among our enemies. Certain species are directly beneficial to man. The honeybee supplies us with honey; the silkworm, with silk. One type of scale insect provides us with the dyestuff cochineal. Another gives lac, from which the commercial shellac is derived.

Insects benefit man indirectly in many ways. If they did not help to pollinate plants, we would be deprived of most of our fruit trees, our beans, peas, and other legumes, and thousands of other plants that are useful to us. Our freshwater fish feed on insects. So do many of our game birds. Insects are also valuable as scavengers. They help keep the earth reasonably unencumbered. It is true that they destroy some of our most cherished plants, but they also attack weeds. They are among our most effective allies in the war against insect pests. Above all, they play an important part in maintaining the balance of nature, upon which our very existence depends.

TYPES OF INSECTS

The class Insecta is divided into two subclasses: Apterygota, or wingless insects, and Pterygota, or winged insects. The winged insects are further divided on the basis of metamorphosis. The insects that undergo incomplete metamorphosis are the Exopterygota. The insects that undergo complete metamorphosis are the Endopterygota.

APTERYGOTA

The wingless insects are primitive forms, probably the first insects to appear upon the earth. They now dwell in almost every land area. The young differ from the adults chiefly in size.

There are four orders of wingless insects. These are the Protura, Thysanura, Collembola, and Diplura.

Protura: The insects of this order were

J. Six

Insects lay eggs, generally many at a time. This potato beetle deposits her eggs on a leaf.

not recognized until 1907. They are whitish and from 0.5 millimeter to 2 millimeters in size. They have no antennae, but use the first pair of legs as feelers. The number of abdominal segments increases with growth.

Thysanura: bristletails, silverfish, slickers. They range up to one centimeter in length. They have very long antennae and often three long tails, or *cerci.* Many of them look like small shrimps. Some are found in lichens, mosses, and rotten wood. Others live in books and under wallpaper. Some live in caves and are blind.

Collembola: springtails, snow fleas. Most species have a curved, taillike appendage that they use as a spring to propel themselves into the air. Some live in ant or termite nests and are blind. Most are dull-colored; but there are several brightly colored species.

Diplura: japygids, campodeids. These insects are closely related to the Thysanura. Diplurans live in soil and leaf litter, bark, and under stones and logs. They are less than six millimeters in length and they lack compound eyes.

EXOPTERYGOTA

The winged insects that have incomplete metamorphosis are classified in 14 orders.

Orthoptera: grasshoppers, crickets, katydids, cockroaches, walking sticks, leaf insects, mantids. The group includes many insect pests, such as the migratory grasshoppers that are often called locusts, and the cockroaches, as well as some very useful species such as the praying mantis which feeds on garden and plant pests.

Orthopterans are able to leap by means of well-developed hind legs. Two wings are usually present, but one or both pairs are absent in many species. Many are good fliers. The rear wings of the flying species can be folded in pleats under the forewings, which are usually known as *tegmina*. Tegmina vary in shapes and sizes. The antennae are either long and slender, or short and comparatively thick. Most species of the Orthoptera are voracious feeders.

Some of the unusual members of the Orthoptera are the *phasmids,* or *walking sticks*. The bodies of the walking sticks are modified to look like sticks or twigs. They are the longest of the insects, attaining a length of more than 30 centimeters in Australia and New Zealand.

Other unusual orthopterans are the *leaf insects* of the family Phyllidae. The wings of the leaf insects closely resemble leaves in shape and color and even in veining. Their legs are modified to look like leaf fragments. They are found in tropical Africa and Indonesia.

The most familiar of the orthopterans are *grasshoppers*. The 18,000 or so grasshopper species are widely distributed around the world. They range up to the snow fringes of mountains and have reached many isolated oceanic islands. These insects are generally large, averaging about five centimeters in length.

The head is rather large and solidly built. The forewings are usually strong. There are two rather large compound eyes, three ocelli, two short antennae, and a powerful set of jaws. The front wings cover the more delicate rear wings when the insect is at rest. The front and middle pair of legs are short and are used for walking. The rear legs, which serve for leaping, are large and strongly muscled.

Along the inner surface of each upper hind leg of many grasshoppers is an elevated, sharp ridge. This may be smooth or notched. There is also a raised vein on each tegmen, or forewing. When the grasshopper scrapes his legs against his tegmina, he produces the familiar rasping noise we associate with grasshoppers.

Most grasshopper species remain comparatively unimportant in man's economy. They furnish food for birds and smaller mammals, and are used for fish bait. Only a few species are damaging pests. The curious life cycle of these insects brings about the terrible "plagues of the locusts" that have tormented man from the earliest times to the present day. In general, a nonmigratory type will live quietly in one place for many generations until it has developed an enormous population. Suddenly there appears a migratory generation quite different in color or size from its immediate ancestors. The great majority will fly off for more abundant pastures, eating all leafy vegetation along the way. The few remaining in the old location will breed a new generation that will have the old color and size. The process of building up a migratory form will then begin again.

One of the most destructive species is the desert locust. *Schistocerca gregaria.* A

The walking sticks have a great camouflage: they resemble twigs.

J. Six

large species, about five centimeters long, it ranges from northern Africa and southern Europe east into Asia as far as India. It breeds in sandy areas with sparse vegetation. The nonmigratory form is yellow; the migratory form, pinkish. *Schistocerca gregaria* may migrate over 1,000 kilometers.

Dermaptera: earwigs. Earwigs are small to medium-sized, brown or black insects. As adults earwigs have two pairs of wings. They have a short elytra, and long antennae. Their distinguishing feature is the pair of pincerlike forceps that adult insects carry at the tip of the abdomen. Earwigs are omnivorous and sometimes damage plants.

Plecoptera: stone flies, salmon flies. These are winged insects one to five centimeters long, with soft, flattish bodies. They are usually dull brownish, greenish, or yellowish, with long slender antennae, and shorter tails or cerci. Stone flies are found in rocky streams and deposit eggs in the water. The larvae live at the bottom of the streams. They eat other larvae and are themselves a food for fish.

Embioptera: web-spinners. Small, slender insects with large heads and threadlike antennae, web-spinners collect in large groups under small stones, debris, and the tangled bases of grasses. During the spring rains, they build shining white webbed tunnels and awnings over the ground between their various retreats. Males have two pairs of slender wings; females are wingless.

Psocoptera: psocids, book lice, bark lice. Very small, fragile insects less than five millimeters long, psocopterans live in straw, debris, fungi, and cereal products. They eat the paste in wallpaper, the glue in books, and the insects in museum collections. Many are wingless, some sightless.

Mallophaga: chewing lice. These are small, flattish, wingless insects, with short legs. They live on the hair, feathers, and dried skin fragments of birds and mammals, excluding humans. Most of them move readily along the body of the host by means of claws. Others cling to feathers or hairs with their jaws. They are very irritating to their hosts.

Anoplura: true lice, sucking lice. These minute, wingless insects, are slender, oval,

J. Six

Grasshoppers, like all orthopterans, show incomplete metamorphosis. This one has just shed its skin.

or crablike. Their mouth parts are modified to pierce the skin of the host and suck the blood. Their legs are fitted with claws for clinging to hairs. True lice are found wherever there are mammals. Those infesting man often carry and transmit such diseases as typhus and relapsing fever.

Zoraptera: Zoraptera are very small — less than three millimeters in length. Wings may be present. Some forms are blind, but other have compound eyes and ocelli. They are found under rotting logs.

Isoptera: termites, or white ants. Termites are small, soft-bodied, pale-colored insects. They live in colonies in wood and soil. Their complicated caste system includes winged, wingless, and short-winged forms. Termites normally live in the soil or in dead trees and stumps. They also build their colonies in wooden foundations, buildings, and fences. Termites are discussed in the article "Ants, Bees, Wasps, and Termites" in The New Book of Popular Science.

Ephemeoptera: May flies. May flies are small to medium-sized, slender, soft-bodied insects. They usually have two pairs of wings, the hind pair being much smaller than the front pair. The larvae live in the water, emerging when they are ready to become adults. The adults never eat; they have weakly developed or no mouth parts. Some live only a few hours.

Odonata: damsel flies, dragonflies. These are medium to large, slender insects, with two pairs of long, narrow, shining wings. They have large, mobile heads and very large, compound eyes. Fossil forms were very large.

The dragonflies are often very large. They are able to catch most other insects on the wing. Damsel flies are smaller, more fragile, and slower in flight. Dragonflies and damsel flies feed on mosquitoes, midges, and other small insects. The Odonata spend their immature life of one to five years under water. They breathe using gills, which are located at the bottom of the abdomen.

Thysanoptera: thrips. Minute, slender insects with sucking mouth parts, thrips are found on vegetation, particularly flowers, throughout the temperate and subtropical world. Many thrips transmit virus and fungus diseases to plants. There may be as many as nine generations in one year.

Hemiptera: true bugs, including stinkbugs, bedbugs, lace bugs, chinch bugs, water striders, water scorpions, back swimmers, water boatmen.

As you look at almost any hemipterous insect from above, you will see a pattern of triangles formed by body parts. Most true bugs have two pairs of wings. The base of the fore wings is hard and leathery, while the outer part is membranous. The Hemiptera range in size from minute to about 20 centimeters. The head can move freely and often has a necklike portion. The compound eyes are unusually large. The antennae are generally long in the land bugs and very short in the aquatic families. The thorax may have various odd adornments.

The mouth parts are typically tubular, beaklike, and fitted for piercing and sucking. All true bugs have such beaklike parts—long and straight, or short and curved. Most bugs are plant feeders. Some of them prey on insects and other animals.

The *stinkbugs* are probably the best-known of all the true-bug families. Their shape is distinctive. Their colors vary from dull to brilliant. Most stinkbugs are shaped like a shield or an arrowhead. Prominent eyes bulge out at the base of a rather long head. The antennae are well developed and so are the wings, although you will seldom see a stinkbug flying. In length the insects range from 0.5 centimeter to 3 centimeters.

The stinkbugs, or pentatomids, numbering 5,000 known species, are one of the largest families in the order of the Hemiptera. There are representatives in almost every part of the world.

Chinch bugs and milkweed bugs belong to the family Lygaeidae. The scientific

A beneficial insect—the praying mantis. Mantids eat insect pests.

J. Six

A male cricket is depositing his spermatophore, or sac of sperm, that will be picked up by a female.

name Lygaeidae refers to the dull, dark color of many of the species. (The Greek word *lygaios* means "shadowy.") However, some of the most beautiful bugs in Europe and North America—the milkweed bugs—are members of this family. The spotted milkweed bug, *Oncopeltus fasciatus*, is a slender insect about 1½ centimeters long. It is orange or bright red in color, with a black thorax and bands across the wings. The eggs and the nymphs are bright red. The insect feeds on milkweed and hibernates on trees and buildings.

Most notorious of the lygaeids is the chinch bug, *Blissus leucopterus*. Less than one centimeter in length, it is dark in color, with reddish or yellowish legs and antennae. Some members of the species have very short, brown wings, barely covering the abdomen and useless for flying. Others in the same brood have normally long, white wings and are able to fly for great distances. Chinch bugs are most injurious in the farm belts of the United States and Canada. Here they gather in huge numbers and suck dry the young plants of corn and other grains. They also feed on hay.

The bedbug family, Cimicidae, has only a few species, all of them small, flat and reddish-brown, with short legs, a small tough beak, and no wings. They suck the blood of birds, bats, and humans. Of the two species that attack man, only one, *Cimex lectularius*, is found in North America. It came to the continent from Asia by way of Europe. The other species is restricted to the tropics of the Old World.

Certain cimicids attack pigeons, wild birds, and chickens. The poultry bug, *Haematosiphon inodorus*, a serious pest in Mexico and the southwestern United States, has recently begun to invade houses. Unlike most cimicids, it has no odor and it is not nocturnal.

The assassin bugs, genus *Apiomerus*, are rather stout and about 1½ centimeters in length. They are black or brown with red or yellow decorations. Legs, head, and thorax are often covered with coarse hairs. These bugs feed on caterpillars, plant lice, and anything they can capture. One species, the bee assassin, is a pest around beehives. It also devours beetles and flies.

Many species of Hemiptera are semiaquatic or aquatic. The semiaquatic species include the water striders and marsh treaders. The aquatic species include the water boatman and the backswimmer.

Homoptera: cicadas, tree hoppers, spittle bugs, leafhoppers, psyllids, white flies, aphids (plant lice), scale insects, mealy bugs, lac insects, cochineal insects. It is very difficult to characterize these insects. They vary widely in shape.

The Homoptera are an important group of insects. Aphids, for example, are serious pests of cultivated plants. Whiteflies and psyllids also damage plants. Psyllids cause plant galls. Whiteflies leave blackish scales.

Scale insects are also serious pests of cultivated plants. However, a few species in the scale insect family—lac and cochineal insects—are useful economically.

ENDOPTERYGOTA

All the rest of the insect orders have complete metamorphosis, passing through four stages: eggs, larvae, pupae, and adults.

Megaloptera: dobson flies, humpbacked flies. Adults are medium to large, with biting mouth parts, slender antennae, and two pairs of wings, held rooflike or flat over the body. The larvae are aquatic.

Neuroptera: nerve-winged insects, lacewings, false mantids. These insects are minute to medium-sized, fragile insects with biting mouth parts, large compound eyes, long, slender legs, and two pairs of

Termites. These isopterans are well known for their wood-eating habits.

A green darner. The largest of the dragonflies belong to the darner family of the Odonata.

large, many-veined wings. In one family (Mantispidae) the adults resemble praying mantids. The larvae of one family are called antlions because they prey on ants. Neuroptera larvae spin silk cocoons in secluded places, and there they pupate.

Mecoptera: scorpion flies. Adults are small to medium-sized, extremely slender insects, with long heads and biting mouth parts. Some have two pairs of long, narrow wings; others have none. Scorpion-fly larvae are wormlike. Both larvae and adults are carnivorous.

Trichoptera: caddis flies, water moths. These are small or medium-sized insects, with long, slender antennae, large compound eyes, husky legs, and two pairs of wings lightly covered with hairs and scales. The larvae are aquatic, living in movable cases, which they construct out of bits of wood or pebbles, held together by silk. The adults live near water.

Lepidoptera: moths, butterflies, skippers.

Coleoptera: beetles, weevils.

Hymenoptera: ants, bees, wasps, gallflies, chalcid flies, sawflies, braconids, ichneumons.

Members of the orders Lepidoptera, Coleoptera, and Hymenoptera are discussed in the articles "Butterflies and Moths," "Beetles," and "Ants, Bees, Wasps, and Termites" in *The New Book of Popular Science.*

Diptera: flies, gnats, midges, mosquitoes, and others. These are minute to medium-sized insects, with sucking, piercing, lapping, or functionless mouth parts. The hind wings are represented by a pair of small stumps, called halteres. There are some wingless species. The larvae of certain species live in decaying animal or vegetable matter and are called maggots. The larvae of mosquitoes and of some gnats and midges are aquatic. Some wingless and eyeless fly species live in the nest of ants or termites or in bee hives.

Midges, or chironomids, are to be found almost everywhere, but especially in the moist parts of the earth and around bodies of water. They range in size from extremely small creatures to insects as large as big mosquitoes. Certain tiny midges, including the sand flies and punkies, are vicious biters. Most midges, however, cannot bite. They annoy people because they swarm around them. The larvae usually dwell in the water. Some of them are bright red and are therefore known as bloodworms. Both larvae and adults are important items of food for freshwater fishes and many birds.

Mosquitoes, or culicids, are the most dangerous insect enemies of man. They are found in almost all parts of the world, from arctic regions, such as Alaska and Greenland, to the tropical swamps of Africa and South America.

The eggs are laid on or near the water of ponds, pools or marshes. They are sometimes deposited in the water that has collected in empty tin cans or discarded automobile tires. After a short time the eggs hatch into the aquatic larvae called *wigglers.* These feed upon minute animals or plants or floating particles in the water.

A giant water bug. These hemipterans live in ponds. They can inflict a nasty bite.

They frequently come to the surface to breathe. The pupae are also active wigglers. Adult males have featherlike antennae, while the antennae of the females are slender and hairy. Only the female bites.

All over the world mosquitoes annoy man by sucking his blood. It is within the tropical and semitropical areas, however, that they cause the greatest amount of human misery. They carry several major diseases that are deadly to man. Malaria is transmitted by certain species of *Anopheles*. Yellow fever is transmitted primarily by *Aëdes aegypti*, but also by other species of *Aëdes* and some species of *Haemagogus*. Dengue is also carried by *Aëdes aegypti* and some of its close relatives. Filariasis is transmitted by various species of *Culux* and *Aëdes*.

Horseflies and their kin—the tabanids—are large, stout-bodied flies, which bite various mammals, including man. Only the females suck blood. They sometimes transmit various diseases, especially tularemia and anthrax. The larvae are found mostly in swamps, ponds, and damp places, where they feed upon tiny animals, especially the larvae of other insects.

The larvae of fruit flies, or trypetids, are destructive to fruits and vegetables. Among the members of the trypetid group are the apple maggot (*Rhagoletis pomonella*), the Mediterranean fruit fly (*Ceratitis capitata*), the Mexican fruit fly (*Anastrepha ludens*), the Cherry fruit fly (*Rhagoletis cingulata*), the olive fly (*Dacus oleae*), and the walnut husk fly (*Rhagoletis completa*). The adults are beautiful little creatures, smaller than houseflies. They have banding or spotting on the wings. Because of the way many of the fruit flies alternately raise and lower their brightly colored wings as they move slowly about on vegetation, they are often called peacock flies.

Vinegar flies, or drosophilids, are tiny creatures, almost always less than one-third of a centimeter in length and yellowish to black in color. The adults are attracted to rotting fruits and vegetables as well as to fermenting substances. The eggs are laid in these materials. The life cycle is always short, usually two to three weeks in length. More than six hundred different species have been described. Of these *Drosophila melanogaster* is the most famous, because it has been used extensively as an experimental animal in research on the laws of inheritance and evolution.

The housefly (*Musca domestica*) is one of the ever-present uninvited guests that live in or near the abodes of man. It does not bite, but its habit of alighting upon the exposed parts of our bodies is most annoying. Houseflies breed upon manure, garbage, and other kinds of decaying matter and feed not only upon such materials but also upon our food. Hence they are efficient transmitters of such diseases as typhoid fever, dysentery, cholera, and anthrax.

ANTS, BEES, WASPS, AND TERMITES

There is no counterpart among other animals to the military, food-gathering, cattle-keeping, and slave-making activities of the ants, or to the perfectly ordered system of the beehive. Ants and some bees and wasps form elaborate social organizations and hence are known, appropriately enough, as *social insects*. Their cooperative enterprises are rivaled only by those of people.

The social insects belong to the order Hymenoptera, which is perhaps the most highly developed group of insects. The order also includes gallflies, sawflies, horntails, and ichneumons (parasitic wasps that lay their eggs within the bodies of other insects' larvae). These are solitary insects; they do not live in communities. More than 100,000 species are known, and probably as many more still remain to be named.

These insects have mouth parts specialized for biting, chewing, scraping, lapping, or sucking. Most members of the Hymenoptera have two pairs of stiff and quite narrow wings. Most ants, however, are wingless.

ANTS

The keynote of the social organization of ants is specialization. In the case of solitary insects, each one must perform all the tasks necessary for its survival—collecting food, seeking shelter, fighting off enemies. Ants live in large nests, or *ant cities,* and divide their labor. Each has its own duties; the rule of one for all and all for one is successfully applied.

The members of the nest are divided into three distinct classes, or *castes,* on the basis of the work they do. There are various subdivisions—some 25 in all—among these three classes.

THREE CLASSES

The largest class comprises the wingless, normally infertile, females, which function as *workers* or *soldiers*. These insects are usually the smallest members of the community. They enlarge, maintain, and defend the nest, gather food, and feed and nurse the queen and young. Occasionally a worker, if carefully fed and attended, may lay fertile eggs. The soldiers are workers with large heads and strong jaws. They are employed in fighting and in crushing hard foods, such as seeds.

The second class is composed of the *queen* ants, which are large, winged individuals. They are the mainstay of the nest, producing eggs almost constantly, except during the cold season.

The winged *males* are the third class, with small heads and small jaws. Some species have both winged and wingless queens or males.

FORMING A COLONY

Ants live almost everywhere—in deserts and fields, in forests and mountains, along the seashore, and in the villages and dwellings of people. Their homes vary from galleries and chambers excavated in the soil to nests built in hollow stumps. Some nests, constructed of paperlike material, are suspended from trees.

An ant colony may be formed either by a solitary queen or by several queens together. Before the nuptial, or mating, flight, each queen has stored immense energy reserves within her body. On the appointed day, which varies with the different ant species, the winged males and the queens swarm together. Mating occurs during or at the end of the flight. It is the queen's only flight. When she alights, she will be ready to begin her egg-laying duties.

After the nuptial flight is over, the males seek shelter under stones, sticks, or fallen leaves. They make no effort to return to the nest. Even if they survive the onslaught of their natural enemies, they die in a matter of days.

When the queen alights after the nuptial flight, she may be surrounded by worker ants of her own species. These workers escort her to their home nest and she becomes a member of a full-fledged colony.

Often, however, the young queen must found her own colony. First she casts off her wings and seeks an appropriate shelter—the underside of a stone or a stick or simply a hole in the ground. She digs until she has hollowed out a burrow with a chamber at one end of it. The insect then blocks the entrance to the burrow and for a considerable period rests quietly, shut off from the outer world. The wing muscles are gradually broken down and converted into fat bodies, for energy.

Finally the queen begins to lay eggs, and in a comparatively short time the first *larvae* hatch. The queen is a devoted mother to the first generation of her off-spring. She forces saliva, containing fat bodies, into the mouths of the young larvae and they grow rapidly. Should her food supply fail, the queen may even eat some eggs and larvae so as to maintain herself and the nest. If mating occurred in late summer, cold weather comes while the brood is quite young. The queen and her larvae then hibernate until spring.

While the larvae of certain species are quite small, they spin cocoons and, enveloped in these wrappings, become *pupae*.

As each pupa becomes mature, it is freed from the cocoon by the mother. It is now a small but perfectly formed worker. In some ant groups, the larva does not spin a cocoon and the pupa is naked.

The workers of the first generation dig their way to the surface of the ground and go in search of food. They bring it to the queen, which is in a rather sad state because she has fed her brood with her own bodily substance. She soon recovers her strength and continues her egg-laying activities. Gradually the workers take over the task of caring for the young. Eventually the queen becomes a sort of egg-laying machine. Workers feed her by cramming food

Right top: an ant colony. The queen, distinguishable by her swollen abdomen, lays eggs. Some soon hatch into light-colored larvae. Workers feed the queen and care for the young. Middle: workers lick the queen and absorb a substance that makes them sterile. Bottom: harvester ant dragging a plant part larger than itself.

Weaver ants use silk produced by their larvae to sew together leaves to form a nest. These ants occur in southern Asia and on some Pacific islands.

A soldier ant. Soldier ants have enlarged jaws. They protect the nest and help obtain food — particularly hard foods such as seeds — for the colony.

regurgitated from their stomachs into her mouth.

As times goes on, the workers become more numerous and larger. Soldiers first make their appearance some time after the queen has established her colony. New queens appear upon the scene last of all. It is the workers that carry on the essential work of the colony. They may move it to a more suitable site again and again. They carry their egg-laying queen, immature workers, larvae, and pupae to the new location. Although most workers are sterile, an occasional worker may lay eggs, which will give rise to male or worker offspring.

BEHAVIOR

Much of a young worker ant's behavior results from imitating its elders. The novice must learn the lay of the land around its nest so that it may return to the nest after a foraging expedition. Perhaps it will lose its way and have to be carried home by the more experienced workers. Certain individuals of a colony seem to learn better than others. These key workers initiate the various activities in the colony and thus set an example to the other ants. If these choice workers are removed from the nest, the activity of the colony is retarded. The colony fails to prosper; the health of the queen and of other workers is bound to suffer.

Modern biologists believe that the seemingly intelligent activities of ants are due in part to inherited instinctual patterns.

Ants also respond to stimuli released by other ants. The stimuli in question are often due to glandular secretions known as *pheromones*. These are found not only in ants but also in various other social insects, including honeybees and termites.

Ants have organs of smell in the form of tiny sensory pits in the antennae. It is with these organs that ants can sense the secretions that have been emitted by other ants. In this way, they can recognize members of their own colony and detect the presence of ants belonging to hostile colonies. In some cases, secretions are laid down as scent trails for other ants to follow, as when a food supply has been discovered by a foraging worker.

Through the release of secretions of different kinds and quantities, ants can bring about alarm reactions, fighting activities, reproduction activities, and nest construction. Certain poisonous secretions serve to repel invaders or to subdue prey. Chemicals secreted by queen ants prevent ordinary workers — infertile females — from developing into reproductive females or rival queens, except in emergencies or at various stages of colony development.

It is generally thought that certain species of ants can hear, because they can make certain sounds that may be a mode of communication with their kind. The ant *Myrmica rubra* produces sounds with a strigil, or file, on its seventh abdominal segment. Other ants (genus *Polyrhachis*) tap on the surface of a leaf with their heads and still others scrape the end of the abdomen on the dry leaves of the nest.

Much of the apparently friendly behavior of ants and other social insects stems from the mutual exchange of pleasing food secretions. The larvae of ants, for example, exude substances that are greatly appreciated by the workers who feed the young and receive these as a reward. The adults of social insects also engage in this practice among themselves. The material may consist of regurgitated food or glandular secretions. The habit is called *trophallaxis,* a name derived from Greek words meaning "food exchange". Furthermore, foreign insects that live in ant and termite nests are tolerated because of their "payment" of food matter desired by their hosts. Certain ants "herd" other insects for the secretions the latter provide.

One of the most amazing practices of some ant species is the keeping and milking of "cows." The "cow" in question is really an insect, the aphid, or plant louse, which deposits a secretion called honeydew on the foliage and stems of vegetation. The ants milk the aphids by gently stroking them with their antennae, thus coaxing the aphids to give droplets of honeydew.

The ants carry their insect livestock to different parts of the plant or to different plants in the garden to make sure that they have enough to eat. Some ants dig tunnels in the soil for the convenience of root-sucking plant lice. Certain ants even take the plant lice into their own nests for the winter.

Aphids are not the only animals that find hospitality in the nests of ants. A great number of beetles, cockroaches, flies, and arachnids take up their abode there. They are allowed by their willing or unwilling hosts to feed on the excretions of the ants or on their food.

Parasitism plays an extremely important part in the life of ants. Different species or groups may live off others temporarily or even permanently. For example, various colonies of the genus *Formica,* in the

Some ants build very large mound nests, such as one below. It is constructed of soil and plant matter carried to the site by worker ants. The mound itself typically has food storage galleries as well as egg-laying sites.

S. G. Bisserot, Bruce Coleman Ltd.

temperate zone, may establish themselves in the nests of other *Formica* communities. A single queen induces the workers of a foreign colony (of a similar species) to care for her and for the offspring from the eggs she lays in their nest. The original queen of the colony may remain, be killed, or be driven off, even by her own workers. As the invading queen's larvae mature, they eventually replace all the original workers, who naturally have been dying off. At this point, there is little evidence that the thriving colony had begun by parasitism.

Ant slavery, or dulosis (from the Greek word for "slave"), is carried out by militant colonies raiding and capturing members of other colonies, often of a different ant species. Workers of *Formica sanguinea* carry off the larvae and pupae of the horse ant. These are tended by ants that have been reared from previously captured young. As the new captives mature, they care for the larvae, forage for food, and so on in the slaveholding community. However, *Formica sanguinea* workers carry on the main duties of the colony.

SOMETIMES PESTS

When ants make their way into our homes, they become unmitigated pests, working havoc with our food. The Pharaoh's ant, *Monomorium pharaonis,* is a notorious culprit, nesting in the foundations and often completely overrunning a house from the cellar to the attic. These tiny yellowish or reddish ants quickly sense the presence of food and rush to consume it.

Carpenter ants, members of the genus *Camponotus,* enter homes to live in ceiling beams, porch columns, and window sills. Here they may do considerable damage by excavating galleries and chambers. Carpenter ants do not eat wood but feed on plant juices, animal remains, and the honeydew of aphids.

SOME INTERESTING ANTS

Among the ants there are many interesting varieties, including gardeners, living receptacles, and slaveholders.

Parasol ants. Members of the genus *Atta* cultivate various species of club fungi, which they use as food. The ants provide a garden bed made of cut sections of leaves. A large ant expedition sets out to obtain this material, generally in the late afternoon. The insects climb up nearby trees and cut off more or less circular pieces of leaves. As the ants return to the colony with their booty, each one carries its piece of leaf overhead like a parasol; for this reason the insects are sometimes called parasol ants.

When the leaf sections have been brought to the nest they are thoroughly chewed and then deposited on the floor of a large chamber. As the leaf layers are built up on the floor, they acquire a spongelike structure. Soon they are covered with the desired fungus growths.

The body of the fungus, called the mycelium, consists of a great many slender branched threads known as hyphae. Small spherical swellings develop on the hyphae, and these swellings provide food for the ants. The smaller workers in the colony rarely leave the nest, but spend most of their time "weeding" the fungus garden — that is, removing unwanted growths.

Black honey ants. Black honey ants have a curious but effective way of storing

Honeybees and some other species of bees have a social organization very similar to that of ants. Below: a queen bee lays eggs in cells of the hive. She is surrounded by workers who will nourish the young.

Jacques Six

ANTS, BEES, WASPS, AND TERMITES

honey. Certain workers, known as *repletes,* gorge themselves until their abdomens are greatly distended. They then serve as animated honey jars, ready to serve the needs of the other ants. When food becomes scarce, the members of the colony stroke the abdomens of the repletes and devour the droplets of honey that are regurgitated.

Repletes are also used to store honey by various species of the genus *Myrmecocystus,* found in the southwestern United States, Central America, and South America. After the abdomens of the repletes have become swollen, they cling to the roofs of underground chambers; they are relieved of their loads of honey as occasion requires.

Tropical driver, or army, ants. These are carnivorous ants whose natural prey consists of insects and other small invertebrates, though they will attack any living thing in their path. Generally, animals such as mammals, including humans, and birds can easily avoid the driver ants. They fall victim to the little carnivores only if they are injured and cannot keep out of the way of the ants.

The two genera—*Eciton* and *Dorylus*—that make up the driver ants are found in widely separated areas. *Eciton* dwells in the American tropics; *Dorylus* in the African tropics. Though these ants are completely blind, they advance in long columns with remarkable precision; their foragers fan out on all sides of the main columns. The ants form temporary bivouacs, consisting of clusters of insects hanging from the branches of shrubs and bushes.

BEES

The bees make up the superfamily Apoidea, comprising some 20,000 species.

The abdomen of a bee is joined to the thorax by a rather slender "waist." Females have an organ called an *ovipositor* at the end of the abdomen. This serves as an egg-layer and also as a weapon; it can inflict a painful sting. In the case of honeybee workers, the ovipositor, or sting, has little barbs that turn inward. If the worker stings a foe, the sting is generally left in the body of the victim and the bee dies. The ovipositor of the honeybee queen is smooth. This bee can sting its enemies repeatedly without harm to itself.

Bees store pollen and honey in their nests to provide food for their young. Pollen is collected by the hairs on a bee's legs and body and also by specialized structures called *pollen brushes,* which are found on the hindlegs and sometimes on the abdomens of female bees.

After pollen is collected, it is brushed off by the insect's head and feet, dampened with dew or some other form of moisture, mixed with honey from the bee's mouth, and formed into tiny pellets. The pellets are then pushed into the so-called *pollen baskets.* These baskets, consisting of long and stiff hairs on the hindlegs, hold a considerable number of pollen pellets.

Honey is a product of nectar, a sweet

Among social bees, each hive contains undeveloped females who serve as workers (left), a mature queen bee (middle), and several males, or drones (right).

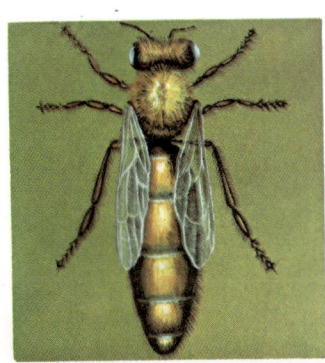

ANTS, BEES, WASPS, AND TERMITES

Some of the cells of a bee hive are used for food storage. Here a honeybee fills a cell with honey, formed from nectar the bee has collected.

liquid secreted by the glands found in certain flower petals. The bee sips nectar and swallows it. The substance is transformed into honey in the insect's crop, or *honey sac,* and is later disgorged. Honey is made up chiefly of the sugars levulose and dextrose and of water. It also contains dextrins, gums, vitamins, enzymes, pollen grains, and various minerals.

Bees also collect a sticky substance called *propolis,* derived from the resinous secretions of various trees. Propolis is used as a cement in the building of nests.

SOLITARY BEES

Contrary to popular belief, most species do not form communities but are solitary in habit. The female of the solitary bee builds her nest cell by cell. She stocks each cell with pollen mixed with nectar and then lays an egg on this food supply. She seals each nest before she goes on to the next one. After the entire nest has been completed, she closes it up and flies away, never to return. The larvae that are hatched from the eggs go through a series of molts, become pupae, and finally emerge from the nest as adult bees.

There are some interesting varieties of solitary bees. The small carpenter bees (family Ceratinidae) and the large carpenter bees (family Xylocopidae) dig through wood and make nests in the resulting burrows. The mason bee of the genus *Chalicodoma* constructs its nest of soil and tiny pebbles, mixed with saliva. Leaf-cutter bees (family Megachilidae) form their nests of pieces of leaves and petals cut from roses and other plants.

Certain bees lay their eggs in the nests of other bees; their young become unwelcome lodgers. These parasitic bees are known as *inquilines,* from the Latin word *inquilinus,* meaning "lodger." For example, the parasitic bee *Stelis* lays its eggs in the cells of *Osmia.* Larvae of both species are hatched. Ultimately the *Stelis* larvae devour those of *Osmia* and take over the nest.

HONEYBEE COLONIES

The social bees—bees that form communities—have developed distinct castes, which correspond more or less to those found among ants. In each hive there is a mature queen bee, a number of males, known as *drones,* and a great many workers, or undeveloped females.

Perhaps the best-known of the social bees is the honeybee, *Apis mellifica.* It builds an elaborate hive with wax secreted from eight wax pockets on the underside of the abdomen. The honeycombs in this hive are set vertically, side by side. Each comb consists of thousands of hexagonal cells. Some of the cells contain eggs, larvae, or pupae. Others are used for the storage of honey and pollen.

At the beginning of the spring, the honeybee hive contains a queen and a comparatively small number of workers. The queen begins to lay eggs. Some of these will develop into workers, others into drones, and a few into queens. The latter are reared in special enlarged cells called *royal cells.* After about three days, the eggs hatch. The larvae, or grubs, are all fed for the first two or three days with *royal jelly,* a secretion from certain glands of the workers. After this period is over, the prospective workers and drones are gorged with honey and pollen. The larvae that are to be reared as queens, or rather *princesses,* continue to be fed with royal jelly.

The larvae grow rapidly. After about six days, the cell is sealed by the attendant workers, and the larva becomes a pupa. The larvae that are to become workers or drones spin a complete cocoon about themselves before entering the pupal stage. The cocoons spun by the larvae that are to develop into queens enclose only the head, thorax, and a small part of the abdomen.

After 12 days or so, the workers cast off the pupal skin and chew their way out of the cell. It takes the drones about two days longer to pass through the pupal stage. Princesses develop from pupae to adults in about seven days. They do not emerge from their cells; attendant bees make a small hole in each royal cell and continue to feed the occupant.

In time the honeybee colony becomes overpopulated and a form of emigration, called *swarming,* takes place. The old queen, accompanied by many of the workers, leaves the hive and seeks a new nest. This may be the hollow of a tree or a man-made hive.

One of the young princesses now emerges from her cell in the old hive. She makes her way to all the other royal cells and slays their occupants. If two or more princesses emerge at the same time, they fight until only one remains alive. The newly established queen then flies from the nest, mates with a drone, returns to the nest again, and begins to lay eggs. The drone dies immediately after mating.

The rest of the drones are tolerated in the hive for a time, though they perform no communal tasks. If the food supply dwindles, however, they are driven out of the nest and are not permitted to re-enter. Since they are dependent on the food brought in by the workers, they are doomed to die once they are excluded from the nest.

Workers have different tasks. Some of them collect pollen and nectar. Others care for the young. Still others attend the queen. A certain number fan the hive by means of rapid wing movements. This is a most important task; if the premises become too hot, the wax cells melt.

DANCE COMMUNICATION

As a result of research in the 1920s by the Austrian zoologist Karl von Frisch, we now know that a worker bee conveys information about a promising food source to the other bees by means of a series of dances. When the food supply is near at hand, the returning bee performs a round dance on one of the vertical combs of the

Below: a worker bee performs a circular dance to tell other members of its hive that a food source is nearby. Right: a bee telling other bees that the food source is some distance away—in the direction of the straight part of the dance. The slower the bee dances, the farther away the food is.

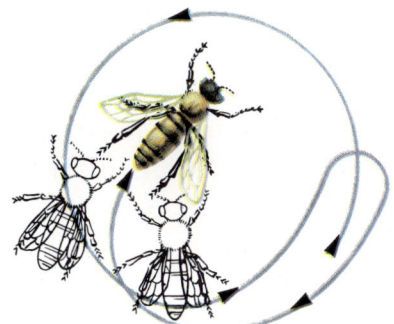
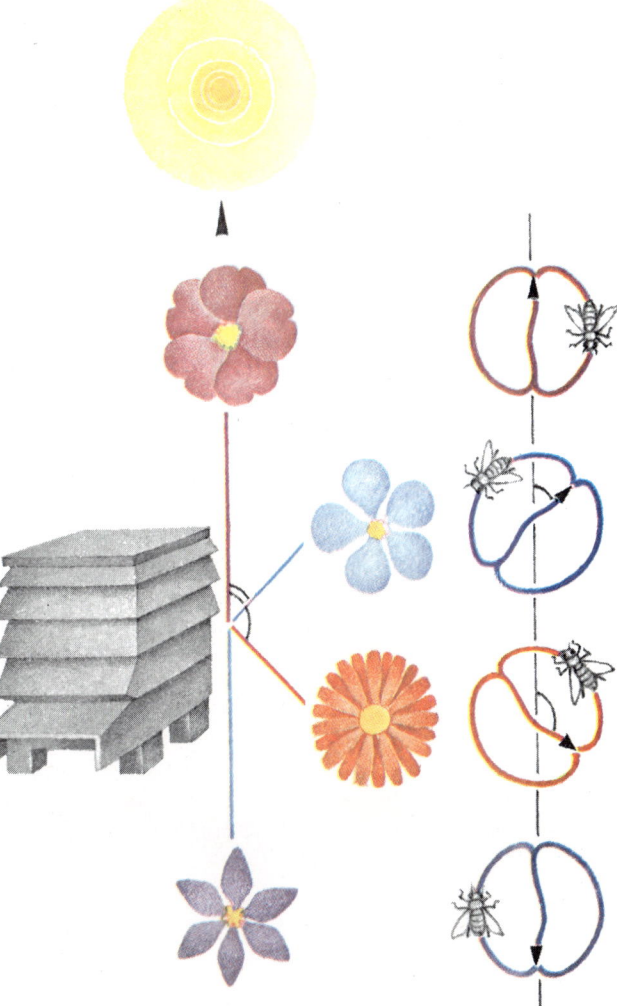

hive. It circles to the right and then to the left, over and over again.

If the supply of food is a hundred meters or more distant, the returning bee performs a tail-wagging dance. First it makes a short straight run up or down the comb, wagging its abdomen from side to side. It then circles to the left. Again it makes a straight run with a tail-wagging motion and this time circles to the right. The bee tells how far away the food is by the speed of the dance; the slower the dance, the farther away the food. It indicates the direction of the food by the direction of its straight run. If the dancer heads directly upward during the straight part of the dance, it means that the feeding place is in the same direction as the sun. If the insect heads directly downward in its straight run, it means that the workers must fly away from the sun to reach the food. Suppose the bee goes 45° to the right of vertical in the straight part of its run. This would indicate that the feeding place is located at 45° to the right of the sun.

OTHER SOCIAL BEES

Bumblebees. These insects, which comprise the family Bombidae, range in length from less than one centimeter to 2½ centimeters or more.

Bumblebee colonies, like those of wasps, must be started anew each year. Each colony is established in the spring by a young fertilized queen, who has spent the winter under brush or debris or in a hole in a log. The queen seeks a site in or on the ground; this is often the vacated nest of a field mouse, chipmunk, or other small mammal. She may have to fight other bumblebee queens for the site.

Once established in her new home, the queen prepares balls of honey and pollen on which she lays her eggs. With the coming of cold weather the old queen, her retinue of workers, and the drone hangers-on all perish. Only the young queens survive to continue the race.

Stingless bees. The Meliponidae are a third important group of social bees; their ovipositors are atrophied. These bees are much smaller than honeybees; some species are only two millimeters long.

Bees also collect pollen to feed their young. The pollen adheres to hairs on the bee's body and is carried back to the hive.

The Meliponidae build combs in horizontal sections. Some make their nests on the ground, others in the hollows of trees. Still others establish themselves in the nests of termites or ants. In constructing their abodes the bees use earth and leaf particles, dung, and other substances in addition to the wax that they secrete. They store honey in fairly large receptacles. Human beings occasionally use the honey as food—a dangerous practice, as this substance is sometimes highly poisonous.

WASPS

Most species of wasps do not form colonies but are solitary insects.

The social wasps all belong to the family Vespidae. The most familiar representatives are the common wasps and hornets of the genus *Vespa*. The young *Vespa* queens mate in the autumn. They alone of all the members of wasp colonies survive the winter season, hibernating until the coming of spring.

Roused by the first warm spring days, each young queen seeks a place for a nest—perhaps under an eave, in a hollow tree, or on a bush. There she rapidly constructs a series of cells of paperlike structure, formed from woody fibers and other vegetable matter. She deposits an egg in each cell.

When the larvae hatch, the queen feeds them with chewed caterpillars, flies, or fruit juices. When the time comes for the larvae to assume the pupa form, they close their cells, undergo metamorphosis, and emerge as undeveloped females, or workers. The workers assist the founder of the colony in making more cells and in feeding the larvae. Tier after tier is made, until the nest becomes a truly imposing structure.

Toward the end of the summer young queens, and after that the males, appear. The earlier generations have all been undeveloped females, or workers. Some of these are capable of laying eggs that may develop into males. The young queens and the young males finally leave the nest, never to return. The workers that remain destroy the larvae that are left in the nest. Then they themselves await death.

The social wasp whose life cycle we have just described eats great numbers of insects that are harmful to people. On the other hand, it is quite destructive of fruit and sometimes attacks human beings, particularly when it is molested or fancies that it is molested. The hornet is a large wasp whose sting is particularly formidable to people.

Not all wasps form their nests of paperlike material. Some of the solitary wasps make nests of sand, clay, or mud — quite efficient dwellings, though less imposing than the beautiful structure fashioned by the social paper wasps. Some of

Bernard L. Gluck, Audubon/PR

A female ichneumon wasp. This wasp is not social. It is parasitic and lays its eggs on or beneath the skins of various insect larvae.

Hugh Spencer

Many social wasps, including house wasps, build paper-walled nests. They chew wood pulp and make a strong paper from it.

ANTS, BEES, WASPS, AND TERMITES 275

the solitary wasps have a rather singular method of supplying their young with food. Since there are no workers among these wasps, the mother wasps have to provide food to which the larvae, upon hatching, may help themselves. This is brought about by the capture of various insects, particularly caterpillars, and spiders. A mother wasp stings the prey in a vital spot and paralyzes it. Then she drags it off to her nest. The victims remain alive, but motionless and helpless, to await the attack of the larvae, as soon as hatching takes place. In this way the young of the solitary wasps are assured of a bountiful food supply.

Other remarkable wasps include those belonging to the genus *Pompilus,* which do not hesitate to attack even such large spiders as the wolf spiders, and the gall wasp, or gallfly, which lays its eggs in plants and causes galls to form.

ICHNEUMONS

The family Ichneumonidae includes some 40,000 known species. Ichneumon flies look somewhat like wasps, but their antennae are usually longer and contain more segments than those of wasps.

The ichneumon fly deposits its eggs on or beneath the skin of various insect larvae. When these eggs hatch, the young feed on the tissues or juices of the victims. In this way, ichneumons kill great numbers of larvae of insects that are harmful to people, thus ranking them among our most valuable allies in the insect world.

TERMITES

The termites, or white ants, are not ants at all but deserve mention as social insects. They form the order Isoptera, a name that refers to the equal size and shape of the wings of those forms having wings. Unlike ants, termites are not slender-waisted and their antennae are not elbowed.

Termite society is based on a caste system. A queen and king form the royal couple. The queen has an enormously enlarged, cylinder-shaped abdomen. During a year, she lays at least 3,000,000 eggs and probably more. The small king is devoted to his queen throughout her life, fertilizing her eggs at intervals. Aside from this royal pair, there are individuals capable of reproducing who are kept in readiness for the queen's eventual decline.

Workers are soft-bodied, pale creatures that labor constantly at constructing the nest, collecting food, caring for the eggs, and feeding the queen, soldiers, and young. With some species, nymphs instead of workers do these tasks.

The soldiers defend the nest against invading ants. Some termite soldiers have enlarged, armored heads with formidable jaws. Others, with small jaws, have a snout-like projection that ejects an acrid or sticky fluid on an enemy. Each termite caste includes individuals of all ages and of both sexes.

Termites nest either below ground or above the surface in mounds, trees, and stumps. They build enclosed mud runways to their sources of food, which consists of dead tree stumps and limbs and the wood of our houses. This wood is broken down in the digestive tract by protozoa that convert the cellulose into a usable food. Termites also eat fungi and grasses.

Termite society is also based on division of labor. Here several greatly swollen queens surrounded by worker termites.

C. A. W. Guggisberg, PR

BEETLES

John C. Pallister

To many people a beetle is a dark, scurrying little insect that arouses disgust and perhaps apprehension. The very name of the insect reflects this attitude, for "beetle" comes from the Anglo-Saxon *bitol,* meaning "creature that bites."

Some beetles, however, are insects that people enjoy watching. Ladybugs, or lady beetles, and fireflies are examples. One beetle, the scarab *Scarabaeus sacer,* was worshiped by ancient Egyptians as a symbol of eternal life.

Some species are very useful because they prey on insects that we consider to be pests. Other beetles are scavengers, burying small dead animals, cleaning up debris, and making the earth more attractive.

But there are thousands of species that are pests that destroy trees, crops, processed foods, clothing, and furniture. A few are parasitic on human beings. Some transmit a tapeworm to rats, other animals, and humans.

Though beetles seldom carry the attack to us, they will sometimes defend themselves when we try to handle them. A few have mandibles, or mouth parts, strong enough to draw blood from the unwary. When handling a beetle of this kind—a "pinching bug," say—it is best to grasp it just back of the head, so that it cannot turn and nip you. The bombardier beetle will shoot out a caustic liquid when it is picked up; this liquid can blister a tender hand. Other beetles' leg joints ooze juices that are evil-smelling and stain the fingers.

The word "beetle" is applied to all the members of the order Coleoptera. The beetles are the most numerous of all the orders of insects. There are some 300,000 species.

GENERAL CHARACTERISTICS

A beetle may be no larger than a pinhead or it may be as big as a man's fist. It may be dull in color or gleam like a precious jewel. It may be slender and graceful or antlike in shape.

Like all insects, a beetle is a boneless animal. Its vital organs and muscles are protected by a jointed, segmented case of hard material called chitin. Its body is divided into three parts: the head, thorax, and abdomen. The head carries the eyes, antennae, and mouth parts, which are very complicated. The thorax, or middle section, bears the six legs and two pairs of wings, and, within, some of the digestive organs. The abdomen is made up of nine or ten ringlike segments of chitin, connected by a softer tissue. It contains the organs for breathing, digesting, and reproducing. The stridulating organs, which produce sound, are also found on the abdomen. Not all beetles possess these organs.

All beetles have one pair of jointed antennae, usually projecting in front of the

A ground beetle, a member of the family Carabidae, which is one of the largest families of beetles.
Edwin Way Teale

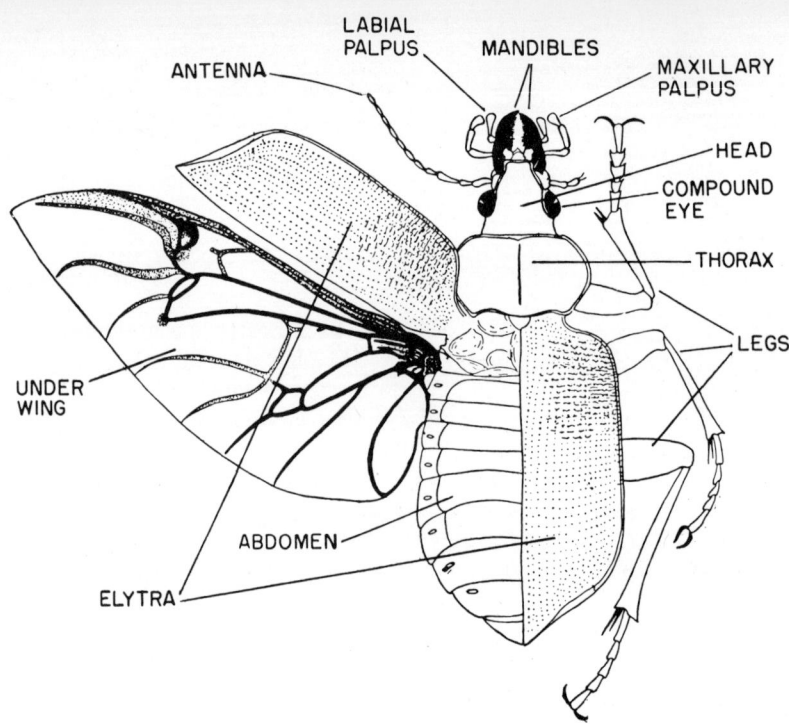

Diagram showing the external structures of a ground beetle. One of the elytra has been spread to expose the abdomen and the underwing. The tiny spiracles, along the margin of the abdomen, are openings leading to the breathing tubes.

eyes. These may be so short that you can hardly see them, as in carpet beetles and lady beetles. Or they may be two or three times as long as the insect's body, as in the long-horned wood-boring beetles. Under a low-powered microscope, the antennae show an astonishing variety of shapes. They may suggest a brush, a feather, a string of beads, a comb, or a club.

The most remarkable structure of beetles, which distinguishes them from practically all other insects, is the outer, or first, pair of wings, called *elytra*. They do not look at all like wings. They are hard and shell-like and serve as a covering for the second, or inner, pair of wings and for the abdomen. The underwings are thin and membranous, and when not in use are folded and refolded under the elytra. The elytra are of little help in flying.

All beetles have elytra or traces of elytra. The name "Coleoptera," which means "sheathed wings," is derived from this wing arrangement. The only other insects that have wing sheaths are the earwigs, some grasshoppers, and some Homoptera. However, their sheaths are not nearly so firm and shell-like as those of the beetles.

A beetle's thorax is composed of three segments, each of which bears a pair of legs. The elytra are fastened to the top of the middle segment, covering it, the back segment, and part or more often all of the abdomen. The top of the front thoracic segment—the prothorax—is as hard and shell-like as the elytra. In some species it fits so neatly against the elytra that when the beetle is resting, its back appears to be in one piece.

During its life a beetle passes through four distinct stages. First it is an egg. This hatches into a wormlike larva, commonly called a *grub*. The grub eats voraciously; this is the only stage of the beetle's life during which it increases in size. (Little beetles are never the young of large beetles; they are always of a different species.)

As the grub grows too big for its skin, it molts, or sheds the skin. It usually molts five or six times before it is full-grown. Having eaten all it needs, the larva seeks a secure resting place. Often it shelters in the plant or tree where it has been feeding. Or it burrows into the ground, where it builds itself a little cell. Some leaf-eating weevils spin cocoons.

Once it is secure in its shelter, the grub becomes a *pupa*, eating nothing and remaining motionless while great changes go on in its body. This period may last only three to four days in the case of lady beetles, when the humidity and temperature are just right. Or it may last all winter, in the case of beetles that become pupae in the autumn months. Finally, the adult beetle emerges from the pupal case. The ordinary beetle lives only two to six months. Certain wood-boring beetles, however, have a life cycle of several years.

Land beetles are found all over the world, except in the extreme northern or southern areas. A few species dwell in freshwater ponds and streams, but none inhabit oceans and seas.

The Coleoptera are subdivided into about 150 families. In the following pages we briefly examine some of these families.

TIGER BEETLES

"Tiger beetles" is an appropriate name for the fierce and swift members of the family Cicindelidae. They feed on other insects, which they stalk along sandy banks and beaches, on roadsides and woodland paths. They are rather small, slender insects, just about the color of the sand or dirt upon which they run. However, one of the commonest species in the eastern United States is a brilliant metallic green, and several related species are greenish or bronze-colored. The elytra are often spotted or banded in lighter tints.

Tiger-beetle larvae are whitish grubs with big, metallic heads and very large mandibles. They are even more dangerous than the adults to small insects. They lurk in small burrows they have dug, with their heads just protruding and their mandibles wide open, ready to seize any insect that walks over them. Hooks on the back of the abdomen serve to anchor the grub to the sides of its burrow, so the intended prey is not able to launch a counterattack by jerking the larva out of its lair.

GROUND BEETLES

If you take a walk in the evening, you may encounter small, dark beetles scurrying in front of you. The great majority of these will be ground beetles, or carabids, belonging to the large family of the Carabidae. Carabids are most numerous in areas where there is plenty of rainfall.

Most ground beetles hunt at night, hiding by day under rocks, logs, or debris. If you raise an old board from the ground, you may catch a glimpse of some carabids, black or brownish, oblong, medium in size, with long, slender legs. They will try to run away rather than fly.

In some species the underwings have atrophied and the elytra have fused together. Other carabids fly very well. One of these—the searcher, *Calosoma scrutator*—preys on caterpillars in trees. It is larger than most ground beetles and is colored a beautiful metallic green with red edges and blue legs. Its relative, the fiery searcher, *Calosoma sycophanta,* is also brilliantly colored and a tree dweller.

Most ground beetles feed on the larvae and adults of other insects, and also upon slugs, snails, and every other creature they are able to capture. A few are seed eaters, and the grubs of a very small number are said to feed on sprouting corn.

WATER BEETLES

Although most beetles are strictly land insects, there are a few families that spend much or all of their lives in ponds and streams. The diving beetles make up the Dytiscidae, the largest family of water beetles. They are oval in shape and generally range from medium to large in size. They are brownish or greenish black. They have long, sturdy hind legs, which they use as oars. Under their wing covers they carry a supply of air for breathing while they hunt or rest on the bottom. Both adults and larvae feed voraciously on all kinds of aquatic insects, tadpoles, and even small fish. Adult diving beetles sometimes migrate at night in great numbers from one pond to another, stopping on the way to whirl around a street light.

If you approach a pond or pool quietly, you may surprise a group of whirligig beetles circling around on the surface of the water. Disturb them and they will quickly dive to the bottom. These small dark insects comprise the family Gyrinidae. Their long, slender front legs are equipped to seize prey, while the middle and hind legs are broad and flat for swimming. Their habits and food are much like those of the Dytiscidae.

The third important water-beetle family, the Hydrophilidae, are sometimes called water scavengers, because they feed on decaying vegetable and animal matter. However, they will eat any living creature they can catch. They somewhat resemble the diving beetles, but they use their legs differently and are not as good divers.

The underside of a water scavenger's body is hairy. Enough air clings to the hairs to give the beetle buoyancy. It also carries air under its wings for breathing.

The insect called the black water beetle, which is found around kitchen sinks and drains and other damp places, is not a beetle. It is the black Oriental cockroach, or kitchen cockroach.

CARRION, OR BURYING, BEETLES

Certain members of the family Silphidae help keep the surface of the earth clean by burying small dead animals. When a pair of these beetles find a dead mouse or small snake, the female deposits her eggs in it. Then the beetles quickly bury the body seven to ten centimeters under the ground. When the larvae are hatched, they feed on the decaying flesh.

There are many species of burying beetles. One of the largest and most conspicuous, *Necrophorus americanus,* is two to four centimeters long. It is a shining black, oblong, heavy-bodied insect, with two large reddish spots on each wing cover. The prothorax is hemispherical and red; the head is almost as large as the prothorax. Other species of burying beetle are oval-shaped or almost hemispherical; they may be black or brownish.

Some burying species dwell in caves and have no eyes. Others, very small, live in decaying fungi and even in ants' nests.

ROVE BEETLES

The rove beetles make up one of the largest beetle families, Staphylinidae. Numbering more than 30,000 species, they are found all over the world. They are usually slender-bodied and range from minute to medium in size. A good many are black. Some are dark red, dark brown, or yellowish. A few shine in metallic greens

Robert C. Hermes, Audubon/PR

Predaceous diving beetle attacking a tadpole. Note the flattened, fringed hind legs used for swimming.

and blues. The larvae are found with the adults, which they resemble.

The elytra of these beetles are so brief that they do not cover the last three to five segments of the abdomen. The underwings are large and thin. When a rove beetle alights it folds these wings and doubles them back. Then, bending up its abdomen, it uses the tip like a hand to poke every portion of the wings completely under the tiny elytra.

Rove beetles are also called short-winged scavenger beetles, an appropriate name, for they feed on decaying animal and vegetable materials. They aid in reducing manure to an available plant food form. Some 300 species live in ants' nests, eating dead ants and cleaning the nests. Some rove beetles are also found in the nests of termites.

LADYBIRD BEETLES, OR LADYBUGS

Unlike most beetles, the ladybirds, or coccinellids, have always enjoyed wide popularity. These members of the family Coccinellidae were long considered omens of good luck; children throughout Europe and North America sing affectionate little verses to them. Nowadays most people know that ladybirds eat the aphids or plant lice that attack house and garden plants. There are coccinellids, however, such as the squash beetle and the Mexican bean beetle, which eat plants instead of other insects and do considerable damage in gardens.

Both the helpful and harmful varieties of coccinellids are attractive beetles. Rather small in size, they are broadly oval or hemispherical. The many species of the true ladybird are enameled in different shades of red with black spots. Usually these are the aphid eaters. Other species are black with red spots; many of these attack scale insects. The herbivorous lady beetles are likely to be yellow with black spots.

When disturbed, ladybirds discharge an evil-smelling secretion that is believed to protect them from insect-eating birds. During the winter months the beetles tend to congregate in large numbers in crevices on the sunny side of buildings. Convergent lady beetles, common in the western United States, collect in masses to hibernate during the winter under leaves and debris on mountain slopes.

SKIN BEETLES

Everybody hates the skin beetles, or Dermestidae (skin-eaters), and with good reason. These beetles love to feed on every animal product that people have processed for their own use. Fur and feathers, woolen cloth, bacon, ham and cheese, flour and meal, dried insects, stuffed birds and animals in museums—each of these products is a food for one or more dermestid species.

Many of these pests are quite small, six millimeters or less in size. Larger ones may be as much as 13 millimeters long. They are usually oval, plump, and dark-colored. They are partly covered with fuzz in a lighter tint, or with scales that rub off. In general, the larger dermestids feed on hides and dead animals; they are called hide beetles. The smaller ones eat pollen in the fields; they also enter homes to feast on clothes, carpets, and dried foods. Many species remain out of doors and never bother people.

Dermestid larvae are small and brownish, covered with many black hairs and bristles. They can move very rapidly. If you watch one closely, you may see it do a curious thing—run a short distance, stop, vibrate its hair rapidly, then start running again. The larvae are more destructive than the adults, eating voraciously when food is available. Yet they are able to live a long time without any food.

CLICK BEETLES

A large variety of extraordinary beetles belong to the family Elateridae. The most amazing characteristic of the click beetles, or elaters, is their ability to hurl themselves up in the air. When disturbed, the elater drops on its back to the ground, feigning death. When it seems safe to move again, the beetle bends its head and thorax back, pushing a special spine on its prothorax almost out of the groove in which it lies. Suddenly the tension is released. The spine snaps back along its groove with a clicking noise, driving the base of the elytra against the ground with such surprising force that the little insect may be shot 10 or 12 centimeters up into the air.

Click beetles are also known as skipjacks and snapping beetles. A conspicuous

A dung beetle. These beetles lay their eggs in dung they have rolled into a ball.

Jen & Des Bartlett/PR

temperate species is the eyed elater, *Alaus oculatus,* a sturdy fellow sometimes five centimeters long. It is shining black, flecked with silvery scales. On top of its large prothorax two big black spots, outlined by a ring of white scales, imitate two glaring eyes. Eyed elaters are found all summer long around old stumps and logs, where their larvae live, feeding on the rotting wood.

Elater larvae are long, slender, smooth, yellowish grubs, so hard and stiff that they are called wireworms. Many feed on decaying wood. Some, however, live in the ground, annoying farmers by burrowing into bulbous roots, tubers, and sprouting corn seed.

The largest and most beautiful elaters live in the tropics. Among these are the luminous fire beetles of the genus *Pyrophorus.* Some of the larger species have a two-way light system. All night, when one of these beetles is resting on a tree trunk, two spots on its prothorax glow with a soft green light; when it takes wing, a bright orange light streams from its abdomen.

METALLIC WOOD BORERS

Nearly all the metallic wood borers, or Buprestidae, are striking insects. Their copper, gold, green, blue, or red backs shine with a metallic luster and often are decorated with intricate patterns in contrasting colors. Since their bodies are hard and the colors do not fade, buprestids are often used as decorations, not only by natives in tropical forests but by art workers everywhere. Two of the most gorgeous species are called jewel beetles; Australians set them in mountings to wear as jewels.

Buprestid species may be short and flat, or oblong and cylindrical. They are large in the tropics—a Brazilian giant is over five centimeters long—and small to medium-sized in temperate regions. Most of them are tropical and nearly all live in forests. The adults like to sun themselves; you may come upon one in some dark woods, glittering in a small patch of sunlight.

The larvae are blind and legless, with a small head and a large flat thorax. The thorax is often mistaken for the head and gives the grub the name of flat-headed wood borer, or hammerhead. The adult female deposits her eggs in cracks in the bark of trees that have been injured by fire or overexposure to sunlight, as when a clearing has been made by storm or lumbering. Curiously enough, forest fires and smoke seem to excite some buprestids, so that they will attack persons working around fires, biting the hands or neck or any exposed skin area, often quite severely.

The family does not restrict its boring to tree trunks. Some species attack herbaceous plants, and some very tiny buprestids, called leaf-miners, bore into leaves.

FIREFLIES

Fireflies, or "lightning bugs," make up the family Lampyridae. These beetles add mysterious charm to summer evenings. Even a city dweller may see a few sparklers on a park or lawn and around hedges.

In the daytime, with its lamps turned off, the firefly is not a particularly notable insect. It is medium to small-sized, elongated or oblong, black or brownish, edged with red or yellow. The elytra and the thoracic covering are not so hard as in most beetles.

The females of many species are wingless, wormlike creatures. They are sometimes three or four times as long as the male. They glow at night from spots along the sides of the abdomen and thorax. The European glowworm belongs to this group.

Each species of firefly appears to have its own code of signals. It has been assumed that the lights are signals between the sexes. But many larvae also are luminous, and so are a few pupae. In some tropical species the larvae have lights but the adults do not. Not all lampyrids are luminous; a great many species are diurnal and have no need of light.

Adults and larvae of nearly all species are carnivorous, feeding on small worms, snails, and the larvae of other insects.

DEATHWATCH BEETLES AND THEIR KIN

Members of the family Ptinidae are

small insects, six millimeters or less in length. They are dull-colored scavengers, living on old dry vegetable or animal material.

The most dreaded member of this family used to be the deathwatch beetle, a tiny brown insect that feeds on decaying wood. It often feasts on the wood of old houses. As it does so it makes a ticking noise that, in the silent night, seems portentous.

In the days when druggists carried large supplies of dried roots and leaves from which to compound their medicines, they were much annoyed by drugstore beetles, which ate their medicaments and also their cigars. Today's medicines are compounded in laboratories and packaged in glass. The drugstore beetle, therefore, has moved to the grocery store, where it can live on anything that is dry and available.

You many find in your bathtub what looks like a little red spider, with a globular body, either smooth and shining or partly fuzzy, and long slender legs. This is really a beetle, one or another of a few species of the Ptinidae. It has fallen into the bathtub, in which it is trapped. It cannot climb the slick sides of the tub and it has no wings for flying. Only in the bathtub is it visible. Elsewhere it would blend completely with its surroundings and would almost certainly escape detection.

DARKLING BEETLES

Darkling beetles, or tenebrionids, form the family Tenebrionidae. They are found mostly in arid regions. They are usually

The highly developed and sometimes branched mandibles of the male give the stag beetle its name.
John Gerard, Audubon/PR

N. Smythe, Audubon/PR
Unlike many other beetles, the ladybug is welcome in the garden because it eats aphid pests.

nocturnal scavengers, feeding on dead or decaying vegetable matter. Some species devour living plants, and sometimes become so numerous that they denude the sparse natural vegetation and damage cultivated crops.

In general, darkling beetles are small to medium, black, and stoutly built. Many species are wingless, and their elytra have been fused together. Their legs are long, but they move rather slowly with an awkward, loose-jointed gait.

The long and slender larvae live in decaying wood and dried vegetable products. Some species of European and Asian origin have specialized on grain and grain products. Now distributed around the world, they cause considerable damage. The meal worm, *Tenebrio molitor,* which attacks cereal products, has become a commercial product itself; it is raised in large quantities to feed pet birds.

STAG BEETLES

Members of the family Lucanidae are large insects. The males have huge mandibles, or jaws—hence one common name, "pinching bugs." The other common name, stag beetles, comes from the resemblance of the mandibles to the antlers of deers.

A scarab beetle. The ancient Egyptians considered some species sacred symbols of resurrection.

It is not known what use the males make of their mandibles. They have been observed fighting each other, not by pinching but by pushing or butting. If you push a small stick in front of the pinching bug, it will grasp it with its mandibles and hold on tight while you lift it into the air. Do not offer a finger; the beetle can draw blood.

The giant male stag beetle, which lives in rotting logs and stumps, is a formidable-looking creature. It ranges in length from 4 to 5.7 centimeters. From the head of its highly polished chestnut brown body protrude two mandibles that are almost as long as the whole body. The antennae are black, elbowed, and end in small combs. The strong black legs are edged with short spines. The female is smaller than the male; her mandibles are short and stout.

Like all lucanids, stag-beetle larvae live in old logs and stumps of oak, maple, and apple, and require two or more years to mature.

SCARABS AND THEIR KIN

The Scarabaeidae are a numerous and famous family. They are so varied in form, size, and habits that about the only characteristics they have in common are antennae that end in a leafy club, large eyes, a large and prominent prothorax, and strong legs.

They include the largest beetles in the world and also some of the smallest. They may be any color from dull black to brilliant hues on metallic or enameled surfaces. Because of these variations the Scarabaeidae have been divided into six or seven subfamilies. Some entomologists have raised the family to the status of a superfamily, called Scarabaeoidea, and the subfamilies to family rank, but the relationship is the same.

The scarab, the ancient Egyptian sacred beetle after which the whole family was later named, is one of the dung beetles. These are useful scavengers that clean up excrement by rolling it into little balls. The females deposit their eggs in the balls and then bury them. The Egyptians held one or two scarab species sacred, as a symbol of resurrection, and placed them in the tombs with their dead. They painted pictures of the scarab on their stone coffins, and made models of it in jewelry.

Another well-known group are the May beetles, or June bugs, big-bodied creatures that bumble around in the early evening. Tropical kinds are brilliantly colored; temperate species are brownish. These beetles are vegetarians, the adults eating leaves and the larvae living underground on roots. The beautiful and destructive Japanese beetle, *Popillia japonica,* is closely related to the June bug.

Pictures and museum specimens have acquainted many people with the giant tropical scarabs and their grotesque horns. The horn projecting from the prothorax of the male Hercules beetle, *Dynastes hercules,* of the West Indies may be more than six centimeters long; the entire insect is often more than 15 centimeters long. The massive elephant beetle, which dwells in the tropical Americas, is the thickest and heaviest of all beetles. It is ten centimeters long; its wing spread of 20 centimeters enables it to fly quite well. The males of the Goliath beetles of Africa and eastern Asia are 12.5 centimeters long. They have no large horns, but the prothorax is enormously swollen and beautifully marked.

LONG-HORNED WOOD-BORING BEETLES

The double-jointed name of the Cerambycidae aptly describes the majority of the species in this large family. The term "horn" refers to the insect's antennae and not to the protuberances on the head or prothorax found on some other beetles.

The cerambycids range in size from the pygmy beetle of central North America, which is about 2½ millimeters long, with antennae of the same length, to the startling *Batocera* of New Guinea, whose 7½-centimeter body carries antennae almost 18 centimeters long.

Cerambycids have large eyes and mouth parts; the mandibles of some males are very large and antler-shaped in some tropical species. The legs are long and slender; sometimes the front pair are nearly twice as long as the others. The insects display all colors and color patterns. Usually they have large, powerful wings, though a few species are wingless. Many have stridulating organs for producing sound; they make a peculiar squeaking noise.

Some species have a pleasant odor. For example, the European musk beetle, a beautiful copper and green insect, smells like attar of roses. Many long-horns are good mimics: some look like bumblebees; others, like wasps. One African species is camouflaged to resemble a piece of velvety moss as it rests on a tree trunk; its antennae appear to be dried twigs.

All cerambycids are vegetarians. The adults feed on fungi, pollen, or green leaves. The larvae live inside a plant or tree, where they may spend from one to three or four years. The larva that lives in the agave, or century plant, in Mexico is a shrimplike creature that is an appetizing addition to a salad.

Cerambycids reach their greatest development in the tropics, where they are avidly sought by collectors; every natural history museum has specimens of the larger and more striking long-horns. Many of the North American species are quite attractive. The common milkweed longhorn, only a little more than one centimeter long, is bright red with black spots. *Prionus im-*

John M. Burnley

When the Japanese beetle was introduced to North America, it caused a great deal of damage to shrubs and lawns.

bricornis, about five centimeters long, is a dark reddish brown with magnificent, heavy plumed antennae; its larvae infest the roots of orchard trees and grapevines.

LEAF BEETLES

Wherever you find green leaves you will find leaf beetles, very attractive little insects and very destructive from our point of view. Third largest of the beetle families, the Chrysomelidae are small to medium-sized and hemispherical, oval, or oblong in shape. Their mandibles, antennae, and legs are usually short. Some are enameled in brilliant colors. Many are striped and spotted. Others are a dull black or brown.

Most leaf beetles live in the tropics, but there are 1,000 species in North America. If you go out on a summer morning after a heavy dew you will find leaf beetles drying themselves on the top of leaves in the sun.

Each species has its preferred plant food. Many of the larvae feed on the outside of the leaf in company with the adults; they are active, bright-colored, chunky little grubs, in contrast to the pale sluggish larvae that live in the ground or inside a plant stem.

Females of the beautiful genus *Donacia* drop down into the water to deposit their eggs on roots and stems of water lilies, pickerelweed, and other aquatic plants. Here the larvae will live until they pupate. *Donacia* beetles are gregarious. You can

BEETLES 285

often see numbers of them flying over or resting on the lush vegetation around ponds and swamps.

The chrysomelids number several bad pests. The famous yellow and black Colorado potato beetle, *Leptinotarsa decemlineata,* has devastated crops in Mexico, eaten its way across the United States, and invaded Europe. Three or four species of brightly marked cucumber beetles do their share of damage to vines. Other species attack sweet potatoes, spinach, asparagus, and other agricultural crops.

WEEVILS

The order of the Coleoptera is divided into two suborders. All the beetles described so far belong in the first group. The second suborder, contains the weevils, or snout beetles. They comprise several families, most of which have beak-shaped heads. In some species the beak is very long, slender, and rigid. Other species have spoon- or shovel-shaped beaks. In one family, the bark beetles, the beak is so short as to be hardly noticeable or is missing altogether.

All weevils are vegetarians, attacking most trees and cultivated crops. They have been a major pest for so long that now almost any damaging insect is called a weevil, and the term "weevily" is used to describe damaged grains and grain products. However, only a few of this enormous group attack people's possessions. The great majority feed on plants with which human beings are not at present concerned, or on weeds, which compete with our cultivated crops.

Brentidae. The weevils of the family Brentidae are odd-looking: their heads are often as long as their slender bodies, or even longer. This is especially true of the female. She uses her head to bore a deep hole in which to deposit her eggs. Sometimes she gets stuck and cannot withdraw her head. The male, who has been standing guard, then tries to pry her loose by pushing down on the end of her upturned abdomen.

Only a few Brentidae are found outside of the tropics. There they are sometimes five centimeters long, but never thicker than a match stick. Some live in large colonies under loose bark. One small reddish species is fairly common on oak and maple in the United States; although small, the males of this species are quite as pugnacious as their larger tropical relatives.

Curculionidae. The curculios are the largest family of beetles. There are more than 40,000 described species and probably many others still unknown to us. The curculios include many formidable pests: the cotton-boll weevil, the apple-blossom wee-

Leaf beetles cause a great deal of damage to cultivated plants, and may transmit plant diseases.

Harry Rogers/PR

Long-horned beetles are so-named because their antennae are at least half as long as their bodies.

Ken Brate/PR

The notorious boll weevil feeding on a cotton blossom. Larvae are more destructive than adults.

vil, the plum curculio, the rose weevil, the granary weevil, and the rice weevil, among others.

Curculios range in size from minute to 7½ centimeters. All have prominent snouts. A great many are colored dull grey or brown. When alarmed, they fold their legs close to their bodies and remain motionless, looking for all the world like seeds or bits of dirt. Some are extremely beautiful. One of these is the diamond beetle of Brazil, which is covered with scales reflecting brilliant blues and greens from minute grooves. It was in such demand for jewelry at one time that it was almost exterminated.

Scolytidae. Bark beetles and ambrosia beetles, belonging to the weevil family of the Scolytidae, have no long snout, are minute in size, and dull in color. They are serious forest pests. The female of the bark, or engraver, beetle excavates a passageway along the grain of living trees, just under the bark, and deposits eggs on either side at regular intervals. The larvae work at right angles to the lengthwise passage through the cambium layer, so that the route of one never intercepts that of any other. This makes a pretty pattern, but kills the tree.

Ambrosia beetles penetrate deep into the wood of dead trees. There the female lines tunnels with a yellow fungus, called ambrosia, probably as food for the larvae. The tunnels spoil the value of the trees as lumber. The beetles spend their whole existence in the tunnels, often remaining after the wood has been cut into lumber.

OTHER BEETLES

The family Meloidae includes the blister beetle, which is well-known to pharmacists. A variety of blister beetle—the Spanish fly, *Lytta vesicatoria*—is dried and reduced to powdered form to yield the pharmaceutical preparation called cantharides.

The smallest of the beetles are those belonging to the Ptiliidae family. All are less than 16 millimeters in length. They are particularly common under loose bark. To the naked eye they appear to be mites. But the microscope reveals that they are true beetles, with elytra and wings.

Its elytra raised high, a heavy May beetle flies clumsily along.

BUTTERFLIES AND MOTHS

by John C. Pallister

The beautiful insects we call butterflies appeal to almost everyone, even to those who fear or dislike other insects, such as wasps, bees, or beetles. Boys and girls often start their insect collections with butterflies. Many adults also delight in collecting them, sometimes exploring remote and isolated areas in search of unknown species.

Closely allied to the butterflies are the moths. Butterflies and moths belong to the order Lepidoptera, which is second only to the beetles in number of species. So far, nearly 150,000 Lepidoptera species have been described; they represent over 1,000 genera and about 200 families.

Moths and butterflies are found throughout the world. A few species inhabit the subpolar regions and ascend to snow line on mountains. There are far more species in temperate regions. In the tropics the insects attain their greatest variety, largest size, and most brilliant coloring.

GENERAL CHARACTERISTICS

The name "Lepidoptera" comes from two Greek words: *lepis,* meaning "scale," and *ptera,*—meaning "wings." The reference is to the minute scales that generally cover the rather broad, usually opaque and membranous wings of these insects. The scales are usually triangular or elongated modified hairs, each fastened by a stemlike base to the wing. They are laid on the wing in regular rows, each overlapping the row below, like shingles on a roof.

Since the scales are held only by the tiny stemlike base, they can be very easily dislodged. Perhaps you have noticed the dustlike particles that coat your fingers when you try to hold a butterfly. Usually scales cover not only the wing surfaces but the rest of the body as well. They are variously colored and form attractive patterns in most butterflies and many of the moths.

In some cases, however—for instance in the so-called glass-winged butterflies of the genera *Haetera* and *Cithaerias*—the scales are extremely limited in number. The few that are present are very fine and scarcely visible except under magnification. The membrane of the wings is so transparent that printing can be easily read through it. A number of other butterfly species and some moths, including the hummingbird moths and clear-winged moths, show scaleless transparent areas.

Many moths of the family Saturniidae have rounded clear spots margined with rows of variously colored scales of blue, red, pink, white, and black. These give an eyelike appearance to the transparent area. Some other moths and butterflies have eyelike spots without transparent centers. The opaque spots frequently occur on only one surface of the wing, the other surface having a different pattern. This type of design probably serves some protective purpose.

The coloring of butterflies and moths is due in most cases to pigments imbedded in the scales. In some Lepidoptera, different colors are produced as light is diffracted from minute, closely spaced parallel lines,

Male Emperor moth, *Saturnia pavonia.* Emperor moths are found throughout much of Europe. Like some other Saturniids, they have eyelike spots on their wings.

Stephen Dalton/PR

Lincoln Nutting, Audubon/PR

Eileen Tanson, Audubon/PR

Eileen Tanson, Audubon/PR

Butterflies and moths have a complex life cycle. The photos above show some of the stages in the life history of a monarch butterfly. The caterpillar (top) is the larval stage that hatches from the eggs. After the caterpillar develops and molts several times, it becomes a pupa, or chrysalis, entering a quiet period from which it emerges (middle photo) as an adult (bottom).

called *striae,* on the scales of the wings. This kind of coloration is called structural. It results in a variety of iridescent hues — violet, blue-green, copper, silver, and gold. The color changes as the surface is tilted.

Most lepidopterous scales are striated (that is, possess striae), but in only a few cases are the striae fine enough and close enough together to produce iridescent structural color. In such cases, the striae may be astonishingly fine. In a Brazilian species of the genus *Apatura,* there are 1,050 striae to a millimeter. In species belonging to the South American genus *Morpho* the striae are even closer together. There are about 1,400 to the millimeter.

There is sometimes considerable variation in the coloring of butterflies of the same species. Where there are two or more broods a season, the spring forms may differ radically in the shade of coloring from those in the summer brood; they are usually much lighter. If there is a fall brood, it may differ from both the spring and summer forms. Variation of this kind is called *seasonal dimorphism*. The two sexes of a given species are often differently colored. The male, which is usually somewhat smaller than the female, is quite apt to show more brilliant coloring. This differentiation is known as *sexual dimorphism.*

Geologically, butterflies and moths are comparatively recent. In North America, butterflies occurred in the Eocene and Oligocene periods of the Cenozoic era, while in Europe small moths were trapped in amber in the Baltic regions during the Oligocene period. It is thought that some European finds may go back more than 135,000,000 years, to the Jurassic period in the Mesozoic era.

LIFE CYCLE

Butterflies and moths rank among the more highly developed insects — those having a complete metamorphosis. This means that they go through four stages of development: egg, larva, pupa, and adult.

Eggs. The eggs are small, round, oval, or somewhat elongate. They are variously colored and delicately sculptured with ridges or pits, according to the species. The

The strange-looking fat caterpillar at top will become a bright strikingly patterned male cecropia moth. The inchworm (bottom) will turn into a moth of the Geometridae family.

female insect usually lays her eggs on or close to a plant whose vegetation will provide the young insects with food. With some exceptions, each species feeds only on one particular plant species or on a closely related group of species.

Larval stage. The creatures that hatch from moth or butterfly eggs are the *caterpillars,* scientifically known as larvae (singular: larva). The young of most butterflies and moths are herbivorous. A few species are carnivorous, feeding on aphids and scale insects. Of the four stages this is the only one in which the insect can grow.

The caterpillar feeds as voraciously as its food supply will permit. As the skin, or covering, of the animal grows tight, it is shed, or molted. The covering that has formed beneath the old one is soft and expands to accommodate the increased growth. There are five or six molts on the average—though some species may molt as many as 20 times.

Caterpillars have three pairs of true legs, one pair to each segment of the thorax. The thorax is the part of the body between the head and the abdomen. On the abdomen there are from one to five pairs of prolegs. These fleshy protuberances are armed with hooks for grasping twigs or leaves of the food plant. Most caterpillars have five pairs of prolegs, one pair each on the third, fourth, fifth, sixth, and last, or terminal, abdominal segment. Caterpillars travel by stretching out the front part of the body, taking hold with the true legs, and then drawing forward the rear part of the abdomen. The eyes are simple and arranged in pairs; there are from two to six on each side of the head.

Pupal stage. After a caterpillar has gone through the required number of molts, it becomes a pupa (plural: pupae). In butterflies and moths the pupa is frequently called a *chrysalis.* It does not in the least resemble a caterpillar. In most cases the appendages are "glued down" to the body—though frequently visible through the pupal skin—giving the insect a compact appearance.

The pupal stage is a quiescent period. It may last from a few days to several months or more, depending on the species. Most butterfly pupae are found fastened to some stable object a little distance from the ground. Some hang head downward from a pad of silk, which the caterpillar spun on a sheltered object before molting. The swallowtail butterflies (family Papilionidae) and the white, sulfur, and orange-tip butterflies (family Pieridae) pupate in a more or less upright position, with the tip of the abdomen in a pad of silk; a silken strand, looped around the middle of the body, is fastened at each end to the support.

Moth pupae are sheltered in various ways. Most of them pupate on or near the ground under leaves, old logs, or loose bark, or in hollow trees. Many moth caterpillars burrow into the ground, where they form a smooth cell in which to pupate. Some line this cell with silk. Others, such as the woolly bears (family Arctiidae), use the spiny hairs from their bodies with a few strands of silk to form a rough cocoon in a sheltered place near the ground.

The most conspicuous cocoons are the beautiful silk ones spun by the larvae of a number of the giant silkworms (family Saturniidae).

Adult stage. After a certain period of inactivity as a pupa — a period that varies in length according to the species — life begins to stir in the insect. With the final molt the pupal case splits down the back and out crawls a butterfly or moth. It is a poor bedraggled creature at this time. Its wings are limp and its body is swollen. But soon the body fluids begin to flow into the veins of the wings, which start to expand and spread out. The spreading of the wings must proceed rapidly. If the air is too dry, the wings may dry out before they are properly spread and the insect will be imperfectly formed. The last transformation of the butterfly or moth usually takes place at night. All the appendages now function and the body has attained its normal size. Soon the lovely creature will be ready to fly off, seek food, and mate.

LARVAE AS PESTS

The larvae, or caterpillars, of the Lepidoptera, particularly some of the moths, are often serious pests because in order to satisfy their voracious appetites they feed on plants and products that are useful to humans. A few of these pests follow.

The larva of the gypsy moth, *Porthetria dispar,* is one of the worst. It devours the leaves of apple, oak, gray birch, alder, willow, and various other deciduous trees. When cankerworms — the caterpillars of the Geometridae — are very numerous, they can strip the leaves of practically all the trees in extensive wooded areas. The codling moth, *Carpocapsa pomonella,* caterpillar feeds on apples and other pome crops. The wild cherry and other food trees are attacked by the larvae of several species belonging to the family Lasiocampidae.

The cutworms — the larvae of various genera of noctuid moths — are so called because they cut the tender stems and leaves of various young plants as they feed. The army worm, *Cirphis unipuncta,* destroys grass, grain, and other crops. The tobacco worm, genus *Protoparce,* attacks tobacco plants; the cabbage web worm, *Hellula undalis,* cabbages and other vegetables. The larvae of several moth species belonging to the family Tineidae feed on woolens, furs, and feathers.

DIFFERENCES BETWEEN BUTTERFLIES AND MOTHS

Many different systems of classification have been proposed for the Lepidoptera. Even at the present time no system is satisfactory to all workers. Even the division of the Lepidoptera into the butterflies, Rhopalocera, and the moths, Heterocera, is much disputed and considered by some to be artificial. However, this division is a practical one and still serves.

The antennae furnish the main point of difference between butterflies and moths. In butterflies the tips of the antennae are distinctly knobbed or enlarged. In moths the antennae assume a variety of forms.

Two types of butterflies found throughout much of the world. Top: a painted lady. Some species of painted ladies are migratory, traveling great distances. Bottom: a fritillary, with typical orange-brown heavily spotted and marked wings.

Hugh Spenser, Audubon/PR

Stephen Dalton/PR

Morpho peleides, from Mexico. Beautifully colored, iridescent morphos are found throughout the tropical Americas.

They may be slender, tapering to very fine points. They may be feathery, or fernlike. Or they may be pectinate (provided with teethlike projections or divisions).

There are exceptions. The skippers (family Hesperiidae), the giant skippers (family Megathymidae), and a few other small families are all definitely butterflies, with knobbed antennae. In most of these insects, however, the knob has a very fine, tapering, pointed tip, which is set at a very sharp angle. The antennae of the Sphingidae, which are moths, are tapering and very finely pectinate, as we might expect of moths, but the tip is curved much as in the skippers.

Another characteristic that distinguishes butterflies from moths is that practically all butterflies fly only in the daytime, unless they are disturbed, while most moths fly only at night, unless disturbed. However, a few moth species, largely of the family Uraniidae, are diurnal. In this family the insects greatly resemble butterflies in their coloring, their broad butterflylike wings, and, in some species, their "tails," or wing extensions.

SOME BUTTERFLY FAMILIES

As already pointed out, the Lepidoptera number about 200 families. In the following pages we shall discuss some of the larger and more conspicuous ones.

Family Papilionidae. The members of this large group are known as swallowtails. Many of the species have "tails," or extensions, on their hind wings. A few even have two "tails"—a feature that accounts for the name "swallowtails." The "tails" of the Papilionidae are not a positive identifying character. Certain members of the family have no "tails" at all. Also, certain butterflies and moths belonging to other families have extensions on the hind wings.

In general, swallowtail species are medium to large, vividly colored or showing contrasting colors of black, yellow, red, and white. There are nearly 1,000 described species of papilionids. Most are found in the tropics, but many have invaded subtropical and temperate regions.

Dimorphism, both seasonal and sexual, is common among the swallowtails. Many of the caterpillars have eye spots on the thoracic region. Many have a **Y**-shaped scent organ that can be extended from a slit on the dorsal (back) part.

Among the most magnificent members of this family are the bird-wing butterflies of the genus *Ornithoptera*. Some entomologists consider them a separate family. Most of the bird-wing butterflies are gorgeously colored. The males, which are generally smaller than the females, have brighter colors. The front wings are very large and elongate; the hind wings are much smaller. These butterflies are very strong fliers. They usually prefer the region of the treetops. For this reason they are difficult to collect. These insects are found particularly in New Guinea, Java, Sumatra, and Malaysia.

Family Pieridae. The whites, yellows, sulphurs, and several other groups make up the family Pieridae. These insects are small to medium in size and worldwide in distribution, ranging from the tropics to the temperate regions. The cabbage butterfly hovering over garden cabbages and the common sulphur that children pursue over clover fields are familiar representatives of this family. The sulphurs of the tropics are medium-sized. They sometimes migrate in vast flocks to some unknown spot many kilometers away.

The European swallowtail, *Papilio machaon*. Swallowtails are found throughout the non-polar world. Many species, like this one, have "tails" on their hind wings.

Family Morphoidae. In the tropics of the Americas, from central Mexico south to southern Brazil, occur the gorgeous butterflies known as morphos. Tropical species show iridescent blues, greens, and purples, for it is in this family that structural coloring is at its best.

Over 100 species of Morphoidae are known. They are much sought after by collectors. Thousands are used for decorative purposes in the manufacture of trays and jewelry. So great has the trade in these colorful insects become that some countries have resorted to strict regulations to prevent the morpho from being exterminated.

It is exciting to see a blue morpho flitting along a jungle pathway. With its slow easy flight, it looks as if it would fly directly into an out-held butterfly net. Appearances are deceptive, however, for the morpho is an excellent dodger. It will dash under the net, or sweep upward and over a treetop or backtrack down the path and lose itself in the foliage.

Family Nymphalidae. The brush-footed butterflies are a very large family. The front legs are greatly reduced in size and are often hairy and brushlike. This character cannot be entirely depended upon to identify the Nymphalidae, however, because several other families, including the Satyridae, Danaidae, and Heliconidae, also have similar brushlike legs.

Nymphalidae are generally medium to large butterflies, garbed in varied patterns and colors. They occupy a wide belt around the world, extending well into the temperate zones, and are easily available to butterfly lovers in the Americas and Europe. Common names have been given to several large groups, including the fritillaries, peacocks, tortoiseshells, and anglewings. Certain species of nymphalids have also received common names, such as the mourning cloak, the red admiral, and the painted lady.

Family Danaidae. The well-known monarch butterfly, belongs to the Danaidae. This family has relatively few species but these number many individuals, which are found all over the world.

Family Heliconidae. These butterflies dwell in the tropics. They flit gracefully from flower to flower along jungle trails on their very long, narrow, brightly colored wings.

Family Brassolidae. These butterflies live in South America, the West Indies, and tropical North America. They are large and brightly colored on the upperside. The underside is brown, with lines and spots. There is a large eyelike spot in the center of each hind wing. When the wings are spread out, the underside, with its brown markings and two eyelike spots, suggests the head of an owl; hence the name owl butterflies.

Male cecropia moth. Cecropias are large, beautiful representatives of the giant silk-spinning moths.

BUTTERFLIES AND MOTHS

Top: a checkered skipper *(Pyragus)*. Checkered skippers are widespread in Europe, Asia, and the Indo-Australian region. Lower: a small copper butterfly *(Lycaena phlaeas)*, found in North America, across Africa, and in central and northern Asia.

These insects are very difficult to collect. When disturbed, they hide in a thicket with their wings folded over their backs, thus covering the bright topside colors; the brown underside markings blend with the leaves and tendrils. If disturbed, they will dash from their retreat, circle a few times, and then make for cover in another thicket.

Family Hesperiidae. These swiftly flying butterflies are known as skippers. The many species are of small to medium size. Their color patterns generally show combinations of brown, yellow, and blue.

These insects form a connecting link between butterflies and moths, though they are classed with the butterflies. The bodies are heavier and the scales more hairlike than those of most butterflies, giving them a mothlike appearance. However, like most butterflies, they are diurnal, flying at night only when disturbed and then merely to find another hiding place. The antennae are knobbed as in the butterflies, but on the end is a tiny tapering extension as in the moths.

SOME IMPORTANT MOTHS

Family Sphingidae. These are the sphinx moths. The name refers to the sphinxlike position many of the larvae assume when disturbed. Many of the adults are swift of flight, dashing on their long slender wings from flower to flower. They are known as *hawk moths*. A few hover like hummingbirds in front of flowers, probing into the depths of the corolla with the long, coiled, springlike proboscis; these are called hummingbird moths. Various species of the Sphingidae pollinate flowers with long corollas.

Family Saturniidae. The giant silk-spinning moths make up an outstanding family. Their generally large size and attractive appearance make them a popular group with collectors. The largest known moths occur in this family. The Atlas moth, *Attacus atlas,* of southern Asia and Malaysia has a wing spread of nearly 28 centimeters. Two other moths of this family—*Attacus edwardsi* of Australia and *Coscinoscera hercules* of Australia and Papua—are just as large but are more rarely found in collections. The Cecropia moth *(Samia cecropia),* spicebush moth *(Callosamia promethea),* luna moth *(Tropaea luna),* polyphemus moth *(Telea polyphemus),* and ailanthus moth *(Philosamia cynthia,* introduced from China) are outstanding representatives of the family in the United States.

Family Arctiidae. The tiger moths are a large family of over 4,000 species, worldwide in distribution. They are also known as woolly bears because the larvae of most of the species are covered with a long and hairlike or thick and woolly covering. They combine this hairy covering with a few strands of silk when ready to pupate.

The moths of this family are generally

small to medium in size, black or white, and decorated with contrasting black, red, brown, or white spots, lines or blotches.

Family Noctuidae. This is one of the largest moth families, if not the largest, in number of species. Known as owlet moths, well over 10,000 species have been described and new ones are constantly being added to our lists. Many destructive lepidopterous pests, including the army worms and cutworms, belong to this family.

The noctuids are medium in size, and many of them are monotonously similar in their brown and gray markings. However, the hind wings of some, especially those belonging to the genus *Catacola,* are most strikingly colored; they are jet black or alternately banded with red, black, orange, or yellow. They are easily attracted to lights or sugar bait.

Family Geometridae. This family, the name of which means "earth measurers," is another very large family. The larvae are known as measuring worms, inchworms, spanworms, and loopers. This family, like the Noctuidae, includes many destructive larva pests; the spring and fall cankerworms are familiar examples. Geometer moths rest with the wings spread out flat.

Family Psychidae. The bagworms are interesting because the larvae construct cases, or bags, of silk interwoven with leaves, bark, and other debris. The entire life of both the caterpillar and pupa is spent in the bag. The female adult, which is usually wingless, also remains in her bag and lays her eggs in it. The adult male, however, has wings to hunt out the female.

Family Limacodidae. The sluglike caterpillars of this family have no distinct legs. Some of these larvae, such as the saddleback, are covered with spines that cause severe irritation to the skin. Some entomologists use the name Cochlidiidae for them.

Family Cossidae. These moths are quite large. The small larvae bore into the wood or stems of trees or large plants. The adult females are sluggish and may frequently be found even in daylight, resting near the light that had attracted them during the night.

Family Aegeriidae. The clearwing moths have wings that are largely transparent; scales occur chiefly along the margins. They are rather small insects. The larvae are borers in certain plants and shrubs.

Other moths. The Pyralididae are a very large family of very small moths. Many of these are destructive to hay, grain, and other crops. In the Tortricidae many of the larvae roll the leaves on which they are feeding into a small case, in which they can hide and be protected.

The Tineidae are chiefly famous because the three species of clothes moths belong here. Most species, however, feed on rotten fungi and other waste products.

Left: an elephant hawk moth *(Deilephilar elpenor),* one type of moth important in pollinating flowers. Right: a luna moth *(Tropea luna),* common in North America. Male luna moths are able to detect the scent of a female many kilometers away.

Adult colonial tunicates. As adults, most tunicates are attached to rocks or seaweed and lack a notochord. The free-swimming larval forms have a notochord.

THE PRIMITIVE CHORDATES

by Yvonne Bonnafous

Animals are usually divided into two general groups: the vertebrates, which have backbones, and the invertebrates, which have none. The vertebrates include mammals, birds, reptiles, amphibians, and fishes, which have backbones. Animals such as insects, spiders, clams, worms, and corals have no backbone. Therefore they must be invertebrates.

Certain animals do not fall clearly in either group. They are the acorn worms, sea squirts, and lancelets, which number some 2,000 species all told. They do not have true backbones, but they develop a structure that is like a backbone. Evidently they are more closely related to the vertebrates than to the invertebrates.

How shall these intermediate animals be classified? Zoologists have solved the problem by putting them in the same phylum, or major group, as the vertebrates. The animals in this phylum—Chordata—are called *chordates*. The acorn worms, sea squirts, and lancelets are ranked as primitive members of the chordates. They are called *protochordates,* or primitive chordates.

CHORDATE CHARACTERISTICS

All the chordates—vertebrates and protochordates alike—have certain features in common. In the first place they possess, at some time in their lives, an internal, rodlike, cartilaginous structure, called the *notochord*. The word "chordate" takes its name from this structure. The notochord extends through the long axis of the body and gives it rigidity. In the higher chordates—the vertebrates—the notochord is replaced early in the life of the embryo by the vertebral column, or backbone, a jointed, bony structure.

In the second place, all chordates at some time or other have *gill slits,* which open into the pharynx, the back of the mouth. In higher chordates such as reptiles, birds, and mammals, the gill slits disappear long before birth. Fish, however, retain them all through life.

Finally, all chordates have a hollow central nervous system, or *spinal cord.* This lies closer to the back than the notochord does. It extends the length of the body.

NOTOCHORD DIFFERENCES

The primitive chordates are commonly divided into three classes: hemichordates, urochordates, and cephalochordates. This division is based on the nature of the notochord.

In the hemichordates ("half-notochord animals"), which include the acorn worms, the notochord extends over part of the body, from the pharynx to the head region. The urochordates ("tail-notochord animals"), or sea squirts, have a notochord in the tail in the larval stage. The cephalochordates ("head-notochord animals"), or lancelets, are so called because the notochord extends into the head. It also extends the length of the body to the tail.

ACORN WORMS

The acorn worms look like elongated worms with an acornlike head. The "acorn" is made up of a *proboscis,* or tubular structure, and a *collar,* into which the proboscis fits. The collar is attached to a long, flat, ruffled *trunk,* tapering toward the end. The animals range in length from 2.5 centimeters to 2 meters; the average length is from 15 to 25 centimeters.

These relatively rare creatures are found buried in the mud and sand of low-tide zones. They are most common in warm waters. Like earthworms, acorn worms pass mud or sand through their bodies. They extract food particles from it and eject the wastes. The spiral castings are deposited outside of the animals' burrows. These castings resemble those of earthworms.

The proboscis of the acorn worm is long and muscular. The animal inflates and stiffens this structure by drawing in water and uses it to burrow in sand and mud. The mouth is located between the collar and the neck. It widens to form a pharynx.

A blind (dead-end) pouch, made of stiff, cartilaginous tissue, extends from the wall of the pharynx up into the proboscis. Zoologists who classify the acorn worms with the chordates believe that this structure is a notochord. They point out that it arises in the same way as the notochord arises in the vertebrates and is made of the same tissue. Other authorities deny that the acorn worm has a notochord. They include the animal with the invertebrates.

The acorn worm's pharynx leads into the intestine, which is perforated with gill slits. The gills lead into a pouch, which opens to the exterior by gill pores. The intestine runs the length of the body.

Acorn worms are hemichordates with the notochord extending only from the pharynx to the acornlike head. They are found—infrequently—in warm waters.

Russ Kinne/PR

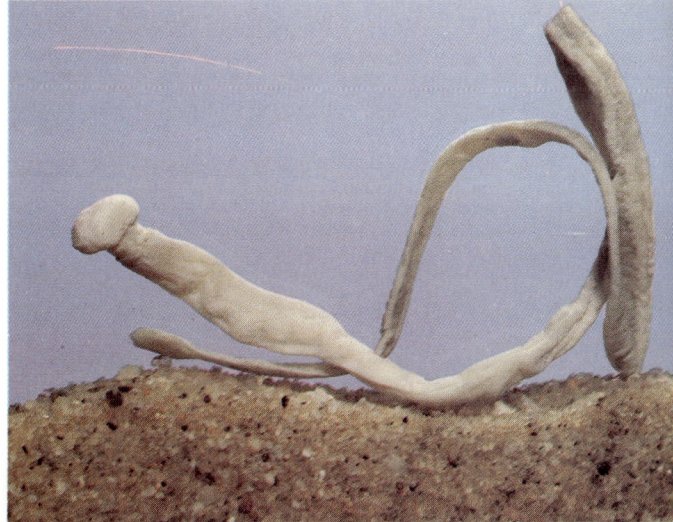

THE PRIMITIVE CHORDATES

It ends in an anus at the tapered end of the trunk.

The reproductive organs are located at the front end of the trunk. The sexes are separate. Males produce sperm and females eggs. Fertilization takes place outside of the body after the sperm and eggs have been ejected. Some deep-sea types, such as members of the genus *Cephalodiscus,* reproduce asexually by budding.

Top: colonial tunicates, or sea squirts. The two openings typical of a tunicate body are visible. Food and water enter the mouth. Waste products leave by way of the atrial pore (lower opening seen on the sides of these tunicates). Bottom: solitary and colonial sea squirts with typical reddish and purplish coloring.

Taronga Zoo, Sydney, Tom McHugh/PR

Gary G. Gibson, Audubon/PR

The larvae of acorn worms resemble those of the echinoderms (sea stars, sea urchins, and their kin) in both structure and development. For this reason many biologists believe that the chordates are descended from the same stock as the echinoderms.

SEA SQUIRTS

In shallow sea water one often finds orange, red, or purple baglike objects attached to rocks, piles, and seaweed. These colorful "bags" are the sea squirts, or tunicates. The name "sea squirts" is very appropriate. If one of these animals is disturbed, it will squirt small jets of water from two openings, called siphons, in the unattached end of the sac, or bag. One of the siphons is the animal's mouth. The other is its *atrial pore*—an opening through which pass wastes, water, and sex cells.

The sea squirt body varies in diameter from 2.5 to 30 centimeters, depending on the species. A material called *tunicin* is secreted on the outside of the body and forms a thick coat, or tunic. This accounts for the name *tunicates*.

In the adult animal, water bearing food particles passes through the mouth and is filtered through gill slits. Food particles are trapped by the gill slits and enter the intestine. The water then flows into a surrounding sac—the *atrium*—and out through the siphon called the atrial pore. Undigested waste materials also pass out of the body by way of the atrial pore.

Tunicates can reproduce by budding. An adult buds again and again, producing new individuals. In time a large colony may develop on a rocky surface. Sexual reproduction also occurs. Tunicates are hermaphrodites. That is, each animal produces both sperm and eggs. Fertilization may take place within the body or eggs and sperm may pass out of the body and unite elsewhere.

Free-swimming larvae develop from fertilized eggs. The link between the tunicates and the vertebrates is clearly seen in the larvae. These young animals, which resemble tadpoles, have a tail. The notochord is contained in the tail. As the animal matures, both the tail and the notochord disappear.

Not all tunicates are permanently attached. The transparent forms called *salpas* are free-swimming. They form colonies, which float about on the surface of the water like rafts. The tiny Appendicularia also remain unattached. They swim about in the open sea.

LANCELETS

The lancelet, also known as *amphioxus*, is closer to the vertebrates than are the acorn worms and sea squirts. It is a tiny creature, rarely exceeding five centimeters in length. Its slender, translucent body, tapered at both ends, looks like that of a fish. There is nothing very fishlike, however, about its long vertical fin. This structure starts at the back of the head and passes along the dorsal (top) part of the body. It widens as it goes around the tail. On the underside, it divides into two branches, which pass along the sides of the body to the head region.

Lancelets are the most widely distributed of all the protochordates. Species are found in tropical and temperate waters around the world. Generally, the animal remains almost buried in sand. Only its "snout" (that is, what would correspond to a snout in a vertebrate) protrudes. The little animal leaves its refuge from time to time and darts through the water. Soon it burrows in the sand again, tail first. It feeds on microscopic organisms.

The snout of the animal consists of a cuplike depression surrounded by *cirri*, or bristles, which extend in front of the head. The cirri direct the flow of water into the lancelet's mouth. The water, carrying bits of food, passes into the pharynx, which is lined with gill slits. The pharynx is surrounded by a sac called the atrium. As water flows through the gills, the food particles are filtered out. The water then proceeds into the atrium and out of the body by way of an opening called the atrial pore. Food passes into the intestine from the pharynx. Undigested wastes leave the body via the anus.

Jacques Six

Lancelets resemble fish but are quite different from them in internal structure and habits.

Blood circulates in a closed system like that of the higher vertebrates. It passes through the gill region. From there it proceeds to the intestine and to the "liver" before it reaches the "heart," which is simply a slightly enlarged blood vessel. From the heart it is returned to the gill region. Oxygenation of the blood takes place through the skin.

The muscles lie in a succession of **V**-shaped blocks along the body wall, with the point of the **V** facing the head. This arrangement is also found in vertebrates.

The notochord of amphioxus extends lengthwise from head to tail. Dorsal to it—that is, above it—lies the hollow nerve cord. The nervous system is very simple; there is no brain. The animal does not possess eyes. However, light-sensitive spots are present on the body and tail. A skin depression at the front end can detect chemicals in the water.

The reproductive organs occur in 26 pairs along either side of the body cavity. They connect with the exterior by the atrial pore. The animals reproduce sexually. Sexes are separate; eggs or sperm are shed into the water in the early summer.

The lancelets are the only members of the protochordates that serve as food for human beings. The tiny animals are netted in the South China Sea. Many metric tons—the equivalent of hundreds of millions of lancelets—are consumed every year in China.

THE PRIMITIVE CHORDATES

J. Burton, B. Coleman

FISHES

by Daniel M. Cohen

From Baja California to the coasts of Norway; from China's Yangtze River to the muddy Mississippi basin; from swiftly moving alpine streams to ocean depths of 2,200 meters—in every part of the world, in every kind of water, there are fishes.

The fishes dwell in a watery world that covers over three-quarters of the earth's surface. As one might expect, therefore, they are the most numerous of the vertebrates, or animals with backbones. They are also the most varied. There are more different kinds of fishes than there are of all the other vertebrates combined. Each of the four principal groups into which the fishes are divided differs more from the other three groups than birds differ from reptiles or than reptiles differ from mammals.

A piranha, one of the most feared of fishes. It quickly attacks almost any animal that enters its home—the rivers of South America.

FOUR GROUPS

The *Agnatha* are the most primitive of the four major groups of fishes. The name is derived from the Greek *a,* meaning "without," and *gnathos,* meaning "jaw." As the name indicates, these fish have no jaws. The mouth is simply a hole in the head and lacks the movable parts—the jaws—that we commonly associate with a mouth. The group is also known as the *Cyclostomi.*

The agnaths are the earliest of the fishes to appear in the fossil record. They existed in the Middle Ordovician period, some 460,000,000 years ago, and were most common during the late Silurian and Devonian periods, from 410,000,000 to 345,000,000 years ago.

The typical agnath at that time was a small, heavily armored creature. It moved clumsily along the muddy bottom of the sea, where it was preyed on by a variety of arthropods. The animal was quite different from the agnaths of today—the lampreys and hagfishes. These are slimy, elongated creatures that completely lack scales or armor. However, we classify them as Agnatha because they lack jaws.

The *Placodermi* ("plate skins") form the second group. They were common during the Devonian period, about 375,000,000 years ago. One kind of placoderm persisted until the Permian period, perhaps 230,000,000 years ago. The placoderms, like the agnaths, had heavy armor. They possessed jaws and paired fins as well. Some of the placoderms were sharklike in appearance, while others were extremely grotesque.

The *Chondrichthyes* ("cartilage fishes") are the third group. These are the sharks, rays, and chimaeras. The earliest fossils belonging to this group were discovered in Middle Devonian formations, going back some 375,000,000 years. Of course, modern representatives of the Chondrichthyes are common today. Sharks

FRESHWATER FISHES

and their kin possess well-developed jaws. They have structures adapted for internal fertilization. They have two sets of paired fins. Most important, though, their skeletons are basically of cartilage rather than true bone.

In several respects, the sharks and their kin are less highly developed than the true fishes, which make up the fourth and largest group. Hence the members of this last group were supposed at one time to be derived from the Chondrichthyes. This relationship has not been borne out by more recent anatomical studies. Furthermore, true fishes appear as fossils as much as 25,000,000 years earlier than sharks. It would appear, therefore, that the Chondrichthyes are actually degenerate, rather than primitive, forms.

The Osteichthyes, or true bony fishes, are the fourth major group of fishes. (*Osteon* means "bone" in Greek.) The earliest known representatives date from the first part of the Devonian period, perhaps 400,000,000 years ago.

Bony fishes have evolved into a bewildering variety of forms. Hence it is not an easy task to name the particular features that distinguish them from the Agnatha, Placodermi, and Chondrichthyes. Here is a list of such features, with some exceptional cases given in parentheses following the description.

1. True fishes have jaws. In this respect, they differ from the Agnatha.

2. They differ from the Placodermi in that the hyoid bone, one of the primitive gill supports, is modified into a prop that supports the jaws.

3. Unlike the Chondrichthyes, the Osteichthyes have true bone. (By way of exception, bone is virtually absent in certain deep-sea fishes and in fishes that mature while very small.)

4. Most true fishes have scales. (Gurnards, sculpinlike fishes, and certain catfishes have very pronounced armor plating. Blennies and some eels and catfishes lack any sort of covering over the skin and are completely naked.)

5. Most true fishes have two sets of paired fins. (Some eels lack any paired fins; fishes of several groups have only one pair. Paired fins may be modified into a sucking disc in gobies, clingfishes, lumpsuckers, and snailfishes.)

6. In true fishes, an outpouching of the gut, or intestine, serves as a lung or a swim bladder. (This feature is absent in sculpins, darters, and others.)

The lamprey, a modern representative of the jawless fishes. Jawless fishes, quite different from those that survive today, were the first fishes to appear, arising more than 450,000,000 years ago.

Aldo Margiocco

Right: a coelacanth. Often called a "living fossil," the coelacanth was long thought to be extinct, but in 1938 the first of several specimens was found in the waters off the east coast of Africa.

CLASSIFICATION

The true fishes are classified in two major subdivisions: Choanichthyes and Actinopterygii. In the Choanichthyes, the nostrils form an opening leading to the mouth—a feature that is also found in human beings and other air-breathing vertebrates. The Choanichthyes form a very small group. They include the famed coelacanth, which was supposed to be extinct for at least 75,000,000 years. Then, in 1938, the first of a number of living specimens was found off the eastern coast of Africa. Among the others in this group are the lungfishes and the early part-fish-and-part-amphibian ancestors of the four-legged, land-dwelling animals.

In the other major subdivision of true bony fishes, the Actinopterygii ("ray-finned fishes"), the nostrils are not connected with the mouth cavity. This subdivision includes the African bichir (genus *Polypterus*), a curious fish with lungs and a series of saillike finlets in a row along its back. The sturgeons, fresh-water gars, and bowfins are also numbered among the Actinopterygii. So are the Teleostei, which are by far the most numerous and most varied of all living fishes. Some authorities estimate that there are as many as 40,000 different species of fishes. If this is so, at least 18,000 of them belong to the Teleostei.

BASIC BODY STRUCTURE

Let us now consider the basic body structure of a teleost fish. As the fish must move through the water, it is a streamlined animal, broadest at the center and tapering at the front and back. The mouth opens at the front and is lined with several rows of teeth. The eyes are set in the side of

Fish are adapted to a wide variety of habitats and life conditions. Right middle: a blindfish that inhabits dark underground streams in North America. Right bottom: a type of "walking fish." It is able to breathe air and to "walk" on marshlands with its powerful pectoral fins.

Most bony fishes and many cartilaginous fishes travel in groups known as schools. Here a school of young salmon.

the head, thus giving the fish wide lateral (side) vision; however, the two fields of vision overlap little, if at all. A pair of nostrils is located on each side of the head on the snout, or region in front of the eye.

On each side of the body, behind the eye, is a bony opercle, or *gill cover,* forming a movable shield over the soft gills. Below each gill cover is a soft, membranous fold supported by a series of bony stays. The two folds are sometimes connected by tissue extending across the bottom of the head. Because a fish has no neck, the head is arbitrarily considered to end at the hind edges of the gill covers.

Two sets of paired fins—pectoral and ventral—are present. The *pectoral fins* are usually set low on the sides of the body and behind the gill covers. The *ventral, or pelvic, fins* are generally behind the pectoral fins and are on the ventral surface (the underside) of the body.

The *dorsal fin*—there are sometimes two and, in a few fishes, three—is set on the dorsal surface (the upper side) of the body and usually near the middle of it. There is an *anal fin* behind the vent, the hind opening of the alimentary canal. Behind the anal fin, a relatively slender part of the body, called the *caudal peduncle,* supports the *caudal fin,* or tail, of the fish. (*Cauda* means "tail" in Latin.) All the fins are supported by bony rods called rays. These are sometimes flexible and sometimes stiff.

LITTLE FISH AND BIG FISH

Fish bodies show a great range of size. They have what is called indeterminate growth. That is, they continue to grow during their entire lives. However, once a fish has reached a certain length, its rate of growth decreases yearly. Different kinds of fishes reach their slow-growth stage when they have attained different sizes. That is why a mature tarpon, say, is bigger than a trout.

There is some debate as to what is the smallest fish. By "smallest" we mean the smallest size at which a fish is sexually mature. Certainly one of the tiniest is a minute goby from the Philippines, *Pandaka pygmaea.* It ranges in length from 12 to 20 millimeters when fully mature. The brain, heart, intestine, kidneys, gills, and other structures and organs are as fully developed in this fish as they are in larger species.

Pandaka is perhaps the shortest of all fishes and indeed of all vertebrates. But another fish claims the distinction of being the lightest. This is *Schindleria,* an odd, almost transparent creature from the central Pacific Ocean. It weighs between two and eight milligrams. It has been estimated that more than 125,000 of these little fishes would be required to make up a kilogram. *Schindleria* was once thought to be the immature form of another fish, the halfbeak. Closer study, however, showed this was not so. The small, larva-like *Schindleria* is sexually mature.

Going now to the other extreme, the acknowledged giant of all fishes is the mammoth whale shark, *Rhineodon typus.* Individuals estimated to be as long as 18 meters have been seen; 13½-meter specimens have actually been measured. The whale shark is a sluggish creature that feeds mainly on small forms of sea life. It would be entirely harmless to people except for its habit of lolling about on the surface of the sea, where it has on several occasions been rammed by the ships of unwary mariners.

The sizes of most fishes lie between the two extremes we have just pointed out. The average fish is perhaps 12 to 15 centimeters in length.

DISTRIBUTION OF FISHES

Fishes are among the most widely distributed of all animals. Of the various factors that have influenced this distribution, temperature is extremely important. Many kinds of fishes are limited to tropical areas. These include butterfly fishes, damselfishes, goatfishes, and the gaily colored specimens found in pet shops. Other kinds of fishes are restricted to temperate areas, while still others are found only in polar regions.

The distribution of fishes at different depths in the sea is due in part to temperature and food supply, but water pressure also has something to do with it. Pressure increases by one atmosphere (about 1 kilogram per square centimeter) for every 10 meters of depth; it ranges from about one atmosphere at the surface to nearly 1,000 atmospheres at the greatest depths in the sea.

The flying fishes are found only very close to the surface—or even above it at times, when they zoom up from the water in order to escape enemies. Fishes such as sardines and mackerels come to the surface but can also thrive at depths down to and sometimes below 30 meters. Still others, including many lantern fishes, thrive in midwater at depths of 150 meters or beyond and rarely if ever come to the surface. Living at still greater depths are the bottom-dwelling rattails and brotulids.

Currents play a significant role in the distribution of weak swimmers and the young of all kinds of fishes. Of great im-

A young flatfish has two normally placed eyes. As it becomes an adult, one eye migrates to lie near the other (top). Then the fish lies on its blind side on the sea bottom, varying its color to match its background (second photo). An angelfish (third photo) also blends in with its background, thus protecting itself from its enemies. A lionfish (bottom) defends itself more actively: it has sharp venom-containing spines with which it can ward off enemies.

portance are upwelling currents, which bring to the surface from the depths water that is rich in nutrient chemicals. These currents influence the growth of plankton, the tiny plants and animals upon which some fishes feed. Thus they affect the distribution of the plankton-eating fishes, and also of course the fishes that devour the plankton-eating varieties.

Various chemical factors also control fish distribution. Salinity (the degree of saltiness) is particularly important in this respect. Certain kinds of fishes, including tunas, parrot fishes, eelpouts, and most sea basses, are restricted to salty oceanic waters. Most of the minnows, catfishes, loaches, and suckers are found only in fresh water. Other fishes spend part of their lives in salt water and part in fresh water; salmon, the common eel, and striped bass are examples. Certain sharks, rays, sticklebacks, pipefish, and toothed carps can apparently move back and forth from salt to brackish to fresh water at any time.

The biological environment of fishes — that is, the animal and plant life that surrounds them — can also play an important part in modifying their distribution. Consider, for example, the case of lake trout in the Great Lakes. Not long ago, the species flourished in this area. But when the sea lamprey entered the upper Great Lakes and began preying on the trout, the latter became all but extinct in some of these waters.

HOW FISHES SWIM

Certain kinds of fishes, including eels and other elongated varieties, swim by pushing themselves in undulating fashion through the water. This is a relatively inefficient and slow type of locomotion. So much of the fish's surface is doing the pushing that there is a considerable amount of friction. A good deal of energy is required to propel the fish.

The swimming movements of the tuna are far more efficient. The tuna is a streamlined fish with a stiff, crescent-shaped caudal fin, or tail. When the fish swims, the tail moves rapidly from side to side in the water, propelling the fish much as a boat is propelled by sculling. The streamlined body of the tuna is fairly rigid and some of the fins fit into grooves in the body when the fish swims, so that there is relatively little resistance.

These are two extreme types of swimming movements. Most fishes do not employ either type exclusively. They use their caudal fins in swimming, but the movement of the body pushing against the water also helps, though not to the same extent as in the eels.

The caudal fin is not the only one involved in swimming. Other fins play an important part in providing proper balance. In general, the pectoral fins of the rapidly swimming oceanic fishes are set almost horizontally on the body. Together with the vertical fins, they can act as stabilizers to prevent the fish from rolling as it moves rapidly through the water.

In fishes that move more slowly, or at least those that do not swim for long distances in a straight line, the pectoral fins are set almost vertically. They can help the ventral fins by acting as brakes and can assist the fish in tricky maneuvering. This is of particular advantage to types such as butterfly fishes and damselfishes. These live in rocky or coral areas, where it is necessary to chase prey around corners or to make abrupt turns in order to escape an enemy.

In most fishes, swim bladders play a vital role in movement through the water. The *swim bladder* is essentially a bubble of gas inside the body of the fish. Because the gas is lighter than the water in which the fish swims, the bubble serves to buoy up the fish. It acts as a flotation mechanism. It enables fishes to use energy to get from one place to another rather than simply to keep from sinking. As one might suspect, many of the fishes without swim bladders are either sedentary bottom-dwellers, such as the sculpins, or else they swim in short, frantic spurts, as is the case with many of the darters.

Many fishes with swim bladders have evolved a very precise sort of mechanism for changing the amount of gas within the swim bladder. To understand the nature of this mechanism, let us see what would happen if it did not exist. At a given depth below the surface of the water, the swim bladder would completely support the weight of the fish in the water. If the fish moved toward the surface, there would be less pressure on the bladder as the water pressure decreased, and the size of the bladder would tend to increase. The buoyancy would be greater than the weight of the fish's body, so that the fish would have to keep swimming downward in order to avoid being carried up to the surface.

On the other hand, if the fish went down deeper in the water, the increasing water pressure would cause the bladder to decrease in size. The weight of the fish would exceed the buoyancy of the gas in the bladder. Hence the fish would tend to sink. It would have to use energy in swimming upwards, or else it would finally find itself on the bottom.

The gas-regulating mechanism prevents these undesirable effects. The swim bladder is divided into two chambers, each with a specialized area. One of these areas secretes gas into the swim bladder when the fish goes down. The other absorbs gas out of the bladder when the fish comes up. Thus the gas bubble maintains a constant volume and the fish is able to remain in a state of neutral buoyancy. In other words, it weighs nothing in the water.

Swimming and the presence of a gas bubble are the main factors that keep a fish afloat. Other, less important factors are also involved. The fats and oils contained in fishes are lighter than water, and they play a part in maintaining a fish's buoyancy in some cases.

The swim bladders of certain deep-sea fishes are modified into fat-storage organs. Fat, of course, is not compressible, and so the volume of the bladder does not change when the fish swims from one depth to another. Many fish eggs contain droplets of oil that help to keep the eggs from sinking.

Certain fish larvae have very long, fine spines, or trailing streamers, or complex outgrowths of one sort or another. Some biologists feel that these odd appendages help to maintain buoyancy by greatly increasing the surface of the fish in proportion to its volume.

In some species that spend their lives in midwater, between the surface and the bottom, the reduction in the amount of calcium in the bones is a factor in maintaining buoyancy. These fishes are flabby and limp; relatively light cartilage replaces heavy calcium salts to a considerable extent.

SOME FISHES WALK

Swimming is not the only type of fish locomotion. Some of the more bizarre species of bottom-dwelling fishes walk along the bottom of the sea and may have completely lost the ability to swim. Among the walking fishes are certain sculpins and also the batfishes, gurnards, and deep-sea lizard fishes.

The lizard fish is perhaps the most

External anatomy of a typical bony fish

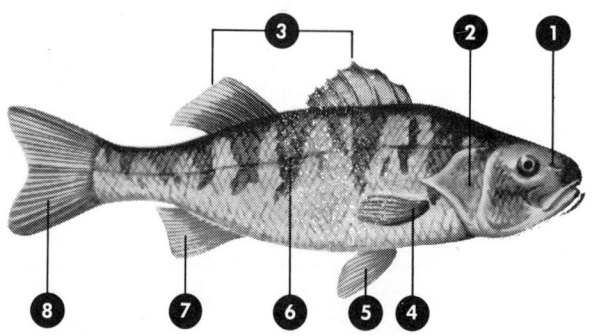

1 NOSTRIL
2 GILL COVER
3 DORSAL FINS
4 PECTORAL FIN
5 VENTRAL FIN
6 LATERAL LINE
7 ANAL FIN
8 CAUDAL FIN

spectacular of all the walkers. This species has long been known from preserved specimens, and ichthyologists (specialists in the study of fishes) have called attention to the excessively long, stiffened rays in the ventral and caudal fins. For many years, the experts were unable to tell what purpose, if any, these rays served. The mystery was cleared up when observers in a bathyscaphe spotted living fishes of this type on the ocean bottom. It was evident that the lizard fishes used the long rays—one in each ventral fin and one at the bottom of the caudal fin—for a tripod support. Thus the fishes were held up over the oozy mud at the bottom. Furthermore, the observers were able to watch these stilt-legged creatures move on their curious appendages. They hopped about like huge crickets.

A few fishes are able to travel on land. The best-known example is the common eel, which is known to move overland in snaky fashion for distances of several kilometers, generally on cool, dewy nights when little water is lost from their bodies. Other land-going fishes include the so-called climbing perch (which does not really climb), the mudskipper of the South Pacific, and the snakeheads of Asia. These fishes walk as the result of various combinations of movements by the tail, paired fins, body, and gill covers. The great majority of fishes, however, spend their entire lives in the water.

RESPIRATION AND CIRCULATION

Fishes require oxygen just as much as land-dwelling vertebrates do. The structures through which a fish draws oxygen from its watery surroundings are called *gills*. They consist of a series of filaments containing many capillaries, or thin-walled blood vessels. The filaments are located on bony supports called gill arches and are protected, as we have seen, by bony plates—the gill covers. The name "gill chamber" is given to the space taken up by the gill filaments and arches beneath the gill cover.

In most fishes, when the mouth opens wide, water is sucked into it and then passes to the gill chamber. At the same time, the

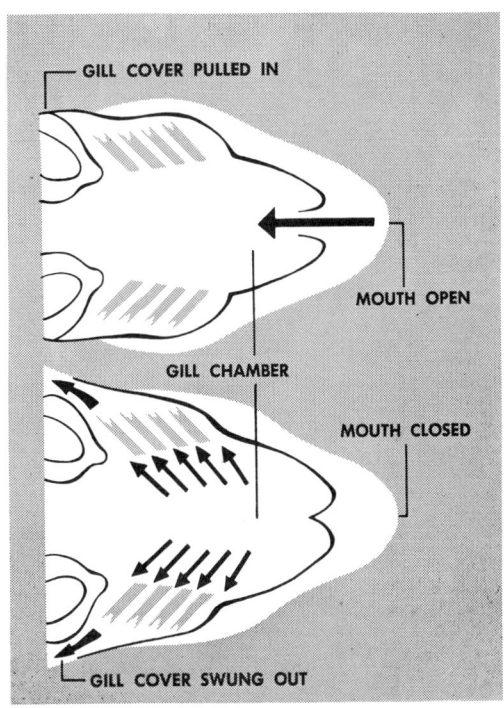

Highly simplified diagram of the breathing apparatus found in most fishes, as seen from above. When the mouth opens, water is sucked into it and passes to the gill chamber. At the same time the gill covers swing in to the side of the head, sealing the hind openings of the chamber. When the mouth is full of water, it closes, and the gill covers swing out. This opens the rear of the gill chamber and water passes out of it. In the meantime, as the water has bathed the gill filaments, an exchange has taken place. The oxygen present in the water has diffused into the capillaries of the gill filaments; the waste product carbon dioxide has diffused out of the capillaries and has become dissolved in the water that will soon pass out of the chamber.

gill cover is pulled in to the side of the head, thus sealing the hind opening of the gill chamber. When the mouth is filled with water, it closes, and the gill cover swings out. This opens the rear of the gill chamber, allowing water to pass out of it. This cycle is continuously repeated.

When the water enters the gill chamber, it bathes the gill filaments. An exchange now takes place. The blood in the thin-walled capillaries of the filaments has just been transported here from other parts of the body. It is very poor in oxygen but rich in carbon dioxide, a respiratory

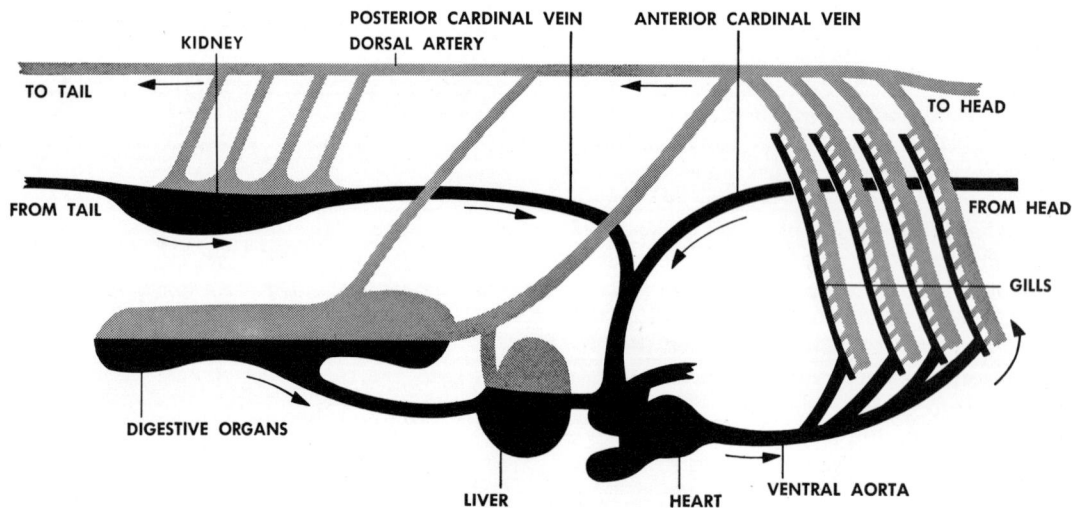

Circulatory system of a fish. Impure, or deoxygenated, blood, indicated in black, flows from various parts of the body through the veins into the heart. From there it is pumped to the gills, passing through the ventral aorta and several other arteries. In the gills, the blood receives a fresh oxygen supply from the water passing through the gill filaments. A large vessel—the dorsal artery (or aorta)—receives the oxygenated blood, indicated in gray. From the dorsal artery the blood passes through arteries in the head, trunk, and tail to the various parts of the body. Moving through the capillary network, the blood gives up oxygen to the body cells and takes on the waste product carbon dioxide. Then it passes through the veins back to the heart.

waste product that must be disposed of. Oxygen diffuses into the capillaries and is quickly taken up by the chemical substance hemoglobin, contained in the red blood cells. At the same time, carbon dioxide diffuses out of the capillaries and is dissolved in the water, which soon passes out of the fish.

Not all fishes have a breathing apparatus exactly like that which we have just described. Certain types dwelling in stagnant waters and some of the overland travelers have accessory respiratory organs near the gills. These organs consist of pouches that are rich in blood vessels. They increase the surface area through which the exchange of gases can take place.

Still other fishes, including the common eel and the mudskipper, are able to take in oxygen through their moist skins. The most remarkable respiratory organs, perhaps, are those of the lungfishes and a few other types. These fishes have not only gills but functional lungs as well. Although all fishes with lungs also have gills, some are so dependent on their lungs that they will drown if held under water.

As one might expect, fishes have a pattern of blood circulation considerably different from that of true land vertebrates with lungs. In the latter, the heart is a four-chambered structure. The fish heart has only two chambers. One of these collects the oxygen-poor blood coming in from the hind part of the body. The other pumps the blood forward, still unoxygenated, to the head and gills. Here the blood picks up oxygen and then circulates back into the body. The air-breathing fishes with lungs have a rather complex blood circulation that combines various features of lung-breathers and gill-breathers.

FISH SENSES

Fishes depend on some of the same senses as land animals to give them information about their surroundings. Most fishes have a well-developed sense of

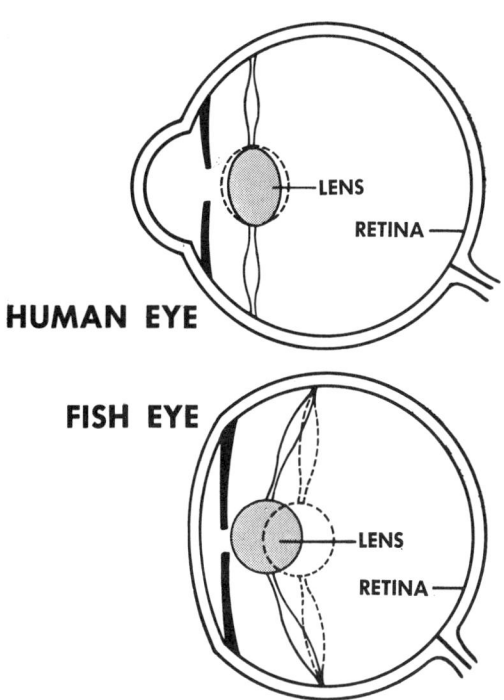

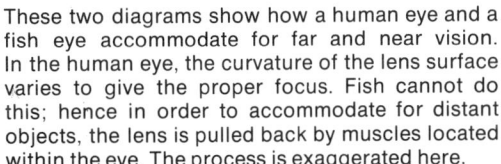

These two diagrams show how a human eye and a fish eye accommodate for far and near vision. In the human eye, the curvature of the lens surface varies to give the proper focus. Fish cannot do this; hence in order to accommodate for distant objects, the lens is pulled back by muscles located within the eye. The process is exaggerated here.

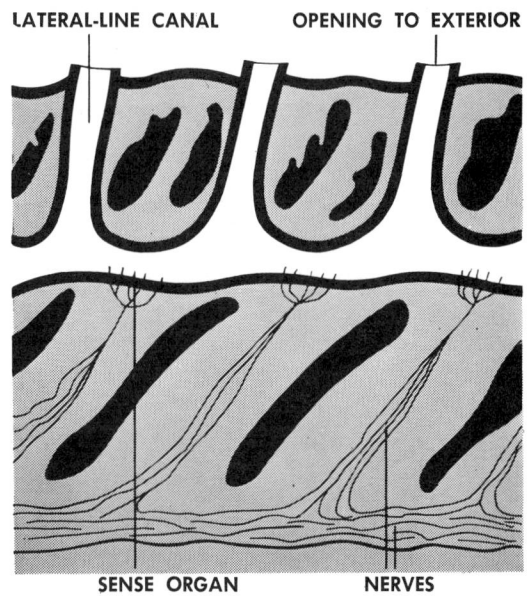

The lateral-line sensory system of fishes, generally running along the side of the body and on the head, has no counterpart in land-dwelling animals. A lateral-line canal under the surface of the skin is connected to the exterior by a series of openings. Along the canal are a series of sense organs connected to nerves; these organs pick up certain waves passing through the water—perhaps pressure waves and sound waves. The system may help the fish to swim, to discern its depth, and to hear.

smell. We have all heard of the frenzy that some sharks may show when they scent even relatively small amounts of blood. An acute sense of smell is also displayed by the salmon that return, in order to spawn, to the freshwater streams where they were born. Recent researches have shown that the salmon can actually smell their way back to these streams.

The sense of taste is closely associated with that of smell and is especially well developed in certain fishes. Carp, sturgeon, cod, and mullet, all bottom feeders, have taste receptors, or organs receiving stimuli, not only in the mouth but also on the lips and other areas. Some catfishes are particularly well supplied with taste organs, being almost entirely covered with them.

Most fishes have well-developed vision. Their eyes are especially modified for sight under water. Few fishes have much in the way of binocular vision—that is, the ability to combine the images perceived through each eye. Hence, they do not have very good depth perception. However, they have the advantage of a very wide field of vision.

Fish dwelling far below the surface have very little light, since light is quickly filtered out in water. Their eyes have been modified accordingly. As one tows collecting nets deeper and deeper in the ocean, one catches fishes with larger and larger eyes. The culmination is reached in the tubular eyes of certain fishes. Among them, the lens is a round ball perched atop a cylinder that is lined on the inside with light-sensitive retinal tissue. This type of eye is probably very efficient in discerning quite small amounts of light. However, it is very poor at focusing or forming images.

In the mid-depths, one also encounters groups of fishes whose eyes become smaller and smaller as we probe deeper into the ocean. In fact, numerous deep-sea fishes have very degenerate eyes or no eyes at all. Certain cave-dwelling types have also lost their eyes.

The fishes have a sensory system that is not possessed by land-dwelling animals. This is the *lateral line system,* which usually runs along the side of the body and on the head. It consists of a series of small pores or specialized receptors, which are connected to underlying nerves. It is believed that the lateral line system picks up certain kinds of waves that travel through the water—perhaps pressure waves and sound waves. Thus the system may help the fish to swim, discern depth, and hear.

There seems to be no doubt that some kinds of fish can hear, for they react to sounds. Some fishes even produce sounds, which are not accidental but are made deliberately and serve a definite purpose. The swim bladder is by far the commonest type of noise-producing apparatus. In many species of catfishes, triggerfishes, puffers, gurnards, drums, and croakers, among others, a strong muscle or ligament is attached to the flexible membrane of the swim bladder, and the membrane can be made to vibrate. This produces a startling variety of croaks, groans, grunts, and drumming noises. Some of the carps and catfishes make noise by expelling gas from the swim bladder through the mouth or anus. Certain fishes produce sounds by rubbing various bones of the body together, the rubbing areas varying with different species. Certain triggerfishes rub together the bases of the dorsal spine and the supporting bones. Other fishes "gnash" their teeth.

FISH LIGHT AND ELECTRICITY

Many ocean fishes produce light—a phenomenon called *bioluminescence.* This has been observed not only in fishes but in a great many other animals (sponges, jellyfish, beetles, flies, and earthworms, among others) and in certain plants (bacteria and fungi).

Some fishes have specialized organs that generate electricity. Among these types are gymnotid eels, the electric catfish, electric rays, some mormyrids, and some stargazers. The electric charges are apparently used for stunning prey and for defense against enemies. Gymnotid eels of the genus *Electrophorus* can produce charges of over 500 volts. In a popular exhibit in the New York City Aquarium, the current produced by these fishes causes light bulbs to glow.

FOOD HABITS

The food habits of fishes are extremely varied. Lampreys attach themselves to fishes and rasp off pieces of flesh with the muscular, toothed area around the mouth. Hagfishes may bore right into the bodies of their victims and leave nothing but skin and bones. Some sharks are voracious creatures that devour other fishes. However, the largest sharks, including the whale shark, and the largest ray, *Manta birostris,* feed on plankton.

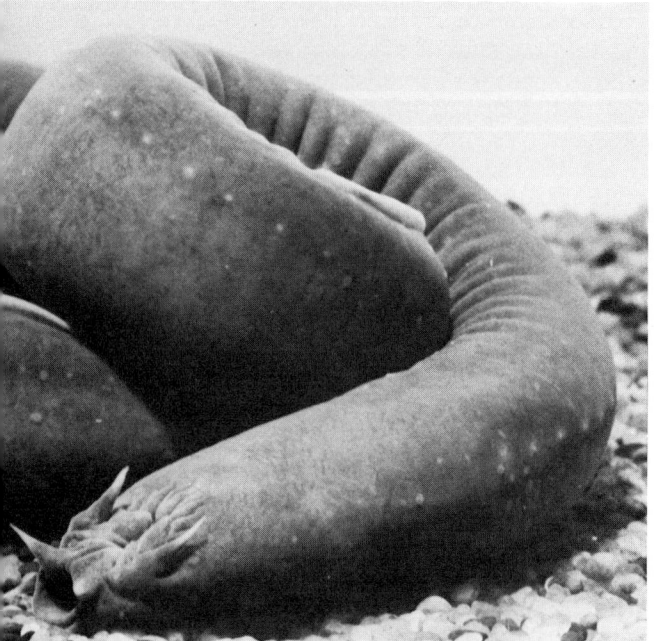

Perhaps the strangest looking of all fishes—a hagfish. This unusual fish, found mostly in cold Pacific waters, has no eyes. It breathes through its nose, has four hearts, and can go for long periods of time without eating.

K. Davidson/PR

A flying fish. This fish's pectoral fins are enlarged and held rigid against air currents, allowing the fish to glide through the air.

The bony fishes are generally provided with small conical teeth to grasp food. A certain number eat vegetation or mud. The herringlike fishes of the family Clupeidae feed on plankton; their teeth are usually poorly developed. Other bony fishes, including the pike and the barracuda, devour other fishes greedily; many of their teeth are large and sharp. Sea breams, which feed on mollusks, have large, crushing teeth deep in their throats. Parrot fishes browse on coral. Their beaklike jaws are extremely powerful.

In the alimentary tract, food passes from the mouth by way of the pharynx and esophagus into the stomach, an organ that is not well defined in some fishes. From the stomach, food enters the intestine. Fishes eating plants or mud usually have larger intestines than flesh eaters.

HOW FISHES REPRODUCE

There are various reproduction patterns in fishes. Perhaps the simplest one is displayed by schooling fishes such as sardines and mackerels. The males and females swim together in schools, and eggs and sperm are released at the same time. The school then swims on, and the union of these sex products is left to chance. Such fishes produce relatively great quantities of eggs and sperm.

Certain fishes build nests for their eggs. Salmon and trout prepare pits in beds of loose gravel. The female then sheds her eggs into the pit, and her attendant male (there may be more than one) releases sperm into it. The nest, or *redd,* as it is called, is buried with gravel. The parents leave the nest.

Other fishes, including midshipmen, sticklebacks, black bass, and sunfish, build nests and then guard them while the young are hatching. Cardinal fishes of some species are known to brood eggs in their mouths; many of the nichlids also brood their young in this fashion. The males of the pipefishes and sea horses possess a special pouch in which they carry the fertilized eggs until they hatch out.

An advanced type of reproduction is internal fertilization, which is practiced by numerous kinds of fishes. Perhaps the best-known are the live-bearing members of the order Microcyprini, a group that includes the guppies and swordtails. In these fishes, the anterior rays of the anal fin in the male are modified into an organ that aids in the transfer of sperm to the female. Sharks, rays, and chimaeras also practice internal fertilization. In these fishes, the pelvic fins of the male guide the sperm from the male to the female in mating.

In most fishes that produce living young, the fertilized egg develops and hatches within the female's body, but the

Lawrence E. Perkins/PR

embryo receives little or no nourishment from any source other than the egg itself. In some sharks and perhaps a few other fishes producing living young, food is actually transferred from the blood stream of the mother to that of the embryo.

RELATIONSHIP TO HUMANS

On the whole, fish are beneficial to mankind. They are, of course, of outstanding value to people as food. The most important of the food fishes are species of the families Clupeidae (herrings, sardines), Engraulidae (anchovies), Salmonidae (salmon, trout, and whitefish), Gadidae (cod, haddock, hake), Scombridae (tuna and mackerel), and Heterosomata (flatfishes). The commercial fishes taken in greatest quantity—the menhaden of the Middle Atlantic and Gulf of Mexico and the anchovetta of Peru—are not eaten but are made into fertilizer.

Fish vary in their reproductive habits. Some, such as the tilapia (opposite page) brood their eggs in their mouths. Others, like trout, simply lay large numbers of eggs in the water (below). Still others—rays, for example—give birth to live young (upper right). In a few the males take an active part. In the stickleback (middle right), the male prepares a nest for the eggs, and in the sea horse (lower right), the male carries the young in a pouch on his abdomen.

Pasquier, Rapho

Gantes, Rapho

Le Cuziat, Rapho   Summ, Jacana

Game fishes, such as trout, bass, pike, marlin, and swordfish, furnish recreation for many a deep-sea fisherman or freshwater angler. Other fishes make interesting pets in home aquaria.

Certain fishes must be reckoned as enemies of mankind. People-eating sharks take a yearly toll of swimmers. The voracious piranhas, found in South American waters, have been known to attack humans, particularly if attracted by blood. Swimmers must be wary of the spines of fishes such as sting rays, stonefishes, lionfishes, and certain catfishes.

When various species of tropical fishes are eaten in certain seasons or places, they are poisonous. Included in this group are barracudas, wrasses, parrot fishes, and some sea basses. Apparently certain kinds of puffers, or blowfish, are always poisonous.

The familiar profile of the shark. This one is a blue shark, a deep-ocean fish.

Tom McHugh/PR

SHARKS AND THEIR KIN

To swimmers and bathers on the seashores of the world, there is nothing more terrible than the sight of a large triangular fin plowing the water's surface in their direction. Today, as no doubt for centuries past, the sight of a shark usually causes a panic.

Some sharks, of course, are very dangerous. The most deadly of all is the great white shark. Next are the hammerhead shark and the tiger shark. The largest sharks—the whale shark and the basking shark—are absolutely harmless.

The danger from sharks is probably exaggerated. Each year more people are killed by lightning than by sharks. And of the roughly 250 species of sharks, only 12 are people eaters. The rest are harmless.

CARTILAGINOUS FISH

The sharks and their kin—skates, rays, and the strange fish known as chimaeras—are called *cartilaginous fishes*. Their skeletons have no bone. They are made up entirely of cartilage. These fish make up the class Chondrichthyes, or fish with cartilaginous skeletons.

The typical shark is a long-bodied fish with a flattened head that slopes to a pointed snout. Skates and rays have wide, flat bodies with long needlelike tails. All have the same basic structure.

The cartilaginous fishes are among the most numerous and important hunters of the sea. They are aggressive and, for the most part, swift swimmers. They feed upon bony fishes, shrimps, crabs, and mollusks such as clams and small squids. A few eat marine plants and many act as scavengers and eat carrion.

Though a few can live in fresh water, sharks and their relatives are almost exclusively ocean-dwellers. The reason for this is probably that, unlike most bony fish (Osteichthyes), which evolved in fresh water and then moved to the sea, sharks appear to have developed directly from earlier salt-water forms.

EVOLUTION

The sharks began their evolution over 300,000,000 years ago. It is believed that their ancestors were very primitive jawed fish that were covered with a shield of bony armor. The sharks developed into a distinct type of fish by losing this heavy armor on the outside. Their internal skeletons did not develop into bone but remained composed of cartilage. The sharks became swift hunters. By relying on attack rather than on defense for their survival, they made up for the loss of the outer bony shields.

In the evolution of a group of animals, many of their body structures, such as teeth, and many of their habits will depend upon the kind of food available to them. In the seas of 300,000,000 years ago there were many kinds of marine animals upon which the sharks could prey. As time went on, different types of sharks would eat only certain kinds of marine animals.

The teeth of one group were arranged as crushing plates for eating mollusks and other hard-shelled invertebrates. The chimaeras descended from this group. Another group of early sharks probably fed on the more agile fishes, for these sharks developed into very rapid swimmers and had sharp, pointed teeth. From this group evolved a later type of shark that had pointed, fanglike teeth in front and rounded, crushing teeth in the rear part of the jaws. These sharks could live on a variety of sea animals and so survived at a time when the shark line otherwise almost disappeared. An interesting descendant of this type of shark is the Port Jackson shark, found now in Australian waters. Besides having the two kinds of teeth, the Port Jackson shark is unusual because of the position of its mouth. It is placed at the front end of the head, or terminally, instead of under the head as with most other sharks.

About 150,000,000 years ago the sharks increased in number and kinds as did the number of bony fishes on which they undoubtedly fed. In time they diverged into the groups that we now know as the modern sharks and the skates and rays. The true sharks developed teeth that were strong, pointed, and sharp. The teeth of the skates and rays remained flattened and specialized for eating mollusks and crustaceans.

SHARKS

Sharks are perhaps the most spectacular of the ocean's fishes because of the great size attained by some and the ferocity shown by others. They have well-developed brains, and their acute senses of

In some sharks and in skates, the eggs are enclosed in a protective case known as a "mermaid's purse." The opaque layers have been peeled away on this one, revealing the yolk and embryo.

Miami Seaquarium

This threatening set of teeth belongs to a sand shark. Despite the menacing mouth, sand sharks do not often offer a threat to bathers. They are common in coastal waters.

smell and sight give them a great advantage in seeking out their prey. Most of them are solitary hunters. Some, like thresher sharks and dogfishes, hunt in packs.

A shark cannot float. It does not have a swim bladder, as does a bony fish, and must constantly move. It swims even when it sleeps. There is another reason why it must always be in motion: to obtain oxygen.

On each side of the shark's body, back of the head, lie openings that lead to the inside of its mouth. These are called gill openings, or *gill clefts*. When the shark "breathes in," water is taken in at the mouth. When it "exhales," water is forced out of the gill clefts. As the water passes out it bathes the gills, which extract the oxygen held in it. Most sharks have five gill openings on each side. The frilled shark has six, and other members of the group to which it belongs have seven openings.

A typical shark has large *pectoral fins*. These are the first fins back of the head. They come straight out from the sides, like airplane wings. On its back, a shark usually has two triangular *dorsal fins,* located one behind the other. Near the rear of its belly, it has a pair of *pelvic fins*. There is a small fin, called the *anal fin,* on the under side near the anal opening. The tail is usually two-lobed. The upper lobe is usually larger than the lower one.

REPRODUCTION

The male sharks—and the male rays and chimaeras, too—have a rodlike struc-

A dogfish and a companion remora. The remora is attached to the dogfish by means of a sucking pad. It feeds on the shark's leftovers and removes parasites from the shark's skin.

ture on the inner edge of each pelvic fin. These unusual structures, which extend back beyond the pelvic fins, are called *claspers*. They aid the male in fertilizing the eggs within the female. The females produce a few eggs having a great amount of yolk.

When the fertilized eggs are laid they are protected by a horny egg case in which the young develop. This case is usually rectangular in shape. At each corner there are tendrils, which catch on to sea plants and serve as anchors. This egg case is the "mermaid's purse" that one finds on ocean beaches.

In many sharks, such as great whites, hammerheads, and spiny dogfishes, and in some of the rays, development of the eggs takes place within the body of the female. In these instances the young are born alive, well developed, and capable of fending for themselves.

TEETH AND SKIN

In modern sharks the teeth grow in several rows. If a shark loses a tooth, a new tooth from the row in back simply moves into the place of the missing one.

A peculiarity of sharks is that something like tiny teeth covers their bodies. Their skin is covered with toothlike scales called *denticles*. Each skin denticle has a central pulp cavity. Around the edge of this cavity lies a layer of cells that secrete a material known as dentine. The outside of the dentine is capped by a layer of enamel, which is produced by cells of the skin.

The skin is extremely rough and abrasive. At one time, cabinetmakers used it for sandpaper. If a shark brushes against a human swimmer, it will most likely take quite a bit of the swimmer's skin with it.

FEEDING HABITS

Most sharks are utterly ravenous when hungry. They will eat anything. They bite and swallow. Everything except the proverbial kitchen sink, it seems, has been found in their stomachs. Scientists have found parts of a horse, the skull of a cow, feathers, a dog's bones, some cans, a raincoat, a rubber tire, a small keg of nails, part of a boat, and an automobile license plate in the stomachs of sharks.

Almost no digestion takes place in a shark's stomach. It is used primarily as a storage bin. Food is digested in the intestine, which contains a spiral valve. This valve looks like the bit in a carpenter's brace-and-bit. Food particles are digested as they pass through the spiral.

The mouth of a shark lies on the under side of its head. Because of this, the shark often attacks its food source from under-

This spotted hide belongs to the leopard shark. The leopard shark is a smaller member of the shark family.

Tom McHugh/PR

A bull, or Zambezi, shark grabs a smaller shark during a feeding frenzy. During a frenzy, sharks often attack fellow sharks in their desire to feed.

Andrew A. Gifford, Audubon/PR

The great white shark. This fish is greatly feared and has a reputation as a man-eater. This shark can reach 9 meters in length and weigh 3 metric tons.

Other representatives of the shark family. Top: the Port Jackson shark. Middle: the nurse shark, a lazy inhabitant of tropical waters. Bottom: the unusually shaped head of the hammerhead shark. Scientists still do not understand why the fish has this strange form.

neath. This gives it a better "angle of attack."

Sharks find their food by means of an almost unbelievably sharp sense of smell. They can detect even a very tiny trace of blood in the water. The scent of blood sometimes seems to have a maddening effect on sharks. As they eat, they may become more and more excited. They may then attack one of the other sharks and eat it. Scientists call this behavior a "feeding frenzy."

Sharks are able to detect bleeding sources even when the blood is moving away from them. In these cases, the sharks cannot smell the blood. Instead, they pick up the vibrations of the bleeding source.

Sharks are especially sensitive to low-frequency vibrations. They can detect a moving or thrashing source by means of their lateral-line senses. The *lateral line* is a series of small pores that lie on a line running along each side of the shark's body and head.

ATTACKS ON HUMANS

Each year about 50 people around the world are attacked by sharks. Half of these people die. The other half are likely to be missing an arm or a leg after the encounter.

Most shark attacks on humans occur off the coasts of Australia. Among shark victims, men outnumber women almost 10 to 1.

No one knows what causes some species of sharks to attack humans. These sharks, furthermore, do not always attack. Sharks are always attracted by blood; bright objects and metal also fascinate them. A shark is almost certain to attack if provoked—a right hook to the snout may cost the swimmer an arm.

A swimmer thrashing or splashing in the water, or swimming unevenly, will attract a shark. The shark will pick up the vibrations and may attack.

Before attacking, sharks will sometimes circle their prey, going around in smaller and smaller circles. No one knows why.

There is no known way to prevent shark attacks. So-called shark repellants have been almost totally ineffective.

Most shark attacks occur in daylight and in shallow water—as shallow as one meter—within 90 meters of shore. Sharks usually single one person out to attack. If a companion tries to help the victim, the shark almost always ignores the rescuer.

Shark scientists at The American Institute of Biological Sciences in Washington, D.C., say that if a shark is seen the swimmer should get out of the water. This should be done quickly but smoothly—no splashing or commotion. The scientists also advise staying out of waters known to be shark habitats. Swimmers, they say, should not enter the water alone or if they have an open wound.

THE BEST-KNOWN SHARKS

We describe below 14 of the best-known sharks, including the dogfishes.

The *nurse shark* of the waters off Florida and the West Indies is a sluggish fish growing up to three meters in length. It can be recognized by the pair of small processes, called barbels, that grow one at each side of the mouth. It is content to feed on crabs, shrimps, and squids, and it may serve as a scavenger. Around Florida, these sharks come into shallow water close to shore. Sometimes adventuresome swimmers climb onto the backs of these harmless creatures and have a spirited ride.

One of the terrors of the sea is the *tiger shark,* which is from 4½ to 9 meters in length. It is a heavy, blunt-headed species

Tom McHugh/PR

Sawfish ray. An unusual form with a snout in the shape of a double saw. It closely resembles the sawfish shark, but it has gills on its underside.

found in tropical seas. However, an occasional straggler will wander into temperate waters. The tiger shark is a voracious fish and will eat carrion as well as almost any kind of sea animal, including other sharks.

The upper part of the *blue shark's* body is colored a deep rich blue. This fish is about four meters in length. Its tapering body terminates in a long, pointed snout. Blue sharks generally are to be found in tropical seas, but some may swim as far north as Nova Scotia and England.

The *sand shark* is a common inhabitant of Atlantic waters. Its brown skin is marked with blotches. It feeds on the small fish, squids, and crustaceans in coastal waters. Sand sharks, which are not considered to be dangerous to people most of the time, may grow to a length of three meters.

The *mackerel shark* hunts mackerel and herring far out at sea in both the Atlantic and Pacific oceans. It is a powerful swimmer and is surpassed in speed perhaps only by the mackerel itself. This shark has a symmetrical, crescent-shaped tail and may attain a length of about four meters.

The dreaded *great white shark,* or man-eater, is related to the mackerel shark. It is probably the most feared of all marine creatures. It is a powerful and voracious hunter of tropical seas, feeding on large fish and sea turtles. It is also dangerous to bathers. It grows to lengths over nine meters and weighs over three tons. One great white shark that was caught was 6½ meters long and weighed 3,220 kilograms.

The great white shark has been seen often in the cool waters off the coast of California, and is believed to be responsible for most of the shark attacks there. The great white has also been seen off the east coast of North America as far north as Nova Scotia. In Australia, it is known as the White Death.

The *basking shark* grows to be from 12 to 14 meters in length. This fish is absolutely harmless to people. It has a large mouth filled with very tiny teeth. The gill clefts are long, and inside the mouth these clefts are screened by long and slender devices called gill rakers. The basking shark nourishes its huge body by eating millions of plankton organisms. Plankton is composed of a floating population of diatoms and algae, protozoa, copepods, and the eggs and immature forms of fish, sea urchins, starfish, shrimps, crabs, and sea worms. These tiny creatures are sucked into the shark's mouth as it draws in water. When it expels the water through the gill clefts, the gill rakers strain out the organisms, which pass on down the shark's gullet. The basking shark likes to float on the surface of northern waters.

The largest living shark, as far as we know, is the *whale shark,* which may reach a length of more than 13 meters and weigh several tons. This shark has a large terminal mouth, small eyes, and extremely small, pointed teeth. Its diet is similar to that of the basking shark. Its dark back and sides are marked by a pattern of light spots combined with crossing lines. Little is known of this huge creature's habits, but probably they are similar to those of the basking shark. Like the basking shark, the whale shark is harmless to man.

The *thresher shark,* which ranges up to six meters in length, has a very long and slender tail that accounts for more than half the fish's length. It is a terrible foe to mackerel, herring, pilchards, and sprats. Nineteen mackerel and two herring were once found in the stomach of a single specimen.

The name "thresher" comes from the fact that the shark beats the water with its long tail as it circles the school of fish upon which it intends to prey. The frightened fish then draw together in a compact mass, and the shark, darting into their midst, has a field day.

The roundel skate, *Raja texana,* has a distinctive eyelike spot on each of its broad pectoral fins. Roundel eggs are being studied by biologists.

Miami Seaquarium

The torpedo ray. This ray has some extra protection—its dorsal side is dotted with electric organs. These fish can deliver 1,600 watts of power in one discharge.

B. Boch from B. Coleman

The *hammerhead shark* is an odd-looking creature. On each side of its head there is a long projection bearing an eye. As a result, the front part of the head resembles the hammer part from which the fish derives its name. The hammerhead, which reaches a length of some 4½ meters, occurs in almost all tropical and subtropical waters, abounding particularly in the Indian Ocean. It is also found along the coasts of Europe and off the shores of North America as far north as Massachusetts and southern California. The hammerhead is a ferocious fish and is among the most feared of the sharks.

The *Greenland,* or *Arctic, shark* is a relentless enemy of the heavy-bodied right whale. The shark rips off the flesh of this huge mammal and so devours the animal practically alive. The Greenland shark is not particularly large as sharks go, ranging from 3 to 5½ meters in length. It is found not only in the waters off Greenland but also as far south as France and as far west as Cape Cod. It has also occasionally been recorded in San Francisco Bay and around Macquarie Island in the South Pacific Ocean not far from Antarctica.

The *bull shark* is known in several areas of the world, but by different names. In South Africa it is called the Zambezi shark. In Australia it is called the whaler shark. In Central America it is called the Lake Nicaragua shark, for that is where it lives. It is, in fact, the only shark that lives in fresh water as well as in the oceans.

The bull shark is a slow-moving fish, but puts on sudden spurts of speed to attack. It is very aggressive and will attack larger fish as well as boats. From the ocean it will travel as much as 160 kilometers up a river. It is gray on top, white on bottom, and grows to a length of about three meters.

To fishermen the prize of them all is the *mako shark,* for it puts up a spectacular fight before it is landed. It grows to a length of four meters and weighs up to 600 kilograms.

The mako is blue on top, white underneath. It is an extremely strong and fast fish. It feeds on mackerel, herring, and even swordfish.

The *dogfishes* are among the smallest of the sharks, but what they lack in size they make up for in numbers. They are often found in large packs, which are the despair of deep-sea fishermen. Dogfishes take bait off fishing lines and sometimes make off with the hooks; they tear fishing nets; they devour large numbers of herring and mackerel and other valuable food fishes. However, these greedy little sharks have proved useful to people in several ways. Some people find their flesh palatable; and the oil extracted from the liver of these fishes is a valuable source of vitamins.

The smooth dogfish is found on both sides of the North Atlantic. On the American side it is particularly abundant south of Cape Cod. The skin of this fish is extremely rough. When dried it is used for polishing wood and other substances. The spiny dogfish is found in especially large numbers off the coast of New England. This fish may be easily distinguished from the smooth dogfish by the presence of a long spine in front of each of the two dorsal fins.

SKATES AND RAYS

Skates and rays differ from sharks in several respects. The body is decidedly flattened, for these fishes have been adapted for living on the bottom of the sea in comparatively shallow waters. Their shoulder fins are greatly enlarged. They extend forward past the gills and are attached to the sides of the head. By moving the skeletal supports within these fins up and down, a ripple, or wave, effect is produced. The rays swim by passing these waves back along the shoulder fins.

The eyes of skates and rays are placed on the upper surface of the head. The gills, on the other hand, lie on the underside of the head. The breathing apparatus of these fishes is well adapted to their life on the ocean bottom. Instead of taking in water at their mouths, as do the sharks, the rays take it in through two apertures called *spiracles,* which are located on top of the head in back of the eyes. Each spiracle is fitted with a special valve. When the fish breathes out, the spiracle is shut off and water is forced out through the gill openings.

Rays have much reduced tails, some being no more than a whiplash. They also have lost the anal fin, and the fins on the back are very small or altogether absent. They mainly feed on invertebrates that live on the ocean floor. The various species of skates and rays are reckoned among the commonest fishes that are found in the sea.

ANGELFISH AND SAWFISH

The curious *angelfish,* or *monkfish,* is a connecting link between the sharks and the rays. The angelfish is a flat fish reaching a length of two meters or more. It is particularly remarkable for the great extension of its pectoral fins, which are supposed to suggest the wings of an angel. It is widely distributed, especially in northern waters.

Another link between the sharks and the rays is the *sawfish* ray. Superficially it closely resembles the saw shark (sometimes referred to as the sawfish shark), for both have an elongated snout with sharp teeth along the edges. The gills of the true sawfish are on the underside of the body and it is therefore classed in the order of rays. The adult may reach a length of 5½ meters and the saw may be as much as 30 centimeters wide at the base and two meters long, with perhaps 32 teeth on each edge. The young are born alive. The eggs hatch inside the female. At that time, the short saws are covered with membranes that they lose soon after birth.

The saw is used for hunting purposes. The fish impales its victim on the teeth by a sideward sweep of the saw. It then scrapes off the speared animal on the sea bottom and chews it, for the mouth contains teeth. The saw can also be used as a shovel for digging up small crabs and mollusks. The sawfish inhabits the tropical Atlantic and

The chimaera is a rare oceanic form. Chimaeras are very distantly related to sharks and rays, and are grouped in a subclass all their own. Like skates and rays, they are mollusk-eaters.

Chicago Nat. Hist. Mus.

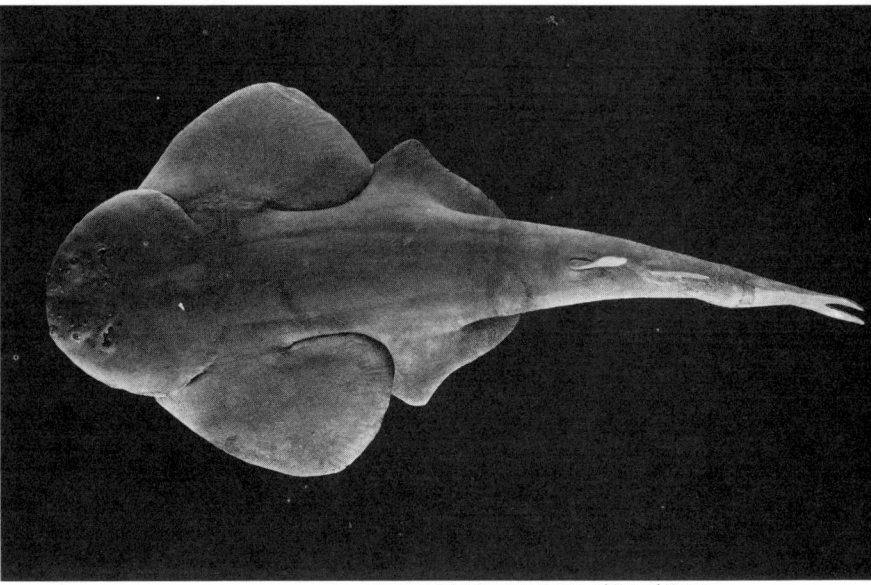

The monkfish represents a link between the sharks and the rays. It is a common fish in northern waters.

Courtesy of the American Museum of Natural History

the waters off India and China. Some species may even live in fresh water.

WELL-KNOWN SKATES AND RAYS

The *guitarfish,* as the name implies, is shaped somewhat like a guitar. This animal is a very ancient form, having existed since the Jurassic period, 165,000,000 years ago. It is a sharklike skate, flatter than the sawfish but not so broad and flat as the true skates.

The *shovel-nosed skate* is a common guitarfish found in many seas. The flattened head and body have large shoulder fins as in the skates, but the tail is powerful as in the sharks. The fish is grayish, with pale spots, and the underside is uniform in color. The adult form is usually about 60 centimeters long. The animal inhabits the warmer seas and is fairly common off the Florida coast. Like other skates, it eats mollusks and crabs. The Indian species is particularly destructive to the pearl-oyster beds of Sri Lanka (Ceylon).

The skates are almost completely flattened and are well adapted to the smooth, sandy bottoms that they inhabit. The disklike head and body are sharply distinct from the slender tail. In some cases the tail is extremely reduced and is almost absent. The skin is armed with spines, which are most prominent on the middle line of the back and the upper part of the tail. The two sexes differ greatly in color and in the formation of the teeth and spines. Some skates can give a mild electric shock, since they have an electric organ in the tail. They prefer cool waters and are not found in tropical seas except in the deeper waters.

The *common skate,* which occurs off the eastern cost of North America, is never more than 60 centimeters in length. The teeth are rounded and set in many rows. They are well adapted for crushing the small crustaceans and mollusks that constitute the diet. Like all skates, the common skate lays eggs that are enclosed in a brown, rectangular capsule with horny strings attached to the corners of the "purse."

One of the largest species of skates is the *barn-door skate,* which occurs in deep waters from Florida to Nova Scotia. It may reach an over-all length of more than two meters. The "wings" of skates make excellent food.

The *torpedoes* are an unusual group of rays. The body is round and disklike and the tail has a large fin at the end. The skin is completely smooth and unarmored. Their only method of defense is their ability to deliver a severe electric shock. The electric organs of the torpedoes are large masses that are located on both sides of the head and extend sideward to the edge of the fins. They are as thick as the entire body and about one third the length of the animal. Actually, the electric organ is modified muscle tissue that is well supplied with nerves that run to it from a special lobe of the brain.

The Pacific torpedo has been known to produce as much as 200 volts of electricity, with a power of 1,600 watts, in one discharge. During several discharges the torpedo may produce enough power to light several household lamps, though only for an instant.

The electrical powers of the torpedo are generally for the purpose of defense, but they can be used in capturing prey. The young are born alive and are capable of producing electricity even before birth. The adult Atlantic torpedo may reach a weight of 14 kilograms. About 23 kilograms has been recorded for the Pacific torpedo. Most of these rays are found in tropical seas. They feed chiefly on fish.

The *stingrays* have a diamond-shaped body and a long, whiplike tail that is provided with a barbed spine at the base. These flat creatures live at the bottom of seas, bays, and rivers, particularly in the tropical regions, wherever there is sand and silt. The tail is used primarily as a defense but it can also paralyze prey. When the ray is alarmed, its tail lashes around the enemy and the barbed spine is driven into the victim's flesh, injecting poison as it does so. The stingray swims by flapping its fins up and down.

The *eagle rays* are typically flattened fishes. In this case, however, the broad shoulder fins common to most members of the ray family are not continued to the end of the snout. They terminate on the sides of the head and appear in front of the snout as two fleshy projections (the cephalic fins). The tail is long and whiplike, with a single dorsal fin near its root. As in the other families, the young of eagle rays are born alive.

The *spotted whip ray* is four meters long and weighs close to 200 kilograms. It inhabits the tropical seas of the western Atlantic.

Cow-nosed rays are found in the Mediterranean and the western Atlantic and off northwestern Africa. The American species may attain a width of two meters.

The largest rays belong to the manta family. They are very similar to the eagle rays, but in the mantas the cephalic fins project in front of the head and resemble ear flaps.

The *devil ray,* or great *manta ray,* is about seven meters wide and weighs over 1,400 kilograms. The brownish skin is rough and covered with knoblike projections. This big animal is generally harmless to people. But when harpooned, the manta may demolish the boats of its attackers, using its great size and muscular powers in a tremendous effort to escape. Generally, it is rather sluggish, often basking in the sun at the surface of the water. It swims by flapping its great triangular wings. The young of the manta ray are very large at birth. They hatch within the mother's body and are born alive.

CHIMAERAS

The chimaeras live in deeper water than either sharks or rays. Like the rays, their tails are often reduced to a long, narrow, whiplike structure. The gills are covered with a flap of skin so that there is only one opening on each side leading to the gills.

The chimaera is generally considered an offshoot of an early cartilaginous fish stock. This group resembles the sharks, for they have a cartilaginous skeleton as well as a similar internal anatomy. They are, however, more like the true fishes, since they have a single gill covering, unlike other cartilaginous fish.

The body is slender and tapered and its first dorsal fin is armed with a spine. The shoulder fins are large. The teeth are flat plates provided with hardened points. They are made for crushing mollusks, but the chimaera's diet consists also of fish and seaweeds. The rough surfaces of the flat plates are well suited for grinding up this varied diet.

A familiar amphibian—a frog. This one is a Cuban tree frog. It has expanded adhesive toe pads typical of many tree frogs. These pads help the frogs be agile tree climbers.

George Porter, Audubon/PR

AMPHIBIANS

by Richard G. Zweifel

You hear him but you can't see him. His distinctive singing is clearest and most cheerful on a damp day, when there is lots of moisture in the air. He sits perfectly still. Sometimes several hours pass before he moves. And because he can change the color of his body, he may match his surroundings, looking like a green leaf or a bit of rock or moss.

He is a tree frog, a small animal perhaps two to five centimeters long. On his toes are suction disks, which help him climb. Some tree frogs even breed in trees. The females lay their eggs on leaves, bending the leaves so that they collect rainwater, which keeps the eggs moist.

Most frogs never climb trees. But they share important features with the tree frogs. They have thin, moist skin. The tongue is attached to the front of the mouth and has a sticky tip. They have well-developed legs. And they lay their eggs in water or in very moist places.

Most frogs spend part of their lives in water and part on land. Thus they are called *amphibians,* which means "both-life creatures." Other members of the class Amphibia are toads, salamanders, and caecilians.

BETWEEN FISH AND REPTILES

The amphibians may be easily identified by their moist, glandular skin. This contrasts with the scaly skin of most fishes and reptiles, the animals most closely related to amphibians.

There are other differences between amphibians and fishes. Generally, amphibi-

The gray tree frog, *Hyla versicolor*, leaps from his perch to a new branch. Its long hind legs are well adapted for hopping and leaping.

ans have legs instead of fins, and breathe by means of lungs instead of gills. However, there are exceptions. Not all amphibians have legs. Some fishes have lungs; all amphibians at some stage of their development have gills, though in most species these are lost when adulthood is reached. Certain fishes lack scales; the amphibians called caecilians have scales.

The leopard frog, *Rana pipiens*, is common throughout North America. These frogs are widely used in biological research.

The dry, scaly skin of most reptiles distinguishes them from amphibians. Also, all amphibians possess functional gills at some time or other. Reptiles never do, though gills exist as a transitory stage in the development of the reptile embryo. Amphibians never develop an amnion, an embryonic membrane found in reptiles and other higher vertebrates. The eggs of amphibians are always laid in water or very moist spots. Those of reptiles are laid on land.

EVOLUTION OF AMPHIBIANS

Primitive fishes called crossopterygians were the forerunners of the amphibians. They developed fins that came to resemble primitive legs. The crossopterygians had lungs, and in various other respects they were similar to the earliest amphibians.

The first amphibians made their appearance some 300,000,000 years ago. The first known amphibian, *Ichthyostega*, possessed limbs much like those of modern amphibians. Possibly it had a somewhat fishlike tail and bore other resemblances to fishes.

We can only guess why and how the earliest amphibians moved from the freshwater habitat of their ancestors to land. The evolution of lungs would enable certain early fishes to live in warm stagnant pools, which would be deficient in oxygen. The fishes would be able to supplement the supply of oxygen by breathing it from the outer air through their lungs. When droughts dried up the pools, primitive limbs would enable the animals to travel overland to seek other water.

The first amphibians that ventured onto land found no vertebrate rivals there. It was a fresh and quite unexploited environment—far more favorable for their development than the aquatic environment, which teemed with competing forms of vertebrate life.

Once they became partly adapted to terrestrial life, the amphibians evolved into a great variety of forms. Some were grotesque creatures with broad bony heads. Others resembled modern crocodiles in size and appearance, and probably also in habits.

The bullfrog, *Rana catesbeiana*, is a large frog native to North America. Its deep-voiced croaks are frequently heard during the summer.

In the Carboniferous and Permian periods, 225,000,000 to 350,000,000 years ago, there were two main groups of amphibians—lepospondyls and labyrinthodonts. It is believed that today's salamanders and caecilians are descendants of the lepospondyls, while frogs are more closely related to the labyrinthodonts. The reptiles also evolved, almost certainly, from the labyrinthodonts.

THREE MAJOR GROUPS

The amphibians are divided into three principal groups—frogs, salamanders, and caecilians.

The name "frogs" is given to various animals, including the toads, that make up the order Anura. These animals are distinguished from other amphibians among other things, by, the lack of a tail in the adult. The frogs are the spriest of the amphibians. They hop or jump from one place to another.

Salamanders and newts make up the order Urodela. They retain their tails as long as they live. Some of them, as adults, dwell on land, walking or creeping about. Other species live in the water. Certain salamanders have no hind limbs.

The caecilians make up the order Apoda. Found both on land and in fresh water, they have no limbs at all.

FROGS AND TOADS

These are the most numerous of the three amphibian groups. Almost 2,000 species are known. The order name, Anura, means "without a tail," and refers to a distinguishing feature of these amphibians. The bodies of the adults are short. The long hind legs are well adapted for hopping.

Frogs are found in most parts of the world. They are almost entirely absent from Arctic and Antarctic regions and from the islands of the sea. However, a very few may be found north of the Arctic Circle and some have been introduced on oceanic islands by people. The vast majority of species, about 80 per cent of those known, live in the tropics.

The common frogs, or ranids, are widely distributed and numerous in many places. The leopard frog, *Rana pipiens*, is the most widely distributed species in North America. It is at home in any place where there is fresh water. Thousands of these frogs are used every year in teaching laboratories. Students dissect them to study the structure of the vertebrate body. In research laboratories scientists use them as experimental animals.

The bullfrog, *Rana catesbeiana*, a large, deep-voiced species native to eastern North America, has been introduced in many other regions of the continent. The legs of this animal are a favorite food item. Misguided persons have tried to take advantage of the steady demand for frogs' legs by establishing "frog farms," where they have raised frogs for sale to restaurants. Unless one owns a swampland with a plentiful natural food supply, the keeping and feeding of bullfrogs on a scale large enough to provide an adequate profit is not practical.

The toads, genus *Bufo*, are rather short-legged and lack the remarkable jumping ability of frogs. They generally move forward with a series of short hops. When the toad finds an insect or worm, he stalks it carefully. Then he flips out his long, sticky tongue—it may stretch five to seven centi-

Blomberg's toad. Toads are very similar to frogs but generally spend most of their time on land and have wartlike structures on their skin.

The hellbender is the largest salamander in North America. It lives in rivers.

meters—and draws the prey back into his mouth.

Most of a toad's life is spent on land. Only in the breeding season does the animal return to the water. The wartlike structures on a toad's skin are actually glands. They have only a superficial resemblance to the warts that develop on human skin. The belief that contact with a toad can cause a human being to develop warts is untrue. However, a few species of toads are poisonous.

Tree frogs are found throughout much of the world, but are particularly abundant in tropical regions. Most species are smaller than true frogs and toads. As a rule they possess expanded, adhesive pads on the toes. These pads make the animals expert climbers.

Not all tree frogs live in trees. The spring peeper, *Hyla crucifer,* so well known to rural residents of eastern North America, remains near the ground. The common Pacific tree frog, *Hyla regilla,* of western North America, likewise seldom ascends more than bush high. The voice of *Hyla regilla* has been recorded and is now used when a movie, radio, or television script calls for frog noises in the background. Thus, in a play whose scene is laid in Burma, Canada, or Brazil, the listener likely hears a common frog that lives near Hollywood.

One of the remarkable properties of frogs, and of tree frogs in particular, is the ability to change color. Many species merely darken or lighten their basic color. Others change from one color to another, as from brown to green.

Color change is accomplished by the contraction and expansion of pigment within skin cells known as melanophores. These cells are under the control of the adrenal and pituitary glands. One would think that a frog would instinctively change its color to match the color of the rocks or leaves upon which it sits. This does not appear to be the case. Color changes seem to be due to such factors as the animal's temperament and its temperature. It is not unusual to see a bright green frog sitting on a drab, gray rock.

SALAMANDERS AND NEWTS

These animals, which have a tail throughout their lives, form the order Urodela—"creatures with visible tails." They look like lizards, but the dry, scaly skin of a true lizard distinguishes it from the salamander, which has a moist, glandular skin.

About 250 species of salamanders have been discovered. The animals are most abundant in the Northern Hemisphere. Only a very few species are found south of the equator.

The red-spotted newt during its pre-adult land stage. During this time it is called a red eft.

One group of salamanders is made up of the newts, which are found in North America, Europe, and Asia. The spotted newt, *Diemictylus viridescens,* of eastern North America is an abundant species commonly found in lakes and ponds. Its interesting life history is described later in this article.

The crested newt, *Triturus cristatus,* is a common species in Europe. It reaches a length of 15 centimeters. During the breeding season, the animal develops a crest along its back.

Salamanders of the family Plethodontidae do not have lungs. In these animals oxygen is taken up and carbon dioxide is given off through the moist skin and mouth lining, rather than through the surface of the lungs. Many of the commonest salamanders of North America belong to this family. They keep pretty well out of sight. About the only persons who see them are collectors, who turn over logs and rocks to find them, and fishermen, who use stream-dwelling species for bait.

The largest salamander in North America, the hellbender, *Cryptobranchus alleganiensis,* lives in rivers in the eastern United States. It rarely ventures on land. It is a flabby-skinned animal, sometimes more than 60 centimeters long. Its relative, the giant salamander of Japan, *Andrias japonicus,* may reach a length of 150 centimeters. It is the world's largest salamander. In past ages, salamanders of this type lived in various parts of the world where they are no longer found. The fossilized remains of a giant salamander discovered many years ago in Europe were at first thought to be the remains of a person who had perished in the Biblical flood.

Although most salamanders have four well-developed limbs, there is one family in which the legs are very tiny and another in which only front legs are present. The congo eel, *Amphiuma,* of the southeastern United States, has four tiny limbs much too small to be of use in locomotion. It spends most of its time in muddy lakes or sluggish streams, and swims in the fashion of a true eel. The sirens, genus *Siren,* found in the same general area, have tiny forelimbs. Both congo eels and sirens are quite large for salamanders, reaching a length of almost one meter.

THE CAECILIANS

The Apoda, or "animals without feet," are a tiny group, composed of about 70 species. These little-known creatures live only in tropical regions in both the New and Old Worlds.

As they are legless, caecilians bear no resemblance to frogs or salamanders. The surface of the body is usually grooved into a series of ringlike segments. Therefore, they look a good deal like earthworms.

An American toad singing. The vocal sac is inflated, amplifying the sounds from the vocal cords.

Some species of caecilians possess scales. These are buried in the skin and can be seen only when the animals are dissected. Most species lay eggs, but some are known to bring forth living young. Some caecilians are burrowers. In all species, the eyes are degenerate, so that the animals are blind, or nearly so.

LIFE CYCLE

Amphibians generally produce a large number of eggs, only a comparatively small percentage of which survive and develop into the adult form.

IN FROGS

Mating. Many more frogs are heard than are seen. During the breeding season, the males of most frog species give forth with songs that apparently attract the females and possibly other males to the breeding site. Frogs frequently gather in vast numbers. A deafening din may be produced by even so small a species as the spring peeper.

Each species has its characteristic call. It is believed that this makes it possible for a frog to find a mate of its own species if several species are contributing to the uproar in a small area.

When calling, most frogs make use of *vocal sacs,* bubblelike pouches of skin filled with air. There may be one or two of these sacs, which are located beneath the chin or at the corner of the mouth. A vocal sac acts as a resonating chamber, amplifying the sound. Air is passed back and forth between the lungs and the sac, passing over the vocal cords to produce the sound. While this happens, the mouth remains closed, so that the frog can call even under water.

Other types of sound besides the mating call are produced by frogs. Often, when one frog is touched by another or when a frog is picked up by a person, a chirping or chucking sound is produced. This seems to be a note of protest. When some frogs are seized by an enemy, they give a loud, piercing scream. In contrast to the mating call, this scream is produced with the mouth open.

Laying eggs. For the vast majority of frogs, the processes of laying and fertilizing the eggs follow a set pattern. The male clasps the female by the back and when the eggs are discharged into the water, they are fertilized by the male. During the breeding season, many kinds of male frogs develop horny, spiny areas on their thumbs. These help the male to grasp the slippery body of the female.

The eggs may be laid in a large clump or diffuse mass, as is the case with many true frogs, or they may be in long strings, as is the habit of toads. Some species may lay their eggs in numerous smaller clumps. Others lay eggs singly. No matter what the shape of the egg mass, each egg is surrounded by layers of jelly that help protect it from injury.

It may take the eggs only a few hours to hatch or it may take many days or even weeks. The time will depend largely on the species but also on the temperature of the water. Generally, eggs laid in temporary pools develop more rapidly than those laid in more permanent bodies of water.

Larger frogs, as a rule, produce greater numbers of eggs. A bullfrog may lay as many as 20,000 eggs. Smaller species, such as the green frog and leopard frog, lay up to 6,000. Toads produce smaller eggs than frogs; these eggs are much more numerous. A toad the size of a leopard frog may lay more than 25,000 eggs. The eggs of tree frogs are usually numbered in the hundreds. The record for the smallest number of eggs belongs to a West Indian form, *Sminthillus,* which lays only one.

The tadpole. The creature that hatches

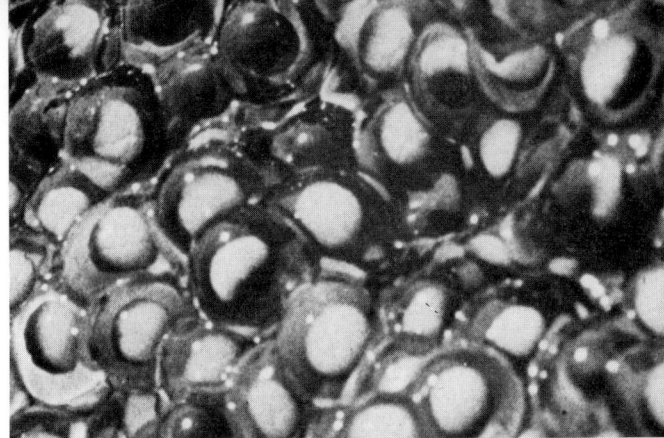

from the eggs—a tadpole—is a fishlike organism that bears little resemblance to the adult frog. It has gills for respiration, no legs, and a tail with which it propels itself through the water.

The tadpole has a horny beak surrounded by rows of comblike horny teeth. It feeds for the most part on plant material, such as algae, which it scrapes from rocks and other aquatic plants. It will also eat meat, usually in the form of dead tadpoles or other animals.

For most of its growth period, the tadpole changes little in external appearance. The gills, which are at first external, are soon covered over by a flap of skin. The hind legs appear at the base of the tail and slowly increase in size.

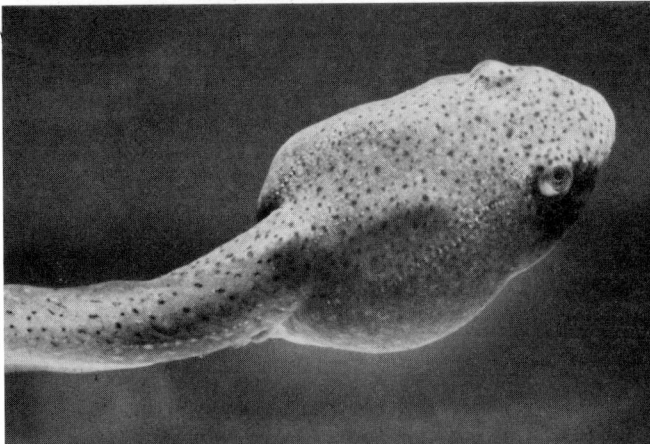

Metamorphosis. When the time for metamorphosis, or the change into the form of an adult frog, arrives, events move rapidly. The front legs, which have been developing out of sight behind the flaps of skin covering the gills, now appear. The tail dwindles away; it is absorbed into the body. The mouth gradually assumes the wide shape characteristic of the frog. Internally, significant changes take place. The long, coiled intestine of the tadpole is replaced by the shorter gut of the frog. Lungs take over the task of respiration from the gills. The tadpole must now dash to the surface of the water to gulp air.

When thyroid extract is fed to tadpoles their development is speeded up. This shows that the hormone produced by the tadpole's thyroid gland is important in starting and bringing about metamorphosis.

The length of time between egg-laying and metamorphosis varies greatly among different species of frogs. Some, such as the spadefoot toads, which lay their eggs in temporary pools in desert regions, may go through the cycle in less than two weeks. At the other extreme, in cold northern cli-

Life cycle of a typical frog. Top: a gelatinous mass of fertilized eggs. 2nd: a tadpole, a form that lives entirely in water. 3rd: metamorphosis occurs at some point, and the tadpole is transformed into an air-breathing form. Bottom: The adult form. The tail has been absorbed.

all photos this page, Audubon/PR; from top to bottom: Lynwood M. Chace, George Porter, W. A. Allen, Jr., and George Porter

mates the bullfrog may take two or three years to transform. For the majority of species, the transformation is completed during a single season. The eggs laid in spring or early summer produce small frogs by the end of summer or early fall.

Unusual development. The development of certain frogs does not follow the general pattern just described. In various species, the tadpole is not a free-living form. It remains in the egg, and metamorphosis takes place before the egg hatches. Such eggs are never laid in the water, but only in moist places on land or in trees.

The male of one remarkable South American species, *Rhinoderma darwini,* keeps the eggs in his vocal pouch. The female Surinam toad, *Pipa pipa,* carries her eggs about on her back. The eggs are set in place by the male. They sink into the skin and each is protected by a covering. The larva that develops from each egg completes its development in its tiny chamber and emerges as a small but fully formed toad.

The female midwife toad, *Alytes obstetricans,* lays two strings of eggs. Her mate thrusts his legs though the egg mass. Thereafter, he carries the eggs, attached to his legs, about with him until the larvae hatch out. By that time the larvae are well-developed.

IN SALAMANDERS

Salamanders have no mating calls like those of frogs. Some species are capable of making a squeaking sound, but most are mute. The sense of hearing is poorly developed, too. Salamanders lack any external evidence of ears, in contrast to the well-developed ears of most frogs. On the other hand, many salamanders show a highly developed courtship behavior.

Except for some primitive species that practice external fertilization in much the same manner as frogs, salamanders have internal fertilization. This is not accomplished in the direct manner of higher animals. The male courts the female by leading her about. He then deposits a gelatinous packet containing sperm, which is called a *spermatophore.* The female picks up the spermatophore with the lips of her cloaca — the chamber through which pass waste products and eggs before leaving the body. The sperm are stored in the cloaca until the time for egg-laying arrives.

The eggs of many species of salamanders are laid singly or in small clumps in water. They hatch into larvae, which spend a period of weeks or months in the water before they are transformed into adults. Some species lay their eggs in moist places near streams. Upon hatching, the larvae make their way to the water. Just as in frogs, many salamanders have given up a free-living larval stage; small salamanders of adult form are hatched. These species never go in the water, either as adults or as young.

No salamander produces the immense number of eggs characteristic of so many frogs and toads. Approximately 500 eggs is the greatest number known to be laid by any North American salamander. Most species produce far fewer than that. Species that lay their eggs on land and have direct development (omitting the larval stage) may lay as few as a dozen eggs.

In contrast to the tadpole, the larva of

This female salamander is guarding her clutch of twenty-four eggs.

Hal H. Harrison, Audubon/PR

the salamander bears a close similarity to the adult. The most conspicuous difference is the presence of gills in the larval salamander. In some wholly aquatic species, the adult retains gills and differs from the larva only in being sexually mature. This condition may occur in species in which adults normally become land animals.

A salamander living in the Alps, the black salamander, *Salamandra atra,* produces live young without the bother of a free egg or larval stage. At the other extreme, the spotted newt of eastern North America has a complicated life history. Eggs are laid in a lake or pond and hatch into yellowish-green larvae. After living in the water for several months, the larva changes its color to bright red, with spots of deeper red. It then crawls out on land. At this stage it is known as a *red eft*. The eft remains on land for a year or more and then returns to the water to breed. Its color has now changed to olive green. It remains in the water for the rest of its life.

HOW LONG DO THEY LIVE?

We saw that most amphibian species produce immense numbers of eggs. Since comparatively few of the larvae that are hatched reach maturity, the mortality among them must be great.

Frogs and salamanders that reach adulthood probably do not live more than three or four years on the average. Animals kept in captivity, however, have lived far longer than that. An American toad attained the venerable age of 31 years; a bullfrog also reached the age of 31. Among salamanders, a large aquatic hellbender lived 29 years, a small spotted salamander 25 years, and a western newt 21 years.

BEHAVIOR

In general the different groups of amphibians are rather similar in their food preferences and adaptations to climate conditions. They all have their particular ways of defending themselves from enemies.

FOOD HABITS

All frogs and very nearly all salaman-

George Porter, Audubon/PR

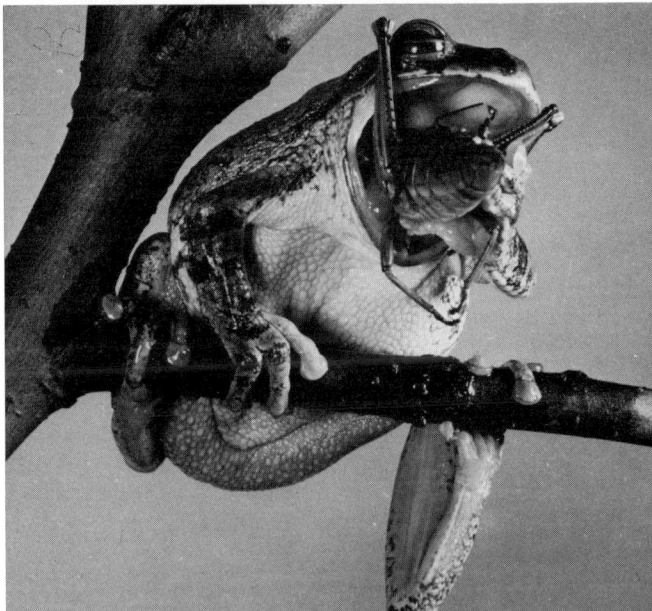
Robert C. Hermes, Audubon/PR

Large species of amphibians, such as the toad (top) and frog (bottom) can swallow relatively large prey. These animals use a long sticky tongue to catch food and get it to their mouths in a fraction of a second.

ders are carnivorous. Some kinds of aquatic salamanders eat plant material, but even these species subsist largely on a diet of smaller animals.

In the case of frogs, the rule seems to be "eat anything that moves and is not big enough to eat you." Frogs and salamanders are stimulated to feed by the motion of their

A spadefoot toad. Spadefoots have a special horny protuberance on each hind foot that is used for digging.

prey. Large species, such as the bullfrog, are capable of remarkable feats of swallowing. Birds, mice, snakes, fish, frogs, and even bats fall prey to this animal. However, most of the food of the bullfrog and of other frogs and salamanders consists of invertebrates—insects, spiders, earthworms, and their kin.

In destroying vast numbers of harmful insects, the amphibians, particularly toads inhabiting gardens and farms, do a service to mankind. One species of toad, the giant marine toad of tropical America, has been introduced in many of the islands of the Pacific in order to help control insect pests.

HIBERNATION AND ESTIVATION

In temperate regions, most amphibians go into hibernation during the winter, with its low temperatures and scarcity of food. Frogs burrow in the mud at the bottom of ponds, or conceal themselves under logs or in crevices. Sometimes toads burrow into the ground to a depth of more than 30 centimeters. Salamanders hibernate under rocks in streams, under logs, in rotting tree stumps, or in damp burrows in the ground.

During the hibernation period all bodily activites, such as digestion, breathing, excretion, and circulation, are greatly reduced. In extremely hot regions amphibians go into a similar state, called estivation, during the hot and dry season.

ENEMIES AND DEFENSES

Frogs and salamanders form a favorite food of many animals. Snakes—particularly water snakes and garter snakes—devour vast numbers of amphibians and are perhaps their deadliest foes. Fishes feast on tadpoles and other larval forms. Frogs dwelling in marshes and streams are often eaten by birds such as the heron. In their nocturnal hunts along streams and ponds, raccoons prey upon any amphibians they may encounter.

People must also be rated as major enemies, not only because of the great numbers of frogs we use as food, but also because we drain swamps and pollute streams in which the animals live.

The great leaping ability of many frogs enables them to escape slower enemies. Many species have no other protection than agility. In others, skin glands secreting ill-tasting or poisonous liquids are found. An animal seizing a frog of this kind will quickly release its prey. A small, brightly colored frog of Central and South America is so poisonous that it has been used as a source of arrow poison by natives.

Even the common toads of North America have poison glands in the skin. Two kinds of giant toad, one found in the southwestern United States and adjacent Mexico and the other in tropical America, are particularly poisonous. A dog or cat mouthing such a toad may be so badly poisoned that it will die.

The poisons of frogs and toads are dangerous to people only when they come in contact with mucous membranes such as those in the eyes or mouth. Ordinarily it is safe enough to handle these animals. However, if you hold a spadefoot toad and, hours later, inadvertently rub your eyes, you may develop itching, watering eyes, and a running nose—symptoms generally associated with a head cold.

The protective device of distasteful or poisonous skin secretions is not limited to frogs. It is also found in certain salamanders. These animals seem to be almost immune from any animal foes, as a result of their skin secretions.

LIZARDS

by James A. Oliver

As the sturdy, fierce-looking Komodo dragon lumbers out of the jungle, it spots a wild pig. The chase is on... and the pig loses. The Komodo dragon leaps on its prey and buries its jaws in the pig's flesh. It tears out huge chunks of meat and gulps them down without chewing—much like the meat-eating dinosaurs must have eaten a hundred million years ago.

The Komodo dragon, a lizard of Indonesia, looks like those terrifying dinosaurs. It isn't as big as the dinosaurs were—but it's big enough to catch and kill pigs and small deer. Some Komodo dragons are more than three meters long. They may weigh as much as 165 kilograms.

Like the pigs and deer of Indonesia, you and I would try to avoid the Komodo dragon. But other lizards, such as the horned lizard, make interesting pets.

More important to people, though, are the many species of lizards that feed on insect pests. In many warm regions of the world such lizards are welcome residents in people's homes. They have even become symbols of good fortune. In parts of southeastern Asia it is considered bad luck to move into a new home until a tokay, one of the gecko lizards, has taken up its residence there.

In certain parts of the world, lizards serve as food for human beings. In Central and South America, iguanas are a valued table delicacy often called "chicken of the tree." Monitor lizards are eaten frequently in Asia and Africa.

The skins of lizards are sometimes used in leather products. Some species have been hunted so intensely for the leather industry that they are now in danger of extinction.

There are approximately 2,500 species of lizards. These reptiles occur in all parts of the world except the polar areas. One species, the viviparous lizard, *Lacerta vivipara*, of Europe, has extended its range to a part of the Scandinavian peninsula

Exxon

The ifkakh, a desert-dwelling lizard. The throat pouch turns a deep blue when the animal is excited.

within the Arctic circle. Most species, however, are found in tropical and subtropical regions.

LIZARDS AND SNAKES

The lizards and their close relatives, the snakes, are the best-known and most widely distributed of all modern reptiles. Lizards and snakes are so closely related that they are usually classified in the same order: the Squamata. The name, a Latin word, means "scaly creatures". Actually, all reptiles are scaly creatures, in a sense, since they are covered with dermal scales or horny plates.

One way in which lizards and snakes differ from other reptiles is in their skull structure. In all reptiles, the lower jaw is attached to the skull by means of the quadrate bone. In lizards and snakes the quadrate bone is movably attached to the skull, permitting the jaws to be more mobile. Also, unlike all other living reptiles, male lizards and snakes possess paired copulatory organs.

In many cases, it is comparatively easy to distinguish between lizards and snakes because of their external appearance. Snakes do not have legs; many lizards do. Most lizards have external ear openings;

LIZARDS   337

The komodo dragon is the world's largest lizard. It may grow to over three meters long—big enough to kill goats, pigs, and small deer.

none of the snakes have. Lizards generally have movable lids and therefore can close their eyes; snakes are not so provided and cannot close their eyes.

There are exceptions, however. Certain lizards have no legs; certain snakes possess remnants of hind legs. Some lizards have no ear openings; some do not have movable eyelids.

Therefore, to distinguish definitely between all lizards and all snakes, we must study internal characteristics. In lizards the right and left lower jawbones are united where they meet in front by a firm bony suture. The brain case is not completely closed with bone in front. And a pectoral girdle—a bony arch supporting the forelimbs—is usually present.

Modern lizards and snakes have descended from the same lizard ancestor. They appeared at about the same time some 135,000,000 years ago, at the beginning of the Cretaceous period. The two groups developed along rather different lines. The lizards known as mosasaurs took up life in the sea. They became elongated, streamlined creatures, with long, powerful tails and paddlelike limbs. Some were large, predatory reptiles, attaining a length of more than six meters. This group of lizards disappeared about 65,000,000 years ago, near the end of the Cretaceous period. The surviving lizard groups are land-dwellers, though the members of certain species forage in the water.

LIZARD ADAPTATIONS

A good many lizard species dwell in trees. The fingers and toes of certain geckos and American chameleons are provided with adhesive structures, enabling them to climb up smooth tree trunks. In the true chameleons, the hands and feet are split so that some fingers and toes oppose the others on the same hand or foot, providing

Chameleons are well adapted for life in trees. Their hands and feet are forked and equipped with adhesive structures and their tails are prehensile—all of which help them climb, grip, and adhere to tree branches. Their very long projectile tongues help them catch prey.

an excellent organ for gripping stems and twigs. Chameleons possess another valuable adaptation for life in trees: a prehensile tail with which they grasp branches and twigs.

The flying lizard is able to glide through the air. The "wings" of this lizard consist of large folds of skin on each side between the front and hind legs. This skin can be extended and held out by means of elongated ribs, thus forming a winglike structure. The flying lizards are accomplished acrobats. They can zoom up at the end of a glide and jump nimbly from one tree to another.

Lizards that dwell on the ground often have elongated bodies and legs that are reduced in size. Some carry this trend to extremes. They have become elongate, limbless, and snakelike. Some of these limbless forms, such as the amphisbaenians, live secretive lives underground or under logs. They generally have no eyes or external ear openings. Certain elongated ground-dwellers have not completely lost their legs. They have small flaps representing the vestiges of limbs.

Few lizards live in either fresh water or the ocean. Those that are adapted for living in or around fresh water resemble crocodilians—superficially, at least. This is reflected in the scientific and common names given to some of these animals, such as the crocodile lizard, *Crocodilurus lacertinus,* and the caiman lizard, *Dracaena guianensis,* both of which live in the Amazon basin.

Marine iguanas are found on the rocky shores of the Galápagos Islands. Here they bask, fight, and reproduce. When the tide goes out, they scamper down the rocks and into shallow water to feed on marine algae. Generally they keep pretty close to the shore. Occasionally, however, they venture quite far out to sea. Marine iguanas are noted for their swimming ability. They use their muscular tails to propel themselves through the water.

LIZARDS NEED HEAT

Like all reptiles, lizards depend on outside sources of heat to warm their bodies. Because of this they are sometimes referred

Phyllis Greenberg, Audubon/PR

The marine iguana lives on the Galápagos Islands off the coast of South America. It is fairly large—about one meter long.

to as ectothermic animals—that is, creatures deriving their heat from without. It is not correct, however, to consider them "cold blooded," as they sometimes are called. When lizards are active, many maintain their bodies at temperatures higher than those of some of the so-called "warm-blooded animals." For example, the desert iguana, or crested lizard, *Dipsosaurus dorsalis,* may tolerate voluntarily a body temperature as high as 47° Celsius. Many species must have a fairly high body temperature to perform most of their functions.

Some lizards bask in the direct rays of the sun and obtain their heat primarily in this fashion; they are called baskers. Others get their heat by contact with some warm substance—the ground, a rock, a log, or the air. Such lizards usually have lower temperatures than the baskers. In general, the species active during the day have a higher temperature than do the nocturnal lizards.

CHANGING COLOR

Much has been written about the ability of lizards such as the old World chameleons to change body color. The popular but erroneous belief is that such color changes always represent conscious efforts to approximate the color of the background on which or against which the lizard appears. Various other factors are concerned besides background color. Among these are age, sexual and seasonal changes, changes in temperature and light, and the animal's emotional state.

Lizards shed their skin periodically as part of the growing process. The skin is made up of dermal scales or horny plates.

The frilled lizard in found in New Guinea and Australia. The frill is usually folded back on the neck — until an enemy approaches.

Usually a decrease in temperature results in a darkening of color. A rise in temperature is apt to produce a lightening in color. The night lizards, genus *Xantusia*, of the southwestern United States and Mexico, form an exception to this rule. Their colors become darker with temperature increase. Intense light generally causes a darkening in color, and faint light the reverse. Color changes resulting from varying emotional states usually produce mottled color patterns.

The tree-dwelling chameleons and anoles exhibit the greatest ability to change color. But even in these animals the range of variations is limited to various shades of green, yellow, brown, and gray. There is no truth in the belief that a chameleon can match a plaid pattern, or any of the infinitely varied nuances of the different colors.

Color variations in lizards are brought about by certain changes in pigment cells in the skin. Most of the cells nearest the surface of the skin are yellowish. Underneath them are great numbers of cells called melanophores, which have dark brown or nearly black pigment. There are whitish cells below the melanophore layer.

Color changes are produced as the melanophores contract or expand. This may be brought about by decreasing or increasing temperatures. In the case of the other factors involved in color variation — sexual and seasonal changes, light, and so on — the melanophores are stimulated by the nervous system or through the action of one of three hormones — intermedin, acetylcholine, and adrenaline. The first two of these hormones cause the melanophores to expand. Adrenaline causes them to contract.

HOW LIZARDS DEFEND THEMSELVES

Lizards have evolved a number of different protective structures and types of behavior to defend themselves against their enemies. The simplest and most widely employed type of defense is to flee from danger. Lizards may escape to a shelter area or nook by gliding, as in the case of the flying lizard. They may run swiftly along the ground on their hind legs. They may swim to the bottom of a stream or pond. They may run upon the surface of the water on their hind legs, as does the basilisk. Or they may burrow underground.

Certain lizards rely on defensive structures, such as spines or armorlike scales. Several lizards, such as the moloch of Australia and the horned lizard, have spines over most of the body. The spines of other lizards, including the false iguanas (*Ctenosaura*) of Central America and the mastigure (*Uromastix*) of Africa and Asia, are restricted to the tail, which is flicked to and fro rapidly as a defensive weapon. Certain large lizards, such as iguanas and monitors, have no spines on their long tails, but lash their tails as a part of bluffing behavior.

One of the most unusual and widespread defensive devices in lizards is the detachable tail. Many species have long and often brightly colored tails that are easily broken or that can be voluntarily detached. When this takes place, the tip of the detached tail wiggles violently, attracting the attention of the enemy, while the tailless lizard crawls away. A special modification of the tail structure provides for the closure of blood vessels in order to prevent loss of blood. The lizard grows a new tail in time. This is never so long as the original tail and it is different in appearance.

In the glass snake, the detachable tail makes up about two thirds of the total length. Some persons believe that the glass snake breaks into pieces when attacked and that, when there is no more danger, the pieces come together again. Of course, once the tail is detached, the severed pieces die. As in the case of other lizards, the glass snake grows a new tail.

Another widespread belief is that many lizards rely on their powerful venom to protect themselves against enemies. Actually, only two species are venomous — the Gila monster and the beaded lizard, both dwelling in the southern part of North America. In these lizards the venom glands are located in the lower jaw, and all of the teeth are grooved, allowing the venom to flow along the sides of the teeth. The venom may be fatal to human beings in rare instances. Fortunately, however, these lizards are very sluggish and it is easy to keep out of their way. Few persons are ever bitten by them, and then only when they insist on playing with or annoying them.

PROTECTING THEIR TERRITORY

Many lizard species have a well-developed social hierarchy, in which large males occupy the dominant position. They maintain definite territories from which they exclude other adult males. Young males and adult females are permitted to remain. Usually the territory of an adult male includes the territories of several mature females. Each of these females has an exclusive territory of its own, from which it excludes other adult females.

To maintain their territory, lizards seek to intimidate intruders, sometimes by feigning to attack them. In most cases threatening behavior suffices to repel intruding lizards. In some species physical contact and biting may be necessary. Male marine iguanas of the Galápagos Islands engage in butting contests with potential rivals. If an intruder refuses to butt, he is given several sharp bites to drive him away or to cause him to fight in the approved fashion. Nocturnal geckos that maintain territories and have social groups give voice to high-pitched squeals, clucks, and clacks to warn off intruders.

In the case of diurnal lizards, the bright colors or "flash colors" of the males serve as a signal to other individuals that they are trespassing. The overall color of the male may be gaudy, as in the agamid bloodsuckers of the genus *Calotes*. In some cases, only a portion of the lizard, such as the dewlap of the anole, may be brightly tinted.

THE LIZARD'S DIET

Lizards eat a variety of foods. Most species are entirely carnivorous, feeding mostly on insects. A certain number eat plants. The moloch of Australia feeds exclusively on small black ants, which it laps

Lizards have adopted some unusual means of defense. This gecko's tail will break off from the rest of the body if a predator grabs it.

Lewis Wayne Walker/PR

up with astonishing rapidity. According to one report, these little animals are capable of devouring an average of 45 ants per minute for 15 minutes — a total of 675 ants — at a single feeding. The horned lizards also feed largely, though not exclusively, on ants.

Many lizards feed on spiders, centipedes, worms, and other small animals. Larger species, such as the Komodo dragon, eat larger prey. The larger chameleons, such as the giant one-horned chameleon, *Chamaeleo melleri,* prey on small birds.

Certain carnivorous lizards eat clams and snails. The caiman lizard of South America and the Nile monitor are well adapted for eating mollusks. In these lizards the teeth in the rear half of both jaws are stubby, low-crowned, and molarlike; they can crack shells effectively. The members of these species are agile swimmers and are well adapted for foraging in water.

The larger iguanid and agamid lizards are primarily vegetarians, eating blossoms, buds, and succulent leaves. Some of these animals feed high up in trees. Others eat plants growing on the ground. The marine iguanas of the Galápagos Islands, as we have seen, feed entirely on marine algae. Certain lizards, such as the desert iguana, *Dipsosaurus dorsalis,* eat mostly vegetable food but devour some animal food as well. Desert-dwelling lizards that eat succulent plants derive the water they need from these plants; they never drink water.

REPRODUCTION

Some lizards are oviparous, reproducing by laying eggs. Others are viviparous, giving birth to living young. For many years it was believed that all the live-bearing lizards merely retained the eggs in the body until they hatched and that the young were not attached to the mother by means of a placenta. However, recent investigations have shown that in some species, such as the European skinks of the genus *Chalcides,* there is a well-developed placenta between the embryo and the mother.

The shells of lizard eggs are generally soft and pliable. In some cases, however, they are hard and brittle. The number of eggs varies from one in some of the small geckos to more than 40 in some of the larger species of monitors and iguanas. Lizards belonging to the large family Gekkonidae have unusually consistent reproductive habits. With few exceptions these animals lay two eggs at a time.

Lizards are not particularly good parents. Usually the mother deposits her eggs in some secluded spot and then goes off about her business, leaving the eggs to incubate unprotected. Some species construct simple nests in the form of excavations beneath logs and stones. A few of these lizards remain with the eggs, protecting them throughout the incubation period and aiding in the process by keeping the eggs warm. They accomplish this by leaving the nest for short periods of time to bask in the sun. When sufficiently warm, they return and coil their bodies tightly around the eggs.

The yucca night lizard, *Xantusia vigilis,* of the southwestern United States and western Mexico gives birth to one or two young. The mother helps to pull the young out when it begins to emerge from her body, and she tears the enveloping membranes from the newborn animal. This behavior indicates an unusual type of parental attention to the young.

The Nile monitor of Africa and the black tegu of South America have solved the problem of incubating their eggs in an unusual fashion. Females of both species excavate holes in large termite nests above ground, and deposit their eggs there. The termites promptly fill in the holes, leaving the eggs securely covered. The eggs then incubate in the uniform warmth of the termite house.

When the embryo of an oviparous lizard is fully developed, it slits the egg shell with a special tooth, called the egg tooth, which is shed shortly after hatching. This tooth is single in all oviparous lizards except the geckos, which have two such teeth. Viviparous lizards usually do not develop egg teeth; the young lizards rip the fetal membrane with their claws.

LIZARD FAMILIES AND SPECIES

There are approximately 20 families of

Fence lizards hatching from their leathery egg cases.

living lizards. Here is a brief description of some important examples.

Geckos, or gekkonids, make up the family Gekkonidae. They are mostly small nocturnal animals with soft skins. They are found particularly in eastern Asia, Africa, and Australia, though a few species are native to the Americas.

The fingers and toes generally have dilated adhesive pads, enabling these lizards to climb easily on smooth surfaces. The geckos can make their way up walls or even walk upside down on the ceiling of a room. They are often found in houses. They also live in trees and on the ground, in crevices, and in burrows. Most geckos have a voice, which ranges according to the species from a faint chirp to a loud squawk.

Skinks, or scincids, form one of the largest lizard families, the Scincidae. They are found in great numbers in Australia, Africa, Asia, and various islands in the western part of the Pacific. There are also a few species in the New World.

These lizards have flat, overlapping scales that give them a smooth and shiny appearance. There are many variations in structure. Some skinks have well-developed limbs with five fingers and toes. Others have no legs at all. There are many intermediate stages between these two extremes.

Certain skinks are fair-sized. For example, the blue-tongued skink, *Tiliqua scincoides*, an inhabitant of Australia, sometimes attains a length of 60 centimeters. Other skinks are small.

Lacertids are common lizards of the Old World. They are land-dwelling animals, with well-developed legs and long tails. Most European lizards belong to this family, the Lacertidae. One of the commonest European species is the wall lizard, *Lacerta muralis*, found in the central and southern part of the continent.

Teiids. The Teiidae are an American family of lizards, resembling somewhat the lacertids of the Old World. Most of them occur in South America. A few are found in the southern part of North America.

One of the largest species is the black tegu, or teju, *Tupinambis nigropunctatus*, which lives in the tropics. It sometimes attains a length of one meter. Other large representatives of the Teiidae are the crocodile lizard and the caiman lizard, mentioned earlier. In some of the smaller teiids, the limbs have been greatly reduced.

Worm lizards (family Amphisbaenidae) are so-called because of their resemblance to earthworms. They have concealed eyes and are without external ear openings. Almost all species lack external limbs. They burrow in damp, sandy places, building underground galleries, and rarely come to the surface.

Monitors are large lizards that comprise the family Varanidae. All members of the genus *Varanus*, they dwell in southern Asia, Australia, Africa, and parts of the Middle East. The limbs have five digits each, provided with powerful claws. Almost all varanids are as much at home in water as on land.

The largest monitor—as a matter of fact, the biggest of all lizards—is the Komodo dragon, *Varanus komodoenisis*. It is found on Komodo, an island of the Indone-

Top: a blue tongue. This skink is a resident of Australia. Middle: a two-legged worm lizard from Mexico. Bottom: the poisonous Gila monster.

sian group, and on several other islands of that group. It has a long forked tongue, strong limbs, and a long tail. Another large varanid is the Nile monitor, *Varanus niloticus*, which may reach a length of almost two meters.

Poisonous lizards are found in the family Helodermatidae. There are two species: the Gila monster, *Heloderma suspectum*, of the southern United States and northern Mexico and the beaded lizard, *Heloderma horridum*, of Mexico. The Gila monster has a heavy body, marked with black stripes against a yellowish or pink background, and small legs. The beaded lizard is somewhat larger than the Gila monster. It is named the "scorpion" (*escorpión*, in Spanish) in the Mexican regions where it occurs.

Anguids are members of the family Anguidae. Some, such as West Indian lizards of the genus *Diploglossus*, have well-developed legs. Members of the genus *Ophisaurus*, which occurs in North America, Europe, and southern Asia, have only slight or no vestiges of limbs and are quite serpentlike. Hence the family name Anguidae: *anguis* means "snake" in Latin. The American and European species of *Ophisaurus* are called glass snakes, because the tail is so fragile that it will break off into several fragments if the reptile is handled roughly.

Iguanids, members of the family Iguanidae, are found chiefly in the New World. They resemble the lizards of the Old World family Agamidae. Most iguanids have small scales, often a crest on the back and tail, and sometimes a pronounced dewlap (a flap of skin hanging from the throat). The common iguana, *Iguana iguana*, which dwells in Central and South America, is the largest species, reaching a length of up to 2 meters. An arboreal creature, it favors trees that overhang the water. The marine iguana, *Amblyrhynchus cristatus*, an inhabitant of the Galápagos Islands, is about a meter long.

The tropical American basilisk, *Basiliscus americanus*, is a rather odd-looking member of the iguanid group. Its body is compressed horizontally, and it has a very

A double-crested basilisk, a member of the iguana group. This lizard can run across water.

long tail that looks much like a whip. In the males there is an unusually large crest. The basilisk is able to run across the surface of the water on its hind legs, using its long tail to balance itself.

Some familiar members of the iguanid family are quite small. Among these is the fence lizard, *Sceloporus occidentalis*, a spiny lizard found in the western United States. It has overlapping scales, with a small spine extending from the middle of each scale.

The horned lizard, *Phrynosoma cornutum*, is popularly known as the horned toad. It is a ground-dweller. Its colors generally blend remarkably with those of the background. The horned lizard occurs chiefly in the deserts of the United States and Mexico, though it is found in various other places. It is well protected by its armor of enlarged scales, or spines. The name "horned lizard" is derived from the hornlike protuberances on the back of its head. When disturbed, this lizard may squirt a jet of blood from the eyes, sometimes for a distance of 60 centimeters.

Other familiar iguanids are the anoles, genus *Anolis*. They are found in the tropical and semitropical areas of North and South America and especially in the West Indies. One species is sometimes called the American chameleon because, like the true chameleon, it is able to change its color. It is not closely related to the true chameleon.

Agamids, or members of the family Agamidae, are widely distributed in the Eastern Hemisphere and include some interesting forms. The flying lizards, or flying dragons of the genus *Draco*, living in southeast Asia and nearby island regions, are able to glide from branch to branch and from tree to tree.

The frilled lizard, *Chlamydosaurus kingi*, is a tree-dweller found in New Guinea and northern Australia. Its frill consists of a large expansion of skin, supported by rods of cartilage, on each side of the neck. Generally the frill is folded back on the neck and shoulders. When an enemy approaches, the frill is erected. At the same time the lizard opens its mouth wide. This awesome display often scares away the foe.

Another curious inhabitant of Australia is the moloch, *Moloch horridus*, a small, short-tailed animal. The moloch has prominent spines, particularly on the head just above the eyes.

True chameleons comprise the family Chamaeleonidae. Particularly noted for their ability to change color, these lizards are found in southern India, Sri Lauka, The Middle East, Africa, and Madagascar. Most members of the family belong to the genus *Chamaeleo*.

A representative species is the common chameleon, *Chamaeleo chamaeleon*. The body of this lizard is compressed from side to side. The eyes are set in turretlike structures and can be moved independently of each other. Each eye rotates in many directions until the lizard sights its prey — generally an insect. Then both eyes are focused on the victim. The long tongue is shaped like a club and has a sticky secretion at the tip. The chameleon approaches its prey slowly. When it is within striking distance, the long tongue suddenly shoots out and is as suddenly withdrawn, with the insect adhering to the sticky tip.

SNAKES

Too many people think of snakes as slimy, ugly, dangerous creatures. But snakes are not slimy. Nor are they ugly. Many are strikingly handsome, with stripes, bands, and blotches of brilliant colors. Many have gracefully tapering bodies.

There are nearly 3,000 known species of snakes. Somewhat less than one third of the snake species possess poison glands, and only a small number of these are dangerous to people. Even these generally tend to shun humans, attacking or biting only when they consider themselves endangered.

Many species of snakes are reckoned among our most valuable allies. Some prey on rodents that destroy grain crops and other vegetation. Others feed on insect pests. Others eat slugs and snails.

A North American pit viper — a diamondback rattlesnake and its young.

HOW SNAKES EVOLVED

Snakes are reptiles belonging to the order Squamata. They are closely related to lizards and more distantly akin to crocodiles, alligators, turtles, and the curious tuatara of New Zealand. Apparently they have descended from some early lizard population.

The oldest known snake fossil comes from Cretaceous deposits dating back some 100,000,000 years. This snake was as specialized as those of today, so it tells us nothing of the early history of snakes.

Some zoologists believe that the snakes' ancestors began to inhabit country where high grasses flourished. Here, legs would be almost useless for quick and easy passage through the dense vegetation. Legs were gradually lost, and the snakes became something like the so-called glass snakes of today — elongated lizards that have no limbs.

According to another theory, the ancestors of snakes were underground burrowers and their legs degenerated because of their mode of life. It is believed that the eyes of these burrowing creatures also degenerated. In time, when the snakes again began to live above ground, their powers of vision re-evolved. In support of this theory, scientists point out that the eyes of snakes are quite different from those of lizards and other animals.

BODY FEATURES

One of the most obvious features of snakes is their lack of limbs. Some of the primitive snakes still possess internal traces of hind limbs and pelvic, or hip, bones. All other serpents, however, entirely lack such vestiges. No snake has a breastbone for the attachment of forelimbs.

The skull is of rather loose construction to allow the upper and lower jaws great freedom of movement. The two halves of the lower jaw do not fuse in front but are jointed by an elastic ligament. Numerous needle-sharp, backwardly curved, replaceable teeth stud the jaws.

As the body became longer in the course of the snake's evolution, the internal organs were considerably affected. The heart, for example, became an elongated organ. The gall bladder, instead of being embedded in the liver as is usual, came to lie far behind it. One lung degenerated to leave room for the other functional one.

Snakes are covered with horny scales. The top of the head may have large scales, called *head shields,* or smaller scales. There may also be large scales, known as *chin shields,* on the underside of the head.

An Indian rock python, one of the largest snakes, grows to a length of over 7 meters. Pythons closely resemble boas.

Big, straplike scales, or *scutes*, run across the belly. Scales may be either smooth or keeled—that is, with a ridge over the center of each.

The color of scales is derived from numerous tiny bodies called chromatophores. Some chromatophores contain pigments that are responsible for blacks and browns. Others lack pigment but break up light into its various colors much as tiny prisms would.

Ground-living snakes that inhabit forested areas often have patterns of blotches or bars that blend with the leaves and sticks of the forest floor. Some arboreal snakes are green. Others are brown or darkly spotted. The desert dwellers generally have an inconspicuous pale hue.

WELL-DEVELOPED SENSES

Because snakes are active predators, or hunters, they need keen senses. Vision seems to be very acute in all but the burrowing types. The lens of the eye focuses accurately on objects at various distances within a comparatively limited range. This is a decided asset for the hunter as it stalks its prey. The eyes have no lids but are capped with a transparent scale. Snakes that hunt by day have rounded pupils. Night roamers have vertically oval pupils, like those of a cat.

The sense of smell is also highly developed. A snake trails prey and recognizes others of its own species by means of this sense. The forked tongue helps the animal to smell. As the tongue flicks out, it picks up tiny particles on its tips. The tongue carries these particles to two small pits lying at the front part of the roof of the mouth. Branches of the olfactory nerve lead to these pits, which are lined with sensitive cells and form part of the smelling part of the nose. The tongue serves also as a delicate instrument of touch but seems not to have taste buds.

Snakes seem to lack a sense of hearing. There are apparently no structures for receiving air-borne vibrations. There are neither external ears nor ear drums. However, the animals can readily pick up vibrations transmitted through the ground.

HOW SNAKES MOVE

The lack of limbs does not seem to handicap a snake particularly in moving from one place to another. It can negotiate fairly smooth or rough terrain, swim well, and climb trees.

A snake may move by throwing its long form into horizontal waves, since its backbone is very flexible. Each curve pushes against obstacles—such as clumps of grass, shrubs, ridges on the ground, small stones, and so on—to force the body forward. Swimming is accomplished in this manner, too; the body curves push against the water.

Another type of motion is the "caterpillar" method: the snake crawls in a straight line by pushing the free edges of its belly scutes, which are powered by strong muscles, against irregularities on the ground. Scutes are useful, too, in climbing. So are the keeled scales found along the lower sides of the body of certain tree snakes.

A third type of motion is a hitching gait. The snake thrusts its front end forward, holds it steady, and then pulls the rear end along.

Desert-living species, such as several African horned vipers and the sidewinder, or horned rattlesnake, of North America, progress by a method called sidewinding. The snake passes with ease over flat, sandy ground in a sidewise direction. As it moves, the animal's body contacts the ground only at several places at any one moment, each part of the body flowing through an S-shaped curve above the ground as the body is transferred from one of these contact points to the next.

The apparent speed of a snake is deceiving. Actually the fastest snakes probably cannot move more than five or six kilometers an hour, and this speed cannot be maintained for any length of time.

ATTACK AND DEFENSE

Snakes capture and subdue their prey in various ways. Sometimes the animal simply grasps the victim in its jaws and swallows it immediately. Such prey as insects and their larvae, worms, snails, fish, salamanders, frogs, and toads are easily handled in this way. Other prey is held to the ground until its struggles cease by a loop of the snake's body.

Some snakes subdue the prey by constriction. They throw several coils around the victim's body and begin to squeeze. As the prey exhales, the coils tighten so that fresh air cannot be breathed in. Death comes quickly to the victim. The constrictor does not break bones nor crush its victim to a pulp.

The most striking manner of killing, perhaps, is by poison. Venom is introduced into a wound by poison-conducting teeth, or *fangs*.

The size of a snake roughly determines the size of the prey it will eat. However, animals several times as large as the serpent's girth are readily swallowed. The snake's teeth firmly seize the victim. Since these teeth are curved backward, the only path the prey can take is inward—that is, down the snake's gullet. The lower jaw of the snake is loosely attached to the skull so that the mouth can be held wide open. The throat is very elastic and each side of the upper and lower jaws can move independently. The snake literally draws its mouth over the prey. Each side of the upper and lower jaws in turn releases its hold, moves forward a little, and then holds fast again. The snake solves the difficulty of breathing while struggling with such a mouthful by thrusting out the end of its windpipe between the separated halves of the lower jaw.

A snake can go without a meal for a considerable period of time. Since it is less active than an energetic mammal, it does not burn up its food so quickly. But it cannot do without water for any length of time. It drinks by thrusting the lower part of its head into the water and sucking up the water by contractions of the throat, in much the same way as a horse does.

If threatened with danger, a snake displays the defensive behavior peculiar to its species. The great majority of snakes will not attack people; cobras are a notable exception. The harmless black snakes will occasionally threaten an attack when disturbed during the mating season.

Snakes produce an evil-smelling secretion in the anal glands situated in the base of the tail. An annoyed serpent may pour out this secretion and smear it about. In this way it discourages a predator from attacking.

HIBERNATION

Unlike a bird or mammal, a snake cannot keep its body at a constant temperature. Usually its body heat is about one degree above or below the temperature of its surroundings. Consequently, a snake does not live long when exposed to the direct rays of a summer sun or when the temperature drops below the freezing point.

Female eastern garter snake and young. The young are born alive. The eggs hatch within the mother's body.

New York Zoological Society

In temperate regions, snakes avoid extreme cold by hibernating. During early autumn they seek out a place where they will be protected from winter's freezing temperatures. This may be in a den on the side of a mountain, reached by fissures in the rocks, or it may be under a pile of rocks or brush or in the base of a hollow tree.

Often, several species hibernate together, as is the case with the copperheads, timber rattlesnakes, and pilot black snakes of the northeastern United States.

MATING AND REPRODUCTION

During the first warm days of spring, snakes begin to emerge from their winter quarters. They do not roam far at first, for the chilly spring nights send them back to their retreats. Before the snake population spreads too far away from the hibernation area, the males seek out the females for the purpose of mating.

It is rather difficult to tell the sexes apart by their external appearance. Generally the females are longer and plumper. The males have a longer tail, and the tail has a stouter base.

Though many snakes are in close association during the mating period, males normally do not try to intimidate one another. By way of exception, however, two male European vipers will often fight over a female. The males face each other; each holds its head and fore part of the body erect. Then they push and thrust at one another trying to force the opponent to the ground. There is much swaying of the body during this contest but no biting. At last, one of the vipers gives up and dashes away quickly, pursued closely by its victorious opponent.

Apparently male snakes discover the whereabouts of a female by the sense of smell. The female gives off an odorous substance from her skin as she crawls along. The male picks up the trail in his wanderings and follows it.

Courtship behavior varies with different species. The male water snake, for example, rubs his chin along the back of his mate; so does the male garter snake. The whip snake male gives chase to the female, which dashes over rocks, through ponds, and over bushes. When the male finally overtakes his mate, each crawls through the coils of the other. Then each animal raises the front part of its body, holding its head close to that of its mate, but keeping the portion below the head well away. The pattern formed thus by the animals is strikingly like the shape of a lyre. The snakes maintain this figure as they move. This dance may last for more than an hour before mating occurs.

Female snakes either bring forth their young alive or lay eggs. Those that produce live young do so during late summer or early fall. Some species retain eggs until they hatch inside the body. Others seem to nourish their embryos, or developing young, more or less after the fashion of mammals. The gestation period, or time between fertilization and the birth of the young, is not known.

The temperature of the female's body, which is approximately that of the environment, influences the development of the embryo. When the European viper lives in the extreme northern part of its range, the period of gestation is two years. It may be a single summer season when the animal lives farther to the south.

Female snakes may produce young several years after mating has occurred by delayed fertilization. Specialized receptacles in the female's reproductive tract store the male's sperm cells until they are needed for fertilizing the ripe eggs.

Each young of live-bearing species emerges from the mother's body enclosed in a thin membranous sac. This sac is ruptured by the newborn snake, which is now completely capable of taking care of itself. The mother shows no interest whatsoever in her offspring.

Egg-laying snakes lay their eggs during the summer in places where moist heat is available, such as manure piles, rotting vegetation, and holes in embankments. Most snakes abandon the eggs after laying them. Female pythons, however, coil about their eggs to incubate them. The python's body temperature is several degrees higher at this time than is normal under other conditions.

Snakes eggs are white or cream colored. The shells are composed of several layers of tough, threadlike fibers. The young snakes usually hatch out during the late summer. Each little snake cuts its way through the shell by means of an *egg tooth* at the front of the upper jaw.

The number of young produced by snakes varies from species to species. The tiny De Kay's snake bears about 14 live young; the red-bellied snake about 7. The worm snake lays 2 to 5 eggs, and the ring-necked snake, 1 to 7 eggs. Certain snakes are exceedingly prolific. A female common garter snake is recorded to have given birth to 78 young at one time, and a common water snake to 76.

GROWTH AND MOLTING

Apparently snakes keep on growing as long as they live. Young snakes generally double their length in the course of the first year of life, but the rate of growth diminishes as the animal matures. The amount of food eaten and the temperature of the surroundings have much to do with growth. Snakes that hibernate grow little, if at all, during the dormant period.

As snakes grow, they periodically shed, or molt, their skins. The serpent actually crawls right out of its skin, which peels off inside out. This skin represents an old layer of cuticle, or outer skin, that has been replaced prior to shedding by a new cuticle layer.

About ten days before molting, the snake's eyes become milky or smoky in appearance and the body colors are dull. Then the eyes clear again. The cuticle begins to peel back from the edges of the lips. The snake rubs its head against stones, bark, or other rough surfaces to help push the skin back over the head. By muscular contractions and more rubbing, the snake quite quickly frees the rest of its body from the skin. It emerges with fresh and brilliant colors because of the transparency of the new cuticle.

The molted skin shows every surface detail of the snake's body, including the caps that covered the eyes. Often the species of snake can be identified by these details.

It is difficult to determine accurately how long a snake lives in its natural environment. Records of captive specimens indicate that the anaconda lives 28 years, the common boa 23 years, the common viper of Europe 22 years, the water moccasin 21 years, the cobra and rattlesnake 13 years, and the garter snake 11 years.

WHERE SNAKES ARE FOUND

Snakes are found throughout the tropical and temperate regions of the world. Only the extreme northern and southern parts of the globe, the Hawaiian Islands, Ireland, and New Zealand are free of native varieties. Snakes are hardy animals and will live wherever they are able to hibernate underground below frost level during winter. The tropics contain the greatest number of serpents, but temperate regions also contain many species.

Some snakes are burrowers, spending the greater part of their lives underground. Their heads are wedge shaped or conical for penetrating the earth. The ground-dwelling snakes may be either heavy bodied or slender in shape. They often go underground by way of crevices in rocks or burrows made by other animals. They can also climb bushes and trees and swim in case of necessity.

The water snakes favor ponds and lakes and never roam far from fresh water. Tree snakes prefer an arboreal habitat. Some kinds have long and slender bodies, almost wirelike, by which they span the gap from one branch to the next. Others are heavy and short and support themselves by a prehensile, or grasping, tail, which can wrap firmly around a branch. The sea also provides a home for some serpents. These are fine swimmers. They have a vertically flattened tail, shaped like an oar.

Though serpents come together when they hibernate, they generally lead a solitary existence except during the courting period. When this is over, each snake goes its own way, roaming over a fairly well defined territory.

NONPOISONOUS SNAKES

Blind snakes. These serpents make up two families. Members of the family Typhlopidae are small, burrowing serpents, widely distributed over the New and Old World tropics. They look much like worms, having a blunt, rounded head and a short, stumpy tail. They possess rudimentary hip bones and rear limbs, which are hidden beneath the skin. The upper jaw carries very tiny teeth. The eyes are small and are concealed beneath scales. The body scales are smooth and shiny. These creatures represent the only serpents on many isolated islands.

Typhlopids live underground, eating worms and insect larvae. Sometimes they invade termite nests and feed on the inhabitants. They lay eggs in underground burrows.

Members of the family Leptotyphlopidae are quite similar in appearance and habits to the typhlopids. However, they have teeth only in the lower jaw. These blind snakes live underground in semiarid areas where some moisture occurs and rocks and boulders are plentiful. Sometimes workers find them while excavating for pipelines and house foundations.

Leptotyphlopids, which use the head to burrow, come to the surface in the early evening and crawl about flicking out the tongue and aiding their motion by pushing with the spine at the end of the tail. They are entirely harmless, never attempting to bite when picked up. Most members of this family are tropical. They are found in the Americas, Africa, and southwest Asia.

Boas. Serpents of the family Boidae are strikingly handsome creatures, varying in size from very large species to small ones not exceeding 60 centimeters in length. They are, for the most part, arboreal creatures with prehensile tails. A few are ground dwellers and burrowers.

Boas are primitive snakes possessing two well-developed lungs and vestiges of pelvic bones. Hind limb bones are also present. They end in spurlike appendages that emerge from the side of the body at the base of the tail. The male uses these spurs to strike the back and sides of the female during courting. Boas have large curved teeth in both upper and lower jaws. They are strongly muscled animals. They kill their prey—birds and mammals—by constriction, coiling around the prey animal so tightly that it is unable to breathe and dies of suffocation. The young are born alive.

The largest boa is the South American anaconda, *Eunectes murinus,* which may grow to a length of 7½ meters. Semiaquatic in habit, this snake preys upon water fowl, young tapirs, and large rodents such as agoutis and capybaras.

The common boa, *Boa constrictor,* of South America is a tree dweller measuring only 3½ meters at most. Close relatives of this serpent inhabit Central America and Madagascar. The Central American species will strike repeatedly and hiss loudly when annoyed.

Another group includes the slender-bodied Cuban boa, the Bahaman boa, and the beautifully iridescent rainbow boa, *Epicrates cenchris,* which lives in Central and South America.

The sand boas *(Eryx)* of North Africa, southeast Europe, and southwest Asia and the rubber boa *(Charina)* of the western United States are burrowing, drab-colored snakes with small heads, tiny eyes, and stubby tails. The harmless California boa, *Lichanura roseofusca,* coils into a ball if disturbed.

Pythons. Snakes of the family Pythonidae resemble boas but have an extra bone in the skull above the eye. This family includes the longest snake in the world—the reticulated, or regal, python, *Python reticulatus,* of southeast Asia and the Malay Peninsula. A length of ten meters has been recorded for the species, but that is far above average. Close relatives are the Indian and African rock pythons, which grow to 7½ and 6 meters respectively. The diamond and carpet pythons of Australia have an average length of 2½ meters.

Africa has a small burrowing python *(Calabria)* and the ball python, which attains a length of about 1½ meters. The ball python curls itself up into a sphere if it is frightened.

Most members of the family Pythonidae prefer to roam at night. They are excellent climbers. They are generally found in the vicinity of water. Females lay eggs, which they brood. The larger pythons are capable of swallowing small goats and antelopes and medium-sized pigs.

Burrowing snakes. The false coral snakes (family Anilidae) of South America and much of southeastern Asia are a small group of harmless burrowing serpents with cylindrical bodies. They have vestigial pelvic and hind-limb bones and clawlike spurs at the base of the tail. These serpents may reach a length of almost a meter.

The shield-tailed snakes of southern India and Ceylon comprise another small family (Uropeltidae) of burrowers. They are related to the false coral snakes but lack hip and hind-leg bones. These are small snakes with moderately stout bodies; small eyes; smooth, glossy scales; and a flat shield at the tip of the tail.

A common water snake at home in the water. These snakes are excellent swimmers. They prey on fish.

Jack Dermid from National Audubon Society

The small ring-necked snake belongs to the Colubridae serpent family. This snake is a burrower that usually lives under stones and debris.

New York Zoological Society Photograph

Another burrowing snake is the mud snake, which is found in southeastern Asia. It belongs to the family Xenopeltidae.

River snakes. Snakes of the family Achrochordidae inhabit fresh and brackish water but often swim out to sea. The nostrils are set on the top of the snout; the body is heavy and rather flabby; and the scales are granular and do not overlap. One species is vertically flattened and carries a fold along the belly that functions as a fin.

These snakes are excellent swimmers and prey upon fish. A single species is native to Central America. The others live in southeastern Asia, the Malay peninsula, and nearby island areas.

"Common" snakes. The largest family of snakes—the Colubridae—includes the typical harmless Old and New World snakes, which occur in all types of habitats. The burrowers are cylindrically shaped, have blunt or cone-shaped heads and, usually, smooth scales. The serpents that roam about over the ground are generally long and moderately slender, with a distinct neck and a long tapering tail. Semiaquatic species are usually stout-bodied and the scales are keeled. Arboreal forms are exceedingly slender with elongated heads and prominent eyes.

The burrowers of the Colubridae family include the rainbow snake *(Abastor),* mud snake *(Farancia),* western ground snake *(Sonora),* hook-nosed snake, *(Ficimia),* hog-nosed snake *(Heterodon),* worm snake *(Carphophis),* and banded sand snake *(Chilomeniscus).* Usually they are secretive animals. They feed largely on salamanders, toads, frogs, earthworms, insects, and spiders. The small ring-necked snake, *Diadophis punctatus,* lives beneath stones and debris.

Among the more active members of the Colubridae are the green snakes *(Opheodrys),* the graceful black snakes, racers, and whip snakes *(Coluber),* indigo snakes *(Drymarchon),* patch-nosed snakes *(Salvadora),* leaf-nosed snakes *(Phyllorhynchus),* rat snakes *(Elaphe),* bull,

The Cuban racer snake. These active snakes are not poisonous.

Jack Dermid from National Audubon Society

The elongated corn snake, climbing up the trunk of a loblolly pine tree. Many snakes are good climbers, and some are wholly arboreal.

New York Zoological Society Photograph

The long-headed tree snake is an expert climber. It preys on young birds.

pine, and gopher snakes *(Pituophis),* and king snakes *(Lampropeltis).* These serpents are terrestrial, though most of them sometimes climb into bushes and trees. Many are nocturnal. All whose breeding habits are known lay eggs. The smaller species eat insects, frogs, salamanders, and lizards. The larger ones prey on rodents. The rat snakes, bull snakes, and king snakes constrict their prey. Bull snakes destroy great numbers of rodent pests; king snakes devour other serpents, both harmless and poisonous varieties. The indigo snake, pilot black snake (one of the rat snakes), and bull snake are the largest harmless serpents in the United States, growing to a length of 2 to 2½ meters.

The whip, or tree, snakes *(Leptophis)* of tropical America are completely arboreal. They appear almost like vines when they loop their coils around a branch. Lizards and tree frogs make up their diet.

The widely distributed Colubridae known as water snakes make up the genus *Natrix*. They keep close to freshwater streams and lakes, where they prey on aquatic animals. The females bring forth their young alive. Closely akin are the common garter snakes *(Thamnophis)* of North America. These are probably the most abundant of all serpents. They are found in a variety of localities, hunting for earthworms and frogs.

The De Kay's and red-bellied snakes *(Storeria)* are diminutive relatives of the water and garter snakes. The little De Kay's snake, which feeds on slugs, snails, worms, and insect larvae, occurs in areas that other snakes have long abandoned, such as vacant lots and backyards.

The Colubridae that we have just described show a variety of behavior when annoyed. The rainbow and mud snakes push the sharply pointed scale at the end of the tail against an attacker. The hog-nosed snake flattens its body, widens its head, hisses, and strikes repeatedly. If it cannot discourage an attacker in this way, the little snake writhes in apparent agony, rolls on its back, and goes limp, letting the tongue hang out of its partially open mouth.

Whip, rat, king, and bull snakes vibrate their tails. Some hiss fiercely. The leaf-nosed snake hisses and strikes from a coiled position. Water and garter snakes flatten the body and smear the ground with the ill-smelling secretion from their anal glands. Tree snakes threaten by opening the mouth and weaving the head from side to side. A number of these species bite, but the scratches they may make are harmless to people unless infection sets in.

Slug-eating snakes. Snakes of the family Amblycephalidae are unique arboreal serpents of tropical America and east Asia. Their heads are chunky and their eyes large. The bodies are slender and the necks exceedingly thin. Since their mouths cannot be opened wide, they feed exclusively on small prey, such as slugs and insect larvae.

The egg-eating snake. Dasypeltis scaber is the only species in the family Dasypeltidae. It feeds exclusively on eggs. An accordian-pleated mouth lining and a stretchable throat allow this little snake to swallow eggs that are three or four times the diameter of its body. Spiny projections, which are modified processes of the snake's neck bones, penetrate the gullet. As the egg goes down the gullet, it is cut open by these projections. The snake compresses its throat muscles and the egg contents go into the stomach. The egg shell is then ejected from the mouth. This African snake eats eggs voraciously during the bird-nesting season. It lives on stored fat the rest of the time.

POISONOUS SNAKES

The venomous serpents have developed a most effective method for subduing prey: they inject poison by means of enlarged grooved or tubular fangs. Modified salivary glands, which produce the poison, are located in the tissues of the head below and to the rear of the eye. Ducts lead from the glands to the poison-conducting fangs. The fangs are sometimes at the back of the mouth, sometimes at the front.

The venom that is injected through the fangs either destroys red blood cells and breaks down blood capillaries, producing hemorrhage, or works on the nervous system, affecting in particular the centers that control breathing and heartbeat. Rear-fanged snakes possess blood-destroying venoms, as do vipers and pit vipers. Coral snakes, sea snakes, and cobras and their allies produce nerve-affecting poisons. The venom of rattlesnakes affects both the blood and the nervous system of their prey.

The rear-fanged snakes, which are widely distributed, are generally only mildly poisonous. Since their grooved fangs, which lie at the rear of the upper jaw, are not easily brought into play, these snakes are not considered dangerous to mankind. Frogs, lizards, small birds, and mice are their principal fare. Prey is seized and held firmly well back in the mouth until the poison has had its effect.

Rear-fanged families. The largest family of rear-fanged snakes is the Boigidae. Certain members of this group closely resemble, in habits and appearance, the nonpoisonous racers, whip snakes, king snakes, and slug-eating snakes. The African boomslang, *Dispholidus typus,* has a very toxic venom. It threatens by inflating its windpipe, thus compressing its neck so that is spreads vertically. The "flying" snake, *Chrysopelea ornata,* of southeast Asia can expand its ribs and draw in the belly. In this way it produces a surface that allows it to glide downward from considerable heights. The tropical American mussurana, of the genus *Pseudoboa,* kills prey by constriction, often attacking and subduing the large pit vipers of the region.

The rear-fanged water snakes (family Homalopsidae) of southeastern Asia and nearby islands are stout bodied and have nostrils on the top surface of the snout. They are similar in habit to the nonpoisonous river and water snakes.

The rear-fanged egg-eating snakes (family Elachistodontidae) of India, like the colubrid *Dasypeltis,* have projecting spines in the throat for cutting egg shells.

Cobras and their relatives. Among the most deadly of the front-fanged serpents are the highly poisonous cobras, kraits, mambas, and coral snakes (family Elapidae). Most of them resemble racers and whip snakes in outward appearance. Their fangs are grooved teeth whose groove edges meet so as to form a tube. New fangs continually replace the older ones. Usually these snakes seize their prey and inject the venom into the wound as they chew.

The king cobra, *Ophiophagus hannah,* of southeastern Asia is considered the world's most dangerous snake, not only because its venom is so potent but also because it is agressive and large, attaining a length of 5½ meters. It feeds almost exclusively on other snakes. The female lays from 21 to 40 eggs in a nest made by pushing leaves or other debris into a pile. She remains close by to guard the eggs.

The Indian, or spectacled, cobra, *Naja naja,* is a jungle or open-field dweller, preying upon rodents, frogs, birds, and other snakes. When annoyed, the serpent erects the front third of its body and expands the larger ribs below the neck upward and outward. The skin stretches over this frame-

The familiar stance of the Indian cobra. When annoyed, the cobra spreads its spectacled hood. It often spits venom into the eyes of an attacker.

American Museum of Natural History

work to form the so-called *hood*. The snake hisses and occasionally ejects two streams of venom, apparently aiming for a foe's eyes. The ability to "spit" venom is due to the fangs, which are adapted for directing the venom forward and out of the mouth.

The African ringhals, *Hemachatus hemachatus,* and the black-necked cobra, *Naja nigricollis,* are even more notorious as spitters. The African water cobras *(Boulengerina)* are stoutly built and lead a semiaquatic existence, feeding on fish.

Kraits make up the genus *Bungarus.* They dwell in southeast Asia and are nonaggressive, never attempting to bite unless stepped on or handled. They have a small head, blunted nose, and smooth, lustrous scales. A ridge runs along the top of the body from neck to tail. These night hunters devour frogs, toads, small mammals, and other serpents.

The mambas *(Dendroaspis)* are very active, swiftly gliding serpents of Africa. The dark olive or green body is elongated and extremely slender. The head is narrow but distinct from the neck. The slender fangs are advanced farther than those of other Elapidae, and the venom is highly potent. Infrequently, perhaps at mating time, these serpents attack people. In habit they are semiarboreal, climbing through bushes and small trees in search of birds and hunting small mammals on the ground.

The coral snakes are shy burrowers, living in the American tropics and subtropics. Similar snakes are found in the Old World. Coral snakes are generally inoffensive creatures, but occasionally will bite. The head is small and blunt and the scales are smooth and glossy. The majority are ringed with black, yellow, and red. Lizards and small snakes form the diet of coral snakes.

The poisonous snakes of Australia all belong to the family of the Elapidae. The smaller forms are inoffensive or secretive, but three of the larger species—the black snake, tiger snake, and death adder—can be exceedingly dangerous. The black snake *(Pseudechis),* a 1½-meter-long racerlike serpent, spreads its neck horizontally when angered. The highly venomous tiger snake *(Notechis)* is quick to take offense and strikes savagely. The death adder *(Acanthophis)* resembles a viper, having a wide head and thickset body. These three serpents prey on lizards, frogs, and mammals, and bring forth living young.

Sea snakes. Snakes of the family Hydrophidae are aquatic serpents related to coral snakes and cobras. Though highly venomous, they are docile, inoffensive creatures. They use their venom to benumb fish prey, some species preferring eels.

The head is indistinct from the neck. The nostrils, which can be closed by a flap, or valve, so as to exclude water, are on the top of the snout. The tail of the sea snake is flattened vertically and resembles an oar.

One species—the yellow-bellied sea snake, *Pelamis platurus*—is found in tropical waters of the Western Hemisphere, occurring as far north as the Gulf of California. All other members inhabit Indian Ocean and western tropical Pacific coastal waters. Most Australian-area sea snakes lay eggs; all others bear live young.

Vipers. Vipers are Old World serpents that make up the family Viperidae. They are, as a rule, comparatively stout-bodied, with wide heads. The tubular poison-conducting fangs are attached to very short, hinged upper jaw bones. When the mouth is closed, the enlarged fangs lie in a horizontal position against the roof of the mouth. As the mouth opens in striking, the jaw bones rotate forward, bringing the fangs into full play. Muscles contract to squeeze venom into the fangs as they sink into the prey.

Of the European vipers, the common viper, or adder, *Vipera berus,* is the only poisonous serpent in the British Isles. It ranges farther north than any other known venomous snake, being found in Norway, Sweden, Finland, and Siberia.

Of the Asian vipers, the 1½-meter Russell's viper, *Vipera russelli,* is considered one of the more dangerous.

Africa is the headquarters of the vipers. The vividly patterned puff adder, gaboon viper, and rhinoceros viper, all of the genus *Bitis,* are outstanding representatives. They are chunky creatures with very broad heads and short, abruptly tapering tails. Other African vipers include the bush vipers *(Atheris),* which have prehensile tails; the slender and secretive burrowing vipers *(Atractaspis),* and the desert-living sand and horned vipers.

Normally vipers are nonaggressive; usually they hiss when threatening. They prey mostly on small mammals and lizards, but often eat toads and frogs.

Pit vipers. Snakes of the family Crotalidae have a special organ that opens by means of a pit, or chamber, on the side of the head below and in front of the eye. This organ, lined with sensory tissue, serves to detect the body heat given off by the snake's prey. Thus, it aids the snake in locating the direction and distance of the prey and directs the snake's strike.

Pit vipers usually strike the prey with a lightning thrust and then draw back to let the venom do its work. If the prey moves off to a distance before dying, the snake tracks it down through its sense of smell. Many pit vipers threaten a foe by coiling the body and arching the front end in an **S**-shaped loop. The water moccasin threatens by exposing its white mouth lining.

The majority of pit vipers inhabit the Americas, but a few species occur in southeastern Asia and nearby islands. The bushmaster, *Lachesis mutus,* which differs from other American pit vipers by laying eggs, is an agile, large serpent of the tropical American forests; it is 2½ to 3 meters in length. The fer-de-lance, *Bothrops atrox,* another tropical American species, is a common, highly venomous snake, which may attain 2½ meters. Relatives of the fer-de-lance include the arboreal palm vipers, the chunky jumping vipers, and the slender hog-nosed vipers.

Copperheads, water moccasins, and rattlesnakes are pit vipers that have received wide notoriety in the United States. The richly colored copperhead and the semiaquatic moccasin of the genus *Agkistrodon* are heavy-bodied serpents of the eastern states. The rattlesnakes *(Crotalus)* and pygmy rattlers *(Sistrurus)* are unique among snakes in possessing a rattle at the end of the tail. This curious device consists of a series of dry, horny segments built up, segment by segment, when the snake sheds its skin. When the rattler vibrates its tail, the segments strike against each other, giving the characteristic buzzing sound, or rattle. This is not intended to give warning, but is a display of annoyance.

TURTLES

Turtle, tortoise, terrapin—how do you tell one from the other? Traditionally, those species that live on land have been called tortoises. Those living in water have been known as turtles. And any species used for food has been called a terrapin. But as far as biologists are concerned, these are false distinctions. All the species are turtles. Period. Which makes life a bit simpler for the rest of us.

A SHELL OF BONY PLATES

The turtles belong to the reptilian order Chelonia. In most species there is a hard shell consisting of bony plates covered by a layer of tough scales. Underneath this plating lie the ribs. Since they serve only to stiffen the shell, they are found fused to the top shell, or *carapace*. They do not encircle the body as they do in most vertebrates.

The spine of the turtle has likewise undergone modification. In the region of the shell many vertebrae have been lost. Others have been fused into a single tube for the length of the carapace. With the development of the shell, the spine lost its supporting function and now serves solely to encase the nerve cord.

The presence of the shell results in other abnormalities. In all other backboned animals, the pectoral and pelvic girdles lie outside the main, or axial, skeleton. In the turtle, the growth of the shell is so rapid, and its influence on the ribs so great, that the normal order is reversed. The bones that support the limbs, fore and aft, are entirely enclosed within the central skeleton.

The rigid shell has also caused a modification in the breathing of the turtle. Reptiles, like other land vertebrates, normally breathe by expansion and contraction of the rib cage. In the turtle this is not possible. The function of expanding the lungs is taken over by the flank (side) muscles. In many aquatic turtles, sensitive patches of skin in the lining of the throat are specialized to absorb oxygen from the water. Although these patches do not supply the entire needs of the animal, they do enable it to stay submerged for hours.

COLD-BLOODED AND SLUGGISH

Turtles vary greatly in size. The weights of adults range from less than one-half of a kilogram for some species to 700 kilograms in the case of the giant leatherback turtles. In general, the heaviest species are aquatic. A huge body that would be unwieldy on land may prove agile when its weight is supported by water. In many species, the female weighs several times as much as the male.

In general, compared to other land vertebrates, the turtle is dull and sluggish. It engages in only a fraction of the activity of a mammal of comparable size. Its food and oxygen consumption are correspondingly small. For example, a 4¼-kilogram South American land turtle can reportedly satisfy its entire food requirement for four weeks with 180 grams of banana. Compare this with the amount of food consumed by a warm-blooded animal of similar weight, such as a small dog.

Turtles feed on the softer parts of plants and on small, slow-moving animal life. Lacking teeth, the turtles cannot eat tough, fibrous plants. And their slowness generally prevents them from actively pursuing fast-moving creatures.

As one might expect of such a sluggish animal, the turtle protects itself from its enemies passively. In some species, the turtle withdraws its body in its shell and clamps upper and lower parts together. The fact that its normal respiratory rate is so low enables it to survive in this suffocating position for hours. It outwaits its enemy.

It should be pointed out that turtles, like other cold-blooded animals, cannot survive in really cold climates. Since the body temperature of the turtle is usually about that of the outside air or water, a very cold environment will cause the life processes to slow down to a point where the animal is incapable of motion.

The mating habits of the turtle follow the usual reptilian pattern. Copulation is generally unaccompanied by a complex ritual, though some male land turtles pound their mates, using their bodies as battering rams. This is accompanied by loud roars, which, in some species, can be heard for almost one-half kilometer.

The eggs are laid shortly after copulation. The female then buries and abandons them. As soon as the baby turtle is hatched, it must be able to fend for itself.

CLASSIFICATION

The order Chelonia includes approximately 250 species of turtles, which are divided into two suborders: Cryptodira and Pleurodira. Most species are classified as Cryptodira. The snake-necked and side-necked turtles comprise the Pleurodira.

SEA TURTLES

Most sea turtles belong to the family Cheloniidae. The most commercially valuable species is the green turtle, *Chelonia mydas,* which is the main ingredient in turtle soup. Like many other sea turtles, green turtles are quite large, sometimes reaching weights of 225 kilograms. They are often found sunning themselves on beaches and rocky ledges. Despite the fact that they occur in virtually all the major ocean areas of the world, not much is known about them. Scientists have yet to determine their breeding grounds, migratory habits if any, egg-laying season, and other basic information.

The largest of the sea turtles is the leatherback, *Dermochelys coriacea,* which makes up the family Dermochelidae. Because of its tough, leathery hide, the leatherback is thought to be the most primitive of the modern turtles. The hide, in which pieces of bone are imbedded, is considered by evolutionists to be closely related to the original bony shell of the ancestral turtles.

Though basically tropical creatures, leatherbacks are found in all parts of the world with the exception of the arctic regions. At home in the water, they are exceedingly clumsy when they venture onto land.

As with most aquatic turtles, the female leatherback is particularly vulnerable when she emerges from the water to lay and bury her eggs. She camouflages herself by heaping sand upon her back as she starts to crawl out of the water and onto the beach. When the nesting spot is reached, a large hole is dug and the eggs deposited in it. After the eggs are buried the female paws over the surrounding area so that they cannot be found readily. She leaves a heavy, musklike odor in the vicinity; this causes the slight odor of the eggs to be masked. Once all this is done, the female heads back to the sea, leaving the young entirely on their own.

SOFT-SHELLED TURTLES

These odd creatures belong to the family Trionychidae. After having taken millions of years to evolve a hard shell and the anatomical peculiarities that go with it, the

Courtesy of the American Museum of Natural History

A common North American turtle—the snapper. This turtle is quite aggressive on land.

Trionychidae proceeded to lose the shell while retaining most of the characteristic turtle features. Why or how this happened is not completely understood, but we do know that the soft-shelled turtles are more active and mobile than the other groups. It is said that they strike with the speed of a snake and move with the agility of a mammal. Perhaps, in the past, the competition for food made speed and activity more important for survival than pure defensive ability, and so the protecting, but encumbering, shell was lost. To bear out this possibility, it should be pointed out that soft-shelled turtles are predominantly carnivorous.

The Trionychidae are freshwater turtles. Various species inhabit North America, Asia, and Africa. They are sometimes called *pancake turtles* because of their flattened shape.

FRESHWATER AND LAND TURTLES

Species of the families Chelydridae, Kinosternidae, Emydidae, and Testudinae are characterized by hard shells, relatively unspecialized limbs, and a straight withdrawal of the head into the shell.

One of the better-known North American members of this group is the common snapper, *Chelydra serpentina,* which is generally more aggressive than most turtles. According to popular belief, once it has seized a victim or tormentor in its jaws, it will not let go until sunset or until the sound of thunder is heard. Oddly enough, this turtle, which is so pugnacious on land, loses all its aggressiveness in water and will try only to swim away.

A close relative of the common snapper is the alligator snapper, *Macrochelys temminckii,* a large creature that may attain a weight of 90 kilograms. It is characterized by a large, blunt head and a particularly rough shell, with prominently peaked plates. Despite its fierce appearance, it is less active and combative than its smaller relative.

Perhaps the best known of all turtles, and the one most commonly kept as a pet, is the box turtle. Two types native to North America are the common, *Terrapena carolina* and the ornate, *Terrapena ornata*. Unlike many turtles, they are primarily active in the daytime. The box turtle is predominantly land-dwelling but seeks out water in very hot, dry weather. It will eat almost anything its jaws and throat can manage, including vegetables, meat, and insects.

A source of amazement to early naturalists were the Galápagos turtles, *Geochelone elephantopus,* found on the Galápagos Islands. Nothing approaching the size of these giant tortoises had ever been seen before by Europeans. They are closely related to the genus *Testudo,* which includes some 50 separate species. Many members of this widely distributed genus are found on a comparatively small number of oceanic islands. Apparently, isolation from the main stream of evolution helped them to survive.

Many *Testudo* species have been wiped out by mankind. Their large size made them a prime source of food and they could easily be caught because of their sluggishness. It was found that a *Testudo* laid on its back would survive for weeks without any attention. This enabled whalers and traders to transport them to markets thousands of kilometers away from their native islands.

SIDE- AND SNAKE-NECKED TURTLES

These turtles belong to the suborder Pleurodira, and are found in Africa, Australia, South America, and New Guinea. When one withdraws into its shell, the neck is pulled in doubled over to one side. The snake-necks are distinguished from the side-necked turtles chiefly by the length of their necks.

Galápagos turtles can weigh over 200 kilograms and live much longer than the average human does.

New York Zoological Society

CROCODILES AND THEIR RELATIVES

Crocodiles, alligators, caimans, and gavials have been on earth almost unchanged in form and habits since the great Age of Reptiles, more than 160,000,000 years ago. They are reminders of that far-distant past, when there were great numbers of fierce-looking armored animals that have long since become extinct.

In that remote, almost unimaginable era, the most important and powerful land animals were the "ruling lizards," or archosaurs. Among these were the dinosaurs, the crocodiles, and the reptilian ancestors of our birds. Although these animals had four legs, the hind legs were long and powerful and the front legs were short. This caused them to develop what is called *bipedal habits*. When they really wanted to get somewhere in a hurry, they would rear up and lumber along on their hind legs. They were heavily protected by thick, horny armor, just as are their descendants in tropical swamps and rivers today.

Modern crocodiles use all four legs to heave their cumbrous bodies along the ground. But their front legs are short, like those of the crocodiles that lived in the Age of Reptiles. When they swim they fold their legs close to their bodies and move through the water by means of powerful strokes of their tails. The hind feet are webbed in most species. Sometimes the forefeet are also webbed.

Crocodiles and other living members of the subclass Archosauria form the order Crocodilia. These animals are found only in temperate and tropical countries. The crocodile itself lives in parts of five continents: Africa, Asia, the tropical parts of North and South America, and northern Australia, as well as in the long chain of islands that extends southeastward from the Malay Peninsula. Alligators live in China and the southeastern United States. The caiman lives in Central and South America. The gavial, or gharial, is found primarily in India.

To the anatomist, the vital organs of the Crocodilia are of particular interest, for, like those of the birds, they are a reminder of the great dinosaurs. This is especially true of the arrangement and structure of the heart and lungs, which are highly specialized in these animals.

Group of American alligators basking in the sun.

N.Y. Zool. Soc.

LAYING EGGS ON LAND

Another characteristic that the crocodile order shares with the birds is that they hatch their young from eggs. A crocodile may lay from 20 to 90 eggs in one clutch. The eggs have a thick, tough shell and the young crocodile is equipped with a strong egg tooth at the end of its snout to enable it to chip its way out of the shell when the time comes. This egg tooth drops off almost immediately after the baby is hatched, and the youngster itself is at once completely able to cope with the world about it without any help from the parent crocodiles.

The laying of tough-shelled eggs on land was one of the very significant steps in the evolution of vertebrates. This habit freed the animals who practiced it from a complete dependence on water, thus making possible the conquest of land. It is true that the early amphibians could leave the water and remain away for some time, but they had to return in order to reproduce — that is, to provide a watery environment in which their eggs could develop. But when pioneering descendants of the amphibians began to lay their eggs on land, a new dimension was given to the life of vertebrates. The reptiles inherited the earth.

The tough-shelled reptilian egg contained its own watery medium, the *amniotic fluid*. This provided for the needs of the embryo just as the aquatic environment had provided for the developing young of amphibians. A large store of food, in the form of yolk, sustained the reptilian embryo. The porous shell allowed for the exchange of oxygen and carbon dioxide, yet kept the amniotic fluid from leaking out. When the development of the embryo was completed, the young reptile, a miniature of its parents, emerged from the egg.

Ironically enough, in the course of evolution, some reptiles returned to the water as a permanent habitat. Consequently, today we find the crocodilians, many turtles, and the sea snakes living out their lives in this medium. However, with the exception of the sea snakes that bear living young, these aquatic reptiles must go to land to lay their eggs.

LIFE IN THE WATER

The amphibious crocodilians are found in tropical rivers, lakes, marshes, saltwater lagoons, and the deltas of large rivers. In these habitats, they do most of their feeding and they are comparatively safe from people. They also spend a great deal of time warming themselves in the sun on river banks or sea and lake shores. Adult females may also be found in the vicinity of their nests, which they guard, presumably, against marauders.

On land these reptiles usually move slowly and rather clumsily. But when charging prey or an enemy or escaping danger, they travel at a good pace. They move with the body well off the ground on extended legs. Only the tip of the tail drags along the ground.

The crocodilian is protected by an armor of horny scales. This covering is reinforced on the back and in some species on the belly, too, by little bony *scutes,* or plates. Such a coat is all but impenetrable to the attacks of other animals and to mankind's more primitive weapons. Unfortunately, however, the high-powered rifle has proved more than a match for the crocodilian's tough hide.

In addition to its use in swimming, the powerful tail of these reptiles is employed as a weapon of defense. A well-placed blow of this appendage will snap a small tree and knock a large animal off its feet. The jaws and teeth, too, are formidable weapons, so strong that they can crush the leg bones of an ox.

In keeping with their aquatic habits, crocodilians are streamlined in shape, with more or less elongated and tapering jaws. The eyes are placed well up on top of the head so that the reptiles can scan the water's surface and edge while being almost completely submerged. The nostrils are elevated high on the tip of the snout. Breathed-in air follows a channel, above the bones of the roof of the mouth, to the windpipe. To keep its mouth open under water while foraging for and tearing up prey and yet not flood the air passage with water, the crocodilian has a flap of skin that closes off the mouth from this passageway. The American alligator can remain submerged three or four hours without drowning.

All kinds of animal food, depending on the environment, may make up a crocodilian's diet. Insects, spiders, crayfish, shrimp, crabs, snails and other mollusks, frogs, snakes, and turtles are preyed upon. The gavial's exceedingly long jaws and sharp, slender teeth are well adapted for seizing fish. Mammals, such as rodents, deer, and hogs, are grabbed by the snout when they come to the water's edge to drink. Ducks and other swimming and diving birds and unwary wading birds are also fair game. The decayed remains of animals may form a substantial part of the food.

The smaller animals are downed in one gulp. Large animals have to be torn apart in a unique way, since the crocodilian's flat jaws and spikelike teeth allow for no grinding or shearing action. The reptile firmly grasps its victim and then rolls over and over until the prey is twisted apart. (The bulk of the prey prevents it from rolling over together with the reptile.) Sometimes a large animal is killed and then hidden. It is eaten later, after aging has softened it so that it is easily broken apart for swallowing.

Both male and female crocodilians possess two pairs of scent glands, one pair located in the lips of the cloaca (discharge chamber), the other pair on the right and left sides of the lower jaw. During the breeding season, these glands produce a musklike secretion that, trailed on the ground or discharged in the water, presumably enables the individuals to find one another.

Mating takes place in the water, at least with the American alligator and probably in the case of other species as well. The eggs are deposited by the female in burrows that are then covered over, or in nests made of vegetation and sand or mud in the form of mounds. The females are said to guard these nests.

TELLING THEM APART

People often confuse alligators and crocodiles. Many times it is difficult to look at one of these animals and decide which it is. But there are some general differences.

The alligator has a broad, blunted snout, whereas the snout of a crocodile is narrow and tapering. When the jaws are closed, the fourth tooth of the alligator's lower jaw fits into a pocket of the upper jaw. This tooth in the crocodile slips into an exposed notch of the upper jaw.

The American alligator and the American crocodile are found together only in the extreme southern part of Florida. The American crocodile is usually much the larger animal, growing to a maximum length of 7 meters but averaging 3 to 3½ meters. The American alligator averages 2½ to 3 meters, though it may attain a length of 5½ meters.

ALLIGATORS AND CAIMANS

The 23 species of crocodilians are divided into three families. The family Alligatoridae includes the alligators and caimans.

Alligators dig burrows, or dens, which may extend back 12 meters from the water's edge. To these "gator holes" the reptiles retreat from danger and retire permanently during the winter. From October to late March, alligators eat little or nothing.

Alligators are not dangerous to people; human swimmers are safe in the waters in

The Nile crocodile is rarely aggressive on land, but it is very swift and aggressive underwater.

J. Six

Some people confuse alligators and crocodiles. The fellow at the top is a crocodile. You can tell by the narrow, tapering snout. The alligator below him has a broad snout. Alligators are smaller than crocodiles.

which these animals dwell. Yet a cornered individual or a female guarding its nest can be provoked into attacking. It is said that when a big alligator angrily crashes his jaws together, the sound suggests a stroke on a bass drum. Certain alligators in zoos have become fairly tame, but their keepers are always careful to keep at a reasonable distance from the animals' jaws.

The American alligator was formerly abundant in the coastal rivers and inlets from North Carolina to the Rio Grande and up the Mississippi and its tributaries as far as northern Louisiana. Now it is largely confined to the rivers and swamps of Florida and the Gulf Coast. Its numbers have been greatly diminished by hunting; fancy leathers made from alligator hides command high prices.

The Chinese alligator is also restricted to a small area: the basin of the Yangtze River. Comparatively little is known about this animal's habits.

The American alligator's breeding habits are rather peculiar. In April or May the female alligator seeks a sheltered spot on a bank and there builds a small mound. The foundation of this mound is of mud and grass; upon it the animal lays some eggs. She covers the eggs with a layer of grass or rotting vegetation and deposits some more eggs upon this. She proceeds in this way until she has laid between 30 and 40 eggs. The eggs are subjected not only to the scorching heat of the sun but also to the heat caused by the decomposition of the vegetable matter in the nest.

After a period of about two months the eggs hatch. It is said that a few hours before the young are ready to come out they make a squeaking noise, whereupon the mother pulls the mound to pieces, making it possible for the young to make their way out as soon as they are hatched.

Young alligators in the egg develop an egg tooth at the tip of the snout. With this they break through the enveloping egg shell. Soon afterward, this egg tooth disappears. As soon as they have chipped the shell, the baby alligators are led to the water by the mother, who provides them with food that she disgorges. At this early period of their existence, they are exposed to many dangers and are a favorite prey of fishes and turtles.

It has usually been stated that alligators grow very slowly. But certain specimens grow in five years to a length of 1.6 meters and a weight of 25 kilograms. The length of life of alligators is not known, but the big reptiles are probably not so long-lived as has been believed hitherto.

Alligators prey upon fish, water fowl, and such small mammals as go into the water or too near its edge, where the reptiles lie in wait. Land animals are dragged under water by alligators, to be drowned and eaten. If the prey is too large to be gulped down at once, it is torn to pieces before it is swallowed.

Caimans live in South America, where they are plentiful. Some species are also found in Central America. Like the alligators, caimans have stub-nosed snouts.

The common spectacled caiman, *Caiman sclerops,* is so named because between its eyes is a curved, bony ridge that resembles a pair of eyeglasses. It lives in sluggish streams where the bottoms and banks are muddy, and feeds on fish, crabs, and snails.

The smooth-fronted caimans form the genus *Paleosuchus*. They inhabit fast-running, rocky-bottomed streams. These caimans have the most complete armor of all crocodilians.

CROCODILES

Members of the family Crocodylidae are more widely distributed than alligators. They are found in North, Central, and South America, Australia, Africa, India, and the islands of the western Pacific. Crocodiles are the largest reptiles. The saltwater species may attain a length of more than nine meters.

Crocodiles are capable of moving quickly on land and can strike powerful and accurate blows with their tails. Young crocodiles feed chiefly on fish. Adult animals prey on water fowl and on land animals that come to the water to drink. In water where the temperature drops below 7° Celsius, these reptiles become inactive and practically helpless.

Certain crocodiles are notorious man-eaters. In 1926, C. F. M. Swynnerton, a game warden of Tanganyika (now Tanzania), exhibited to a London gathering of scientists the contents of the stomach of a crocodile that he had killed. Among other things there were eleven heavy brass arm rings such as natives wear, three coiled-wire armlets, a glass-bead necklace, fourteen human arm and leg bones, three spinal columns, and a length of fiber cord that had probably bound some bundle borne on the head of a Tanganyikan who had become a meal for the crocodile.

The 19th-century Scots explorer David Livingstone had many unpleasant experiences with crocodiles in the course of his African travels. He wrote how one of his bearers was suddenly seized by a crocodile while swimming in a river. The man kept his head. When the reptile pulled him down to the bottom of the river, he whipped out a knife and stabbed it behind the ear. Writhing with pain, the reptile released the man and swam off at a great rate. The man carried the marks of this encounter on his thigh for the rest of his life.

Ancient Egyptians worshiped the crocodiles that infested the waters of the Nile River. They tamed the animals and kept them in tanks at the temples, particularly at Thebes. When these sacred crocodiles died, they were preserved in mummy form as holy relics. Today, crocodiles are objects of veneration in certain parts of India.

THE GAVIAL

The family Gavialidae contains only one living species, *Gavialis gangeticus*. This animal has a very long, thin snout that is very useful in catching the fish that constitute its diet.

Gavials inhabit rivers in India and can also be found in Burma, Borneo, and Sumatra. It spends almost all of its life in the water, though the female will climb onto land to lay eggs—40 or more at a time—on the sandy river banks.

A gavial. The gavial has a long, thin snout that is particularly well adapted for catching fish.

THE LIFE OF BIRDS

The birds are among the liveliest creatures in nature. With hustling energy, they seek food, court their mates, build nests, care for their young, and sometimes make long migratory journeys. Since the time when they first evolved, they have invaded land and water areas in every quarter of the globe. They are now found in all continental regions from tropical rain forests to arctic tundras, from lowlands to mountaintops, on deserts, polar ice fields, and remote islands, and also far out at sea. In even the largest cities, we find pigeons, house sparrows, and various songbirds.

Birds are vertebrates, or backboned animals. They belong to the class Aves. There are roughly 9,000 known species.

EVOLUTION OF BIRDS

Birds have sometimes been called feathered lizards. To understand how appropriate this name is, we must consider the origin of birds in the remote past.

About 230,000,000 years ago, in the early Triassic period, there flourished a population of reptiles called archosaurians. They had developed the striking habit of walking on their hind legs, which were longer and more strongly developed than the front ones. Later, they evolved into different lines. One group took to walking on all fours and became amphibious types—that is, creatures equally at home in the water and on land. The modern crocodiles are descendants of this group. Other branches from the archosaurian stock gave rise to dinosaurs. Still another branch evolved into the birds.

Apparently the reptiles from which birds descended were small, carnivorous animals with teeth along the edges of the jaws. The narrow skull had the suggestion of a beak—a feature that was to become characteristic of birds.

The earliest "birds" that we know of were found as fossils in Bavarian limestone deposits going back to the Jurassic period, approximately 150,000,000 years ago. They belonged to the genus *Archaeopteryx*. These lizard-birds were about as large as pigeons and had an elongated body similar to that of a lizard. The body was covered with feathers, except for the head and neck, which had scales. (Reptilian scales still persist on the legs and feet of modern birds.) The tail was long and had a row of feathers down each side. There were teeth in both jaws, and the shape of the skull was more like that of a reptile than of a bird. The front limb, which apparently was used as a wing, ended in three clawed digits, or fingers. Possibly *Archaeopteryx* could fly to some extent, or at least glide.

Some theories have been proposed for the origin of flight and development of feathers. One is that the earliest birds were ground-living creatures that used their modified forelimbs to aid in running and eventually in taking off into the air. Another theory holds that primitive birds were tree climbers that adopted the habit of gliding from one tree to another or from

An albatross—often called the "king of the sky" because of its unsurpassed powers of flight.

G. Holton/PR

Although the ability to fly is characteristic of most birds, there are some that cannot fly. Ostriches, among the most primitive of birds, walk and run along the sandy grasslands of their African homeland.

a tree to the ground. The fact that *Archaeopteryx* had a long tail (probably used as a rudder) and clawed digits on the wings suggests that it lived in forests where it climbed, jumped, and glided among the branches. Another theory is that dinosaurs were warm-blooded and that feathers actually developed as a heat-protective device for small forms.

The aquatic birds called *Hesperornis* and *Ichthyornis* flourished in the Cretaceous period, some 80,000,000 years ago. Both possessed teeth, like the earliest bird forms, but their tails were short. *Hesperornis* was a large, loonlike diving bird that could not fly. *Ichthyornis* looked more like a gull. It had well-developed wings and a keeled sternum, or breastbone.

Other Cretaceous birds included a cormorantlike form and a primitive flamingo. By the end of this period, birds had evolved more typical bird features. The tail was short, the digits on the wings were reduced, the sternum of flying birds was strongly keeled, and teeth were absent.

In the Eocene epoch, about 50,000,000 years ago, there were herons, ducks, cranes, rails, grouse, sandpipers, and owls—types from which our modern birds were derived. It was not until 30,000,000 or 40,000,000 years later, however, that the truly modern forms made their appearance.

ADAPTATIONS FOR FLIGHT

Since many birds live at least part of the time in the air, their streamlined bodies and their wings have obviously been adapted for flight. The skeleton is extremely light. For one thing, many of the bones are hollow and are filled with air. Also, many bones found in other vertebrates are fused together in birds or are altogether absent.

Yet, light though it is, the bird's skeleton is very strong. The bony framework is rigid and suggests a ship's hull. The keeled sternum, below, is connected by the ribs to the fused backbone above. At the front of the sternum, several bones unite to form the shoulder girdle, which gives firm support for the strong wingbeats. The hipbones are fused with some of the

Birds of prey, including the osprey, are usually powerful fliers and quick to swoop down on prey.

vertebrae to form the pelvic girdle, or *synsacrum.*

The wing bones are similar to the arm bones of other vertebrates. In birds, we find the humerus (bone of the upper arm) and also the radius and ulna (forearm bones). The bones of the wrist and hand have been reduced in number, and some are fused. The outermost flight feathers—the *primaries*—are borne on several bones of the hand. The secondary flight feathers, or *secondaries,* are borne on the ulna; the so-called *alula,* or bastard wing, on one of the digits.

Each feather is made up of a central *shaft,* to which a great number of parallel *barbs* are attached on either side. The barbs are held together by rows of *barbules* jutting out at an angle from each barb. The barbules of one barb overlap and interlock, by means of tiny hooks, with the barbules of an adjacent barb.

HOW BIRDS FLY

Bird flight is based on the same principles as the flight of propellered, winged aircraft. A bird's wing is really an airfoil. That is, it has a rounded leading edge tapering to a thin trailing edge. The upper surface is convex, while the lower one is flattened or more or less concave.

When the wing is moving through the air, the airstream is deflected from the convex upper surface so that airflow is increased in speed and the air pressure directly above the wing is decreased. The airstream below the wing is only slightly deflected, and air pressure on the lower surface is therefore greater. As a result of these unequal pressures, the wing is forced upward, or given lift. When the leading edge of the wing is tilted upward a little, air hits the bottom surface more directly and lift is increased. However, if the wing is tilted too much, the lifting force ceases and the wing stalls. When a bird takes off or lands, the wings are tilted in just this way, but stalling is prevented because of the bastard wing. This is raised so that a slot forms between it and the main wing. The slot serves to increase the speed of the air flowing over the wing, sustaining its lift.

The secondary flight feathers, on the inner half of the wing, are principally responsible for lift. In order to maintain lift, however, the wing must be moving through the air. Forward thrust is provided by the primary flight feathers. The tail serves as a steering mechanism. It can move up, down, or sideways. When fanned out, it gives additional surface for lift. Fanning the tail also acts as an air brake when a bird comes in for a landing. Both the tail and wings maintain the bird's balance when in flight.

Flapping flight is the most common type. Usually, the wing beats downward and forward and then, more rapidly, upward and backward.

Most birds glide to some extent and some use this type of flight almost exclusively. Gliding is usually accomplished by holding the wings stiff and almost fully stretched out. The bird slowly loses altitude but maintains its flying speed because of the accelerating effect of gravity.

Soaring birds, such as vultures and albatrosses, can not only glide for hours at a time on extended wings but can even soar upward. This is because they make skillful use of rising air currents. We should point out that gulls, pelicans, gannets, albatrosses, hawks, and other soaring birds have a comparatively light body, supported by a large wing area. Hence birds such as these can take advantage of even moderate ascending currents.

Hovering flight is used by smaller birds, such as hummingbirds, when they seek or collect food. Over water, the hovering kingfisher holds its body almost vertical and beats its wings rapidly back and forth. The sparrow hawk will remain suspended over a patch of meadow, scanning it for grasshoppers. With head to the wind, the bird holds its slightly quivering wings outstretched and uses its tail for balance.

Both takeoff and landing are skillful operations. To take off, a bird must gain enough forward speed for normal flight. Large land birds such as vultures and eagles move rapidly along the ground to get their lifting speed. Swans, coots, and diving ducks run some distance on the water. On the other hand, pheasants, quail, and surface-feeding ducks, such as mallards, rise almost vertically with great speed. Many seabirds launch themselves from cliff edges. Large land birds often take off from treetop perches or the tops of rocks. A great many birds bend the legs and spring into the air.

When landing, birds often lessen their flying speed by gliding. Just before dropping, there are several forward wingbeats, the tail is fanned out, and the legs are low-

Feathers are not only adaptations for flight, but also serve other purposes—including ornamentation. Below left: the structure of a typical feather. Below right: closeup of a bird of paradise with its graceful feathers, and of beautiful eye-spotted feathers of a peacock.

Pelicans are well-adapted for their aquatic life. Their feet are webbed, making them more facile swimmers, and their long beaks enable them to probe for and pierce fish.

such as ducks and geese, travel 45 to 95 kilometers per hour. Certain birds, including peregrine falcons, swifts, and sandpipers, are capable of speeds of over 160 kilometers an hour in straightaway flight. Even greater speeds — up to 300 kilometers per hour — can be attained in "power dives" by certain hawks and eagles.

IMPORTANT EXTERNAL FEATURES

Feathers. The feathers of birds serve other purposes besides those of flight. The lightweight and relatively waterproof plumage provides a durable covering for the bird's body. It is very useful, too, in regulating heat loss from the body. On cold days, birds fluff out their feathers. The dead-air spaces within the fluffed plumage serve as additional layers of insulation in preventing heat loss. When the feathers are held close to the body, in warm weather, there is less insulation, and heat escapes.

Feathers play an important part in the courtship ritual of birds. The male often erects or spreads out various feathers as he seeks to dazzle a prospective mate. The

ered to take up the impact. Many birds slow down by using the energy of flight to carry them in a glide from a lower position to a landing spot above their line of flight.

Most small birds travel at speeds of 25 to 50 kilometers per hour. Larger ones,

VARIOUS KINDS OF BIRD'S LEGS

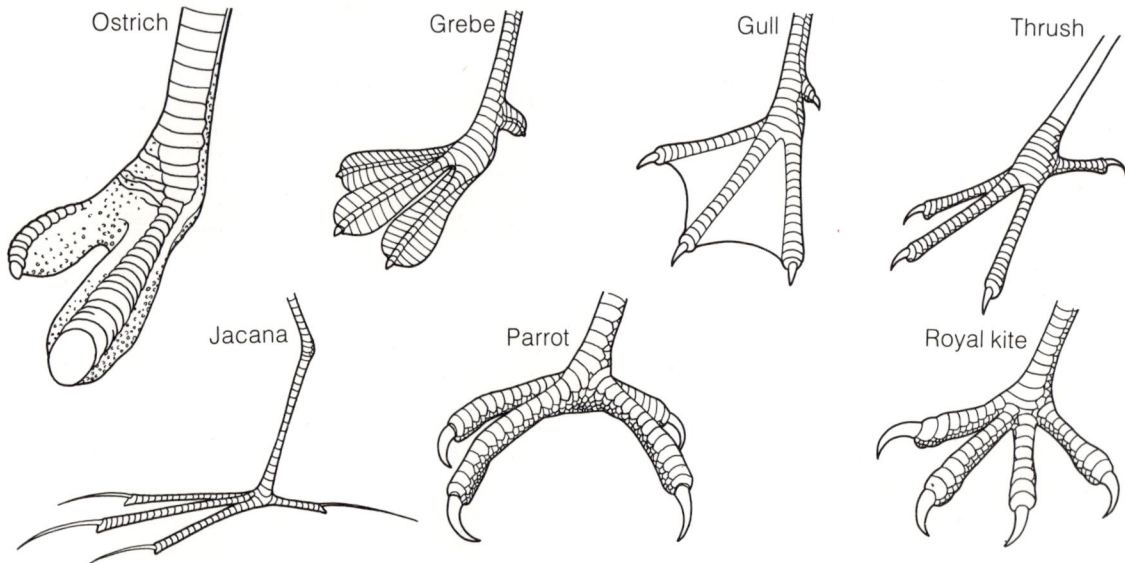

370 THE LIFE OF BIRDS

colors of bird feathers serve as a bond of attraction (and also recognition) between males and females of the same species. These colors are exceedingly varied and often brilliant. Feathers may also serve a protective function, helping the bird to blend with its surroundings or breaking up the bird's outline.

Soft down feathers, which have no central shaft, cover fledgling birds and occur as an undercovering on such birds as ducks. The feathers of young birds are replaced in time by adult plumage. Adult feathers are shed and replaced annually, usually in the fall. Many species also have different seasonal plumage.

Feet. The feet of birds tell us a good deal about their way of life. The feet of perching birds are adapted for grasping branches or other perches. In such birds, a tendon runs down from the thigh to the underside of each foot. As the bird perches on a branch, its weight causes the legs to be bent and holds the tendon taut. As a consequence, the toes are locked to the perch and the bird does not lose its hold even when it is asleep. When it is ready to fly away, it relaxes its claws by raising its body.

Woodpeckers have two toes forward and two toes to the rear, an adaptation suitable for clinging to and climbing vertical surfaces. The strong feet of birds of prey are provided with curved, pointed claws for killing and grasping prey. Ground dwellers, such as quail and grouse, have strong claws and fairly long toes, which enable them to scratch for food and walk on the ground. In the winter, grouse develop fleshy fringes on their toes to give a snowshoelike effect.

Ptarmigans, which live where snow often perpetually covers the ground, have their toes enclosed in feathers. Webbed or lobed feet are common among swimming and diving birds. Marsh birds that walk on mud or lily pads have extremely long-toed feet.

Beaks. The beaks, or bills, of birds are also modified in many ways. They are used, often together with the feet, for food gathering, nest building, and fighting.

Peter and Stephen Maslowski, Audubon/PR

Karl Maslowski, Audubon/PR

Many species of birds change plumage with the seasons. Top: the starling in its spring plumage; bottom: in its winter plumage.

THE LIFE OF BIRDS

DIFFERENT TYPES OF BEAKS

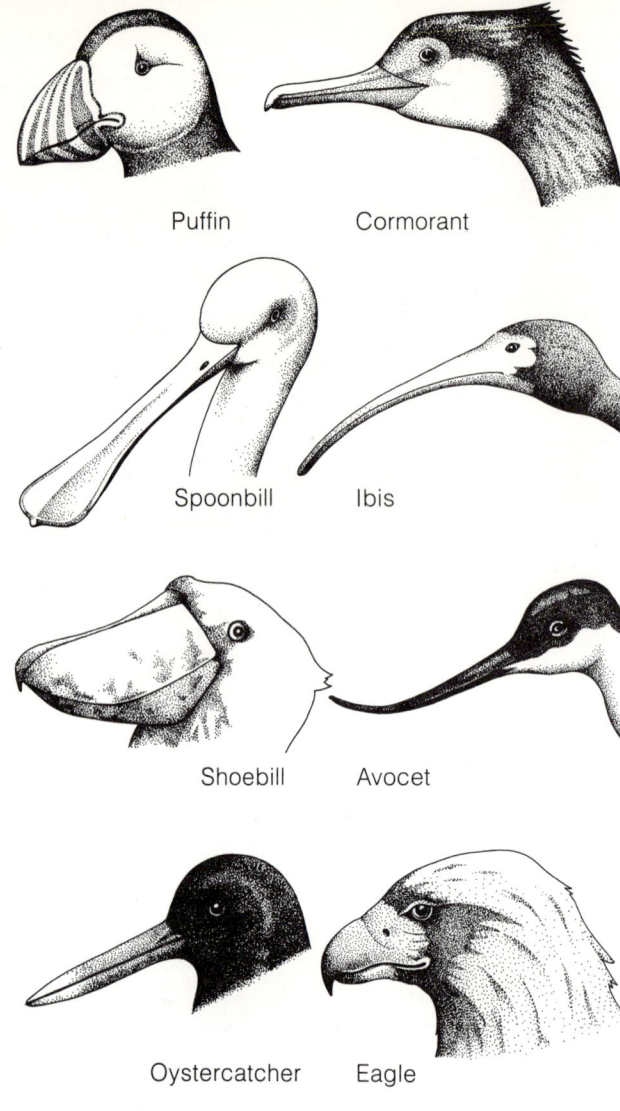

Egrets have long legs and long necks and dagger-like beaks—characteristics common to many marsh birds, and all adaptations for movement and food-getting in deep swamplands.

McWaren/Jacana

The short, curved, and sharp beaks of hawks and owls rend the prey. The seed-eating grosbeaks, finches, and sparrows have short, stout conical beaks for crushing seeds. The pickax beak of a woodpecker digs out wood-boring insects and excavates a nesting hole. The crossbill extracts seeds from pine cones with the crossed bill from which it derives its name. Hummingbirds have needlelike bills with which they take insects and sip nectar from flower blossoms. Shorebirds probe into water and mud with their long, slender, and often curved bills. Strainer plates in the broad, flat bills of ducks sift food from mud and water. Slippery fish are held by the saw-toothed edges of the merganser's bill.

INTERNAL FEATURES

The digestive system. In order to maintain its high body temperature, which ranges from 40° Celsius to 45° Celsius, and its high energy level, a bird eats a great deal of food. This is passed quickly through its digestive system. All that is useful is absorbed, so that only a minimum of waste is left. Since birds have no teeth, the digestive system is responsible for completely breaking up the food.

After food is swallowed, it passes by way of the esophagus to the *crop,* where it is stored and softened up. Below the crop is the stomach, which is composed of a true stomach, called the *proventriculus,* and, beyond it, a muscular *gizzard.* In the proventriculus, food is mixed with gastric juice before it passes to the gizzard. The gizzard grinds up the food by means of its horny lining, sometimes aided by sand grains and pebbles swallowed by the bird.

The food then passes through the intestine. At the junction of the small and large intestines are two blind ("dead-end") pouches, or caeca, in which water may be absorbed from the food. In vegetation-eating birds, cellulose is decomposed by bacteria in the caeca.

From the large intestine the waste products of digestion enter a chamber called the cloaca and then pass out through the cloacal opening.

The respiratory system. A bird's respiratory system is similar to that of other vertebrates. There are certain modifications, however, to guarantee a large supply of oxygen for the rapid burning of food. Within the lungs is an intricate branching system of air capillaries that act as membranes for the exchange of oxygen and carbon dioxide gases.

The hollow bones are connected with the lungs by means of a series of air tubes. A number of thin-walled air sacs, also connected with the lungs, are found in the body cavity and sometimes right under the skin. When these air sacs are inflated, the bird's body becomes very buoyant. The sacs make it possible to make the fullest possible use of the oxygen that is breathed in. They are also important for getting rid of excess body heat. Birds can perspire into them and so get rid of extra heat by way of the lungs and mouth.

At the base of the trachea, or windpipe, is a structure called the *syrinx*. Within it are membranes that are capable of vibrating, thus producing voice — that is, a bird call. The pitch can be changed through the action of various muscles.

The circulatory system. Birds have a very efficient circulatory system, shared only by the mammals, for quickly moving oxygen and nutrients to the tissues and getting rid of waste products. The comparatively large, four-chambered heart pumps blood throughout the body by way of the right aortic arch. (In mammals the left aortic arch serves this purpose.) The bird's heart beats rapidly. There are about 400 beats per minute in a crow, from 400 to 600 in a sleeping chickadee, and over 1,000 in a canary in a state of excitement.

The nervous system and sense organs. A bird's brain is similar in function and structure to that of other vertebrates. It is larger in relation to body size than the brain of all but the mammals.

The size and structure of the brain reflects the nervous activities that take place in birds. For example, a bird's sense of smell is poor; the olfactory area of the brain, having to do with smell, is poorly developed, as one would expect. The brain's

In most birds, the eyes lie on the sides of the head. Each eye has wide, independent, and very acute vision.

optic lobes, which receive impulses relating to sight, are large, indicating keen vision. The cerebral hemispheres, which are concerned in part with mating and nesting behavior, are well-developed, especially in crows, rooks, and parrots — birds that seem to show a considerable amount of intelligence.

The sense of touch is not acute, probably because of the protective plumage that covers the bird's body. The bill, however, is supplied with numerous nerve endings, making it a sensitive probing and exploring organ.

Only the nocturnal kiwi, which has poor vision, uses the sense of smell to locate hidden prey. Other birds depend on their eyes and their sense of hearing to detect food and enemies.

Taste buds are sparingly scattered over the palate and tongue. Many birds bolt their food, apparently relying little on taste. On the other hand, a number of birds, especially insect eaters and those having a varied diet, are quite particular about what they eat. This indicates that, in some cases at least, taste may be a governing factor.

Hearing is an important sense in birds. The hearing of owls is particularly acute, enabling these birds to locate their prey in the dark. The woodcock and robin seem to pick up by ear the underground whereabouts of earthworms. Birds can detect sounds varying quite widely in pitch. Mock-

Top: a blackbird giving food to its young. Bottom: a robin feeding a cuckoo. Cuckoos lay their eggs in the nests of other species and the foster parents feed the cuckoo young.

ingbirds, catbirds, and starlings must be able to distinguish between simple tunes, for they are good mimics of other birds' songs. Unlike the ear of mammals, which has three bones in the middle ear, the bird's middle ear contains only one vibration-transmitting bone.

Birds rely particularly on vision, which is perhaps better developed in them than in any other animal. The curvature of the cornea and the shape of the lens can be rapidly altered by special muscles so that the eye almost instantaneously focuses sharply on near or very distant objects.

Birds have good color vision. This enables them to locate food, to recognize other members of the species, and to distinguish the sex of each individual. Probably the development of colored plumage and color vision evolved together in bird evolution.

The eyes of most birds lie at the sides of the head. Therefore, birds have monocular ("one-eye") vision. Each eye has a wide, independent field of vision for scanning the ground, water, or foliage. However, there is a narrow range of vision toward the front, in which the bird's eyes function together to give binocular vision ("vision through two eyes"). This allows a bird to gain better distance perception for pursuing prey that is either flying or moving along the ground. Since the eyes of owls are directed forward, these birds may use binocular vision exclusively.

A unique membrane called the *pecten* projects from the retina of a bird's eye, forming a second eyelid. The pecten, a tissue plentifully supplied with blood vessels, apparently nourishes the eye. A third eyelid, the *nictitating* ("winking") *membrane,* can be drawn across the eyeball. These membranes are semi-transparent. They probably clean the surface of the eyeballs as they cover them. Perhaps they also shield the eyes of birds flying into the wind, protect the eyes of aquatic birds when swimming under water, and lessen the glare of light on the eyes.

SOCIAL ORGANIZATION AMONG SOME

When not engaged in reproductive activities, a number of birds of a species will often flock together. Food is frequently more easily and quickly found by group action than by a solitary individual. Escape from danger is more successful when there is a good chance that several members of the flock will spot the enemy and give the alarm reaction.

In some social species, there is a rigid system of dominance. The most aggressive individual lords it over the others of the group. The next bird in line dominates the rest of the flock but is submissive to bird number one, and so on. The term *pecking order* is applied to this sequence.

LEARNING ABILITY

Because of their highly developed nervous system, birds make use of past experiences in facing new situations. In other words, they have the ability to learn. This is true especially in situations concerning food. Birds will remember a locality where food is particularly abundant and will return to this place until the supply is exhausted; then they explore new localities.

Some birds have learned to utilize unusual sources of food. Birds of prey sometimes feed on dead animals killed by automobiles on the highways. Catbirds and mockingbirds will eat insects off the radiators of parked automobiles. Sparrow hawks may even watch pigeons feeding on bread and then join and feed with them.

In spite of this capacity to learn, much bird behavior depends on innate patterns handed down by heredity.

COURTSHIP AND BREEDING

At the beginning of the breeding season, males establish territories, which they jealously guard from other males of the species. Such territories apparently guarantee that the individuals of a given species will be evenly distributed over a suitable locality. Furthermore, a territory offers an isolated area where male and female can court and mate without interference. It gives the paired couple a place to find food, during the critical period of incubation and rearing of the young, without going far from the nest.

The territory-holding male sings lustily, or displays himself in specific ways, or both. Song or display is a threatening signal that warns other males away from the territory. At the same time, it is a signal for attracting unmated females.

Once a female is attracted, the male proceeds with its courtship ritual, which prepares the female for mating. Courtship, or sexual display, involves all kinds of movements and much flashing of color. Males dance, strut, and bow before their prospective mates. They fluff out their feathers, drag their wings, erect plumes and tail feathers, show off brightly colored feathers, and open the mouth to flash its highly colored interior. Some birds, especially meadow-nesting species, sing while performing a courtship flight. Gulls and terns feed the females in courtship; others present nesting material.

Courtship is not a one-sided affair. After the male's display causes a certain response in the female, this response brings a new reaction in the male and so on. Once they have mated, the birds set about constructing a nest. Perching birds sometimes make elaborate nests, using the bill and feet to gather, carry, and weave together nesting material. Other birds dig burrows into embankments, excavate a nesting hole in a tree, pile up a heap of sticks, scoop out a depression on the ground, or make no nest at all.

Birds lay hard-shelled eggs. These

Many species of birds use the sun's position to orient themselves as they travel long distances in their seasonal migrations.

Chaumeton/Jacana

range in number from one in certain species (gannets, petrels) to as many as 15 in others (quail, coots). Certain species raise only one brood a year. Others raise more.

Once the eggs are laid, the female does much of the incubating, but in some species the male may assume full responsibility. Often a special nest-relieving ceremony occurs when one mate relieves the other.

In the hatching process, the young birds peck their way out of the eggshell with the help of an *egg tooth* at the end of the bill. This is later shed. Some young birds, such as baby chicks, leave the nest an hour or two after hatching and never return to it. They follow the parent about and instinctively pick up insects or other kinds of food. The young of most species, however, are almost or entirely naked at birth and are helpless. They are carefully fed and protected by the parents.

BIRD MIGRATION

A fascinating and complex part of the life of birds is their migration—their seasonal movements from one place to another. Some mountain-dwelling species travel to valleys and slopes during the winter and return to the heights at the breeding season. Other birds migrate in an east-west direction. But by far the most impressive migrations are made over the land areas of the Northern Hemisphere—northward in the spring, southward in the fall.

Birds that migrate seem to do so because of an inherited instinct. We may speak of an impulse to migrate for want of a better term. The stimulus that arouses this impulse may be an internal factor, acting alone. It may be combined with various environmental factors such as the increase (or decrease) in the amount of daylight per day, the increase (or decrease) of humidity and rainfall, the rise (or drop) of air temperature, and the supply of food.

It is not known how birds maintain their sense of direction during migration. It seems almost incredible that such a bird as the greater shearwater can freely roam the whole Atlantic, going as far north as Greenland, and yet return unerringly to a tiny island, Tristan da Cunha in the South

Jacques Six

Three widely different types of birds: (above) a cassowary of South America; opposite page: (left) a spotted sandpiper; (right) a penguin.

Atlantic, to breed. It has been suggested that a bird's orientation (sense of direction) is based on the sun's position, on the earth's magnetic lines, on an inherent sense of direction, or on the memorization of landmarks such as coastlines and mountains.

Generally, when a bird migrates, it travels to a place where weather conditions are favorable, where food is plentiful, and where there is room to raise its young.

Of course, some birds do not migrate at all but remain as permanent residents in one more-or-less widely extended area.

CLASSIFICATION OF BIRDS

There are various classification schemes for birds. According to the classification that we adopt here, there are 26 orders of living birds in the class Aves. Two other orders—the elephant birds (Aepyornithiformes) of Madagascar and the moas (Dinornithiformes) of New Zealand—became extinct during historic times.

Five orders of living birds make up a superorder known as the Palaeognathae. They include the more primitive forms. They differ from other birds, among other things, in the arrangement of some of the skull bones. They are flightless, or almost so, and have undeveloped wings and sternum. Their legs are long and their feathers curly. Some species are very large. Following are the five orders of the Palaeognathae:

Struthioformes: ostriches. Largest of living birds, flightless, confined to deserts and plains of Africa and Arabia.

Varin/Jacana

Russ Kinne/PR

Rheiformes: rheas. Large South American birds, ostrichlike in form and habits.

Casuariiformes: cassowaries, emus. Similar to ostriches. Cassowaries, which have a horny helmet on the head, live in Australia and New Guinea. Emus inhabit grasslands of Australia.

Apterygiformes: kiwis. Much smaller nocturnal, flightless birds of New Zealand, having a long beak and tiny eyes.

Tinamiformes: tinamous. Partridgelike, weak-flying, ground-living birds of grasslands and forests of Mexico through South America.

The remaining 21 orders make up the superorder Neognathae. They include birds with well-developed wings and sternum. They are as follows:

Sphenisciformes: penguins. Stout-bodied, flightless, aquatic birds living on the Antarctic ice, islands, and barren coasts of the Southern Hemisphere. They use their forelimbs, which are modified into paddles, to swim.

Colymbiformes: grebes, loons. Ducklike water birds that have a narrow head and neck and a pointed bill. Excellent divers and swimmers. Found in many parts of the world.

Procellariformes: albatrosses, shearwaters, fulmars, petrels. Chiefly cosmopolitan, oceanic birds with long, narrow wings and a hook-tipped bill covered with horny plates.

Pelicaniformes: swimmers having all four toes united by a web—tropic birds, pelicans, boobies, gannets, cormorants, snakebirds, frigate birds. Chiefly long-winged, oceanic diving birds. Most are nearly cosmopolitan, though the tropic birds are confined to tropical oceans, the snakebirds mostly to inland lakes and swamps of the tropics and subtropics.

Ciconiiformes: long-legged waders—herons, egrets, bitterns, storks, ibises, spoonbills, flamingos. Usually large, long-necked, long-billed wading birds that capture prey in shallow marsh and lake waters. Most are cosmopolitan.

Anseriformes: waterfowl (ducks, geese, swans), screamers. Waterfowl are cosmopolitan. The South American screamers are semiaquatic birds with webless toes, a slender and short bill, and spur-bearing wings.

Falconiformes: vultures, kites, hawks, eagles, ospreys, falcons, secretary birds. Powerfully winged birds of prey generally of worldwide distribution.

Galliformes: mound birds (megapodes), gallinaceous birds (grouse, ptarmigans, quails, partridges, pheasants, peacocks, jungle fowl, chickens, guinea fowl, turkeys, curassows), hoatzins. Largely ground birds (curassows and hoatzins are arboreal) with plump bodies, short and stout bills, and short wings. The legs and feet are specialized for running and scratching the ground. Quails, turkeys, curassows, and hoatzins are New World species; grouse and ptarmigans inhabit the Northern Hemisphere; the others are of the Old World.

Columbiformes: pigeons, doves, sandgrouse. The Eurasian and African sandgrouse are grouselike with very short legs. Pigeons and doves are worldwide in distribution.

Gruiformes: "marsh" birds—cranes, rails, coots, gallinules, button quails, trumpeters, sungrebes, cariamas, bustards, limpkins. A diversified group adapted for ground-living and often semiaquatic habits. Many are poor fliers or are flightless. They are often plump and long-legged; the bill varies in shape and length. Rails, coots, gallinules, and cranes are almost cosmopolitan; bustards inhabit the Old World; button quails,

Africa and Australasia; sun-grebes, the tropics; limpkins, southern United States through South America; trumpeters and cariamas, South America.

Charadriiformes: shorebirds, gulls, and relatives—jacanas, oystercatchers, plovers, sandpipers (including snipe and woodcock), avocets, phalaropes, pratincoles, coursers, seed snipes, sheathbills, skuas, jaegers, gulls, terns, skimmers, auks, murres, puffins. A mixed group more or less specialized for aquatic habits. Typical shorebirds are usually long-legged waders, often have long bills, and for the most part inhabit marshes and shores of open water. Gulls and terns are strong flying birds of the sea, though some may be found far inland. Skuas and jaegers are hawklike, oceanic predators. Auks murres, and puffins are somewhat ducklike in appearance; they frequent open seas. Pratincoles and coursers inhabit the Old world; seed snipes, South America; jacanas and skimmers, the tropics; sheathbills, Antarctic shores; skuas and jaegers, the cold oceans; auks, murres, and puffins, oceans of the Northern Hemisphere. The rest are cosmopolitan.

Psittaciformes: lories, macaws, cockatoos, parrots, parakeets. Tropical arboreal birds with strong, hooked beaks, and with two toes directed forward and two behind.

Cuculiformes: turacos, cuckoos, anis, roadrunners. Mainly long-tailed arboreal birds. Turacos are African; cuckoos inhabit the Old and New World; anis and roadrunners are American.

Strigiformes: owls. Large-headed, large-eyed cosmopolitan night birds of prey.

Caprimulgiformes: oilbirds, frogmouths, owlet frogmouths, potoos, goatsuckers (whippoorwill, nighthawk). Mainly nocturnal, insect-eating birds, having an exceptionally wide mouth, small bill, and weak, tiny feet. Oilbirds and potoos are South American; frogmouths, which have a broad, flattened bill, and owlet frogmouths frequent the Australian region; goatsuckers are cosmopolitan.

Micropodiformes (sometimes *Apodiformes*): swifts, hummingbirds. The long-winged, wide-mouthed swifts are cosmopolitan; hummingbirds belong to the New World.

Coliiformes: colies (mousebirds). Long-tailed African birds that creep along branches.

Trogoniformes: trogons. Stout-bodied, brilliantly colored birds of the tropics.

Coraciiformes: kingfishers, todies, motmots, bee-eaters, rollers, hoopoes, hornbills. A varied group. The cosmopolitan kingfishers, which have a strong, pointed bill, plunge into

A wild cock. Many game birds have been widely domesticated and are important food sources.

water for prey or capture small animals on land. Hornbills, bee-eaters, and rollers are birds of the Old World tropics; hoopoes live in Africa and Eurasia; todies inhabit the West Indies; motmots are tropical American.

Piciformes: jacamars, puffbirds, barbets, honey guides, toucans, woodpeckers. Birds chiefly specialized for climbing and digging into wood. Two toes are directed forward, two backward. Most have a straight or curved tapering bill. While the woodpeckers are cosmopolitan, toucans (which have a large, long, and vertically broad bill), jacamars, and puffbirds are tropical American. Barbets occur in all world tropics; honey guides, in tropical Africa and Asia.

Passeriformes: perching birds. Generally small or medium-sized birds with varied habits. The toes are specialized for gripping a perch. About half of the known species of birds belong to this order. Representatives common to North America (many are common to other lands as well) include: flycatchers, larks, swallows, crows, jays, titmice, nuthatches, creepers, dippers, wrens, thrashers, mockingbirds, thrushes, kinglets, gnatcatchers, pipits, waxwings, shrikes, starlings, vireos, wood warblers, blackbirds, meadowlarks, orioles, tanagers, weaver finches (house sparrows), grosbeaks, finches, sparrows. Some exotic forms are cotingas, antbirds, woodhewers, honeycreepers (New World tropics); ovenbirds, tapaculos (South America); manakins (Mexico to Argentina); cuckoo shrikes and Old World orioles, warblers and flycatchers (Old World); bulbuls, sunbirds, babblers (Old World tropics); pittas, drongos (Africa, Asia, Australia); white-eyes (Africa, Asia); flowerpeckers (Asia, Australia); lyrebirds, bowerbirds (Australia); birds of paradise (New Guinea).

FLIGHTLESS BIRDS

Kiwis looking for worms in a New Zealand forest. Penguins shuffling across the Antarctic ice in search of fish. A band of ostriches picking at lizards and mice on an African grassland. A pair of cassowaries feeding on plums in a New Guinea rain forest. If an enemy approaches or danger threatens, these birds lack a major escape mechanism common in most birds: they cannot fly. Their wings are small and underdeveloped.

Flightless birds are not, however, defenseless. Ostriches do not bury their heads in the sand if faced by an enemy. Rather, they deliver powerful kicks and inflict deep wounds with their toenails. Or they may run away—they can run faster than a horse, and so can escape most pursuers. The cassowary defends itself with its long, sharp toenails. The rhea is a fast runner. The tinamou, grayish or tawny in color, escapes detection by blending in with its environment.

A number of bird orders have flightless representatives. Generally, however, the term "flightless birds" refers to members of six orders: Struthioniformes—ostriches; Rheiformes—rheas; Casuariiformes—cassowaries and emus; Apterygiformes—kiwis; Tinamiformes—tinamous; and Sphenisciformes—penguins. Because they cannot fly, these birds are obviously ground dwellers.

THE OSTRICH

Some 50,000,000 years ago, about half a dozen species of ostriches were distributed through the continents of Africa, Asia, and Europe. Today, there is only one species, *Struthio camelus*.

The ostrich lives in Africa, south of the Sahara, typically in dry savanna or brushland. The birds usually feed in groups of six to ten, although occasional bands of 50 or more may be seen. Ostriches often associate with grazing mammals such as antelopes and zebras. The two groups of animals have a mutually beneficial alliance. The grazing animals disturb small rodents, lizards, and insects, which are then eaten by the ostriches. The tall, sharp-eyed ostriches can spot an enemy a long way off, and warn the mammals of danger.

Ostriches are the largest living birds. A male may be 2.5 meters tall and weigh as much as 135 kilograms. No wonder they are flightless—imagine how much energy would be needed to lift so heavy an animal.

The ostrich has only two toes, which are very unequal in size, on each foot. This gives the foot greater strength and thrust, which is useful to an animal adapted to a walking and running way of life.

The male's body plumage is black; the tail feathers and wing quills are white. The smaller females are dusky gray in color. The heads and long necks of the birds are sparsely covered with down.

Adélie penguins in their rookery protecting their dark-plumaged young. These aquatic birds are adapted to life in the harsh environment of the Antarctic.

George Holton/PR

Toni Angermayer/PR
A male ostrich showing off his plumage. The male ostrich is the largest living bird.

Frank Stevens, Audubon/PR
Darwin's rhea from South America. Rheas have long strong legs and are swift runners.

Ostriches are nomadic animals, wandering to wherever food is available. They are indiscriminate eaters. In addition to invertebrates and small vertebrates, they eat all sorts of plant material—and will even gulp down shiny objects such as bottle tops.

These birds are polygamous. A male has a harem of three to five mates. The females in the harem generally all lay their eggs in the same shallow, sandy pit that serves as the nest. Each female lays six to eight eggs. Though comparatively small in relationship to the ostrich's size, these are the largest eggs laid by birds. One ostrich egg may weigh more than one kilogram, and be as big as 20 or 24 chicken eggs.

The females in a harem incubate the eggs during the day; the male sits on them at night. The large clutch of eggs makes a grand meal for predators such as vultures and jackals. But the predator must worry about those muscular ostrich legs, which can deliver a maiming kick, and those toenails, which can disembowel a jackal.

THE RHEAS

These South American birds are similar to, but much smaller than, the ostrich. There are two species. The common rhea, *Rhea americana,* may be 1.5 meters tall and weigh 25 kilograms. Dull gray in color, it is found on the grassy plains of Brazil and Argentina—a range that is constantly made smaller by the spread of human civilization. The second, smaller species, Darwin's rhea, *Pterocnemia pennata,* lives in the eastern Andes.

The legs of a rhea are long and powerful. Each foot has three toes. The animal is a swift runner and a good swimmer.

Like ostriches, rheas roam in groups for most of the year. When the breeding season begins, the males fight each other for control of a harem. After a victorious male acquires his harem, he constructs a nest—a shallow lined pit in the ground. After he has mated the hens, he leads them to the nest. All the hens in a harem lay their

380 FLIGHTLESS BIRDS

Violet-necked cassowary. Cassowaries inhabit dense rain forests where they live in small groups.

A kiwi and its egg. Relative to its own size, the kiwi lays the heaviest egg of all birds.

eggs in the one nest, and the male then incubates the clutch.

CASSOWARIES AND EMUS

Cassowaries and emus are swift runners and usually can avoid an enemy. If, however, they cannot avoid an enemy, they can be very aggressive, delivering vicious kicks and inflicting wounds with their long, sharp claws.

Another distinguishing characteristic of these birds is their hairlike feathers. Each feather has two shafts, and does not interlock with other feathers, as is the case in most birds.

Cassowaries inhabit dense rain forests in northern Australia, New Guinea, and nearby islands. There are three species, all in the genus *Casuarius*.

These birds live in pairs or small family groups. The male incubates the three to six large eggs and also takes care of the young chicks.

The largest and most common species, the double-wattled cassowary, *Casuarius casuarius*, may be more than 1½ meters tall. It has black, glossy plumage that is hairlike in appearance. There are no feathers on the blue neck and head. Two red wattles, or folds of loose skin, hang from the neck. Atop the head is a flat horny casque that looks like a tall crown; it may be 15 centimeters high.

The emu, *Dromaius novaehollandiae*, is found only in Australia and is the largest bird of that continent. It averages 1.5 to 1.8 meters in height and weighs up to 55 kilograms. Female emus are larger than the males.

The drab, brownish-gray emus live in a variety of environments, from scrubland to rain forest. They feed mostly on plant matter but will also eat insects.

Emus breed in the winter. The female lays eight to ten dark-green eggs in a nest on the open ground, usually beneath a tree. The male incubates the eggs and cares for the young.

FLIGHTLESS BIRDS 381

The Emperor penguin is the largest of all penguins. It lives and breeds on the Antarctic ice. The female lays only one egg per season.

THE TINAMOUS

There are about 50 species of tinamous, which live in a variety of habitats from southern Mexico to the tip of South America. The tinamou can fly, but it is clumsy in the air. Its short, rounded wings let it fly only short distances.

Tinamous have compact bodies with long necks, short tails, and strong legs. They range in length from 20 to 50 centimeters. The plumage consists of beautifully patterned browns and grays and blends in well with the environment. These are solitary birds, except during the breeding season, when they are polygamous. The males incubate the eggs and care for the young.

THE PENGUINS

The order Sphenisciformes contains 17 species of fascinating, highly specialized birds. The order name comes from the Greek word *spheniskos,* which means "a small wedge" and refers to the penguin's flippers. These are hard, narrow structures covered with scalelike plumage. They are used to propel the penguin underwater at speeds of 40 kilometers per hour or more.

Penguins have other adaptations to a marine life. Their bodies are streamlined; their feet are webbed; and their feathers form a dense, waterproof insulation.

These birds live in the Antarctic and on islands as far north as the Peruvian coast. They eat fish, squid, and shrimp. The largest species, the Emperor penguin, *Aptenodytes forsteri,* is more than one meter tall and weighs 35 kilograms. It never sets foot on dry land, but lives and breeds on the Antarctic ice. The smallest penguin, the little blue penguin of Australia, *Eudyptula minor,* is three tenths of a meter tall.

Most penguins are monogamous and remain with their mates throughout their long lives. They breed in colonies that may contain many thousands of birds. Most species lay two or three eggs. The parents share the tasks of nest building, incubating the eggs, and caring for the young. During the nesting period, most species undergo a long period of starvation, living off fat reserves in their bodies.

THE KIWIS

Kiwis live in the damp forests of New Zealand. There are three species, all in the genus *Apteryx.*

These birds have grayish or reddish-brown plumage. The bill is long and slender, and well adapted for probing the soil in search of insects and their larvae, worms, snails, and other small animals that comprise the bird's diet. The nostrils are located at the tip of the bill, and the sense of smell seems to be highly developed. The eyes are small and weak, which perhaps explains the kiwi's nocturnal habits.

The kiwi has long legs and well-developed leg muscles. It is a swift runner. The wings are very tiny—only five centimeters long—and there is no tail. Kiwis range from 35 to 55 centimeters in length.

The bird uses its heavy claws to scratch on the forest floor and to excavate the nest. Female kiwis lay the heaviest eggs, relative to their size, of any bird. The weight of a kiwi egg may be as much as one-fourth the weight of the bird. Most kiwis lay only one egg, though some lay two. The male incubates the eggs for 75 to 80 days.

G. R. Austing and Donald Koehler, Audubon/PR

Birds of prey hunt other animals for food. In these photos, a saw-whet owl moves in on a mouse. It grabs the prey in its sharp talons.

BIRDS OF PREY

On a hot, sunny African plain, a zebra dies. Soon, the vultures come. With their sharp beaks, they tear the hide and flesh off the dead animal.

In a dimly lit South American rain forest, a harpy eagle kills and eats a young monkey.

When night comes to the woods of Europe, tawny owls begin their hunt for mice and other rodents.

Vultures, eagles, and owls are birds of prey. The birds of prey include some of the most powerful fliers and fiercest hunters, for their size, to be found in the animal world. They are graceful on the wing and ever alert for unwary prey.

Birds of prey may be divided into those that hunt by day—diurnal birds—and those that hunt by night—nocturnal birds. The hunters that fly by day include vultures, hawks, eagles, kites, falcons, and ospreys. Most owls, on the other hand, hunt their quarry by night.

The daytime hunters are all placed in the order Falconiformes. They are grouped together because of their habits rather than because of close physical similarities. Owls make up the order Strigiformes.

GENERAL CHARACTERISTICS

All these birds generally have stout hooked beaks with a soft area, called the *cere*, at the base. The beaks have wavy edges. Most of the birds of prey have three toes set forward and one directed backward. The owls and ospreys have a reversible outer toe—that is, it can be turned either to the front or to the rear.

Birds of prey range in size from the condor, which measures over three meters from wing tip to wing tip, to the pygmy falcon of India, which is scarcely larger

Cruickshank, Audubon/PR

Illinois Natural History Survey

Top: a marsh hawk and her young. These birds live in wetlands. Bottom: a barn owl. Owls have very large eyes set immovably in their sockets.

they tear it to pieces with their hooked beaks. The vulture lives on carrion and so does not need taloned feet.

The eyes of the day hunter are placed at the sides of its head so that the two eyes never look in the same direction. In owls, the eyes, which are larger than those of the diurnal birds of prey, are set in front of the rather flat face, so that both look together in the same direction. The owl, therefore, has binocular vision, while the other birds must survey an object first with one eye, then with the other.

The eyesight of birds of prey is extremely keen. From a hundred meters overhead they scan the ground and are able to see the tiniest mouse or lizard. In the instant required for them to drop from that height and pounce on their victims, their eyes change focus so quickly that they never lose sight of the prey.

The eye color varies according to the family, genus, or species. It may be yellow, ruby-red, gray, or brown. The eye color may also vary according to the bird's sex or age.

Soft shades of brown and gray mark the plumage of most species. Some, however, show striking patterns of blue and reddish-brown. Some eagles and vultures have ornamental crests. Young hawks are marked on the underparts with perpendicular stripes, which become horizontal (adult marking) after the first molt. It is easier to distinguish between species by size or shape than by color.

Perhaps the most outstanding characteristic of the birds of prey is their fearlessness and truculence. The goshawk will courageously attack any person coming near its nest. The European sparrow hawk, although less than one third of a meter in length, will fall upon any other creature in the air, no matter how large. Many hawks will attack animals as large as themselves, or even larger.

The voices of the day-hunting preying birds are harsh, discordant screams. The short-winged species that lie in wait for their prey are usually silent. The others call frequently. The owl's voice is perhaps best described as blood-curdling—at least to small animals.

than a sparrow. The females of the birds of prey are larger than the males.

Long, well-curved, sharp *talons* characterize all birds of prey except the vultures. With these truly formidable weapons they seize and dispatch their prey. Then

All birds of prey are exclusively carnivorous. Most kill their food. The scavenger species, however, feed on decaying animal matter. The larger birds eat other birds and small mammals. Small ones feast on insects, mice, frogs, lizards, and snakes. Only the short-winged hawks and occasionally the marsh hawks are destructive to game birds and domestic birds.

Certain birds of prey, especially ospreys and eagles, repair and use the same nest year after year. A number of hawks and eagles mate for life, but are quick to get another mate when the old one has died. Most species, however, get a new mate and a new nest each year. Many build nests of sticks and branches high up in trees. Some nest on cliffs and ledges, or on the ground in prairie or marsh land. All lay strong, heavy-shelled eggs with granular surfaces.

HUNTERS BY DAY

The order Falconiformes is made up of five families. The family Cathartidae contains the New World vultures. The family Accipitridae includes the Old World vultures, hawks, eagles, and kites. The falcons are placed in the family Falconidae. The osprey by itself makes up the family Pandionidae. The secretary bird of Africa is also the only species in its family, the Sagittariidae.

THE VULTURES

Vultures are heavy birds that may weigh 10 kilograms or more. They have weak legs with feet adapted for standing or walking and a long middle toe that helps the birds balance their ungainly bodies.

Vultures feed almost exclusively on dead animals. They have an absolute immunity to the poisons produced in decaying flesh—poisons that would kill many other creatures.

Vultures hunt in a wide range of habitats. With their remarkable eyesight they can detect the smallest dead animal from tremendous heights. When one vulture swoops down on a find, its descent is seen by others for kilometers around, and it takes only minutes for a flock to gather on the carcass. When the birds are gorged, they cannot fly away and are easily captured. It is interesting to note that captive vultures that become accustomed to food that is not putrid will, in some cases, refuse tainted meat.

Vultures lay their spotted eggs on the ground under a log, in a hollow tree, or in a cave. The young of these birds are covered with whitish down and are helpless for a long time. Young vultures are quite noisy. Adults make no sound except a hiss.

Vultures display some interesting behavior. In North America they have learned to patrol the highways for run-over animals. The Old World lamb vulture carries the bones of large animals aloft and drops them on rocks. The bones break open and the bird can easily get at the marrow.

There are six species of New World vultures and 14 species in the Old World. The turkey vulture, *Cathartes aura,* is the most widely distributed of the New World vultures. Its range extends from Canada to the southern part of South America. This vulture can be distinguished by its reddish head and neck and its long wings. The South American condor, *Vultur gryphus,* is one of the largest living birds, with a wingspan of up to three meters. This vulture is found in the Andes Mountains from Venezuela to Patagonia.

An osprey, or fish hawk, brownish above, white below. It typically dives for fish, seizing the fish with its long curved claws.

G. Ronald Austring, Audubon/PR

TRUE HAWKS AND EAGLES

Broad-winged hawks. The broad-winged, fan-tailed hawks include the red-shouldered, red-tailed, broad-winged, Harris', Swainson's, and rough-legged hawks and the bald and golden eagles. With extended wings and spread tails these birds circle high overhead, as do the vultures. However, with the exception of the eagles, they are smaller than the vultures.

In general, these birds live in wooded districts but do most of their hunting about open fields. Most of them feed principally upon small rodents and the larger insects and thus are beneficial to mankind.

These hawks are best identified by the dark marking on their underparts. The broad-winged hawk has the under surface of the wings pure white with the exception of the dark tip formed by the dark first primary feathers. The under surface of the red-shouldered hawk's wings is barred but with no distinct black patch. The red-tailed hawk has a distinct crescentlike black patch at the second joint of the wing.

The eagles are usually recognizable by their large size. From ancient times the eagle has been regarded as a symbol of bold strength and courageous character. It is often found in mythology. Its form has been used in art, heraldry, and religion.

A bald eagle on its nest. This bird is not actually bald—adults have snowy white head feathers.
Karl H. Maslowski, Audubon/PR

The bald eagle, *Haliaeetus leucocephalus,* of North America, was chosen as the national bird of the United States because of its majestic appearance rather than its habits, for it feeds principally on fish cast up by the waves or those that it steals from an osprey. In the adult the head and tail are white; they are brown in the nestling. The young require four years to develop adult plumage.

Young bald eagles are very similar to the dark brown golden eagle except in the feathering of the legs. The lower leg of the bald eagle is bare. That of the golden eagle is feathered all the way to the toes.

The golden eagle, *Aquila chrysaetos,* lives in mountainous areas throughout the Northern Hemisphere. It is a much more active bird than the bald eagle and preys upon rabbits, grouse, young lambs, and fawns. Extensive killing by human beings has greatly diminished their numbers.

Short-winged hawks, genus *Accipiter,* include three common North American species: the sharp-shinned hawk, Cooper's hawk, and the goshawk. They lie in wait for prey and their wings are adapted to sudden bursts of speed from a stationary position. These three are responsible for most of the killing of poultry and game birds for which the hawk family generally is blamed. The sharp-shinned hawk itself has sometimes been killed for food by people, but this bird is now protected by law.

Two thirds of a meter or less in length, the short-winged hawk seldom wheels aloft on the lookout for food. It flies swiftly from place to place, flapping its wings rapidly a few seconds and then gliding noiselessly, ready to drop like an arrow among a flock of poultry. It seizes the victim in its talons and leaves quickly. After it has reached some favored branch or log, it plucks the bird before eating it. Short-winged hawks diving for prey may reach a speed of more than 300 kilometers an hour. They are known to pounce upon game birds that have just been shot before the surprised hunter has a chance to retrieve them.

KITES AND MARSH HAWKS

The kites, the lightest and most graceful of the hawks, are almost swallowlike in their flight. This is true especially of the New World's swallow-tailed kite, *Elanoides forficatus*. It is a strikingly marked white and black bird about ½ meter long, whose swallowlike appearance is augmented by its deeply forked tail. A troop of these birds skimming the water like a band of swallows, darting after one another in playful sport, rising abruptly high into the air, diving, and sailing is a beautiful sight.

One interesting bird in this group is the Everglades kite, *Rostrhamus sociabilis*, which is found from Florida south to Argentina. It depends entirely on a certain large freshwater snail for its food supply. Its bill, adapted to draw the snail from its shell, is very long and slender.

The marsh hawk, *Circus cyaneus*, lives in Eurasia and North America. In spite of its light body and long wings, it does not look like the kites. It can be identified by a white patch above the tail. Except during migration, it is usually found near marshes, although it often feeds in open fields where mice are abundant. Most of its food consists of mice, frogs, snakes, and other small creatures.

The marsh hawk nests on the ground, usually in the marshes but sometimes by stumps in upland pastures, and lays from five to seven pure white eggs. During the mating season the male performs curious evolutions in the air, sometimes turning somersaults from a considerable height toward the ground or, again, "looping the loop" on an angular course across the marsh.

THE OSPREY

The fish hawk, or osprey, *Pandion haliaetus*, is most abundant along seacoasts, but is found inland near most large bodies of water from the tropics to the Arctic Circle. With slight variations, it is found all over the world.

The osprey sometimes nests in small colonies where food is abundant and protection is afforded. It builds an enormous nest of sticks, usually at the top of a broken tree, although under protection it sometimes descends to nest on the ground.

N.Y. Zoological Society

Falcons are adapted for great speed. The sparrow hawk is the smallest of the North American falcons.

Ospreys may be confused with bald eagles because of their great wing expanse and the large white areas on the head. Unlike eagles, however, their underparts are white and their tails dark.

Ospreys feed entirely on fish, which they locate near the surface of the water while they hover overhead. With a plunge, often from some height, the fish hawk catches the fish with its talons. The fish frequently weigh one or two kilograms.

THE FALCONS

Approximately 50 species of falcons are known. Placed in the same family, the Falconidae, are some nine species of caracaras. The latter are common in Central and South America.

The falcons have pointed wings adapted for great speed rather than for soaring. They pursue and strike their prey in full flight. These powerful birds, vigorous and fierce hunters, were tamed and trained for the chase in China as early as 2000 B.C.

By the ninth century, falconry had spread to England. The females, called "falcons," were prized above the smaller males—the "tiercels." The most favored species was the peregrine, which was fierce and yet tractable.

The falcon is captured either as an *eyas,* or nestling, or during the annual migratory flight. Training has changed little in 4,000 years. The captive is hooded and fettered and it is taught to respond to the feeding call. Only when it is tame and can be handled is the hood removed. Well-handled falcons show a good deal of attachment to their owners and a keen interest and intelligence in their work. They are trained to kill the quarry and remain upon it until the falconer calls the bird to his wrist.

A common group of falcons are the kestrels, of the genus *Falco,* which are worldwide in distribution. The New World kestrel *(F. sparverius),* is sometimes called the sparrow hawk because it eats small birds—though its more normal fare consists of rodents and insects.

HUNTERS BY NIGHT

The owl, unlike the other birds of prey, usually hunts at night. A hardy bird, it can adapt to desert, mountainous, or swampy terrain. In most cases it migrates only short distances.

There are some 300 species of owls. Owls are found in all countries and under all conditions. These birds range in type from small, rather mild 15-centimeter-long insect-eaters to fierce birds more than 60 centimeters long. The latter are capable of taking over an eagle's nest.

In many countries and in all ages, the owl has been the object of quaint superstitions. There seemed to be something uncanny about a bird that hid during the day and only came out at night. Few people are aware, however, that when this bird comes near houses, it is to rid the barn or garden of mice and rats.

GENERAL CHARACTERISTICS

Coloration. In owls that live in woodlands, the predominating colors are browns and grays, streaked, mottled, and barred in shades that imitate the colors of dead wood and bark. Their hues hide them effectively during the day, when they are hidden in hollow trees or perched on branches close to the trunk. The heads of many species bear tufts of feathers, called horns or ears, which break up the outline and make the protective coloration more effective.

The short-eared owl, dweller of open marshes, is pale buff in color and finely streaked with brown, which makes it incon-

Tropical monkey-eating eagles. Top: the rare *Pithecophaga jeffreyi* of the Philippines. Bottom: the harpy of Central and South America, shown with another favorite prey, a paca.

New York Zoological Society

James R. Simon, Audubon/PR

spicuous in the dead and tangled low vegetation. The burrowing owl is of a dull sandy color. The snowy owl of the Arctic is almost entirely white. None of the owls show any brilliant colors.

Eyes. An owl's eyes are set immovably in their sockets and directed toward the front. Therefore an owl, in order to look at a particular object, must move its whole head, and not simply turn its eyes toward that object.

The eyes of owls are very large. The size of the retina is further increased by an appendage, called the *pecten,* situated in the middle of the eyeball. Thus equipped, the eye is able to register the faintest light rays, a necessity for birds that hunt during the night.

It is commonly believed that owls cannot see during the day. As a matter of fact, they can see very well. And, in fact, certain species hunt regularly by day and some occasionally, particularly if they have young to feed. The snowy owl, for instance, lives most of the year in the Arctic, where the days are six months long. Obviously it must adapt itself to hunting in daylight.

The owl has well-developed eyelashes, which are uncommon in birds. When the owl closes its eye, the upper eyelid moves downward. In other birds, the lower eyelid moves upward.

Ears. Owls are not dependent on their eyes alone for hunting, for their hearing is very keen. The owl is the only bird having an external ear suggesting that of the mammals. The ear is concealed and partly protected behind the radiating feathers of the bird's facial disk. In some species a funnellike arrangement of feathers serves as a sort of ear trumpet. The slightest rustle in the leaves or grass is heard by the owl, which swoops down on its prey noiselessly.

Feet. The owls differ from all the birds of prey, except the osprey, in the arrangement of the toes. The outer toe may be turned forward, backward, or outward. When it is turned back in perching, the owl gets a much stronger grip. This outer toe is also turned back when the owl is about to strike its prey, thus making its foot a most efficient weapon.

Feathers. The feathers—especially the wing feathers—are soft and fringed with down. Therefore these birds make very little noise as they fly. The North American Indian name for the owl—"hush wings"—was bestowed because of the bird's quiet flight.

THE DIET OF OWLS

Nocturnal small animals form the bulk of the owl's diet, small rodents being the chief food. Some species also feed on insects, snails, frogs, or larger animals. The great-horned owl is the only species that preys on poultry to any extent. A fierce, bold hunter, it will also attack skunks, turkeys, and domestic cats.

The rest of the owls are among the most beneficial of birds, and represent an extremely effective natural check on the increase of small rodents. The meadow mouse, for example, which feasts on stored grain and attacks the bark of young fruit trees, has from five to eight young in a litter and from three to six litters a year. If their numbers increased unchecked, they could do untold harm. It is fortunate, indeed, that they are sought after so greedily by owls.

The roving habits of some owls make them particularly efficient protectors of grain crops. When, for example, meadow or field mice become abundant in a locality, a flight of short-eared owls often follows. They remain in the area, nesting if need be, until the rodents once again become scarce.

REPRODUCTIVE ACTIVITIES

Most owls make no pretense at building nests but lay their eggs in cavities of trees and other convenient places. They sometimes use the old nests of crows, hawks, and even squirrels. Some owls, however, including the screech owl and the short-eared owl, build crude nests on the ground.

Owls breed very early. All lay white or whitish, spherical eggs. There are generally from three to five eggs in a single clutch, or brood, but the number may vary considerably.

Owls usually begin to incubate as soon as the first egg is laid. Sometimes both par-

The snowy owl and its eggs. This bird lives in the Arctic, but is sometimes seen in Canada and in the northern United States.

ents sit on eggs side by side. The first eggs may be hatched much sooner than the last.

Baby owls are thickly covered with down when they emerge from the eggs. At an early age they display certain instinctive defense reactions. For example, when molested they fluff out their feathers and either hiss or make clicking sounds with their bills. It takes the young bird a comparatively long time for its flight feathers to develop.

BARN OWLS

Most owls belong to the family Strigidae. The barn owls are placed in a separate family, the Tytonidae. The furculum, or "wishbone," in barn owls differs from that in other owls.

There are ten species of barn owls. One found in most parts of the world is *Tyto alba,* the common barn owl. *Tyto alba* is about 45 centimeters long. Its back is mottled buff and grayish. Its underparts show buff speckled with black. The feathers of the face radiate outward, forming a heart-shaped disk.

A hollow tree, a barn, a steeple, or even a hole in a bank may serve this odd-looking owl as a egg-laying site.

OTHER SPECIES

The great grey owl, *Scotiaptex nebulosa,* is one of the handsomest members of the family Strigidae. It haunts the forests of Canada and Alaska but in winter it migrates south into the northern part of the United States. It is a large bird—measuring 75 centimeters in length—with grayish brown plumage.

The great horned owl, *Bubo virginianus* is a magnificent North American bird measuring 60 centimeters in length. It has very prominent ear tufts about 5 centimeters long. This owl is colored sooty brown or dusky and is streaked or mottled with grayish white. A related bird, the eagle owl, *Bubo bubo,* is the largest European owl, averaging 65 to 70 centimeters.

The barred owl, *Strix varia,* ranges from southern Canada to Mexico. It hunts in the deep solitudes of dense forests and forbidding swamps. It is about 50 centimeters in length, having a brown plumage with whitish bars. It has no ear tufts.

The snowy owl, *Nyctea scandiaca,* is white, lightly barred with brown, and about 55 to 65 centimeters in length. It spends the summer on the Arctic tundra and winters from the Arctic shores southward, feeding on rabbits, lemmings, and ptarmigan.

The long-eared owl, *Asio otus,* lives in coniferous forests throughout Europe, Asia, and North America. It has ear tufts like those of the great horned owl. It is much smaller and more slender, however, and does not have the white throat patch characteristic of the larger bird.

The short-eared owls, *Asio flammeus,* are found near grassy marshes or pastures throughout the Northern Hemisphere and in South America. Numbers of these birds spend the day together, usually on the ground or in tangled places.

The commonest owl in North America is the little screech owl, *Otus asio,* not much larger than a robin, but much heavier. It is found even in large cities in hollows of trees or in crevices about buildings, for the mice upon which it feeds are everywhere. Different individuals may show varying shades of tan, gray, or brown without regard to age or sex. The call of the screech owl is really not a screech, but a low tremulous whistle. Other species of screech owls are found in various parts of the Americas.

The saw-whet, or Acadian, owl, *Aegolivs acadicus,* is only 20 centimeters long. Its range extends from Canada to Mexico. It spends the daytime in evergreen thickets or brush and often sleeps so soundly that it does not stir even if taken in one's hands.

WILD FOWL

There's no mistaking this wild fowl. The peacock has a large fan of showy, beautifully colored plumes. The female peafowl, or peahen, is rather drab.

Quails, grouse, turkeys, and pheasants are wild fowl. They are also know as upland game birds for two reasons: they are usually found in grasslands and forests, rather than in marshy areas at the edge of water, and they are fine game for hunters.

Sadly, the numbers of these splendid birds have been greatly reduced during the years. Hunters value them not only for the sport but because they are good to eat. In addition, farms, houses, and other human developments have encroached on their lands.

These birds are members of the order Galliformes. The order contains more than 250 species, placed in seven families:

The family Phasianidae includes pheasants, partridges, peacocks, quails, and the domestic chicken.

The family Tetraonidae includes the grouse, ptarmigans, and prairie chickens.

The Meleagrididae are the turkeys.

The Megapodiidae includes junglefowl, brush turkeys, and other fowl found only in Australia and on nearby islands.

The Numididae contains the guineafowl of Africa.

The Cracidae contains curassows, chachalacas, and guans—forest-dwelling birds found only in Central and South America.

The family Opisthocomidae contains only one species, the tropical hoatzin, *Opisthocomus hoazin*.

There is a considerable amount of confusion in the usage of such common names as "grouse," "partridge," and "quail." These names are applied to quite different birds in some parts of North America and interchangeably in others. For example, the harlequin quail is also known as the Montezuma quail, massena quail, fool quail, fool hen, and black quail. Similarly, the mountain quail is often called the plumed partridge or mountain partridge.

This confusion of names can probably be traced back to the early settlers in the United States and Canada who first named the birds. They often mistook the American birds for similar species native to their European home. For example, an immigrant from central Europe, who might have been familiar with a species of European partridge native to his home, would call any similar birds he saw in America by the name of that partridge. In many cases, the bird to which he was referring was not a partridge at all, but a turkey, quail, or grouse.

GENERAL CHARACTERISTICS

The Galliformes are mostly ground dwellers. Turkeys and many of the grouse keep to the deep woods. Some of the grouse, however, frequent open prairies and deserts. The bobwhites and ring-

Quail are ground-dwelling wild fowl. Top: the California quail has a crest of feathers that give it its other name: the helmet quail. Bottom: North American quail—the bobwhite.

necked pheasants often live about farm lands, and ptarmigans, above timberline.

They may be rather large birds, with heavy bodies, small heads, and stout, short bills. They have short, rounded wings that are strong, but the birds do not fly long distances. Their stout legs are moderately long, and the three front toes are often slightly webbed.

Most upland game birds are not conspicuous, but, rather, blend in with their surroundings. However, parts of the plumage of turkeys and pheasants are richly colored. And the order contains the most spectacularly colored of all birds, the male peafowl—better known as the peacock.

The upland game birds flock together at mating season and then go their separate ways to make their nests in various natural hiding places on the ground. From 6 to 18 eggs are laid. When the young are hatched, they are already covered with down and are able to care for themselves. These birds have a varied diet depending upon the season of the year: weed seeds, buds, berries, insects, snails.

THE QUAILS

Quails are fowllike in form. New World species have large, heavy bodies and small heads.

Old World quail, belonging to the genus *Coturnix,* are much smaller than their American cousins. Many of them are no larger than sparrows.

The bills are stout, short, and convex. The outer edge of the lower mandible is distinctly serrated, or notched, in the New World quail. In the Old World quail, the edges of both the upper and lower mandibles are smooth.

The legs and toes of quail usually lack feathers and are scaly. Occasionally the three front toes are webbed. The hind toe can be elevated. Most American quail lack the tiny feathers inside the nostrils that are found in the English quail. The quail wings are normally short, arched, rounded, and quite strong.

In habit and habitat, most quail are strictly ground-dwelling birds. Some favor open and cultivated fields. Others prefer wooded areas and mountains.

The mating season usually occurs around February. Most quail tend to be rather promiscuous during this season. The males go through elaborate courtship procedures and usually collect harems. The females lay 10 to 15 eggs in May or June. Both parents are devoted to their young; the female constantly cares for the eggs while the male guards the nest. In some species, the males have been known to purposely attract attention to themselves when danger approaches so that the nest will not be harmed. When the eggs hatch, the young quail are quite well-developed and are able to forage for food for themselves within a few hours. They may even be able to fly when one week old.

The best-known North American quail is the northern bobwhite, *Colinus virginianus*, which is found from southern Canada to Mexico. Other bobwhite species, also members of the genus *Colinus*, are common in Mexico and Central America. All are known by their call: "bob-white."

The northern bobwhite is extremely popular for a variety of reasons. It has always been a favorite game bird and its call is considered to be quite musical. It is now protected in some parts of the United States because it feeds upon the destructive cotton boll weevil.

Unlike most of the gallinaceous birds, the bobwhite is not polygamous. It is a devoted mate and a conscientious father, helping to incubate the eggs and care for the young. It selects a protected area under a fallen branch or log, or in high grass that can be bent to form a roof. It then scratches out a depression where the female lays eggs and incubates them until hatched. Bobwhite eggs are the whitest and most pointed of any of the gallinaceous birds.

At the close of the breeding season, the birds gather in large flocks, or *covies*, which may consist of all the members of only one or perhaps two families. The covies generally are to be found in an open field or garden where food is plentiful. They remain together, unless scattered by hunters, until the wintertime, when they retire to more wooded areas. At night, the covies form circles, suggesting the wagon-train arrangement of early North American settlers. Each bird in the circle stands with its tail pointing to the center and its head pointing outward. When the birds are disturbed and take to the air, the circle seems to explode.

The California quail, *Lophortyx californica*, ranges from northern California south into Mexico. It is slate blue, with a crest of club-shaped feathers that remain erect and curved forward, giving the appearance of a helmet. For this reason it is sometimes called the helmet quail. The feathers are bare at the base and swollen at the tips. The male's plumage includes areas of soft grays and warm browns; the female's coloring is plainer.

The scaled quail, *Callipepia squamata*, lives in semi-desert regions of the southwestern United States and Mexico. The scaled quail is gray in color. The feathers of the neck and breast are edged with black, giving the appearance of scales—hence the name of the bird. There is an inconspicuous tuft of white feathers on its crown; this accounts for another of this bird's names—the "cottontop."

The common European quail, *Coturnix coturnix*, is a solitary animal except during periods of migration. It lives in pasture and crop lands and is a difficult bird for hunters to flush.

THE PHEASANTS

The pheasants are all native to the continent of Asia and the nearby islands.

Most of the male members of this family are brilliantly colored so as to attract their mates. The coloring of the females is dull by comparison. For example, the male English pheasant, *Phasianus colchicus*, has a peacock-blue head and neck that are

Male ring-necked pheasant. Pheasants are native to Asia but now flourish in many other areas.

Leonard Lee Rue III, Audubon/PR

glossed with green, purple, and bronze metallic reflections. The sides of the head are scarlet. The feathers of its back are orange-brown, with streaks of green, buff, and black. The tail is olive-brown with black bars. And the breast is a glossy copper brown edged with purplish hues. The female, in great contrast, is a very plain brownish color.

The common pheasant was successfully introduced into England many years ago—according to some accounts, by the Romans. It is still raised there for hunting purposes. The Chinese ring-necked pheasant was introduced into the United States in 1880. However, the well-known ring-necked pheasant now found in the United States is not the same species that was brought over in 1880. It is a hybrid—a cross between the original Chinese ring-necked variety and the English common pheasant. This hybrid has flourished in the New World.

The pheasants possess strong legs with which they can run to cover quite rapidly. They can fly for short distances, but only with some effort because their wings are relatively short. The tail is certainly the outstanding feature of these birds. One species, the argus pheasant, *Argusianus argus,* of Southeast Asia, is about 2.5 meters long, of which 1.8 meters is tail.

THE GROUSE FAMILY

The family Tetraonidae may be divided into three distinct groups: (1) the ptarmigans, (2) the grouse proper, and (3) the prairie chickens and heath hens.

The birds of this family are generally larger than quail, measuring 35 to 50 centimeters in length, as compared with the 10 to 30 centimeters of the quail. The legs are fully feathered in most species, because they live in areas of snow. In species found in the far north, even the toes are feathered; this enables the birds to walk more easily on snow. The leg feathers help distinguish the grouse from quail.

Ptarmigans live in arctic and subarctic regions. They change their plumage from season to season: gray and brown in summer; white in winter.

Charles G. Summers Jr./PR

Leonard Lee Rue III, Audubon/PR

The ptarmigans inhabit arctic and subarctic regions and rocky areas on the peaks of lofty mountain ranges. They are the only birds in the entire gallinaceous order that change their plumage from one season to another. As is common with animals that spend a large part of their lives in snow, the ptarmigan has a pure white winter coat; this blends admirably with the snow background and helps conceal the bird from its foes. The summer coat of mottled gray and brown also provides excellent camouflage; it blends beautifully with the lichen-covered rocks amid which the ptarmigan makes its home.

All ptarmigans belong to the genus *Lagopus*. There are many species, each native to certain specific areas. *L. rupestris* is found in Greenland; *L. hemileaurus*, on the polar island of Spitsbergen; *L. leucurus*, in the Canadian Rockies and the Sierras of the United States. Another species, the red grouse of Great Britain, is actually a ptarmigan, *L. scoticus*. It is the only member of the genus that is confined to the British Isles.

There are a number of species in the group to which the name "grouse proper" is applied. These birds are inhabitants of wooded areas in the temperate zone. Perhaps the best-known is the ruffed grouse, *Bonasa umbellus*, of North America. It is about 45 centimeters in length. It has reddish-brown plumage, occasionally mixed with gray and yellow. The name of the ruffed grouse is derived from the tufts of black feathers on the sides of its neck. These feathers can be raised and spread so as to form a ruff.

The courtship procedure of this bird is very curious. During the breeding season, the male bird mounts a hollow log, struts back and forth with its tail spread and ruff erect, and rapidly beats its wings, producing a hollow rumbling sound known as "drumming."

The black grouse, *Lyrurus tetrix*, of Europe and Asia has a similar mating ritual, during which the male dances and sings. It nests on the ground but may be found in trees. It inhabits moors, rocky heather-covered areas, and peat-moss regions.

Harry Engels/PR

A. W. Ambler, Audubon/PR

Top: a sage grouse displaying. The male inflates air sacs on his neck to impress onlookers. Bottom: the guineafowl, a game bird of Africa and Madagascar.

The prairie chickens of the genus *Tympanuchus* inhabit the open plains and semi-arid regions of North America. Normally about 45 centimeters long, a prairie chicken weighs almost a kilogram. It is reddish-brown in color with small markings of red and black. On the sides of its neck are air sacs that can be inflated at will. During courtship, the male inflates these sacs and then rapidly expels the air with a loud booming sound calculated to attract females. The male also does a characteristic dance as part of the courtship routine.

The heath hen, *Tympanuchus cupido*

WILD FOWL

Wild tom turkey. Wild turkeys have been widely hunted and are now quite rare.

cupido, also called the eastern prairie chicken, was once found in large numbers throughout the coastal plain of New England and the Middle Atlantic states. Unfortunately, it was so relentlessly hunted by people that its numbers decreased sharply. Finally there was only a small flock left on the island of Martha's Vineyard, where they were protected by rigidly enforced laws. This protection did not avail. The last heath hen to be seen alive disappeared on March 11, 1932. Bird authorities now agree that the heath hen is extinct.

WILD AND DOMESTIC TURKEYS

Turkeys belong to the family Meleagrididae. The name "turkey" is the result of mistaken identity. An Old World relative of the turkey, an African guinea fowl, was imported into Europe by way of Turkey. The guinea fowl then became known as the turkey cock in Europe. When the first Europeans settled in the New World, they confused the American bird with the turkey cock. Consequently they gave the American bird the name "turkey."

The original North American wild turkey, *Meleagris gallopavo,* has five subspecies distributed through eastern Canada, the United States, and the Mexican plateau. There is one other species of North American wild turkey, the ocellated turkey, *Agriocharis ocellata,* which is restricted to the Yucatan peninsula of Mexico. It gets its name from the ocelli, or eyelike spots, on its tail feathers.

The turkey that the Pilgrims ate at the first Thanksgiving was a northern subspecies of *Meleagris gallopavo.* At that time, wild turkeys were plentiful throughout the wooded areas of the land. Unfortunately, they were hunted down mercilessly. They are now quite rare and possibly near extinction. The male stands almost a meter tall and weighs about nine kilograms. The plumage shines with a metallic luster, reflecting browns, reds, greens, blues, and blacks.

The North American wild turkey may be distinguished from the ocellated turkey by a peculiar tuft of hairlike feathers that hang down from the breast and are sometimes known as a "breast beard." The wild turkey has powerful legs and wings. Although it prefers to remain on the ground and run from danger, it is quite capable of flying well over short distances.

Male turkeys of both the wild and domestic species are polygamous and are constantly seeking to add other females to their harems. Consequently, during the mating season, most males pompously strut about trying to attract prospective mates. Often their promenading is interrupted as they fight with other males.

The original domestic turkey is a Mexican subspecies domesticated by the Aztec Indians in Mexico long before Europeans ever voyaged to the Americas. This species was brought back to Spain in 1519 and then distributed throughout Europe. It was introduced into England some time between 1524 and 1541.

The original Mexican species was crossbred with many other varieties after it was brought to Europe. Many of the modern varieties of domestic turkeys resulted from this intensive crossbreeding. A number of these domestic breeds were reintroduced into North America.

PIGEONS AND DOVES

Wherever one goes, there are pigeons or doves. Hordes of common, or domestic, pigeons have firmly established themselves in large cities around the world. Mourning doves are found in great numbers in small cities and towns and in many rural areas. Only the arctic and antarctic regions are excluded from the range of pigeons.

Pigeons and doves make up the family Columbidae, one of two families in the order Columbiformes. The other family, Pteroclidae, contains the sandgrouse, which, depending on the species, inhabits areas of Europe, Asia, and Africa.

The flightless dodo, now extinct, also was a member of this order. This bizarre bird was larger than a turkey. It had a comparatively large head and an enormous beak, which was hooked at the tip. Dodos flourished in the islands of Mauritius and Réunion, in the Indian Ocean. When Europeans, with their domestic animals, moved to the island homes of the dodos, the big birds were doomed. By the end of the seventeenth century they were extinct.

GENERAL CHARACTERISTICS

For practical purposes, the terms "pigeon" and "dove" can be used interchangeably. However, the name "dove" is more properly applied to the smaller species of the family Columbidae.

Pigeons and doves have small heads and are decidedly plump and full-breasted. Their feet are so adapted that they can live with equal facility on the ground or in the trees. Their bills are horny at the tip and have a swelling, called the *cere,* at the base. The powerful wings are pointed. The notes of most wild pigeons and doves are soft, cooing sounds.

These birds feed chiefly on fruits, grains, and weed seeds, although insects also form a part of their diet. When a pigeon or dove drinks, it immerses its bill to the nostrils and sucks in continuously. As you may have observed, other birds drink in small gulps, hold up their heads, and swallow with the aid of gravity.

Some pigeons and doves live in large flocks. Others are solitary creatures. The male and female share the duties of nest building, sitting on the eggs, and feeding the young. The nest, made of twigs, is a very flimsy structure, in which one or two immaculately white or buff eggs are laid. The young are fed on pigeon milk, a substance secreted by the lining of the parent's crop. The adult bird pumps it into the mouth of the young. Pigeons and doves mate for life. When one of the mates dies, it is some time before the other accepts a new partner.

None of the surviving species of doves and pigeons approach the dodo in size. They dwell in the temperate and tropical zones throughout the world. The greatest number of species is to be found in the island chains off southeastern Asia. The plumage of these Asian birds is particularly handsome, showing brilliant shades of red, green, and purple.

THE DOVES

The dove has long been a symbol of gentleness, love, and purity. It figures in poetry and legend, both ancient and modern.

The mourning dove, *Zenaidura macroura,* which is about 30 centimeters in length, is easily identified, for in flight the tips of the outer tail feathers show a distinctive white band. The bird displays a black spot on the side of the head, behind and below the eye. The breeding range of the mourning dove extends through southern Canada and the United States southward into Mexico. Sometimes as many as three or four broods of young are raised in a single breeding season.

Mourning doves are among the first migrants to return north each year. Their mournful calls on a spring morning are in sharp contrast to the bubbling lay of the

The mourning dove, which is found from southern Canada to Mexico.

song sparrow. When the dove flies, the whistling sound produced by the wings can be distinctly heard. The mourning dove feeds principally on weed seeds and grasshoppers and thus is an ally of mankind.

The turtle dove, *Streptopelia turtur,* of Europe is very similar to the North American mourning dove. It is fawn-colored, with a larger black mark on the sides of the head. The attention that the turtle dove pays its mate is legendary. The paired birds sit and coo to each other for hours at a time. They inhabit woodland and open, bushy countryside.

The Inca dove, *Scardafella inca,* is distributed from the southwestern United States through Mexico and into Central America. About 20 centimeters long, it is grayish-brown above, grayish-red and buff on the underside. It nests in bushes, often very close to human dwellings. In farming areas, Inca doves can often be seen in barnyards, sharing a grain dinner with the chickens.

WILD PIGEONS

The most colorful pigeons inhabit areas of the eastern hemisphere. There are many attractive species. The topknot pigeon, *Lopholaimus antarcticus,* of Australia has gray feathers—and a bright rust-red topknot on its head. The crested pigeons of the genus *Goura* are named for a crest of rigid feathers on their heads. Their feathers are slate-blue, with areas of brick red. These are the largest of all pigeons. They equal large chickens in size.

The band-tailed pigeon, *Columba fasciata,* is an American species with a range from southern Canada to Nicaragua. It is about the size of the domestic pigeon. It has brownish and grayish upperparts, and a purplish-pink head and underparts. There is a white collar on the back of the neck, and a broad pale gray band on the end of the tail, which is black above. The band-tailed pigeon feeds largely on acorns, young sycamore balls, and wild berries.

The passenger pigeon, *Ectopistes migratorius,* now extinct, was a native of North America. It was bluish-gray above and reddish-fawn below; it had a broad white tip on its tail. The bird measured about 40 centimeters in length. The last one died in the Cincinnati Zoological Garden, in Ohio, on September 1, 1914.

Passenger pigeons lived together in enormous flocks, sometimes numbering over 2,000,000,000 birds. Migratory flocks sometimes formed columns 350 kilometers long. It took them many hours to pass a given point. In the places where the pigeons roosted, the vegetation became covered to a depth of several centimeters with their

Left: a common pigeon. These birds are found everywhere and are common in parks in large cities. Right: many people raise and train pigeons as a hobby. These birds are in a race. They will travel from the start at the Nevada border and reach San Francisco in six hours.

excrement. The ground was strewn with massive branches broken off by the weight of the birds roosting on them.

Mankind is entirely responsible for the passenger pigeons' complete disappearance. For years, people mass-slaughtered the birds on their nesting grounds, shooting them, clubbing them, catching them in huge nets, and choking them by producing sulfur fumes beneath their roosts. People kept track of the flocks by telegram and overtook them by train. From one nesting ground in Michigan, over 1,000,000 birds were sent to market in a single year. When the market was surfeited with the pigeons, people continued to slaughter them and drove hogs in to fatten on the victims.

After 1900, rewards of several thousands of dollars were offered for a pair of living passenger pigeons, but nobody ever collected any of these rewards. The last known nesting took place in Michigan in 1881.

DOMESTIC PIGEONS

The parent species of the domestic pigeon is the rock dove, *Columba livia*. In Europe this bird inhabits the rocky seacoasts, but in Asia and North Africa it lives inland. There are about 15 subspecies, all of which possess the grayish coloration we see on pigeons of city streets. It is interesting to note that ornamental domestic pigeons will revert, in a few generations, to this type of coloring if they are not carefully selected for mating.

In the domestic state, the pigeon has been carried to all parts of the world. More than 150 different kinds that breed true are now recognized by fanciers. Among the most interesting of the domestic strains is the homing pigeon. It is capable of flying 95 kilometers per hour and can find its way back to its home loft when released up to 2,000 kilometers away. It must be said, however, that this is not an unusual bird performance; as a matter of fact, many wild birds have even a stronger homing instinct and will find their way to their home loft over even greater distances.

Other strains developed from the wild stock include the fantail, whose tail may have 42 quills instead of the normal 12, and the pouter, which is able to inflate its gullet to great size. The pompous swagger of the pouter pigeon is a veritable parody on the ordinary male pigeon's inflated posture when cooing and strutting.

Hawk-headed parrot. Parrots have strong hooked bills and grasping feet. Members of the parrot group are usually found in tropical and sub-tropical environments.

Russ Kinne/PR

OTHER NON-PERCHING BIRDS

How are birds named? If you consider a variety of birds, you notice that their common names are based on many different characteristics. Some names are based on physical features. The hornbills, for example, are named for their horn-shaped beaks.

Some birds are named for the songs they sing. "Poor-will, poor-will," cries the poor-will. And you know who calls "cuccoo."

Hummingbirds are also named for the sound they made — not by their voices but by their wings. The wings move so rapidly that a humming sound is produced.

Some birds are named for their behavior. The roadrunner is named for the rapid way in which it can travel down a highway, making a sharp turn off the road only if some one chasing it is gaining.

And many birds are named after the foods they favor. The diet of bee-eaters consists largely of insects — especially bees. Goatsuckers received their name because superstitious people often saw the birds near goats and assumed — incorrectly — the birds milked the goats.

In this article, we will discuss these and other non-song birds. These birds belong to very different groups.

The order Psittaciformes contains tropical birds such as macaws, parrots, parakeets, and cockatoos.

The order Cuculiformes includes the cuckoos, roadrunners, and some tropical birds.

The order Caprimulgiformes includes birds with very wide mouths, small beaks, and tiny feet. Whip-poor-wills, nighthawks, and frogmouths are examples.

The order Micropodiformes contains the swifts and hummingbirds.

The order Coliiformes consists of the colies, or mousebirds.

The order Trogoniformes consists of tropical birds called trogons.

The order Coraciiformes is a varied groups that includes hornbills, kingfishers, bee-eaters, and others.

The order Piciformes contains birds that are adapted for climbing and digging into wood. Woodpeckers, toucans, puff birds, and honeyguides are examples.

PARROTS AND THEIR KIN

Members of the order Psittaciformes have short, hooked bills that are strong, enabling the birds to crack nuts and hard fruits. Each foot has four toes—two forward and two backward—and is well adapted for climbing and grasping. Their long, pointed wings are generally strong. Most parrots are able to fly long distances in search of the plant matter or insects that usually make up their diets. An exception is the kakapo, *Strigops habroptilus,* of New Zealand, which cannot fly at all.

These are noisy birds, with harsh, shrieking voices. In captivity many species can be taught to speak, and throughout the world, these birds are popular pets and zoo animals.

Members of the order occur in the tropics and much of the sub-tropical areas of the world. Australia and nearby islands contain the greatest variety of these birds. In fact, 16th-century mapmakers called Australia *Terra Psittacorum*—the "Land of Parrots." Unfortunately, advancing human settlements have decreased the parrots' numbers and threaten the existence of some species.

Approximately 320 species make up the Psittaciformes. Most have brightly colored plumage. Most are gregarious, except during the breeding season. They build their nests in holes—usually in trees but also in buildings, rocks, cacti, ant nests, and so on.

Cockatoos. Cockatoos are medium- to large-sized parrots; some are 50 centimeters long. They can be distinguished from other parrots by the erectile crest atop the head. Their tails tend to be relatively short and square-ended. The plumage is

Phillip Gendreau, Audubon/PR

Photo Ylla, Audubon/PR

Top: Wild macaw. Macaws are brightly colored birds that live in the tropical forests of Mexico through South America. Bottom: An Australian cockatoo. These birds range from Indonesia to Australia.

OTHER NON-PERCHING BIRDS 401

largely white, gray, or black. Cockatoos inhabit Australia, New Guinea, and the islands of Indonesia.

Lories. Lories are distinguished by the brush-like tip of the tongue. This, together with the narrow bill, is useful in obtaining the nectar and pollen that form the major part of a lory's diet. These are brilliantly colored birds that live in Australia, New Guinea, and nearby islands. Smaller species are called lorikeets, a common name also given to some other parrots.

Lovebirds. Lovebirds are small African parrots of the genus *Agapornis*. They are renowned for the preening and affection showered on one another.

Macaws. Macaws are large, gaily colored parrots that inhabit tropical forests from Mexico south into South America. The largest is the scarlet macaw, *Ara macao*, which may be more than 90 centimeters long.

CUCKOOS AND THEIR KIN

Most birds of the order Cuculiformes are insect eaters and render us valuable service by feeding on many pests. The bodies of these birds are rather long. The tail is also relatively long; it has eight to ten stabilizing feathers. The bill is slender and curved down slightly at the tip. The four toes on each foot are arranged in pairs: two toes point forward, two backward.

Cuckoos. The cuckoo family, Cuculidae, contains over 200 species and subspecies. These birds range from sparrow size to pheasant size. They are found all over the world except in the coldest spots where insects are few. They are particularly numerous in the African and Indo-Malayan tropics. Cuckoos are shy, furtive birds and are more likely to be heard than seen. In general, they have dull brown, gray, or black plumage, although some of the tropical varieties are brilliantly colored and have plumes. The bird has a long tail and short wings.

Some species of cuckoos have plumage closely resembling that of various birds of prey. The European cuckoo and the hawk cuckoos of Asia look so much like the sparrow hawk that they alarm all small birds when they approach. The drongo cuckoos of India resemble the drongo shrikes, whose nests they use, even to the extent of having the outer tail feathers curved outward after the fashion of their hosts. These cuckoos tend to be aggressive, in keeping with their assumed identity.

The cuckoos' nesting habits are of particular interest. The New World varieties build their own rather haphazard nests and care for their young, only rarely absent-mindedly dropping eggs into each other's nests.

The Old World species are parasitic. They lay their eggs in the nests of smaller birds and depend on the foster parents to raise the young, gaily going about their business and forgetting their offspring. Only one cuckoo egg is laid in each parasitized nest. The cuckoo's egg has a shorter period of incubation than the eggs of the host. When the egg is hatched, the young cuckoo grows more rapidly than its nestmates. The cuckoo seems to have an in-

White-crested turaco. This brightly colored bird is an inhabitant of southeastern Africa.

A. W. Ambler, Audubon/PR

stinctive urge to push the rightful occupants out of the nest. The parasitized birds never seem to realize what has happened. Apparently they take parental pride in filling the cavernous mouth of the strange young bird.

Young cuckoos do not have a juvenile plumage. They are almost naked when hatched and are black-skinned little creatures. Their growing feathers remain in the sheaths until fully developed so that for a long time the little birds' bodies seem encased in curious coats of mail. When the feathers are grown, the sheaths burst open at about the same time so that the change to adult feathering takes place in only a few hours.

The notes of the European cuckoo, *Cuculus canorus,* so familiar to everyone because of the cuckoo clock, are always associated with the coming of spring. These cuckoos are migratory, spending the winter in Africa and returning to Europe among the first of the spring birds.

The migrations of the bronzed cuckoo, *Chalcites lucidus,* of the southwest Pacific are nearly incredible. At the bird's nesting sites in New Zealand, the eggs are laid in the nests of flycatchers. When fledged, the young birds fly over 2,000 kilometers of water to Australia, and then another 1,500 kilometers to the Solomon and Bismarck Islands, where they join their parents, who preceded them there.

Anis. Anis are fascinating jay-sized cuckoos of South America, the West Indies, and tropical North America. They are typical pasture birds. They follow cattle, pecking ticks off their backs and catching insects that the cattle stir up. The anis are black with a metallic sheen and have large convex bills. They are weak flyers. Their long tails are composed of but eight feathers, the smallest number that any bird possesses.

These birds seem to prefer a communal way of life. When the nesting season arrives, the entire flock builds a nest. All the females lay their eggs in it, and, if there is not enough room for all the eggs, green leaves are placed over the first layer and the rest of the eggs are laid on

Karl H. Maslowski/PR

John H. Gerard, Audubon/PR

The cuckoos are a large, diverse group of birds. Top: the yellow-billed cuckoo lives in wooded areas in the United States. Bottom: the roadrunner, a cuckoo that would rather run than fly.

top. The lower layer does not usually hatch. The females incubate and feed the young in a group.

Roadrunner. Another curious bird of the cuckoo family is the roadrunner, *Geococcyx californianus,* which lives in the region of cacti, mesquite, and sagebrush near river valleys in Mexico and the southwestern United States. This bird has brownish-streaked feathers tipped with white, bright alert eyes, a very long

OTHER NON-PERCHING BIRDS

Two members of the goatsucker family. Top: a frogmouth, a sluggish bird that feeds on large insects. Lower: a nighthawk guarding its eggs. Nighthawks do not build nests, but lay and incubate their eggs on gravelled ground.

bronze-green tail, and strong legs. It has blue and orange skin around its eyes. The feathers on its crown can be erected to form a ragged crest.

The roadrunner dashes along the road at speeds clocked at 25 kilometers per hour. It prefers to travel by running and by fast, gliding jumps. It flies only as a last resort. The roadrunner is an amusing sight as it races along before a horse or slowly moving automobile with wings outspread, head and neck forward, and its tail jerking up and down. When the bird tires, it abruptly turns aside into the brush and brakes itself to a stop by throwing its tail over its back.

Nicknamed the snake-eater, the roadrunner is more than a match for the rattlesnake. It also eats lizards (of which it is especially fond), horned toads, young birds, and various insects. It builds a nest of sticks, snake skins, and grass in a cactus bush and lays from four to nine white eggs.

Turacos. Turacos inhabit central and southern Africa. Tree dwellers, they live in forests and wooded savannas. Fruits make up most of their diet. Some are called *plantain eaters,* because they seem to eat mostly plantains, or bananas. Other species are called go-away birds, after their call: "go awaa."

These pigeon-sized birds have short, round wings that restrict them to brief flights. They are well adapted to climbing and running along branches. Turacos have a crest atop the head that is usually made of red or white-tipped feathers.

Members of two genera are brilliantly colored by green and red pigments found in no other animals, as well as by tints of blue and purple. Other species are gray or brown.

These birds nest in trees, building a pigeonlike platform of twigs. The female lays two eggs, which are incubated about 18 days.

THE GOATSUCKERS, OR NIGHTJARS

The order name for these birds, Caprimulgiformes, is derived from the Latin words *capri mulgus,* meaning "milker of goats." As is so often the case, this name is due to a very old legend. Even the ancient Greek philosopher Aristotle thought that the birds used their unusually large mouths to suck milk from goats. Actually, almost all the goatsuckers feed on small flying insects, which they catch on the wing, using their gaping beaks as scoops. Since goats ordinarily attract myriads of insects, the goatsuckers often take advantage of their presence for a feast.

These birds are also called *nightjars,* because their booming calls "jar" the night. The order contains five families, which are distributed over wide areas of the world.

True goatsuckers. About 70 species spread throughout the world make up the family Caprimulgidae. They include the nighthawk, whip-poor-will, poor-will, and chuck-will's-widow, the last three named for the sound of their calls. The poor-will, *Phalaenoptilus nuttalli,* displays a phenomenon absolutely unknown in any other bird—hibernation. In western North America, it is often found in protected niches, hibernating through the winter.

The long feathers of the goatsuckers make the birds appear much larger than they really are. For example, the nighthawk, whose body is smaller than that of the robin, appears, on the wing, to be about the size of a sparrow hawk.

The goatsuckers build no nest whatever. They lay their eggs on the bare ground—not even in a depression to keep them from rolling. The eggs are whitish or cream-colored, marked with gray and purple. The young are hatched blind and helpless. They are soon covered with long grayish or brownish down, not very different from that which covers young owls.

Nighthawks are usually birds of the pasture or prairie country. They are seldom found in heavily wooded districts. They spend the day perched on a rock or post. At dusk, they begin seeking out their insect prey. They consume great quantities of gnats, mosquitoes, and other flying insects. Over 500 mosquitoes were found in the stomach of one nighthawk, and 1,800 flying ants in the stomach of another. Occasionally these birds will devour insects crawling along the ground. Nighthawks are sometimes seen at dusk along country roads, busily pursuing grasshoppers. They are also found, in increasing numbers, in cities, where they have little competition from other birds for the host of flying insects that are attracted by lights.

The whip-poor-will is a bird of the woodlands, spending the day on the ground under the trees and coming out into clearings or along the forest borders at night to feed. Whip-poor-wills feed on larger insects than nighthawks. They are particu-

Hugh M. Halliday, Audubon/PR

The whip-poor-will gets its name from the bird's call. The bird nests on bare ground.

larly fond of the large night-flying moths, whose larvae are very destructive to the tree foliage.

The whole family of goatsuckers is without exception beneficial to mankind, since its members feed almost entirely on insects. They prey particularly on night-flying insects, which have few other bird enemies and which include some of the most destructive species known.

Potoos. The woodland nightjars, or potoos of the family Nyctibiidae, are large, owl-like night birds of Central and South America. The mouth of the potoo is so large that it can close over a tennis ball.

Guacharo. The family Steatornithidae contains only one species: the oilbird, or guacharo, *Steatornis caripensis.* It lives colonially in caves in isolated parts of the Caribbean and South America. Natives trap the oilbird and melt down its body grease to make lamp oil and butter substitutes.

Frogmouths. The owlet frogmouths are a small family (Aegothelidae) found in Australia and New Guinea.

In the east Asian and Australian tropics we find a dozen different species of the true frogmouths (family Podargidae)—rather odd nightjars with wide, froglike mouths, and very sluggish habits. The frogmouths do not exert themselves to fly after insects, but confine their efforts to the large ones that can be caught by hopping about on the tree limbs.

HUMMINGBIRDS AND SWIFTS

Even close inspection fails to reveal much similarity among the adult hummingbirds and swifts—except in the shape of the long narrow wings and the tiny feet. Nevertheless, the kinship exists. Much of the internal structure is very similar. One can see the resemblance more clearly by noting the resemblance of the swifts to newly hatched hummingbirds with their small bills and large mouths.

Hummingbirds. The smallest bird in the world is the bee hummingbird of Cuba. It measures 5.7 centimeters in length and weighs less than 25 grams. Not all hummingbirds are tiny. The giant hummingbird of the Andes is over 20 centimeters long. However, the majority of the 580 odd species and subspecies are under 10 centimeters in length—of which half consists of bill and tail.

Hummingbirds are noted for the brilliance of their colors. John James Audubon, the noted 19th-century American ornithologist, called them "glittering fragments of the rainbow." Sometimes extremes of color, in wonderful combination, are found on the back or breast of single birds. Interestingly enough, the colors are not derived from pigments. They are the result of light diffraction, because of the peculiar structure of the feathers. The birds may actually appear quite somber if viewed in a certain light. They are most brilliant when they are flitting about from flower to flower in ever-changing light.

Many species of hummingbirds are curiously ornamented with tufts of feathers in various places.

Hummingbirds are found only in the New World. All efforts to introduce them elsewhere have failed. They undoubtedly originated in the Andes, in Colombia or Ecuador, where the majority of the hummingbird species are still found. The bird has spread as far as Patagonia to the south and Alaska to the north. Most species are quite local in range, some restricting themselves to a single valley.

These birds live on the nectar of flowers and the insects within them. Wherever they dwell, one will always find multitudes of flowers. The bills of many hummingbird species have become adapted to particular flowers. In all species the bill is probelike and the tongue tubular for sucking the nectar. In certain species the bill has become curved downward, even sickleshaped. Others have upcurved bills to help them get nectar from flowers having pouchlike or liplike corollas.

One species has a bill more than 12 centimeters long. The bill of another species measures scarcely 6 millimeters. Curiously enough, the two species feed from the same long, tubular flowers. The long-billed variety sips the nectar in the conventional way. The other drills a hole through the base of the flower into the nectary and inserts its bill in the hole. Many flowers depend on hummingbirds for cross-fertilization. The pollen is usually carried on the head or bill as the bird moves from flower to flower.

The voice of the hummingbird is seldom heard. It is the sound made by the rapidly beating wings that gives the bird its name. However, during the mating sea-

N.Y. Zoological Society

son the male of most species gives vent to its feelings with excited chirpings as it passes, with flashing wings, back and forth past the female. A few species have melodious songs of surprising volume.

The nests are skillfully constructed of plant down or catkins and are fastened in place by spider webs. The outside of the nest is usually ornamented with lichens and bits of moss. It looks much like the knot of a tree. Usually the nest is set upon a branch. In some species it is regularly fastened to the underside of a large leaf or even to a projecting cliff or overhanging rock. Invariably only two eggs are laid and these are always pure white without spots.

Swifts. Watching chimney swifts as they dart back and forth over housetops or circle in dark clouds before making their way into some unused chimney, one would never suspect their relationship to hummingbirds.

There are nearly 100 species of swifts (family Micropodidae). With few exceptions, they are sooty-black birds, sometimes with white on the rump or underparts but often with no marks whatsoever. The chimney swift, for example, is entirely sooty and only a little lighter in color below than above. In the East Indian tree swifts, however, the plumage has a metallic gloss and the feathers are quite silky. Most swifts have short, drab plumage.

The most interesting thing about the swift is its method of nesting. The nest is built of sticks, straws, feathers, or other material in the form of a shallow saucer. The materials are cemented together and fastened to the wall of a cave, hollow tree, or chimney by means of the bird's saliva, which is like glue. In one group of swifts inhabiting the islands off the east coast of Asia, the nests are made entirely of this saliva. These are the nests from which the Chinese make their famous bird's-nest soup.

The feet of swifts are small and very weak, and cannot support the weight of the body. Therefore, all feeding, nest-material gathering, and contacts between the birds must take place while in flight. When forced to rest or to climb the walls of its home, the swift remains vertical, grasping the wall with the aid of its sharp, curved claws and the spiny protuberances of its tail feathers. It rarely perches or stands on the ground.

During the mating season, a single unmated swift often attaches itself to a nesting couple and assists them throughout their labors. This cooperation may continue for two seasons, but ordinarily the helper acquires its own mate the following year and establishes a new family.

The swifts—as their name implies—are fast flyers. Even the smaller ones reach

Hugh M. Halliday, Audubon/PR

Left: emerald hummingbird perched on a branch. Hummingbirds can not only fly quickly, but also can hover and fly backwards. Right: chimney swifts climb walls and anchor themselves by spikes under their feathers. They attach their nests to walls using gluey saliva.

Mousebird nesting. Mousebirds are small dull-colored birds with unusual feet adapted for running along tree branches.

speeds of almost 100 kilometers per hour. This is a fair speed for a small bird, but not uncommon. The larger swifts of Asia and the Middle East have been clocked at 275 to 320 kilometers per hour, and together with certain falcons are the fastest living birds.

MOUSEBIRDS

The mousebirds, or colies, are small, long-tailed birds that resemble mice both in appearance and behavior. Their legs are short. The feet have sharp, hooked claws and an outer toe that pivots and so can point either backward or forward. This enables the mousebirds to run along tree branches—in a mouselike fashion. The plumage consists of grayish feathers that look like hair. The head has a crest.

Mousebirds live in Africa south of the Sahara. They feed mainly on fruits. They are graceful fliers, though they generally do not fly farther than from one tree to another.

These are gregarious animals. At night they gather in small groups in a large bush or tree and fall asleep cuddled against one another.

TROGONS

The 35 species of trogons, which comprise the order Trogoniformes, are beautiful, brightly colored birds. The male plumage often includes vivid shades of red, pink, orange, or yellow on the breast and belly. Some are a bold green. These colors contrast sharply with the glossy colors of the throat and neck. The square tail has a black-and-white design. The female plumage is generally more subdued.

Trogons are tree-dwellers that inhabit tropical areas around the world. Most species feed on fruit and large insects. They range from 22 to 35 centimeters in length. The bodies are stout and heavy, with a long tail and short, rounded wings. The beak is short, broad, and curved. The trogon's feet are unique: the two inner toes are turned backward.

Most trogons excavate nests in soft, decaying wood. Some nest in termite mounds or wasp nests. The female usually lays two or three eggs, which take 17 to 19 days to hatch. The parents share the tasks of incubation and caring for the young.

The most spectacular-looking trogon is the quetzal, *Pharomachrus mocino,* which inhabits dense mountain forests of Central America. It is colored in brilliant shades of green, blue, and crimson, and has long tail feathers. The quetzel is the national bird of Guatemala and has long been sacred to the Aztecs and other civilizations of Central America.

KINGFISHERS AND THEIR KIN

Members of the order Coraciiformes have long tails, strong beaks, and plume crests. The three front toes on each foot are partly joined.

The order includes kingfishers, hornbills, bee-eaters, todies, motmots, rollers, and hoopoes.

Kingfishers. The kingfisher family, Alcedinidae, contains 250 species and subspecies. Only 11 of these brilliantly colored bird species are found in the Americas. They are most numerous in the Malay Archipelago and New Guinea.

The kingfisher is most readily identified by its characteristic top-heavy appearance. It has a large crested head, heavy body, short tail, long straight bill, and feeble feet, with the second and third toes united. The colors usually seen in its plumage are metallic greens and blues, satiny whites, russets, and reds. The males and females are very similar. Sometimes the females may have the brighter plumage.

The Alcedinidae are divided into two natural groups, the wood kingfishers and the water kingfishers. The former live in woodlands, open country, and gardens, where they feed on insects, snails, frogs, lizards, and sometimes young birds and mammals. The water kingfishers are found about streams and lake shores. Here they live almost entirely on fish and other water creatures.

All the New World kingfishers belong to the fish-catching group. These birds prefer freshwater. They are seldom seen on the seacoast except where there are estuaries.

The water kingfisher stakes out a fishing territory along the edge of a stream or lake and defends it boldly against other feathered trespassers. The bird's call, a harsh penetrating rattle, is often heard as it goes about its fishing. Sometimes it sits patiently on an overhanging branch and waits for fish to swim past. At other times it will fly over the water until it sees a fish swimming near the surface. When the kingfisher sights its prey, it makes a rapid, head-foremost plunge with closed wings, and the fish is grasped or speared with its sharp bill. The bird is carried entirely beneath the surface of the water by the force of the fall. When it returns to its perch, it beats the fish against the branch to kill it. Then, juggling its prey about, the bird swallows it head first. The size of the fish the kingfisher can swallow is surprising. Indigestible bones and scales are later spit up in the form of pellets.

The water kingfisher nests in a hole that it digs, using its bill as a shovel and its feet as rakes. It begins by tunneling straight into a sandy bank for from 1 to 5 meters. At the end of the tunnel it digs out a rounded chamber in which it lays from five to eight white eggs.

The bones of the fish the kingfisher has eaten often line the cavity of its home. It is not a very clean housekeeper. It does not remove the excrement from the nest as most birds do. The kingfisher makes two grooves in the bottom of the tunnel as it waddles in and out. The presence of these grooves tells us the nest is in use.

Young kingfishers are naked when they are hatched. The first feathers are those of the adult bird. As with cuckoo young, the feather sheaths do not break open until the feathers have nearly matured.

The belted kingfisher, *Ceryle alcyon*, is the commonest North American species. It is about 33 centimeters long. It has grayish-blue upper parts and a band across its chest. Its wings are tipped with white. Its tail is banded in the same color. The white of its underparts extends around its neck in a broad collar. This species is found from the Arctic to the Gulf of Mexico. It winters in the more southern part of this range.

The common kingfisher of Europe and Asia, *Alcedo atthis,* has habits much like those of the American species. It is only 18 centimeters long, however, and is brighter colored—bright blue above and rusty orange-red beneath. Many beautiful legends are told about this kingfisher, or halcyon, as the ancient Greeks called it. It was believed that the bird built a floating nest of fish bones upon the sea, and that during its two-week brooding period, or "halcyon days," the god of winds kept the waters calm and peaceful.

The largest and most unusual kingfisher is the Australian laughing jackass, or kookaburra, *Dacelo gigas*. It is a brown and grayish-white colored bird about 50 centimeters long. Its cry is a discordant, abrupt laugh. One of the wood kingfishers, it feeds on crabs, reptiles, mice, rats, and young birds.

Motmots. Motmots are solitary forest-dwellers that occur from Mexico to Argentina. Most live in the northern part of this range. They feed on fruit and insects.

The plumage is boldly colored, especially around the head. The motmot has a serrated bill, short, rounded wings, and a tail with long central feathers.

Motmots build nests that are difficult to find. They use their bills to dig a nesting tunnel—that may be more than 1.5 meters in length—in a bank of soft earth.

Most species lay three or four white eggs, which take three to four weeks to hatch. The parents share the task of incubation.

Hornbills. Hornbills are Old World birds found in Africa and Asia. They inhabit dry wooded savannas and tropical forests. The diet consists mainly of fruit, though some hornbills also eat insects, lizards, and small mammals.

A hornbill is distinguished by its huge bill, which is usually topped by an equally large—or larger—boldly colored casque, a helmetlike, usually hollow structure.

Hornbills have an interesting nesting procedure. The nest is made in a tree hole. After the female settles into the nest, the opening to the outside is closed up with mud and other materials until there is only enough room for the female's beak to emerge. During the entire period of incubation—some 30 to 50 days—the female is fed by her mate. After the eggs hatch, the female remains with the chicks in the enclosed nest—perhaps another 30 to 50 days. Only when the young are able to fly is the wall finally broken down.

WOODPECKERS AND THEIR KIN

The woodpeckers, toucans, honeyguides, jacamars, barbets, and puffbirds make up the order Piciformes. All members of this order have feet with two toes pointing forward and two pointing backward. This makes an effective pincer for grasping the bark of trees. The wings are usually rounded. The feathers are hard and brightly colored, with differences in the coloring of the male and female plumage.

Piciformes are tree dwellers. They inhabit tropical and temperate woodlands and forests throughout the world except for Australia, New Zealand, and the island of Madagascar.

Woodpeckers. Woodpeckers (family Picidae) dwell not only in forests and woods but also in city parks and rural areas. The North American woodpeckers known as flickers are often seen on lawns and in gardens, diligently searching for ants and grasshoppers.

As a rule, woodpeckers are solitary birds. Family groups break up after the young are able to care for themselves. The Picidae do not migrate great distances as do many other kinds of birds. Certain species live the year round in the north temperate areas of the world.

Woodpeckers are highly specialized for their tree-climbing and grub-hunting activities. Their feet are strong and equipped with sharp, curved claws. The stiff tail feathers end in spines. These spines are pressed against the ridges in a tree's bark and help prop the bird as it digs for grubs or excavates a nesting site.

The woodpecker's head is large and its neck short and powerful, enabling the bird to deliver rapid and forceful blows with its stout beak. This beak, with its chisel-shaped tip, is an effective woodcutting tool. With it, the bird penetrates the bark and wood of trees, where wood-boring grubs, hibernating insects, and insect eggs are to be found. Once a small hole is made, the woodpecker's tongue dislodges the insect prey. The tongue is long and slender and can extend a considerable distance from the mouth. Its tip is usually pointed and barbed and is covered with an adhesive secretion.

The woodpeckers have no true song. Their call notes tend to be rather harsh and unmusical. During the mating season, the males go through an acrobatic courting ritual and replace the mating song with the characteristic woodpecker tattoo.

Woodpeckers have remarkably uniform nesting habits. They drill a hole in the tree they have selected and then lay their eggs on the chips in the bottom of the cavity. The size of the hole varies from about 3 to 6.5 centimeters in diameter. The hole is drilled into the trunk for a short distance and then dips down, in some species, to more than one-half a meter. It is enlarged at the bottom and covered with chips so that the eggs will not roll about. The various species of woodpeckers lay from 2 to 12 eggs, which are invariably glossy white and unspotted.

Most woodpeckers excavate new nesting cavities each year. Both male and female assist in the drilling. It is said that woodpeckers shrewdly remove the chips to a safe distance from the tree so that the nest's location will not be revealed. This is not always the case. As a matter of fact, woodpeckers sometimes lose their nests to starlings.

Sapsuckers. The North American woodpeckers known as sapsuckers, genus *Sphyrapicus*, get their name from drilling rows of small round holes in the bark of many species of trees and drinking the sap that collects. One bird usually taps several trees in several places and then makes the rounds. The sapsucker's tongue is modified for collecting sap, having a brush on the end instead of barbs. Sap, however, is not the main part of the bird's diet. It is primarily an insect eater. The sap holes serve, among other things, to attract insects. The bird feasts on them as it makes its sap-collecting rounds. The holes drilled by the sapsucker often mar timber.

Piculets. The smallest woodpeckers are the piculets. They are 7 to 10 centimeters long and usually lack the pointed, stiff tail-feathers found on other woodpeckers. These species are found in South America, Africa, and Asia.

Toucans. Toucans are noted for their large, highly specialized, and very colorful bills, which are several times larger than the birds' heads. Because the bill is filled with air chambers, it is very light. But it is also very strong.

These birds range in size from 30 to 60 centimeters. In some species, more than half of this length is bill. The plumage is mostly black or green, with bold patches of vivid colors, especially on the head and neck.

There are 37 species of toucans. They are gregarious, noisy birds that live in tropical forests from Mexico south to Argentina, sometimes as much as 1,500 meters above sea level. Toucans feed mostly on fruit, though they will also eat insects and the eggs and young of other birds.

John H. Gerard, Audubon/PR

Kenneth W. Fink/PR   R. Van Nostrand, Audubon/PR

Top: firmly anchored to the side of a tree by its clawed feet and spined tail feathers, this red-bellied woodpecker enters its nest hole. Middle: quetzal, the most spectacular of the trogons. Bottom: sulfur-breasted toucan. The very large, colorful bill is light in weight but very strong.

Geese in flight in a characteristic V-pattern as they travel in their seasonal migration to better climates.

WATER AND SHORE BIRDS

It is a memorable experience to hear the cry of the loon for the first time. It begins on a low note. Then it rises in pitch and increases in volume until it ends with a terrible spasmodic gasp. It has been variously described as "sinister," "wolflike," "defiant," "suggesting hysterical laughter." The owner of this distinctive voice is a large, handsome diving bird.

Other water and shore birds also have distinctive calls. There's the loud honk of the beautiful trumpeter swan, the "ker-wacky-wack" of the sooty tern, the grunt of the Cape Bareen goose of southern Australia—which is, as a result of its call, sometimes called the pig goose.

Birds that live on or near the water are placed in seven orders:

The order Colymbiformes contains the grebes and loons.

The so-called shore birds—gulls, plovers, skimmers, sandpipers, terns, puffins, and so on—belong to the order Charadriiformes. This order also includes woodcocks, snipes, lapwings, and many other species.

The marsh birds—cranes, rails, and coots—make up the order Gruiformes.

Waterfowl, such as ducks, geese, and swans, make up the order Anseriformes.

Stilt-legged birds, such as herons, storks, ibises, spoonbills, and flamingos, make up the order Ciconiiformes.

Seabirds—albatrosses, petrels, and shearwaters—form the order Procellariformes.

And the order Pelicaniformes contains the cormorants, anhingas, gannets, boobies, frigate birds, tropic birds, and, as the name tells us, the pelicans.

In the following pages, we will look at representatives of these orders.

LOONS AND GREBES

These birds, which spend almost all their time in the water, share certain characteristics. Both loons (family Gaviidae) and grebes (family Podicepedidae) are diving birds. They have elongated bodies and necks and are able to submerge their bodies so that, as they swim, only their necks and heads are above water.

The toes are webbed in loons and broadly lobed in grebes. Wings and tail are short. The plumage is dense and the body contains an extensive layer of fat.

THE LOONS

The loon is an expert diver and swimmer and can overtake the swiftest fish. It actually uses its powerful wings to "fly" under the water. In contrast, it is almost helpless on land. If danger threatens there, the bird must awkwardly flop to the water's edge, for it cannot get into the air from the ground. Once in the water, escape is possible by either diving or rising in flight.

The loon is a solitary bird, usually found alone or in pairs. Occasionally, however, several pairs may nest fairly close together if the marsh or lake is unmolested and the fish supply is plentiful.

The nest, which is never far from the water's edge, is merely a depression in the ground, sometimes crudely lined with bits of weed stalks or marsh vegetation. Two mottled, olive-colored eggs are laid. The newly hatched young, which are covered with down, are capable of swimming. Usually, however, the female carries them on her back for several days.

The common loon, *Gavia immer,* has a wide distribution in North America. During the nesting season it ranges from Alaska across arctic North America to Greenland and, southward, from northern California across the United States to New England and Labrador. In winter the loon migrates to southern California and Mexico on the west coast and to Florida and nearby on the Atlantic coast. It occurs inland along the largest rivers. The loon's winter plumage is a dull black or gray above, without white spots.

In summer the common loon is glossy black above, spotted with white, while the underparts are white. The head and neck are velvety black except for patches of white streaks on the lower throat and sides of the neck. The feathers are thick and compact, making an effective cloak for shedding water.

The loon's three front toes are completely webbed. Its wings are short, narrow, and pointed, and the tail is short and stiff. The heavy beak is a sharp-edged

Alvin E. Staffan, Audubon/PR

A common shore bird: the herring gull, *Larus argentatus.*

spear, making an excellent weapon for capturing and holding the fish that make up the main part of the loon's diet.

THE GREBES

The grebes measure under 60 centimeters in length. Like the loons, they excel as divers and swimmers, but are helpless on land because their feet are placed too far back on the body to be very efficient in walking.

Grebes have softly patterned plumage of white, gray, black, or brown tones. They have pointed bills and very short tails—nothing more than tufts of feathers. The toes are individually webbed, or lobed. The grebe's stomach invariably contains a ball of its own feathers, which it has swallowed to prevent sharp fishbones from continuing into the intestine.

The diving ability of the grebe has led to its being nicknamed the "hell-diver." It is able to stay under water for lengthy periods. When it does come up after having been alarmed, it rises to the surface very quietly, and only until the bill shows above the water. The grebe can dive either head foremost with a flip of its feet, or it can settle backwards so carefully as to leave scarcely a ripple on the surface. It is able to dive so rapidly at the flash of a gun that it disappears before the shot reaches it. The grebe seems to prefer to use this method of escaping enemies rather than to fly. It seems to take much effort for the bird to fly. Ordinarily it must patter along the surface of the water for some distance before it is able to get up enough speed to lift itself.

The commonest grebe is the pied-billed, *Podilymbus podiceps,* which ranges from Canada to Chile and Argentina. It is an inconspicuous, brownish little bird even when in its breeding plumage. It is most easily identified by its chicken-shaped bill encircled by a black band. It is usually found on reed-bordered ponds and marshy lakes, where it builds a floating nest and anchors it to the reeds. The nest is only a pile of debris, and the grebe's eggs, white when laid, are soon discolored by the decaying vegetation.

Young pied-billed grebes have black and white striped down and are able to swim as soon as they are hatched. The first-hatched ones soon slip into the water and follow their father while the mother incubates the remaining eggs. When tired or cold, the young snuggle beneath the wings of one of the parents and onto the back.

SHORE BIRDS

These birds are most commonly seen in marshy areas and along shores. They are well adapted to flying and many migrate long distances in the spring and fall. Most species have long, slender beaks that are well-adapted for seizing the insects, shellfish, and worms that are their primary foods.

THE GULLS

Around the seacoast there is no bird more familiar than the graceful, long-winged sea gull. Most people think of it only as a sea bird, but several of the approximately 50 species live on inland bodies of water as well.

The gull's skill at wheeling and dipping flight, and its strong curving bill fit it admirably for its career as scavenger of port and

Peter Slater/PR

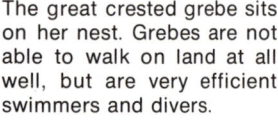

The great crested grebe sits on her nest. Grebes are not able to walk on land at all well, but are very efficient swimmers and divers.

harbor. Wherever it is found, the gull follows the ships, ready to pounce on any offal thrown overboard. It also congregates in large flocks in harbors. This bird is absolute master of the air and can survive the fiercest storms. It often rests by sleeping on the waves, and it has webbed feet like a duck's. It flourishes in the severest climate.

The gull is about 60 centimeters long, and is usually white, with the back and upper surfaces of the wings (called the mantle) washed with gray, black, or brown. The head is often black during the nesting season and black is sometimes seen on the flight feathers. The tail is square.

A gull never plunges into the water; rather, it is a surface feeder. It will eat almost anything. This bird is often seen carrying bivalve mollusks into the air and dropping them in order to break the shells. It occasionally preys on other water birds' eggs and young. However, the gull certainly renders mankind a valuable service by consuming large amounts of refuse.

The gull nests in large colonies and lays three heavily marked eggs in a crude nest of seaweed. It is sociable but quarrelsome, and it is quite likely to be harsh toward other gulls' young. The baby gulls announce their hunger by tapping on the grownups' bills.

THE TERNS

The tern is another graceful flier of the seacoast. It is more slender than the gull. It has a forked tail, which gives it the nickname "sea swallow." The long, straight, slender bill equips the tern to fish for its living. It dives for small fish and catches insects while on the wing. In appearance, the tern is typical of this tribe of water-loving birds—it is white below and gray above and has a black or gray cap.

Helen Cruickshank/PR
Audubon/PR
PR

Top: the least tern is the daintiest of the terns, with a body length of about 20 centimeters. Middle: the black skimmer is a bird of the Western Hemisphere with an unusual scissor-like bill. Bottom: the spotted sandpiper is a shore bord that is easily identified by its jerky walk.

The migrations of certain terns are famous. The arctic tern, for instance, has an annual round-trip migratory flight of 35,000 kilometers—the longest of any known creature. Some terns do not go on these extended migratory flights. They favor coastal regions in warm latitudes.

Terns nest in colonies. In slight hollows scraped in the sand, they lay one to three white eggs mottled with dark brown. Some species build nests of seaweed in bushes. The fairy tern balances its single egg on the gnarled bark of a horizontal limb.

SKIMMERS

The skimmer differs from all other birds in having a vertically flattened knifelike bill, which gives it the nickname "scissorbill." The lower part of the bill is much longer than the upper part. The skimmer also differs from other birds in eye structure—it has a vertical slitlike pupil. Flying low over the water with its bill cleaving the surface, the skimmer is able to seize small marine organisms and fish.

Three species of these long-winged birds make up the family Rynchopidae. The American black skimmer, *Rhynchops nigra*, nests along the U.S. east coast and south to Argentina. Its plumage is black, with the forehead, sides of the face, and underparts white. It has red legs. The skimmer lays buff eggs with black blotches in a hole scraped in the sand.

Another skimmer species is found in Africa; still another, in India.

SANDPIPERS

The sandpipers family Scolopacidae, are so called because most live on sandy beaches and have a piping cry. They have long, flexible bills with which they probe the sand for food. Their sizes vary from the least sandpiper, no longer than a sparrow, to the curlews and godwits, standing 30 centimeters tall. However, aside from size they all look much alike, plumaged with similar patterns of gray, brown, chestnut, and white.

There are about 100 species of sandpipers, living in various parts of the world. They are seen mainly in the temperate and northern portions of the Northern Hemisphere during their nesting seasons, many of them nesting within the Arctic Circle. Sandpipers are birds of the seacoasts mainly, but some are found inland on the shores of bodies of fresh water.

The sandpipers are great travelers. Some traverse the entire length of both the Americas in their migrations. The majority of species spend the summer in the Far North and in the fall travel across the Caribbean to northern South America. Some, however, stop on the North American Gulf Coast while others continue as far as Chile and Patagonia.

When the sandpipers leave their summer homes they have stored up thick layers of fat. By the time they reach their winter quarters, however, the majority have grown thin. This is particularly true of those that have undertaken the long flight from Nova Scotia to Venezuela or from Alaska to the islands of Hawaii without a single stop. A non-stop flight of 4,000 kilometers seems almost incredible, but such a feat is not at all unusual for these birds. Some of them have even been observed at Cape Horn—about 14,500 kilometers from their nesting grounds. During the nesting season they are solitary, but at migration and during the winter they may be seen in large flocks.

Common figures on the North American continent: a pair of mallards. The male has the bright green head. The female is duller and blends in with her nest site.

Phillipa Scott, Audubon/PR

With the exception of the solitary sandpiper, all the members of this group nest on the ground. The solitary sandpiper, however, utilizes the old nests of other birds such as robins and grackles. The sandpipers lay three or four eggs, which are quite large for the size of the bird, and which are sharply tapered at one end so they will fit together like the pieces of a pie. Otherwise the parent bird would be unable to cover them while incubating. The eggs vary greatly in color, but are most often spotted with black, brown, or lavender.

The young sandpipers are covered with down when they are hatched. Many of them have a striped pattern. They are able to run about and follow their parents or even to swim across streams. The first plumage is similar to that of the adults in the fall, and in the following spring all molt into the breeding plumage. If there is a bright plumage, females wear it as well as males.

The food of the sandpipers includes many mosquito and fly larvae. Among some species, grasshoppers and other destructive insects form a large part of the diet.

The commonest species of sandpiper in North America is the spotted sandpiper, *Actitis macularia*—"tip-up" or "teeter tail," as it is variously called. In summer it is found along almost every stream and lake from northwestern Alaska to Louisiana, and in winter from Louisiana to southern Brazil. About 19 centimeters long, it has light brown upper parts and white underparts spotted with black. In the fall it loses its spots, but is still easily identified because of its jerky walk. Several other species jerk their heads when they walk, but the spotted sandpiper teeters its tail or its whole body as though it had difficulty balancing on its slender legs.

THE PLOVERS

The members of the plover family Charodriidae resemble sandpipers somewhat. They are about 25 centimeters long and have plump bodies, long pointed wings, and short necks and tails. Their bills are short and shaped rather like a pigeon's, with an enlargement at the tip.

Golden plover. This species is famous for its very long migration—over 4,000 kilometers.

All the plovers' movements are quick and energetic. They run rapidly over the ground—much like mechanical toys. There are about 75 species in this family which also includes the lapwings, turnstones, and surf birds.

The plovers are most remarkable for their migrations. The lesser golden plover, *Pluvialis dominica,* for example, covers 13,000 kilometers in its round-trip migratory flight. It passes south from Nova Scotia to the northeast coast of South America—a distance of 3,900 kilometers. This flight over long stretches of open ocean is often made without a stop. In the spring the bird returns to its northern nesting grounds by a different route—by the northwest coast of South America and the Mississippi valley. The two routes are fully 2,500 kilometers apart.

The golden plover has upper parts spotted with golden-yellow and black, and its underparts are uniformly black in summer and grayish-white in winter. A white stripe from the forehead down the side of the neck and breast is conspicuous in the summer plumage and is set off against the black underparts.

The black-bellied plover, *Pluvialis squatarola,* is very similar to the golden plover. It also has a change of plumage with the seasons, but lacks the golden-yellow spots of the upper parts. This bird is found over nearly the entire world. It, too, is a great traveler.

The golden and the black-bellied plovers have quite similar habits. They rarely enter the water. They feed on sand bars and mud flats exposed by the falling tide. Here they run along the beach in search of stranded aquatic insects and crustaceans, which they pick up with a vigorous tilt of the body. They are also seen in plowed fields or pastures.

An interesting member of the plover family, found in the Eastern Hemisphere, is the Nile plover, noted for its life habits in relation to the crocodile. The bird picks leeches and other parasites from the mouth of the reptile and feasts on them. In return, not only does it rid the crocodile of these pests, but it also acts as a sentinel, warning the reptile of danger.

Plovers nest in depressions on the ground. Sometimes the depression is lined with moss or grass. More often it is not. Like sandpiper's eggs, the plover's eggs are much larger at one end than at the other. If the wind blows them, they will roll away. The egg color is neutral, with specklings of dark brown, black, lilac, and so on.

OTHER SHORE BIRDS

Phalaropes, family Phalaropodidae, are ocean-going birds. Wilson's phalarope, *Steganopus tricolor*, spends many months at sea during the winter. It is well adapted for this type of life, with its weblike toe lobes and dense, gull-like plumage. However, it often nests in the interior marshes of North America. The female phalarope assumes the bright plumage and does the courting at breeding time, and it is the male phalarope that incubates the eggs and cares for the young.

Avocets and stilts, family Recurvirostridae are long-legged wading birds with long, upcurved bills. They build their nests on sandy banks, in muddy flats, or in marshes. The black-winged, or common, stilt, *Himantopus himantopus*, of southern Europe is easily recognized by its exceptionally long pink legs, which project some 17 centimeters beyond its tail when the bird is in flight.

The jacanas are tropical birds, with long legs and long toes that are adapted for walking on floating water plants — a habit that has earned them the nickname "lily-trotters."

Auks and puffins, family Alcidae, often live together in huge groups that also include murres and gulls. They inhabit cliffs and islands in the northern seas. Most migrate south in winter, returning north in the spring to breed. Auks and puffins do not build nests, but lay their eggs on ledges or niches in the rocks. Each female lays only one egg. She and her mate take turns guarding it.

The woodcock, *Scolopax* species, is a shy bird. It ventures into the open after dark, but spends the day in thickets. The woodcock has a protective coloring of grayish-brown, reddish-brown, and black. Its long, thin bill takes up a third of its length. The large eyes, set far back in the head, give the bird a very distinctive appearance. It likes to feed in the lowlands, where it can bore into the soft mud for worms and other small invertebrates.

MARSH BIRDS

The members of the order Gruiformes are not well adapted for flight, but they are very well adapted for life on land and in the water. The order's three major families are the Gruidae, or cranes; the Rallidae, which includes rails, gallinules, coots, soras, and wekas; and the Otididae, or bustards.

CRANES

These birds are found on every continent except South America. The best-

The black-necked stilt of the Western Hemisphere pretends injury to distract prey from its nest.

Robert J. Erwin, Audubon/PR

WATER AND SHORE BIRDS

The common crane, *Grus grus,* found throughout much of Europe and Asia.

Jacana, a tropical bird that typically "trots" over floating vegetation.

known species is probably the American whooping crane, *Grus americana,* because of the tremendous efforts that have been made to try to save this beautiful animal from extinction. The whooping crane summers in western Canada and winters on islands off the coast of Texas. It flies from one site to another—a total distance of more than 3,700 kilometers—at an average rate of 160 kilometers a day.

The common crane, *Grus grus,* also migrates great distances. It summers in north-central Europe and Asia and winters in northern Africa. The migrating birds move in a "V" or line formation, flying with their long necks and legs extended.

RAILS, GALLINULES, AND COOTS

The members of the family Rallidae are all "thin as a rail," for nature has given them slender, compressed bodies which enable them to slip easily through bunches of marsh grasses. These secretive birds are mostly freshwater marsh dwellers.

There are about 225 different forms of Rallidae. They live everywhere but in Antarctica. The family characteristics are short rounded wings, long necks, and muscular legs, with long slender toes that help them to scamper easily across the marshes. The coot has lobes on each side of its toes to assist in swimming. It is much more aquatic than the other species and often gathers in large flocks on the water like ducks.

Many species have a chickenlike appearance, with their short erect tails and short wings. Like chickens, too, they have clucking voices. The gallinules, coots, soras, and yellow and black rails have short, thick, pointed bills. The Virginia clapper and king rails have rather long, slender, and somewhat downcurving bills.

In the water, both coots and gallinules are very ducklike except for their small heads, which they continually pump as they swim. Strutting and running along the border of the marsh, they look more like little hens, as they peck at seeds and insects on the ground. Some of the most startling sounds that come from the marshes can be traced to these birds. Their ordinary calls sound somewhat henlike: *cut-cut,* or *tuka-tuka.* But occasionally they give vent to loud, angry screeches that shatter the air. Unlike the rails, they rarely call at night, and are heard mostly in the early morning or in the evening toward dusk.

The coots and gallinules build their nests of dried rushes, hidden in clumps of marsh vegetation, and sometimes lodged in tall, growing stalks, one half to one-one half meters above the water. Some are floating platforms that can rise and fall with the floods. The eggs are buff, the coot's being evenly spotted with black and brown, and the gallinule's, more sparsely spotted with brown and lavender.

The young of almost all the coots, gallinules, and rails are covered with glossy

WATER AND SHORE BIRDS 419

black down when hatched, the coots being ornamented with an orange, beardlike fringe around the throat. The coots and gallinules lay their eggs at the rate of one a day, and begin to incubate each one as soon as it is laid. The young are able to swim and run almost as soon as they are hatched, and leave the nest very shortly.

Rails in general are the most secretive members of this elusive family, and seem to take wing only when an escape by dashing through the weeds is impossible. Their nests are built on the ground, hidden in the grassy marshes. The sparrow-sized black rail, smallest and rarest of the family, conceals its nest very effectively in a small depression.

THE WATERFOWL

The waterfowl include swans, geese, mergansers, and ducks—all members of the order Anseriformes and the family Anatidae. Over 200 species of waterfowl have been recognized.

These birds are adapted for life on the surface of the water. They have broad, flat bodies that are lightened by air sacs and hollow bones. They keep their feathers well oiled to prevent water from reaching the skin. Their swimming ability is aided by webbed feet and strong leg muscles.

Waterfowl have long necks with which they reach to the bottom of streams and ponds for food. Their center of gravity lies forward. Hence it is not necessary for them to tip up as the bill searches for food on the bottom. The bill in most species is broad and flat and has grooves at the side, forming an effective strainer when the bird dabbles after small water organisms.

One interesting characteristic of waterfowl is that they lose all the wing quills at the same time that they molt their nuptial plumage after the breeding season. As a result, they are unable to fly until new quills have grown in. They must, therefore, depend on swimming for a while to escape their enemies.

THE SWANS

These elegant birds have much longer necks than the other waterfowl—even longer than their bodies. Their beauty and grace has given them a position in the poetry and legends of many nations. Of the eight species of swans, two are found in North America—the trumpeter and whistling swans. Both species are pure white except for the black bill and feet. The adults are more than one meter long. These two species closely resemble the common domesticated swan, which has been derived from the European mute swan and which can always be identified by the tubercle or knob on its bill.

One unusual species is the black swan, *Cygnus atratus,* of Australia. A Dutch explorer who saw these beautiful birds in 1697—at the mouth of what is now called the Swan River—captured some of them and carried them back to Europe, where they had never before been seen.

THE GEESE

There are about 30 species of wild geese. The Canada goose, *Branta canadensis,* is the most abundant and the best known of the species in North America. It has been introduced to Europe, where it may often be seen in parks. It is about one meter long. It is gray-brown above, grayish beneath, and has a black head and neck. A white patch runs under its chin and up both cheeks like an incomplete chinstrap.

The Canada goose nests from the northern United States northward to the limit of trees. It winters from the Great Lakes southward into Mexico. The travel-

The Black Swan of Australia and Tasmania was introduced into Europe in the 18th century, and recently into New Zealand.

Russ Kinne, PR

ings of the Canada geese are among the most conspicuous of bird migrations. One hears their loud honking long before they become visible as they travel high overhead in a great wedge or "Y." These birds migrate both by day and by night. On their migrations they are great vegetarians and are fond of grazing on young wheat, both in the spring and in the fall. In the South on their wintering grounds, however, they seem to prefer to feed in the shallow water of the bays and lagoons, tipping for aquatic plants and organisms like the dabbling ducks.

The snow geese, *Anser hyperboreus,* are easily recognized because they are white except for their black flight feathers. Sometimes, however, the head is stained orange. Found in both North America and Europe, these are colonial birds that generally breed on lake islands and tundra.

The graylag, *Anser anser,* is the best known of the European wild geese. This ashy-brown bird is the species from which the domestic goose has been derived. It has the same nasal, reedy voice as the domestic bird.

THE MERGANSERS

Mergansers are fish-eaters. They are easily distinguished from other members of the family Anatidae by their crested heads and narrow saw-toothed bills, which end in a strong hook to aid them in grasping slippery fish. They seldom walk far from the water's edge and must patter along the water's surface before flying.

Mergansers pursue their finny prey under water by the drive of their powerful feet, which are set far back. They first locate their prey by lowering their heads until their eyes are beneath the surface film. They can often be seen swimming along in this position. These birds are not valued as food, because their fishy diet gives their flesh a coarse, rank taste.

Three of the nine species are found in North America. All of them nest either on the ground or in hollow trees. Six to eighteen eggs are laid, and the young enter the water as soon as they are hatched. The mergansers breed in the Far North and migrate southward as far as Mexico on

Plomion

Canada geese. These are the most abundant and well-known of all the North American species.

the approach of the colder season. The females of all three species are grayish birds with conspicuously crested, reddish-brown heads. The males are boldly marked black and white birds.

THE DUCKS

Ducks are divided into two groups — the surface-feeding ducks (subfamily Anatinae) and the diving ducks (subfamily Nyrocinae).

It is true that surface-feeders sometimes dive for food and that diving ducks sometimes feed on the surface. In the main, however, the surface-feeders live in marshes and on lake shores where they can feed in shallow water by tipping their bodies. They seek food mostly at night or on dark days, and are also frequently seen feeding on land.

The diving ducks, on the other hand, feed in deep water, often far from land, for they dive readily and secure their food of mollusks, roots, and buds of aquatic plants in water up to 45 meters deep. They seldom feed on land, since they do not walk as well as their surface-feeding relatives. These ducks also migrate later than the surface-feeders, for they are not influenced by the freezing of the marshes and shallow water. And, since they are less exposed to enemies while foraging than the surface-feeders, they feed more during the day than at night.

The male pintail duck has handsome plumage, black and white except for the brown-feathered head.

They are better adapted for diving than the surface-feeders. They have larger feet and stockier bodies, with shorter necks and shorter wings. Also, the hind toes of the diving ducks have lobes which make them resemble paddles. These lobes are not present in the surface-feeders.

While swimming, diving ducks rest lower than the surface-feeders and do not hold their tails up from the water. A beauty spot, formed by the iridescent color of the secondary feathers of the wing, is a conspicuous feature of most surface-feeders, while it is absent from nearly all divers. When rising from land into the air, the surface-feeders spring directly up without a preliminary run, while the diving ducks must run over the surface of the water to work up enough speed for flight.

One interesting characteristic of all ducks, both wild and domestic, is that around midsummer, the drakes (male birds) shed their bright plumage for two months or longer. The bright feathers are replaced by duller plumage, resembling that of the females. This is known as the eclipse feathering, and takes place at the same time that the bird sheds its flight feathers. The reason behind this periodic change of dress remains somewhat of a mystery to the scientist, for it only occurs among ducks native to, or transplanted to, the Northern Hemisphere.

Most ducks pair and nest the spring following hatching. The wild duck is monogamous, but the domesticated male is polygamous. This is probably due to living in captivity under unnatural conditions. The drake is not very much devoted to its family, even under the best of circumstances—it usually abandons the female as soon as incubation is well under way.

The best-known of the surface-feeding ducks regularly nest on the ground, usually near water. The nests are crude affairs of grasses and weeds. As incubation proceeds, the female plucks down from her breast; she uses the down to cover the eggs, thus keeping them warm when she goes off to feed.

The mallard, *Anas platyrhynchos*, is the most common and one of the most widely distributed species of surface-feeding ducks. It breeds readily in captivity as well as in the wild state. The males have bright green heads and white rings around their necks. The females are uniformly streaked yellowish or grayish-brown.

The pintail, *Anas acuta,* is also widely distributed throughout the Northern Hemisphere. The male has a chocolate-brown head and neck, with a conspicuous white streak up each side of the neck. Its tail is long and pointed. The female looks much like the mallard female.

Diving ducks commonly gather in large flocks—thousands of scaup ducks, for example, are not an unusual sight on larger bodies of water. The diving ability of these ducks can hardly be exaggerated. Some members of the family, especially the old squaws, are repeatedly captured in gill nets set for fish 30 to 45 meters below the water's surface. The old squaws, scoters, and eiders are believed to use their wings as well as their feet in diving. The rest of the diving ducks, however, use only their feet, which are set very far back—this is the reason that these birds can walk only with difficulty.

The diving ducks prefer the marshes for nesting. They can slip from their nests directly into the water without having to walk on dry land.

The canvasback, *Aythya valisineria,* so called from the white back of the male, is easily recognized because of its long, flat, gradually sloping profile. The male has a reddish head and neck. The female's back is gray; her head and neck are cinnamon brown.

Eider ducks, *Somateria mollissima,* inhabit northern regions around the world. Eider ducks are the largest ducks and are very hardy. The male is the only duck that has a black belly and a white back. The female is brown.

Like various other ducks, the female eider duck covers her eggs with down plucked from her breast. This down, taken from the nest by humans, is the eider down of commerce.

STILT-LEGGED BIRDS

Members of the order Ciconiiformes are well adapted for their life as waders. In addition to their long legs, they have long necks and long bills, which help them catch fish, snakes, frogs, and so on.

The approximately 120 species are widely distributed. These birds are strong fliers, and many of the temperate-zone species migrate.

HERONS, BITTERNS, AND EGRETS

The long-legged herons are wading birds that live in marshes of salt-water lagoons, freshwater lakes, and rivers. No matter how much they eat, they always look emaciated, with their long necks and legs and light, bony bodies.

Herons are patient fishermen and often can be seen standing quite still in the water waiting for prey to come near. Some herons wander about in search of prey, staring at the water and lifting each foot and putting it down so carefully that there are no ripples to warn the water creatures. The heron's bill is shaped like a stiletto and is a most formidable weapon. The bird has a vora-

Left: a purple heron. These birds live in colonies and nest in bushes or trees near the shore. Middle: a white stork. White storks are common in Europe, Asia, and Africa. Right: an egret.

Top, George Holton, Audubon/PR; bottom, B Tollu—Jacana

Sea birds. Top: giant petrels. This courting pair nests in Antarctica. Bottom: a pair of courting albatrosses. Albatrosses have a very large wingspan.

cious appetite and consumes fish, frogs, tadpoles, crayfish, small snakes, meadow mice, shrews—all with a spicy relish.

The voices of the heron family are hoarse croaks and squawks. The birds vary in size from the least bittern, whose body is not much larger than that of a robin, to the great white heron, which is more than one meter long. Colors range from the streaked brown plumage of the bittern through various shades of blue, gray, and chestnut to the snowy white of the egrets. Herons are ornamented with elongated feathers on crown, foreneck, or middle of the back.

Herons are gregarious birds, roosting and nesting in colonies. The nests are usually built in trees or large bushes in swamps. The bittern, however, nests on the ground in treeless marshes. Herons' nests are bulky platforms of twigs and sticks lined with grass. Three to six pale greenish-blue eggs are laid. The old birds double up their long legs under them when brooding.

When the young are hatched, they are covered with down. If alarmed, the young herons stretch up their long slender necks and remain perfectly quiet so that they look more like sticks than birds. Should the nest be disturbed, they may greet the intruder with thrown-up stomach contents of partly digested food from the parent's crop.

The name "egret" is usually applied to the white members of the family that wear long nuptial plumes, or aigrettes, on their back during the breeding season. These plumes, which look as if they were made of spun glass, have nearly led to the egret's extinction, for they were once in great demand as a decoration for women's hats. They could be obtained only by killing the adult birds; the fledglings would be left to starve. As a result, the egrets all but disappeared. Thanks to the efforts of legislators and environmentalists, most species are no longer in danger of extinction.

Bitterns are smaller than herons. Like the other members of this family, they live in reedy marshes. Bitterns are difficult to observe, however, as they are shy and skulk about among the reeds. When approached closely, a bittern seems to freeze into position with head, neck, and bill pointing to the sky and the streaks along the throat blending with the surrounding reeds. The bittern has a booming call in the springtime and is a skilled ventriloquist.

FLAMINGOS

With plumage of rosy pink or bright, vivid scarlet, the flamingos are among the most beautiful of birds. They have very long necks and legs. Their bills are distinctive. The lower part is like a broad, deep box. The upper part fits into the lower part, like a box lid. The sides have gill-like structures that serve as sieves. When the bill is thrust into mud, the sieve strains out shellfish, frogs, plant matter, and other food from the mud.

The most widely distributed species is the greater, or scarlet, flamingo, *Phoenicopterus ruber*. In the New World it ranges from the Bahamas through the Caribbean and into parts of Central and South America. In the Old World it is found around the Mediterranean Sea, Caspian Sea, and Persian Gulf. It favors shallow coastal waters.

The female lays one or two eggs in a conical nest made of mud. The young birds have dull grayish-brown plumage. Several years pass before they attain the vivid adult colors.

SPOONBILLS AND IBISES

These birds of the family Threskiornithidae are, for the most part, tropical or sub-tropical, though some summer as far north as Scandinavia and southern California.

Except for their bills, spoonbills and ibises resemble each other, as well as herons and storks. The spoonbill has, as you have guessed, a spoon-shaped bill. It is very well adapted to catching the frogs, shellfish, insects, and small fish on which a spoonbill feeds. The ibis has a long, slender, down-curving bill that is more or less cylindrical. It seeks prey similar to that of the spoonbill.

Spoonbills are gregarious birds, often forming very large colonies during the breeding season. They build nests on the ground, in bushes, or in low trees. Some ibises are colonial and are often found together with spoonbills and herons. Other species are generally solitary birds. They, too, build nests in a variety of sites: on trees, bushes, in holes in rock ledges, or on the ground.

Plumage of these birds ranges from white to shades of pink and rose. The white ibis, *Eudocimus albus,* has pure white plumage, with a few large feathers tipped in black. The glossy ibis, *Plegadis falcinellus,* has green, purple, and bronze tints on its plumage. The roseate spoonbill, *Ajaia ajaia,* has pink plumage, pink legs, and a yellow bill.

SEA BIRDS

Four families of sea birds make up the order Procellariiformes: the albatrosses; storm petrels; diving petrels; and shearwaters and fulmars. These birds spend almost all their time in the air. They come ashore only to breed.

These birds are commonly called tubenoses, because the nostrils open through tubes that extend along the top or sides of the bill. The tip of the bill is hooked and enlarged. The upper half of the bill is longer than the lower half, and is curved downward.

Every part of the world is visited by

A large number of herons, egrets, and spoonbills aggregate on the shores of a bird sanctuary.

Audubon/PR

Spoonbills, found in temperate and tropical regions, use their bill to strain for crustacea.

Procellariiformes. Individual species may be very wide ranging. The sooty shearwater, *Puffinus griseus,* for example, makes its homes in the oceans of the Southern Hemisphere but, in summer, flies as far north as California and northern Europe. It returns every year to the same nesting site.

The albatrosses are the largest members of the order. The wandering albatross, *Diomedea exulans,* may attain a length of more than one and one third meters and a wingspread of over three meters. Storm petrels are the smallest sea birds, looking rather like swallows.

The plumage usually consists of gray, black, or white feathers. There are no brilliant colors. Males and females have the same plumage; seasonal variations, if any, are slight.

PELICANS AND KIN

All members of the order of Pelicaniformes have long bills and an expandable membrane, or throat pouch. In pelicans, this pouch is very noticeable, being 15 centimeters or more deep, and may be used to scoop up and store fish. In frigate birds, the pouch has a function during courtship. The male attracts a female by inflating his normally invisible pouch, which gradually turns scarlet, and by rattling his bill.

The order consists of six families of birds: pelicans; cormorants; anhingas; gannets and boobies; frigate birds; and tropic birds. All feed exclusively on fish.

PELICANS

These are gregarious birds that breed in large colonies. Their large nests, composed of sticks, reeds, and mud, may be built on the ground or in trees. The two to five eggs are a bluish-white.

Pelicans are among the largest of birds, ranging in length from 1.25 to 1.75 meters. Some white pelicans, *Pelecanus onocrotalus,* have a wingspan of close to three meters. White pelicans are found in temperate areas in both the New and Old Worlds, along coastal lagoons and marshes and large inland bodies of water.

The brown pelican, *Pelecanus occidentalis,* is a New World species, most commonly found from the Gulf of Mexico to South America. Occasionally, it will wander north as far as Canada. It is a spectacular sight to see this bird as it dives headlong from high in the sky into the water, emerging soon thereafter with a fish.

THEIR KIN

Cormorants have long necks and thin, hooked bills. The plumage is usually a metallic black. Occurring throughout the world, they dive after their prey, and are used in the Far East to catch fish. One species, the Galápagos cormorant, cannot fly. The nests, of sticks and seaweed, are built on the ground or in trees. Usually there are two to four eggs.

Anhingas are also called snakebirds or darters because of their very long necks. They are found on all continents except Europe. They live in colonies and lay three to six eggs in tree nests.

Gannets occur in cooler waters and boobies in warm waters. Both are colonial nesters and excellent divers.

Male scarlet tanager *(Piranga olivacea)*. Scarlet tanagers are found throughout much of eastern and central North America and, in winter, as far south as Peru.

Karl & Steve Maslowski, Audubon/PR

PERCHING BIRDS

The birds most familiar to people are perching birds. Most species are also known as songbirds. These birds have been more successful than others in adapting to human environments. Almost anywhere you travel you may see or hear such birds as sparrows, finches, swallows, and thrushes.

Most species of birds—some 5,500 of the 8,600 known to us—are perching birds. They comprise the order Passeriformes. The great majority of these are in the suborder Oscines; these are the songbirds.

GENERAL CHARACTERISTICS

Perching birds have four toes on each foot, one of which points backward. This arrangement enables the bird to easily perch on everything from leaf stems to thick branches. Tendons in the feet enable the bird to flex its toes and grasp its perch tightly. Any movement that threatens the bird's equilibrium will cause the feet to tighten their grip. This explains why perching birds don't fall off thin telephone wires—even when they are asleep on a windy night.

There is great variety in the appearance of perching birds. They range in size from tiny wrens only 7½ centimeters long to the lyrebirds, which may be more than 100 centimeters long—counting their 75-centimeter-long tail. Some, such as the Baltimore, or northern, oriole and the cock-of-the-rock, are brilliantly colored. Others, such as the woodwarblers, have dull-colored plumage.

The beaks vary, depending on the food eaten by the bird. For example, wrens have long, slender, slightly downcurved bills, which are useful in picking insects off leaves. Cardinals have large, heavy bills adapted to crushing seeds. Drongos have strong bills that are hooked at the tip and

both photos, Paul Schwartz, Audubon/PR

Barred antshrikes *(Thamnophilus doliatus)* of Central and South America. The female (top) is reddish-brown, the male (lower) boldly barred with black and white.

slightly notched, the better to hold insects captured while in flight.

Almost every songbird species can be recognized by its distinctive song or songs. A few, such as crows and blue jays, imitate the calls of other birds.

A number of different classification systems are used for the Passeriformes. In general, they are divided into more than 50 families. Over 40 of these families comprise the suborder Oscines—the songbirds. Members of the other families are sometimes called sub-Oscines.

In the following pages we will look briefly at some of the better known and more interesting perching birds.

SUB-OSCINES

Many of these perching birds can sing. However, their vocal organs are not as well developed as those of the Oscines.

WOODCREEPERS

About 50 species of woodcreepers make up the family Dendrocolaptidae. These solitary tree-dwellers are found from Mexico and some Caribbean islands south to Argentina.

Woodcreepers range in length from 13 to 38 centimeters. Most species have olive plumage with reddish wings and tail. Their legs are short and strong. The bill is stout and long. It is adapted for probing bark to find insects and spiders, which are the main foods of a woodcreeper.

OVENBIRDS

The family Furnariidae consists of approximately 220 species of ovenbirds. These are small to medium-sized drab brown South American birds. They live in a variety of habitats, from the sea coast to the Andes, from swamps and jungles to grasslands and rocky mountain slopes. Most have narrow beaks and feed on insects.

Ovenbirds get their common name from members of the genus *Furnarius*. These birds build large, elaborate mud-and-cow-dung-nests that look like old-fashioned ovens.

AMERICAN FLYCATCHERS

The American, or tyrant, flycatchers make up the family Tyrannidae. There are more than 350 species, found throughout the Americas but most numerous in the tropics. They are not to be confused with the true flycatchers, which are members of the thrush family (Muscicapidae), and which are Old World birds.

In temperate zones, American flycatchers often live in open country, thin woods, orchards, or gardens. There they perch on branches and poles, from which they quickly dart after insects, their chief food. Other species live in deserts and tropical rain forests.

These birds have small, weak feet, short necks, and large heads. The bill is broad and flat at the base, then tapers down to a hooked tip. The mouth, which has bristles at its corners, can open wide.

These birds are called "tyrants" because of their aggressiveness. They drive off larger birds, such as hawks and crows, from their territory.

Among the most fearless Tyrannidae are the kingbirds, genus *Tyrannus*. The eastern kingbird, *Tyrannus tyrannus,* is a common species. A little smaller than a robin, it is white, black, and gray, with an orange crown. It nests in rural areas in eastern North America. It spends the winter in Central America and northern South America.

One of the most colorful members of the family is the many-colored tyrant, *Tachuris rubrigastra,* of South America. Its plumage includes greens, oranges, and reds, as well as black and white.

The family also includes the small phoebes, of the genus *Sayornis,* and the wood pewee, *Myiochanes virens.* They are the size of sparrows, gray-brown or olive above and whitish below. Pewees prefer woodlands; phoebes, human habitations.

COTINGAS

The family Cotingidae contains 90 species. These birds all live in the New World, from the southern United States down into South America. Most are forest dwellers. They have short, sturdy legs. The beaks are large and hooked. Cotingas usually eat insects and fruit.

The birds in this family differ greatly in appearance from one another. Most have dull-colored plumage, but some are brightly colored. The male cocks-of-the-rock of the genus *Rupicola* are bright red or orange. Atop the head is a flattened, disklike crest. The female cock-of-the-rock also has a crest, but she is brown in color and smaller in size.

LYREBIRDS

The beautiful lyrebirds make up the family Menuridae. They are Australian birds known for the graceful tail of the adult male. It has two large outer feathers. Between these are 12 filamentary feathers. All are silvery on the underside. The feathers may be 75 centimeters long. When the male courts a female, he spreads the tail feathers and brings them forward over his back. He also dances and sings to the female, mimicking the songs of other forest birds.

The lyrebird is somewhat henlike in appearance. It has a large head and a long neck. The wings are short, rounded, and not very strong. The legs are long and the feet bear heavy claws.

OSCINES

LARKS

Only one of the 75 species in the family Alaudidae is native to the Americas. The other species are found in Europe, Asia, and especially in Africa. In general, larks are sparrowlike birds with small, rounded bills, rounded shanks, and greatly elongated hind toenails. Many have very lovely songs.

The horned lark, *Eremophila alpestris,* is widely distributed throughout North and South America. It averages 18 centimeters in length and has curious black markings about the face and little tufts of erectile black feathers on the head. These birds frequent open farming country and prairie land, where they run along roads or soar aloft with a cheerful whistle. They feed largely on the seeds of weeds except during the nesting season, when they consume a great many harmful insects.

The horned larks begin to nest very early in the spring. Olive or grayish speckled eggs are laid in a nest placed in a depression in the sod. During the breeding

Vermillion flycatcher *(Pyrocephalus nanus),* found from southern North America to Argentina. It is one of the few brightly colored tyrant flycatchers.
Summer Hays, Audubon/PR

season, the males perform daring aerial evolutions.

In former years, horned larks were netted in tremendous numbers and sold as game birds. Today, they are protected by law. Farmers realize their value to agriculture in destroying great numbers of harmful insects and weeds.

SWALLOWS

The swallows, family Hirundinidae, are among the most pleasant and well-loved of birds, and have long appeared in literature as symbols of spring and its promise of happiness. The nesting swallows are often the first to greet the early sun of a summer morning with pleasant warbling. They are day birds, and birds of the air. Their bodies express this adaptation beautifully. They have a thin, streamlined shape; long, tapered wings; and graceful tails that are never round but always forked or notched. The swallow's smooth glide over the surface of a pond, as it quenches its thirst in flight, is an inspiring sight. It rarely walks on the ground as its feet are small and weak, and fit only for perching.

The swallow family is worldwide in distribution. There are 75 species. All are insectivorous, and, like all such birds, have short, flat bills.

The migrations of swallows are linked to the abundance of insects in the air. As insects begin to disappear with the approach of cold weather, thousands of swallows assemble in flocks, preparing for their journey to warmer lands. In the spring, the swallows dependably return with the first warm weather and its increase of insects. In nesting, swallows are remarkably diverse.

CUCKOO-SHRIKES

Members of the family Campephagidae are related neither to cuckoos nor shrikes. There are 70 species, distributed from Africa through Asia to the Philippines and Australia. They are tree-dwellers, living in forests and woodlands and feeding on insects and small fruits. They range in length from 12 to 33 centimeters.

The black-faced cuckoo-shrike, *Coracina novaehollandiae,* is common in south-

Scarlet, or flamed, minivet *(Pericrocotus flammeus)*, found in the tropical forests of India, southeastern Asia, and the Philippines.

ern Australia and Tasmania during the summer months. It flies to warmer places for the winter. It usually nests on a thick, horizontal branch of a eucalyptus tree. Like most cuckoo-shrikes, it builds a shallow, frail nest. Both parents share the tasks of nest-building, incubating, and caring for the young.

Minivets, also members of the Campephagidae family, are a group of slender, brightly colored birds found mainly in tropical forests of Asia. They are 15 to 20 centimeters in length. The bill is sharply hooked and well-adapted to picking insects from leaves or catching them in midair.

SHRIKES

The shrikes (Laniidae) are an Old World family containing 70-odd species. Only two of these, the loggerhead and the northern, are found in North America.

The shrike, which is about as large as a robin, often preys upon other birds. It well deserves the name "butcher bird" that is often given to it. The shrike has a strong, sharp, hooked bill. It has ordinary songbird feet, however, rather than the sharp and strong talons of birds of prey, such as owls or eagles. Therefore, it does not attack birds with its feet. Instead, it uses its powerful bill to impale its victim on a thorn or on a barb of a wire fence, or else to wedge it into a fork in a branch. Then it perches on one side of the victim, braces itself, and begins to feed on its prey.

The shrike feeds not only on small

birds, such as sparrows and chickadees, but also on insects, mice, and snakes. In time of plenty, it may kill more than it needs and leave the surplus impaled on thorns.

Shrikes usually select some prominent fence post or the top of a tree from which they can survey the country and attack their prey. They can adjust their eyes for different distances and thus can keep their victim in focus all the time. It is said that a shrike can fly directly to a spot several hundred meters away and catch a grasshopper.

When the nesting season arrives, the shrike builds a substantial open nest. Four or five mottled grayish eggs are laid. The shrike is bold in defense of its family.

The shrike's song is a subdued "shek-shek," sometimes prolonged into a magpie-like rattle—a curious mixture of harsh and musical notes. The bird's anxiety cry is a grating "jaaeg."

WAXWINGS

The waxwing is known for its dignified carriage and its quiet coloring and conduct. This feathered aristocrat's family name, Bombycillidae, refers to the silky texture of its brownish plumage. The word *bombyx* means "silkworm" in Latin. A reddish-chestnut crest, yellow-tipped tail feathers, and red, waxlike appendages on the inner wing feathers further distinguish this bird. Adults have a black eye stripe and throat patch.

The waxwings are one of the smallest families of birds, containing only three species. However, they have a wide distribution throughout the Northern Hemisphere. Eastern Asia is the home of the Japanese waxwing, which has a rosy-red tail band instead of yellow. The Bohemian waxwing is found in North America, Europe, and Asia. The cedar waxwing ranges throughout the United States and Canada.

The waxwing is no singer. Its call-note is a weak and high-trilled "zhreee," sounding like the whistle belonging to a vendor's peanut-roaster. It has been suggested that the waxwing gave up music because it hindered its success in eating cherries. Indeed, this bird is liable to be unwanted around

Karl Maslowski, Audubon/PR

Stephen Dalton, Audubon/PR

Top: the common mockingbird (*Mimus polyglottos*), one of the most gifted songbirds throughout its North American range. Lower: song thrush (*Turdus ericotorum*) a European songbird, here seen near its nest.

sweet cherry trees, especially where native fruit or mulberries are scarce. However, it prefers wild fruits. Therefore, the best way to protect a cherry orchard against the waxwing is to plant plenty of native fruit about the orchard to supply the bird's food needs. The waxwings more than pay for the orchard fruit they eat by devouring insect pests found in orchards and shade trees.

Waxwings love company. They travel in compact flocks until the nesting period. Their direct, even flight can be recognized at a distance.

It is not uncommon to see the members of a small flock arrange themselves on a branch where only the one at the end can

PERCHING BIRDS 431

reach the fruit. It plucks the fruit and very politely passes it to its neighbor and thus on down the line until the last bird is reached. This last bird then swallows the morsel. This may go on for some time before the birds scatter and commence feeding by themselves.

The waxwing nests late, builds a bulky nest 2 to 7 meters above the ground, and lays three to five spotted, bluish-gray eggs.

WRENS

The energetic, small brown wrens belong to the family Troglodytidae. There are about 350 species, found in both hemispheres. The majority of the species, however, dwell in the tropics of South and Central America.

In spite of their numbers, wrens are remarkably uniform in their plumage, wearing browns and grays in very inconspicuous patterns. The birds are from 10 to 15 centimeters long, with rounded wings and short tails, which they characteristically hold erect or tilt forward over the back. Their small, plump, brown bodies and their habit of haunting brush piles or walking on the ground give them a strikingly mouselike appearance. They constantly flutter about as though on springs, scolding trespassers in a loud voice.

Individual species can be identified by their songs, which are distinctly different. The male wren has a song of surprising volume and sweetness. The female seldom sings. In some tropical species, however, the mates sing "duets."

The house wren, *Troglodytes aedon,* is one of the commonest species. It is found throughout North and South America from Quebec to Argentina. It is uniformly dark brown above, faintly barred with black, and brownish-gray below.

The house wren somewhat resembles another quite common species, Bewick's wren, *Thryomanes bewickii.* The latter, however, has a light line over its eye and light spots on the corners of its tail.

Both the house wren and Bewick's wren are fond of the habitations of people and are quick to avail themselves of nesting boxes put up for them.

The long-billed marsh wren *Telmatodytes palustris,* of North America frequents the cattails and sedges of marshes bordering lakes, creeks, or sloughs. Here, its incessant song will always be heard. Even during the hours of darkness, when most birds are quiet, the marshes will resound with their chorus. Often the wrens seem to be carried away by the exuberance of their own songs and, springing from the cattails, seem actually to explode upward. With their feathers shaken out, their short

Robins are familiar birds in many parts of the world. Top: a red-cap robin *(Petroica goodenovii)* of Australia, shown here with its young. Bottom: the American robin *(Turdus migratorius),* found throughout much of North America.

Peter Slater, Audubon/PR

Gerard, Audubon/PR

wings vibrating, their cocky tails tilted far forward, their plump little bodies look like cotton balls.

The long-billed marsh wren is noted for its amazing nest-building habits. It typically builds several elaborate nests each season—why is not known.

In the arid regions of the western United States dwells the largest and most unwrenlike of all the wrens, the cactus wren, *Campylorhynchus brunneicapillus*. It is a gray bird with a white-spotted breast whose large, retort-shaped nests are among the most characteristic sights of the cactus country. Its song is less musical than that of any other wren, although it is given in characteristic wren fashion with the tail drooping and the head thrown back.

Although at present the cactus wren is found primarily in arid, uncultivated, desert land, its dietary preferences may, at some future date, be put to very valuable use by mankind. These birds feed on insects injurious to valuable crops, and their cultivation in agricultural areas would bring a powerful ally to the side of farmers in their war against harmful insects.

MOCKINGBIRDS, CATBIRDS, AND THRASHERS

Members of the family Mimidae are mimics. They imitate other birds. There is a record of one mockingbird that imitated 32 different species in ten minutes of continuous singing. Not all members of the family are good mimics, or mockers. Many confine themselves to their own brilliant notes.

There are 34 species in the family Mimidae. They are New World birds. Included in the family are mockingbirds, catbirds, and thrashers.

Mockingbirds. Mockingbirds are about the size of thrushes. The best-known species is the northern mockingbird, *Mimus polyglottos*. In places along the Gulf of Mexico it is the most abundant bird, and its rich songs are heard on every side.

The northern mockingbird is a slender, ashy-gray bird about the size of a robin, with white marks in the darker wings and tail. It is found wherever there is a thicket in which to hide and an exposed perch from which to sing.

Catbirds. The catbird, *Dumetella carolinensis*, is a common species of Canada and the United States. It resembles the mockingbird in being a long, slender, gray bird, but it is darker and it does not have the white bars in the wings and tail. Its only marks are a black cap, black tail, and reddish-brown feathers at the base of the tail. It gets its name from the harsh catlike notes with which it scolds every intruder and with which it ruins an otherwise melodious song.

Top: Stellar's jay *(Cyanocitta stelleri)* of the western United States and Mexico. Bottom: the great tit *(Parus major)* of Europe, Asia, and northern Africa.

Bill Reasons, Audubon/PR

J. A. Hancock, Audubon/PR

Many learn to imitate the notes of other birds with great skill.

Catbirds are either very sympathetic to the troubles of all the bird world or very inquisitive, for whenever a bird is in difficulties and gives an alarm cry, all of the catbirds of the neighborhood assemble to stare at and to scold the disturber. In the defense of their own nests, they are seldom excelled for bravery.

The catbird is very largely insectivorous and therefore beneficial. However, together with many other birds, it shows a partiality for cherries and other small fruits in their season. Where mulberries and wild fruits are available, cultivated varieties seldom suffer.

Thrashers. Thrashers are the most numerous of the mockingbird family. Their center of distribution is in the southwestern United States. They extend southward through Mexico and westward through southern and lower California.

Only one species, the brown thrasher, *Toxostoma rufum,* is found east of the Rocky Mountain region. It occurs throughout eastern North America as far north as Quebec and occasionally somewhat farther. The brown thrasher is often confused with the wood thrush, although it differs in its much longer, slightly curved bill, its long tail, and its streaked, rather than spotted, underparts. It is a shy bird, much more often heard than seen, for it keeps to the undergrowth, where it scratches among the leaves on the ground or digs holes with its bill in its search for larvae.

The sound produced as it apparently blows the soil from its nostrils is almost like a snort. When singing, the male mounts to the topmost branches of a tree, from which its loud, ringing notes can be heard for long distances. The song is a rich, ringing medley. Though limited in its range and confined to one air, it rivals the mockingbird's in the exuberance of its tones.

Occasionally the thrasher lives about gardens, especially if some effort is made to develop a tangle of shrubbery in which it can find seclusion and safety from enemies. Like the mockingbirds and catbirds, it will come to a food shelf for suet and crumbs and sometimes becomes quite friendly. When the bird is angry or in active defense of his nest, it vigorously switches and thrashes its tail about — hence its name.

THE THRUSH FAMILY: ROBINS, WARBLERS, AND OTHERS

The family Muscicapidae is a very large family. Its members are found in most parts of the world, though the greatest numbers live in the Eastern Hemisphere. The family includes many birds that are very well known: the robin, bluebird, and wood thrush of North America; the garden warbler and blackbird of Europe; the fantail and tomtit of New Zealand; and so on.

The birds of this family are generally insect-eaters. Many capture the insects in

The capped wheatear *(Oenanthe oenanthe),* a wide-ranging Eurasian member of the thrush family.

Vanguard, Audubon/PR

The eastern bluebird *(Sialis sialis).* This small, beautiful North American thrush is often used as a symbol of happiness.

Alvin E. Staffan, Audubon/PR

the air. Others hunt for insects in foliage, on branches, or on the ground.

Thrushes. Thrushes make up the subfamily Turdinae. They are found almost everywhere. The more than 300 species include the robins, wood thrushes, bluebirds, wheatears, bluethroats, rock thrushes, and nightingales. They inhabit parks, gardens, fields, and woodlands.

Thrushes are medium-sized songbirds averaging under 30 centimeters in length. They have strong wings and legs and long, slender bills slightly curved at the tip. Their color is uniform rather than streaked. Most thrushes are dull brownish to grayish, but some species are more brightly colored. The underparts are white, with spots. Most thrushes are beautiful singers.

Except during the nesting season, thrushes travel in scattered flocks, hunting food. They usually eat insects during spring and early summer, and wild fruits or berries in late summer and fall.

Some thrushes often winter in cool areas, but others spend their winters in tropical regions.

The American robin, *Turdus migratorius,* with its black head and reddish breast, is one of the best-known thrushes. It is famed as a herald of spring, although some robins remain in one area during the entire winter. The robin builds its mud nest wherever it finds a sheltered ledge about the house or garden.

To shape the nest, the female robin uses her breast, turning round and round. The eggs are the familiar "robin's egg blue." The robin's cheerful phrase, "cheerily, cheerily," and the brisk, businesslike manner with which it dispatches earthworms and insects makes it one of the best-known and most popular of all wild birds in North America.

The "robin redbreast" of children's rhymes is the European robin, *Erithacus rubecula.* It is smaller than the American robin. The upperparts are olive brown. The forehead, throat, and breast are bright orange-red. It has a lovely song, which it uses to define its territory and warn other birds away.

Several small North American thrush-

Canada warbler *(Wilsonia canadensis),* found throughout southern Canada and northern and central United States.

es of the genus *Sialia* are commonly known as bluebirds. The males have more blue in their plumage than do the females. The bluebird is not always readily identifiable unless it is seen in strong light, for only then does the blue in its plumage show up. This is because it is a diffraction color and not due to pigments. This beautiful bird is very devoted to its family, as befits a symbol of happiness. It nests in an abandoned woodpecker hole or cavity in a tree or fencepost. It is also quick to take advantage of a bird house which has been put up for it. A most welcome guest in the garden, the bluebird destroys enormous numbers of insects.

Few birds have sweeter songs than that of the hermit thrush, or American nightingale, *Catharus guttata.* One seldom hears the bird, for it is very shy and frequents woodlands rather than gardens. It is easy to identify, for its tail is a much brighter brown than the rest of its back. In true thrush fashion, it white breast is dark-spotted. It nests in Canada and in the hills and mountains of the northern United States above an altitude of 150 meters.

The name "blackbird" is commonly given to several species of thrushes. That of "four and twenty blackbirds" fame is the European blackbird, *Turdus merula,* which has also been introduced into Australia and New Zealand. Some 25 centimeters in length, the male is entirely black, except for a yellow bill. The female is dark brown, with a paler throat and a brown bill. It lives in many different types of environments. A ground feeder, it favors insects and fruit.

Flycatchers. These birds range from

PERCHING BIRDS 435

Top: Yellow warbler *(Deudroica petechia)*, found throughout much of the Americas. Middle: black-capped chickadee *(Parus stricapillus)*, a friendly, inquisitive bird of the woodlands of North America. Bottom: brown, or tree, creeper *(Certhia familiaris)*, a wide-ranging, stiff-tailed bird found in cool, wooded areas of North America and Eurasia.

Europe and Asia through the islands of the Pacific to Australia and New Zealand. Some 50 species are found in New Guinea.

There is great variety in the appearance of these birds. Some have brightly colored plumage. Others are dull in color. Some have wattles on the face. A few have crests. Most are small-sized, with long pointed wings and relatively short tails. A typical species is the spotted flycatcher, *Muscicapa striata,* of Europe. It is brownish-gray with a creamy breast. Its song is a monotonous chirp.

Warblers. Small birds, warblers generally have thin, pointed bills and feed on insects. The common name comes from the warbling, or trilling, song that these birds produce. Many have very melodious songs, while other species sing monotonous or even unpleasant songs. The 275 species are distributed throughout the Old World.

Two species, known as kinglets, are found in North America. The kinglets sometimes follow the winter flocks of chickadees and nuthatches. More often, however, these small grayish-olive-green birds keep to themselves among the evergreens, searching for scale insects and aphids. Their movements are quick, their wings constantly flitting. Their bright gold or red crowns are partially concealed except when they raise their crown feathers.

TITMICE AND CHICKADEES

The family Paridae contains some 60 species of small birds that are common in woodlands throughout the Northern Hemisphere.

As a group, these birds have the common characteristics of small size—most weigh less than 20 grams—short cone-shaped bills, blunt-ended tongues, rounded and well-developed wings, and rounded tail feathers. Their plumage is never spotted, streaked, or barred. The predominant colors are gray, brown, and olive on the upper parts, dull white and gray on the lower ones. Some may have black, gray, or yellow around the head, or else they may have protruding tufts, as in the case of the tufted titmouse.

Although the titmouse will eat almost

anything, it prefers a diet of insects. When these are not available, it will eat various seeds and the larvae and eggs of insects.

Only a few species migrate south for the winter. After the nesting season, members of most species of these sociably inclined little birds join forces in loosely organized groups and spend the cold winter months foraging for food together. Seeming to enjoy their society, other winter birds, such as the downy and hairy woodpeckers and the golden kinglets, often follow these flocks. Not only do these birds get along well together, but they seem to have little fear of people and gather about suburban dwellings wherever food is offered to them.

In North America, the best known and most beloved titmice are the chickadees. Chickadees are dull grayish birds, lighter below, with conspicuous black crowns and throat patches. Although they differ slightly, all ten species can be easily distinguished as chickadees.

The songs and call-notes of the different species vary considerably, but all have a family likeness. The call of the common species gives the name to the family, for it is a clearly enunciated "chick-a-dee" or "chick-a-dee-dee-dee." In other species the call is less clear, more highly pitched, or more nasal. In addition to this note, the chickadee has a song of two or three whistled notes resembling the syllables "phe-be" or "phe-be-be" so exactly that amateur bird students are often led to believe that it is a phoebe calling. When flying through the woods, the chickadees have a variety of conversational notes that are rather difficult to describe. They also utter a hissing sound when protecting their eggs or their young against intruders.

Chickadees are friendly, inquisitive birds, and it is not only at the winter feeding stations that they become tame. They are always ready to answer an imitation of their "phe-be" call and will greet the person uttering the call. They will perch on the branches above the person's head. Sometimes they will even drop to the person's shoulder or hover in front of his face in a vain effort to discover the whereabouts of the other "chickadee."

Michael Ord, Audubon/PR

Dan Sudia, Audubon/PR

Charlie Ott, Audubon/PR

Top: the Hawaiian honeycreeper *(Hemingnathus wilsoni)*, a beautifully plumaged, nectar-feeding bird found in Hawaii. Middle: painted bunting *(Passerina ciris)*, a small, brightly colored North American bird related to the finches. Bottom: the cardinal *(Cardinalis cardinalis)*, a brilliantly plumaged, crested relative of the buntings, that is one of the most familiar birds of the eastern United States.

Western tanager *(Piranga ludoviciana)* of western North America. It winters as far south as Brazil.

Baltimore, or northern, oriole *(Icterus galbula)*, a well-known North American member of the blackbird family.

NUTHATCHES

The nuthatches (family Sittidae) are largely confined to the Northern Hemisphere. There are about 70 species.

The nuthatch is an acrobatic little inspector of the trunks and larger branches of trees. It climbs with equal facility whether its head is upwards or downwards, devouring vast numbers of insects. In clinging to trees, it does not use its tail as a balancing prop, but relies on its large, strong feet. These are of the ordinary perching type, but the toes and claws are much better developed than in most other perching birds. In fact, the nuthatch has been known to sleep hanging head downward, clinging to the bark beneath a jutting limb. It is a lively little creature, always on the move. The "hatch" in its name comes from the Middle English *hake,* meaning "hack." It refers to the fact that the bird wedges nuts or seeds into crevices in bark and then hacks away at them in order to get at the contents.

TREECREEPERS

Members of the family Certhiidae are weak flyers. Instead, they "creep" up and down tree trunks in search of insects. They have large feet. The tail feathers are strong and, when pressed against the tree, help to support the bird. The beak is slender, rather long, and curved—well adapted for probing into crevices in search of insects.

The only North American species is the brown creeper, *Certhia familiaris.* It is a slender bird about 14 centimeters long. Its streaked brown back makes it resemble a bit of animated bark as it climbs about the trunks of trees, using its stiffly spined tail feathers as a prop. The brown creeper has a slender, curved bill and dines on such insects and larvae as it can find in the crevices of the bark. This bird's song consists of a few weak notes.

The largest creeper, *Tichodroma muraria,* is about 17 centimeters long. It lives in mountainous parts of Europe and Asia. Its plumage is dull gray, with black on the chest and throat, and patches of red on the wing that are visible when the bird is flying.

There are six species of Australian treecreepers, which are similar to other treecreepers but do not have strong, rigid tail feathers. They are placed in the family Climacteridae.

SUNBIRDS

Sunbirds are very small, very brilliantly colored birds comprising the family Nectariniidae. The family name comes from these birds' eating habits. The birds have long, slender bills that are curved downward in many species. The tongue is partly tubular. These two adaptations help the bird reach into flowers to gather nectar and insects.

Sunbirds are distributed from Africa

Goldfinch *(Carduelis carduelis)* of western Asia, Europe, and northern Africa, feeding its young.

House finch *(Carpodacus mexicana)*, a sparrowlike bird native to western North America but now established throughout much of the continent.

and the Middle East across Asia to New Guinea and Australia. The greatest variety is found in Africa. These birds live in many different habitats, from mountains to coastal lands, from tropical forests and swamps to grasslands and semidesert areas. Some are common near human abodes.

The males of most species have metallic-colored plumage on the back and breast and nonmetallic belly plumage. The colors are bright greens, purples, and bronzes. The female plumage is usually dull shades of olive or brown.

BUNTINGS

The buntings make up the family Emberizidae. They are closely related to the finches. Buntings are found throughout the world. They are generally ground birds, and are more common in grasslands than in forests. The bills are short and cone-shaped, and adapted to a diet largely composed of seeds. The tail is often rather long.

In North America, many buntings are commonly called sparrows or finches. "True" sparrows and finches, however, are members of other families (Fringillidae, Ploceidae). Among the best-known buntings of the Americas are the grosbeaks and the cardinals.

The cardinal of the eastern United States, *Cardinalis cardinalis,* may be more than 23 centimeters long, with a tail of up to 12 centimeters in length. The male is scarlet, except for black around the stout, red bill. The female is brownish, with some red in the plumage and a red bill. Both sexes have conspicuous crests.

TANAGERS

The brightly colored tanagers are New World birds. Most of the more than 200 species of the family, Thraupidae, are confined to South and Central America. The greatest number live in tropical areas.

These are stout little woodland birds. The tail is shorter than the wings, and the heavy bill has a large, naked nostril. The males are the brilliant members of the family, having much red, green, yellow, blue, and black in their plumage. The females have a subdued yellowish-green plumage. This blends with the leaves of the tree tops where they nest and feed. Males may assume a quieter garb in the late summer.

Tanagers generally have wheezy or squeaky voices. The scarlet tanager of North America has perhaps the most pleasant song of the family. It sounds much like a hoarse robin giving a hurried performance. These birds live on insects and fruits. Most of them are 17 to 18 centimeters long.

The scarlet tanager, *Piranga olivacea,* is the most familiar North American member of the Thraupidae. In the summer the male is bright scarlet and has a black tail and wings. In winter these change to olive

Top: red-eyed vireo *(Vireo olivaceus)*, one of the most common North American forest birds. It is widely known for its intermittent song. Middle: spotted breasted oriole *(Icterus pectoralis)*, a beautifully colored and beautifully voiced oriole native from Mexico to Costa Rica. Here it is shown with its young in nest. Bottom: yellowhammer *(Emberiza citrinella)*, a European bunting having the same habitat and habits as New World sparrows.

green. The female is olive green throughout the year. The scarlet tanager spends summers in the woodlands and orchards of the eastern United States and Canada. It winters in South America.

WOOD WARBLERS

The American wood warblers offer a challenge to anyone seeking to learn about them. Almost all warblers are colored alike. They are smaller than sparrows and hard to notice.

The wood warblers form the family Parulidae, the second largest family of American birds, after the finches. Warblers have thin bills and attractive plumage with pleasing color patterns. Yellow is the dominant color, combined with markings of white, black, chestnut, olive green, or gray-blue. The female's feathers are generally drabber than the male's. Young birds often have different colors.

Many warblers dwell in the trees of parks, gardens, and forests, where they hunt insects. Others inhabit low trees and bushes. Still others live on the ground, on the shores of streams or ponds.

Except for the myrtle warbler, most northern species head south for the winter. Many fly from the northern United States and southern Canada to the southern United States. Other warblers winter in Mexico, Central America, the West Indies, or South America.

The warblers travel mainly at night in their southern migrations, though they continue their journey slowly during the day, feeding as they go. They make long flights across bodies of water, occasionally by day but much more often by night.

The wide variety of plumage and of feeding and nesting habits among the warblers is probably the result of a relatively recent evolution. The woodlands and forests contain so many species of warblers that competition would be extremely severe if each species had not adapted itself to a slightly different manner of life.

Though almost all the warblers are alike in feeding on insects, they differ in the manner of catching them. Some snap them up in flight, while others glean them from the bark of trees. Some snatch insects up as they flit along the ground. Others search them out of the debris on the forest floor. This insect diet is often supplemented with other foods, including the juice of various fruits and aquatic organisms.

Unfortunately, the name warbler is a misnomer for most of these birds. Their ordinary calls are shrill, unmusical, wheezy sounds, such as the ovenbird's *teacher-teacher,* the myrtle's *tchep-tchep,* the blackpoll's rising *east-east-east-east,* and the sewing-machine trill of the worm-eating warbler. A few, though, have songs of subtle beauty, such as the Swainson's clear, rich whistle, the Nashville's eight full-voiced notes and rolling twitter, and the myrtle's "sleigh bell" trill.

The trees of the warblers' forest home are in constant danger from leaf-eating insects and caterpillars. It is indeed fortunate that they fall prey so often to insect-eating birds such as the warblers.

VIREOS

Vireo is a Latin word meaning "I am green." Green is the basic color of the birds called the vireos (Vireonidae). It is combined with gray and yellow in attractive combinations. The vireos often travel in the company of warblers. They can be distinguished from the warblers by their softer color combinations, slightly larger size, proportionately larger heads, and heavier bills. Their movements are much more deliberate than those of the nervous little warblers. Even their song is more unhurried. The vireo sings from dawn to dusk. It even sings after the exhausting molting season has caused other birds to cease.

The vireos form an exclusively American family containing about 40 species. They are, with few exceptions, arboreal birds frequenting shade trees of city streets, brush, and woodlands. Vireos feed greedily on insects and insect larvae. In the fall, berries are added to the diet.

These birds are found from sea level to 3,000 meters. Very few species breed in South America. Most breed in the cold parts of the Northern Hemisphere and then migrate to the tropics in the winter.

Dick Hanely, Audubon/PR

J. A. Hancock, Audubon/PR

Delbert Rust, Audubon/PR

Top: male satin bowerbird *(Ptilonorhynchus violaceus)* of eastern Australia. Bowerbirds are architects and artists. They build complex bowers and decorate them, often using tools. Middle: male house sparrow *(Passer domesticus),* a weaverbird widely distributed throughout Europe and North America and also known for its building abilities. Bottom: song sparrow *(Melospiza melodia),* a well-known North American finch.

PERCHING BIRDS

Top: common mynah *(Acridotheres tristis)* of southeastern Asia, a member of the starling family. Lower: Eurasian, or common, jay *(Garrulus glandarius),* a brownish bird with vivid barring on the wing coverts. It is shown here with its young on the nest.

One of North America's most common forest birds is the red-eyed vireo, *Vireo olivaceus.* Its intermittent song is heard wherever there are trees in any number. It is olive-green above, white below, and has a slate-gray cap bordered on either side with black. There is a white line over the eye, which is red in the adult and brown in the young.

AMERICAN BLACKBIRDS AND THEIR RELATIVES

The varied family Icteridae is found only in the Americas. Included in this group are blackbirds, bobolinks, cowbirds, grackles, meadowlarks, and orioles. There are about 95 species. The majority are tropical birds. This family includes birds of widely differing habits.

All these birds have stout bills with exposed nostrils, strong perching feet, pointed wings, and tails of 12 feathers, the tips of which are square or rounded—never forked. Black is usually prominent.

The Icteridae eat seeds, insects, and fruits. During the summer, when their diet is principally composed of insects, they are very valuable birds. In the fall, however, many species of blackbirds assemble in large flocks and often do considerable damage to grain fields. This family has the dubious distinction of containing the only American parasitic bird species, the cowbird. It lays eggs in nests of other species.

The red-winged blackbird, *Angelaius phoeniceus,* is about 25 centimeters long. The female is streaked gray and black like a large sparrow, but the male is glistening black with red epaulets. This bird summers in cattail marshes of eastern North America and flies south for the winter.

Grackles may attain lengths of 40 centimeters or more, and are particularly large blackbirds. Most have a metallic bronze and dark purple sheen to the plumage. The tail is shaped like the keel of a boat. The grackle walks jerkily over lawns, "conversing" in its harsh, squeaky voice.

The bobolink, *Dolichonyx oryzivorus,* is a valuable field bird with ground-nesting habits. The male is black with buffy or yellowish-white markings, while the female is streaked brown. Its joyous song is most delightful to hear. Rice growers in the southern regions of the United States dislike the bird because it feasts on the growing rice.

There are about 40 species of orioles, distributed through temperate and tropical America. (The Old World orioles belong to another family—the Oriolidae.) They are not great singers but have sweet whistling calls. The males are quite grand in their orange, chestnut, yellow, and black plumage. Orioles devour great numbers of destructive insects.

A well-known oriole is the Baltimore, or northern, oriole, *Icterus galbula,* so named because the orange and black colors of the male resemble the colors on the coat of arms of the British House of Baltimore. The bird summers throughout eastern

North America and winters from Mexico south to Colombia. The adults are about 20 centimeters long.

THE FINCH FAMILY

The large finch family, Fringillidae, has representatives all over the world, except for some Pacific islands and Antarctica. But they are most abundant in the Northern Hemisphere. They range in length from 10 to 25 centimeters. Among their distinguishing characteristics are a thick-based, conical bill and the presence of only nine primary flight feathers. These birds feed mostly on seeds but they also eat other plant matter and some insects. They build open, cup-shaped nests.

Many members of the Fringillidae are commonly called sparrows. (Other birds that are called sparrows belong to the weaver family, Ploceidae.)

The typical sparrows are rather dull-colored, brown and gray birds, usually heavily streaked to blend in with the pattern of the grasses, for the majority of them are terrestrial birds. Others, such as the goldfinches, are dressed in striking colors.

Many species have beautiful songs and are well known as cage birds—the canary and European bullfinch are examples.

Sparrows adapt themselves readily to their environment. They are hardy birds, with many species not migrating at all. An excess of light and aridity tends to make them pale in color. They are darker if there is excessive humidity and shade.

WEAVERS

The 300 species in the family Ploceidae resemble the Fringillidae but have ten primary flight feathers and build dome-shaped nests. These are Old World birds, though some species have been introduced into the Americas. The Old World sparrows belong to this family.

The English, or house, sparrow, *Passer domesticus,* is well known around the world. In some countries many people consider that this pugnacious, hardy little bird has become too well naturalized, since it has displaced several native species. After a few house sparrows were brought to the East coast of the United States, it took less than 30 years for the species to spread across the continent to the Pacific coast.

The house sparrow is about 14 centimeters long. The male adult has a dark gray crown, chestnut neck, black throat, and whitish cheeks. His back is brown and he has a grayish white belly. The females and young males have duller plumage. They are brown above and whitish below. This bird is noisy—it "cheeps" a lot.

STARLINGS

Until 1890 there were no starlings in North America. At that time, 60 common starlings, *Sturnus vulgaris,* were released in New York City; 40 more were released in 1891. From these 100 birds have descended all those that now swarm over most of North America.

There are about 60 species of starlings—family Sturnidae—native to Europe, Asia, and Africa. As with the house, or English, sparrow, America was warned against the starling by European ornithologists. The warning came too late.

It is true that studies of the starlings' food throughout the year show that economically they do much good. But this domineering bird replaces much more valuable birds, such as blackbirds, swallows, and flickers, by driving them away from their nesting sites and feeding grounds. Since starlings withstand northern winters, they appropriate the best nesting and feeding sites before migratory birds return.

After the nesting season, starlings gather in enormous flocks, whereupon they either roost with the blackbirds in the marshes or gather about steeples or cupolas. Before retiring and upon arising, the entire flock performs a series of aerial maneuvers with all the precision of trained soldiers. In the air, starlings can be recognized by their strong, direct flight, their pointed wings, and square tails. On the ground they walk like blackbirds but have longer bills, which are yellow during the nesting season. Their much shorter tails are quite distinctive. During the fall and winter their iridescent black feathers are spotted with buff, but this wears off as spring approaches.

BIRDS OF PARADISE

These birds, which make up the family Paradisaeidae, are famous for the beautiful colors of their plumage and for the elegant, filmy feathers that are displayed by the males during courtship. In some species, males and females have similar coloring. In others, the female is a drab brown.

Birds of paradise live in the forests of New Guinea and nearby islands. A few species are found in the Moluccas and northern Australia. These birds are primarily fruit eaters but will also eat insects, frogs, lizards, and other small animals.

The birds of paradise may reach lengths of 45 centimeters. The largest species is the great bird of paradise, *Paradisea apoda*. The male has a yellow head, green throat, and metallic yellow plumes projecting from the sides of his body.

THE CROW FAMILY

The crows, jays, ravens, magpies, jackdaws, rooks, and choughs make up a single family—that of the Corvidae. There are over 400 species and subspecies of Corvidae. They are found all over the world except in New Zealand. All of them make interesting but mischievous pets and can often be taught to articulate a few words. In their natural state, they are disliked by farmers and feared by other birds.

All members of the family have stout, heavy bills with thick tufts of bristles at the base, concealing the nostrils. They have strong legs and toes, which are adapted for walking and perching, and strong, rounded wings. The American crows and ravens are uniformly black with metallic reflections. The jays and magpies are brilliantly colored; blues, greens, blacks, and whites predominate. The crows and ravens have short, square tails. Jays and magpies have tails that are long and tapered.

Corvidae eat nearly everything—insects, fruit, grain, nuts, crustaceans, fish, and the eggs and young of other birds. Whatever is most easily procured always suits the taste of crows and jays. Consequently they are often of considerable value during insect outbreaks, because they then feed on insect pests to the exclusion of nearly everything else. However, when grain or eggs are more readily available than other foods, these birds do considerable damage. Their food habits make it possible for many to winter in northern climes.

Jays. These are noisy woodland birds. They bully smaller birds and rob their nests. These mischievous creatures also delight in mobbing a sleepy owl, following and tormenting a dog, cat, or fox, or mimicking a hawk—to the consternation of smaller birds. Jays store up grains and seeds for winter use and so are important factors in the natural dispersal of nut, oak, and fruit trees.

Jays nest early in the spring and build bulky nests of twigs lined with fine grass and feathers. They lay four or five eggs, mottled with gray and brown. Both male and female take part in building the nest and feeding the young. These birds are pugnacious when their nest is disturbed.

The plumage of many jays includes blue. Some species are exceptionally colored. An example is the green jay, *Cyanocorax yncas*, which ranges from the southern United States to Peru. Its back is green, its belly and outer tail feathers yellow, and it has a blue crest.

Crows. Crows are much more destructive than jays. In addition to thieving, egg-destroying habits, they are larger and bolder than jays. They often come into the poultry yard after young chickens and eggs and they regularly feed on corn and other grain fields. It must be said in their favor that they also destroy large numbers of harmful insects.

Ravens. Several large, dark crows of the genus *Corvus* are known as ravens. These birds, found throughout the Northern Hemisphere, have been immortalized by many famous writers. They are intelligent animals and make good pets.

The common raven, *Corvus corax*, is about 63 centimeters long. Its plumage is sable brown. It feeds on the ground and will eat a variety of foods: fruits, small birds and mammals, and carrion. It inhabits forests and wilder areas. With the spread of human habitation, ravens have been replaced by crows in many areas.

Observing birds can be fun and informative. Here are two bluejays feeding at a station that anyone can make at home.

W. A. Schwarz

BIRD OBSERVATION

by Robert W. Howe

Birds are among the most familiar animals of our environment, and through the years people have observed their activities with keen interest. Bird observation may begin in one's backyard and may result in rewarding experiences. You will find it fascinating, for example, to watch a robin tugging at a worm in the soil and then flying off with its prize to the nest in order to feed its young. Listening intently on a summer day, you may be able to hear the mockingbird, often including in its song the calls of various local birds. Later, strolling in a meadow or in the woods, you will add greatly to your bird-watching experiences. Your observations need not end with the coming of winter. You will be able to note many different types of bird behavior at feeding stations set up for birds.

A few hours spent in learning how to observe, where to observe, and how to attract birds will greatly increase your enjoyment. In this article, you will find various suggestions to help you become a more skillful observer.

PREPARING FOR OBSERVATION

By way of preparation, you should obtain a good illustrated pocket field guide, which will identify the common birds in your area.

Another helpful source of information is a local bird list. Such a list can usually be obtained from a local naturalist organization or from an ornithologist at a college or university in your state.

Using the field guide and the local bird list, you should be able to identify the common birds in your area. Among the useful characteristics which will help you distinguish one bird from another are size, color, markings, and shape. By carefully noting such features, you will be able to identify birds more readily. Is the bird you are observing as large as a crow or as small as a sparrow? Is it predominantly red, yellow, or blue? Does it have markings on its wings, breast, or tail? Does it have a stocky body or a slender body?

Experienced bird watchers find a field notebook and field glasses extremely valuable. In your notebook, you should set down the names of birds, the locations where they were observed, the dates of the observations, and the weather at the time the observations were made. These data will provide clues for future observations. Field glasses make it possible to see birds

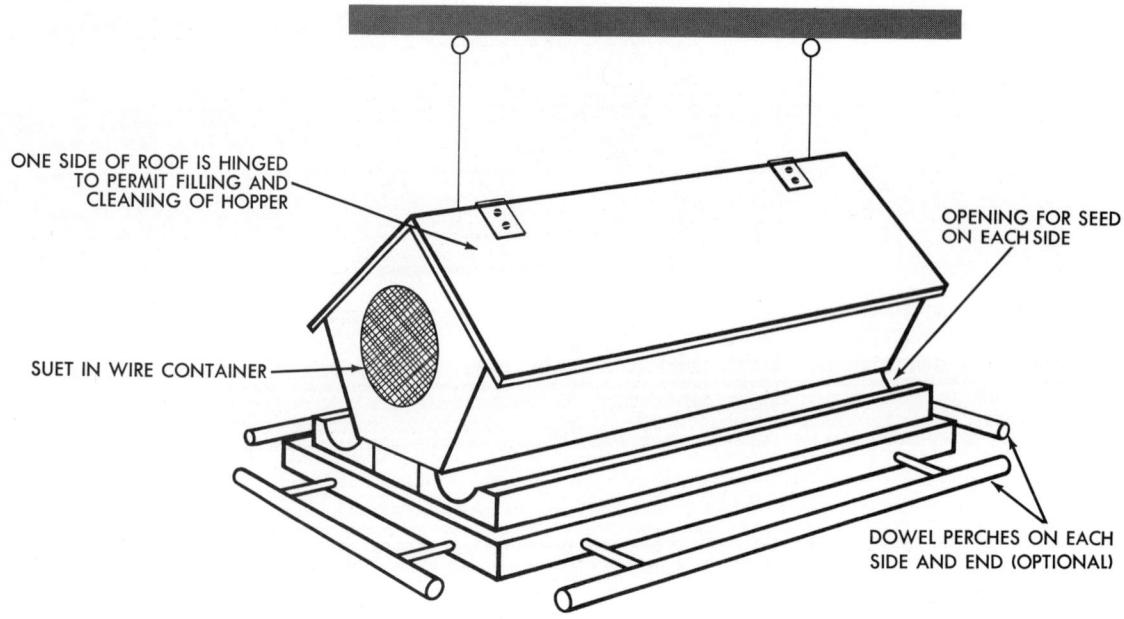

A hopper-type feeder. The roof of the feeder is hinged and the seed is supplied from the top.

which are too far away to be made out clearly with the naked eye. Many observers have found six-power or seven-power binoculars to be a good selection. Glasses with six-power magnification would make a bird appear to be six times larger than if observed with the naked eye.

FIELD OBSERVATION

Many birds can be observed only if you go to the locations where they live. Woods, fields, marshes, and swamps will provide observations of many different species.

By adopting the following procedures, you will be able to see and hear more birds. Plan your hike for the early morning — 4:30 to 5:00 A.M., or as soon as it is light — or late afternoon. Birds are usually most active during these hours. You should wear clothes that are not bright and that will blend with the landscape. However, you should not be in the field in such clothing during hunting seasons. Be sure to take your field guide, notebook, and pencil and also your field glasses, if you have a pair.

Plan your observations so that you will have the sun behind you as much as possible. You should observe while walking west in the early morning and east in the late afternoon. Walk as slowly and quietly as possible through the area you are investigating. Stop occasionally. Animals that might sound an alarm at your presence and cause birds to fly off will not be disturbed if you observe quietly.

Often you will be able to hear a bird singing even though you will not be able to see the bird itself. If you learn the common birdsongs, you will be more aware of what birds are present in a given area. With a little practice, you will be able to attract birds by whistling their songs, or by producing a squeaking sound as you blow against the back of your hand.

Different birds will be found in different areas and also on different types of plant life. In planning observations, consider the location, vegetation, and season.

If you live near a forest, you will find several good areas for observation. This forest may be made up of deciduous trees, such as beech and maple, which lose their leaves each winter, or else of evergreen

A catbird at a feeding station. This easily constructed structure is set at the top of a post.

Hal. H. Harrison from Monkmeyer

trees, such as the many species of pine, which retain their needles. The edge of such a woodland will usually contain more birds than the interior. Openings in the forest — glades — will also provide good observation places. Birds common to forest edges and glades include woodpeckers, crows, thrushes, vireos, robins, titmice, warblers, and sparrows.

Areas where logging operations have taken place often become covered with brush and young saplings. Cardinals, chats, sparrows, goldfinches, warblers, and catbirds are often seen in these areas. Where woods give place to open fields or meadows, you will find other birds, including meadowlarks, various sparrows, blackbirds, and doves. Still other types of birds will be seen in places such as marshes, lakes, seacoasts, or deserts.

In towns and cities, parks and cemeteries provide excellent areas for observing. A great variety of birds can usually be seen in places where natural woodlands and planted areas are combined.

The season of the year will have an important bearing on the kinds of birds you will observe. Spring and early summer generally offer the greatest variety. There is usually the least variety in the winter. The observer who wants to see the greatest possible number of birds will plan trips throughout the year.

ATTRACTING BIRDS

Many people like to observe birds that they are able to attract. You can attract birds by providing them with food, shelter, and water. If you maintain feeding stations or set out plants that will supply natural foods, you can have bird visitors throughout the year.

Food should be provided in the early spring, late fall, and throughout the winter if close observations are desired. If you begin feeding birds in the fall, be sure to continue the practice throughout the winter months. If you were to stop, many birds which had come to depend on you for their food might die.

You will be able to attract seedeaters such as juncos and sparrows, insect eaters such as woodpeckers and warblers, and also birds, including blue jays and chicka-

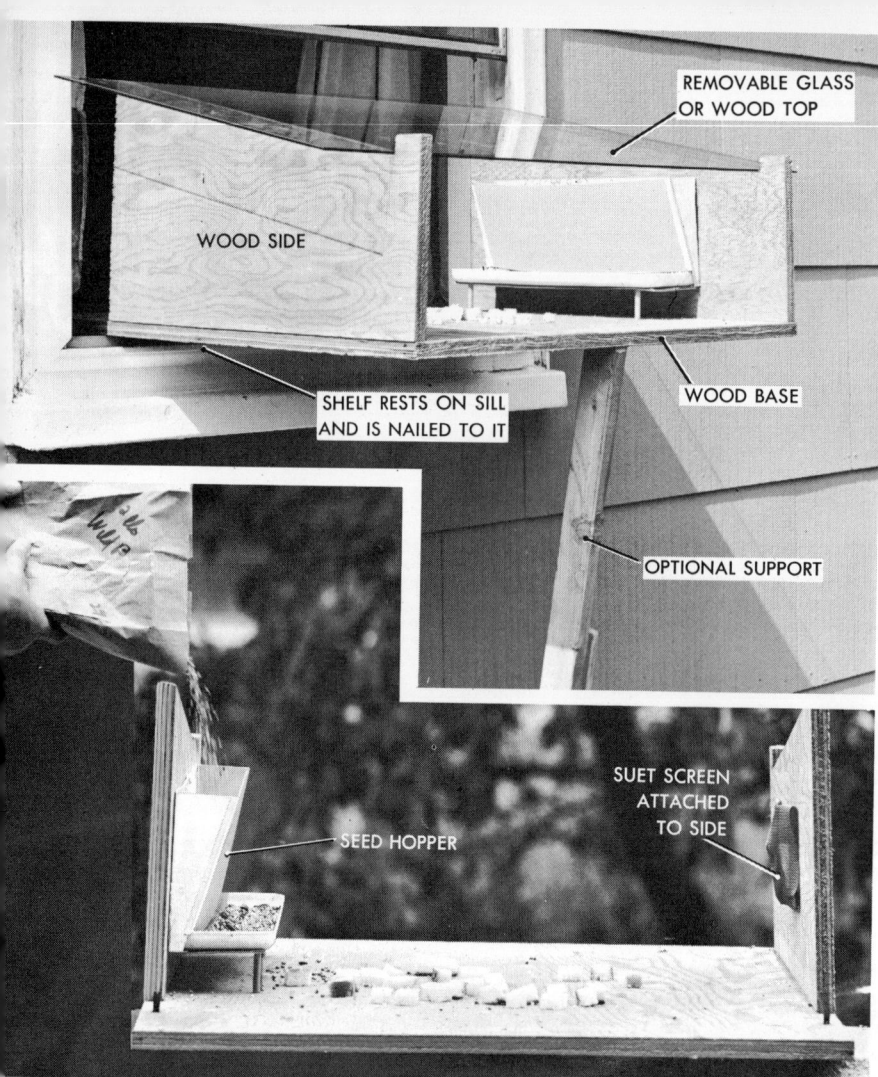

Two views of a shelf feeder attached to a windowsill. The upper photo shows the general appearance of the feeder. The lower photo shows the hopper filled with seed.

W. A. Schwarz

dees, that eat both seeds and insects. While some of the insect eaters will eat seeds in the winter, their special needs should be considered.

Three bird-feeding techniques have been found to be successful in many areas. An easy way to feed a number of birds is to use a food hopper, such as the one shown on page 444. It can be suspended from a tree, from the eaves of a house or garage or from an archway. Several hoppers can be used effectively in one yard. Mixed seed for your hopper can be purchased at many garden stores. Suet, which will attract a number of the insect eaters, usually can be purchased from a meat market or garden store.

Shelf feeders are also very popular. They can be attached to a windowsill, placed on a pole, or set on a box. A typical shelf feeder may be attached to a windowsill. A box shelf of this type can easily be constructed from an old apple box or from similar materials. The shelf should be large enough to accommodate several birds and contain a variety of foods. It should be set in a place where it will be readily visible. A variety of seeds, suet, and small pieces of fruit can be set out. It would also be helpful to include some fine sand or other gritty substance. If you live in an apartment house, setting up a shelf feeder is an excellent way to attract birds. You should be sure, naturally, that your shelf is secure.

A third feeding technique is to attach food to trees. Suet can be placed in a screened container, which is fastened to a tree trunk. Peanut butter can be spread on either a trunk or a branch.

Birds will often be attracted to natural

nesting sites in your vicinity if you make nesting materials available to them. These materials might include cotton yarn, narrow strips of cotton flannel, feathers, wool, horsehair, twigs, and straw. They should be set out in the spring. They can be placed in a net potato or onion sack. You will have many hours of enjoyment watching birds obtaining this nesting material and taking it to their nests.

You can supplement natural nesting places by supplying suitable nesting quarters such as those shown on page 448. They can be used to attract a variety of birds, including woodpeckers, chickadees, house wrens, bluebirds, purple martins, flickers, house finches, house sparrows, robins, and swallows.

Materials for constructing a nesting box should be dull-colored or weathered boards, rustic cedar, or wood with bark still attached. The box should be at least 13 centimeters by 13 centimeters by 20 centimeters in size. It should be somewhat larger if you want to attract the bigger birds. The entrance should be about five to eight centimeters from the top of the box. An entrance of about three centimeters in diameter will be large enough for small birds, such as house wrens. For larger birds, such as flickers and purple martins, an opening of at least nine centimeters is needed. A drainage and ventilation hole, about one-half of a centimeter in diameter, should be provided in each corner of the floor. An exit cleat should be placed outside the entrance. Boxes should be placed at least eight to ten meters apart. Each box should be set in open sunlight or bright shade, about five to fifteen meters above the ground.

Birds need water during all seasons of the year. When the temperature is above freezing, water may be provided in a birdbath or a simple pan. Another way to supply water is by dripping water into a pan. The dripping and splashing of the water will aid in attracting birds.

When the temperature is below freezing, water can be provided for the birds by a heated bath. The heat from the electric bulb will keep the water from freezing unless the temperature is extremely low. The light can be connected to an inside or outside electrical outlet. For your outside cord be sure to use heavy-duty wire, designed for exterior use.

You can provide natural food, shelter, and nesting materials for birds by planting selected shrubs and trees. It is important to choose a combination of plants that will give food and shelter throughout the year. Flowering dogwoods, American and English hawthorns, various evergreens (junipers, pines, and cedars), and cherry trees are examples of plants that would provide for the needs of many birds. Suggestions for planting in your area can be obtained from a local plant nursery. There you will receive advice concerning selections of early and late fruiting species to provide food for a large portion of the year. You will also learn what kinds of shrubs and trees not to plant. These include barberries and certain types of currants, which are links in various plant diseases.

INTERESTING ACTIVITIES

As you attract birds to your home or as you observe them in the field, you will be able to investigate many interesting physical characteristics and behavior patterns. There are various observational techniques. The most common method is to observe with binoculars and to record in a notebook your observations concerning

Special foods for birds are not always necessary at a feeding station. Here, a starling is attracted to peanut butter spread on a branch.

W. A. Schwarz

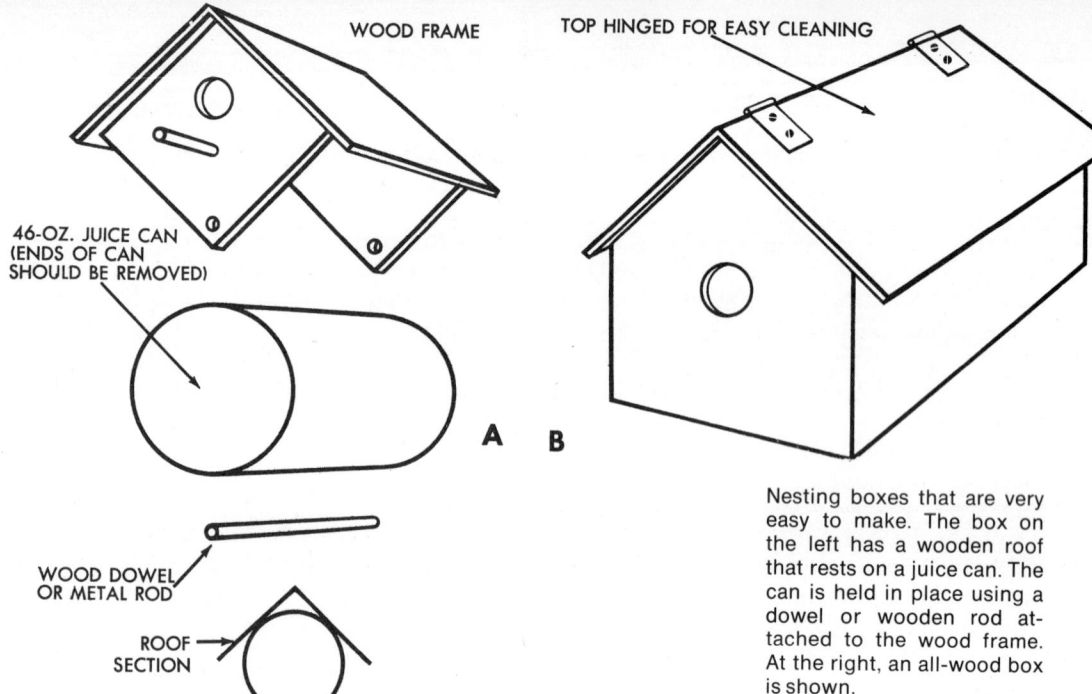

Nesting boxes that are very easy to make. The box on the left has a wooden roof that rests on a juice can. The can is held in place using a dowel or wooden rod attached to the wood frame. At the right, an all-wood box is shown.

different types of birds and their behavior. Many bird watchers take still photographs with black-and-white or color film. Color film provides more data for future study. Also, your color transparencies and prints will be a constant source of viewing pleasure. Thirty-five-millimeter cameras are particularly useful for photographing birds. When equipped with a telescopic lens they are remarkably versatile. Motion-picture photography is excellent for the study of bird behavior.

Two other techniques that have been used with success are tape recordings of birdsongs and time-lapse photography of nesting sites. A microphone for a tape recorder which operates on 110–120 volts can easily be attached outside a window. Battery-operated tape recorders will serve effectively in the field. Time-lapse photography can be used at open nest sites. The resulting pictures are often extremely rewarding. The technique, however, requires adequate equipment and planning and it is not recommended for beginners.

Many bird watchers develop particular interests. Some concern themselves especially with birdsongs. Others are concerned with bird coloration. Still others are interested in the social habits of birds. As you increase your observations, you will develop your own interests. By way of introduction to the field, here are some activities that have proved fascinating to many people.

Daily cycles. Most birds tend to awake during certain hours and they follow more or less regular cycles of activity throughout the day. The robin is generally an early riser in most areas, while blue jays and crows normally do not begin the day's activities until later. Robins sing before feeding. Other birds may feed before singing or feed while singing. Activities of most birds decrease during the later morning hours and usually are at a minimum during early afternoon. They again increase during the late afternoon. Some birds are extremely active during the later part of the day.

Which birds in your area are early risers? Which ones are late risers? When are particular birds most active? Do the daily cycles of birds change from one season to the next? If they do, can you think of possible reasons for such changes?

Feeding habits. To investigate the feeding activities of certain birds, it is best to observe in the field. You will note that woodpeckers drill into trees with their beaks in search of grubs, hibernating insects, and insect eggs. Other birds, such as chickadees and nuthatches, probe the crevices of bark for eggs and hibernating insects. Flycatchers, swallows, and nighthawks seek out insects in flight. You will find, too, that many birds feed mainly on seeds. Bird diets may also include fruit, live animals or dead animals, and refuse. On what does the English sparrow feed? The barn owl? The kingfisher? The sea gull?

Certain types of observations are more effective at bird feeding stations. You might, for example, want to investigate the kinds of seeds that different birds prefer. You will be able to carry out this investigation at your own bird feeder. Count out 50 of each of three types of seeds. You might select from the following: white millet, buckwheat, cracked corn, and sunflower. Mix the seeds and place them in the bird feeder. After a specific species of bird has been at the feeder, count the number of seeds of each type that remain. Use the same procedure for several species of birds. By analyzing the information you collect in this way, you can determine the seed preferences of various types of birds. You can use this information in buying the particular seed mixture preferred by the birds that you wish to attract. You can carry out similar investigations using suet, broken dog biscuits, and bits of raisins, oranges, and bananas.

Birdsongs. Birdsongs have long intrigued people. Investigations have revealed that birds have specific songs for different purposes. By using either a standard or a portable tape recorder, you can record birdsongs for later study. Once you have learned the songs of a certain number of bird types, you can investigate various factors related to their singing. Do birds sing more in some months than in others? Do males sing more than females? Is this related to courtship and nesting?

Social behavior. Observations in the spring present many interesting opportunities to observe territorial defense, courtship behavior, and nesting habits of birds. Male birds usually arrive first in the spring, announce their arrival with considerable singing or display, and defend their territories. During the spring, try to observe the male birds. Are they conspicuous? Do you find several males together or are the males observed singly? Note the different types of courtship patterns and nesting activities.

Flocking by many species may be observed in the late summer and early fall. Starlings, grackles, and purple martins often form such groupings after breeding. Swallows, sparrows, and some other birds may also flock. Groupings may be observed among resident birds throughout the winter. How many birds are there in a flock? Is the behavior of the males toward each other in the flocking period similar to their behavior toward each other in the spring?

If you observe birds at your bird feeder, you will probably be able to note the so-called *pecking order,* in which certain birds dominate others. The dominant birds establish their position by fighting with other birds in the flock or group, and then they may peck a bird of a lower order without being pecked back. The dominant bird may exert its rights when feeding or drinking. At your bird feeding station, you should be able to observe the pecking order in a particular flock or feeding group. Can you identify a dominant bird in the group? A submissive bird? If a group of birds

A wren and its prey perched outside a wooden wren house.

H. Armstrong Roberts

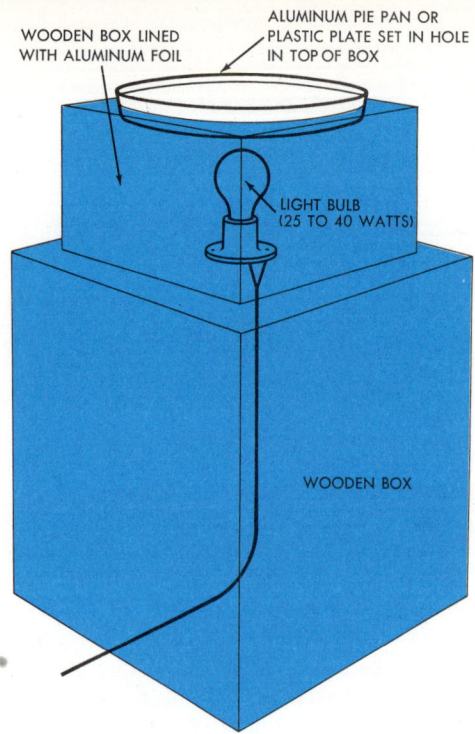

For bird baths that will operate only in warm weather, water from a perforated can trickles down into a receiving pan.

This type of bird bath can provide water to birds even during the winter. The water is kept from freezing by the heat from a lightbulb.

gathers at your feeder over an extended period of time, try to identify any changes that may happen to take place in the relationships of the birds.

Flight patterns. There are many different flight patterns among birds, but three general kinds of flight may be recognized. These are gliding flight, flapping flight, and soaring.

Gliding flight is typical of ducks over water and swallows in the air. Nearly all birds engage in flapping flight at some time or other, though there are many variations. Thus hummingbirds have a much different flapping flight from that of robins or sparrows. Soaring utilizes air currents and updrafts, so that flapping flight is not necessary for a considerable period at a time. Soaring may be observed in birds, such as hawks and gulls, with a large area of tail and wing surface compared to body surface.

Can you observe the three flight patterns in the birds you see? Which technique do they use when they come in to land? How do flight patterns vary for the different birds you observe?

Color changes in birds. Many people use bird coloration as an aid in identification. One factor to consider is whether or not a given bird has reached maturity. Many birds become mature only after two or three years. Before maturity they differ from the mature bird in color. Can you observe the several differences in coloration that exist between a mature robin and a young robin?

Birds also renew their coloring by shedding and replacing old feathers. This process is known as *molting,* and it commonly takes place in nearly all birds in the late summer after nesting. Some birds also molt, in various degrees, in the spring before the breeding season. Hence a bird may present a different appearance at different times of the year.

Observe specific birds in the late fall or winter. How does their color at this time differ from their color in spring and summer? If there are any American goldfinches in your vicinity, you will note that they undergo a remarkable seasonal change.

Structural similarities and differences in birds. There are many structural similarities and differences in birds. Bills, feet, and legs are relatively easy to investigate and to relate to bird behavior.

ANIMAL BEHAVIOR

by M. A. Freiberg and J. A. Rozé

An opossum closes its eyes and lies very still. A chameleon suddenly changes its color. A chimpanzee punches keys on a computer.

All are startling and fascinating observations of animal behavior. Is the opossum dead? How can an animal change its body color? Is there such a thing as an intelligent chimpanzee?

The observation of how animals behave is probably one of the oldest sciences. Indeed, the survival of early peoples depended to a large extent on their observations of animals. They needed animals for food, for clothing, and for shelter. And they had to protect themselves from attacking animals.

Today, people are becoming more and more aware that they must live in harmony with all forms of life and with their natural surroundings. For their own survival and for the survival of many animals, human beings must prevent further destruction of the environment. Many species of animals have already become extinct because their environment has been changed. The passenger pigeon is an example. The feeding and migrating habits of the passenger pigeon were interfered with. It became extinct: the last individual—named Martha—died in the Cincinnati Zoo in Ohio on Sept. 1, 1914.

INSTINCT OR LEARNING?

The study of animal behavior is called *ethology*. For many years this science has been divided into two main schools of thought—one stresses heredity, the other emphasizes learning.

Early in the twentieth century some observers noticed that every species has definite ways of acting in certain situations. Each species seems to have inherited a certain way of acting. These ways are called *fixed action patterns*.

Ethology, the study of animal behavior, is also the name most often used for the school of thought that stresses the importance of heredity in animal behavior. One of the most important scientists in this field is the Austrian biologist Konrad Lorenz.

Ethology is based largely on the idea that an animal must be observed in its natu-

Many species of birds engage in complex courtship rituals. Top: male frigate bird proudly displays his bright red inflated throat pouch for a female. Bottom: two albatrosses in their courtship ritual.

George Holton/PR

ral surroundings. A duck must be observed in a pond, a penguin in the Antarctic region, a camel in the desert. Only when an animal is in its natural surroundings, or habitat, can we see how that animal responds to its environment. How does a duck act when it can no longer find food in its pond or when other ducks are put into the pond? Ethologists believe that the way an animal acts—its fixed action patterns—is as important a characteristic of the animal as its shape or how its body functions.

Other scientists take a different approach to the study of animal behavior. They stress the importance of learning in animal behavior. This school of thought is called behavioral psychology or comparative psychology. Behavioral psychologists explain how an animal learns to act in certain situations in its environment. Like ethologists, they also study animals in their natural environment but will also study them under laboratory conditions.

The Russian physiologist Ivan Pavlov did much to further this way of thinking. In a famous series of experiments made early in the twentieth century, Pavlov rang a bell each time he put food out for dogs. After this process was repeated several times, the dogs began to connect the sound of the bell with food. Then whenever the bell sounded, the dogs expected food. In fact, they even salivated, waiting for food. The dogs continued this response to the sound of the bell even after food was no longer offered. The dogs had developed a *conditioned reflex*. The dogs had learned, or were conditioned into, a certain way of behaving by the repetition of a certain stimulus—the bell—and a certain reward—food. Later, other experiments showed that similar conditioned reflexes could be developed in other animals using other stimuli.

The two schools of thought try different approaches to explain behavior. Both present accurate points and must be considered together to get a full picture of behavior. For the ethologists, behavior is "written down" before birth. It is inherited, not primarily learned. The animal acts by instinct in a certain way that is determined before its birth.

For behavioral psychologists, behavior is learned. External stimuli determine the reflexes and development of behavior.

Today we know that both automatic reflexes and learning are important in the development of behavior.

CERTAIN COMMON ACTIVITIES

Behavior differs with each species. As some observers were quick to point out, certain patterns of behavior can be as important a characteristic of an animal species as the animal's body form.

However, certain types of behavior occur in many species. Animals communicate with each other, defend themselves, mate, and produce young. A brief look at how some animals go about these activities will show us some common features of behavior.

ANIMALS "TALK" TO EACH OTHER

Animals communicate, or "talk," to each other through signals. Each animal uses certain signals to communicate with other members of its species. It uses these signals to recognize other members of its species, to attract mates, to call for food, to warn of danger, and for other reasons. The signals differ with the species.

Calls and songs are important ways of communicating. A hen "talks" to her baby chicks with several types of calls. Among fur seals, cows, goats, rabbits and many

The brightly striped lionfish has long, sharp spines with which it fends off enemies.

Russ Kinne/PR

other species, mother and young communicate by voice signals. Other animals "call" other members of their species when they find food. They also warn others of danger. A prairie dog, for example, will give an alarm call if it sees danger, thus telling other prairie dogs to take cover.

Vocal noises are also a very important way of communicating between the sexes. Some species have definite vocal calls or songs in courtship. Male frogs, for example, peep and croak during the mating season. Females recognize the croaks of their own species and go to find the male. Many birds "sing" to attract mates.

Odor is another way of signaling. Many mammals "mark off" a territory, or home region that they consider theirs, by odor. They urinate around the boundary of their region. The odor tells other animals not to intrude. Odor is also often a sexual message. The male silkworm moth has an extremely sensitive sense of smell. He will pick up the odor of a female at a distance of three kilometers and follow the odor.

Some species have developed very unusual ways of communicating. Bees, for example, use various "dances" to tell other members of their hive where food is. A bee who has found a food source returns to the hive and performs a "dance." The form of the dance tells other hive members the direction of the food and how far away it is. Dances are also a part of courtship for many species.

TO LIVE ALONE OR IN A GROUP

Some animals live alone, avoiding contact with other members of their species, except in the mating season. Other animals typically live in a group. Some animal groups—a school of herring, for example—may include several thousand individuals, none of which is the leader. They all just live and travel together. In other groups, such as a flock of sheep, there is one leader, and all other members of the group follow. In a flock of chickens, there is a "pecking order." The hen leader pecks all members of the flock without being pecked back. The next ranking hen pecks all but the leader and so on.

James H. Charmichael/PR

Land hermit crab emerging from its borrowed house—a snail shell.

In other types of animal groups, there is a complicated social organization. Ants, for example, live in large colonies that have a complex organization. Each member has a special job and rank.

Baboon group life has been the subject of much study. Generally, one older male baboon is the acknowledged head of the group. He is supported by a core of "central males"—elder statesmen, so to speak. Younger males and adolescent males fight for rank and for the females. The females are protected by the males and spend most of their time caring for and playing with the young. Each member knows its place within the group. Rank is recognized through different types of rituals, such as the process known as grooming.

TO BUILD A HOME OR TO WANDER

Some animals build permanent elaborate homes. Bees, for example, live in large colonies and build complicated hives. The hives may have many connecting rooms and passageways, with some of these rooms used as nurseries and some for honey.

ANIMAL BEHAVIOR 455

Some animals prefer not to put up a fight. This hognose snake rolls over and plays dead to fool its enemies.

Beavers are well known for their home-building talents. They typically use sticks and mud to build a dam and a comfortable lodge along a stream. They are able to fell trees with a diameter of 45 centimeters to use for their homes. They hide the entrance to their lodge and store food in the lodge for use when they hibernate.

Other animals make use of "temporary" homes. A hermit crab, for example, uses the empty shell of a sea snail as a home. As the crab grows too large for the shell, he simply moves to another, larger, temporary home.

Still other animals, most typically fishes and birds, move about almost constantly, stopping only for nest building.

Many vertebrate, or backboned, animals and at least some lower animals establish a home territory even if they don't build a home. For some species, the territory is a certain area that the animal or animals living together consider their own—where they live, eat, sleep, reproduce, and raise their young. Other species establish a territory only during the breeding season. For them the territory is the area for courtship, mating, and the care of the young. For a lizard the home territory may be simply a small area around a tree log. For a mountain lion the territory may be very large—almost 40 square kilometers. Most animals defend their territories from intruders, using various forms of attack to do so.

Some animals have two widely different home regions, like winter and summer homes. Birds such as the Canada goose, sandpiper, and arctic tern often travel tremendous distances and have remarkable abilities to sense direction between their two homes. Certain bats, antelopes, and other mammals also migrate between two home regions.

HIDING FROM AN ENEMY

Every animal must protect itself. For some animals this means hiding from an enemy or trying to fool it. Some play dead. The opossum, for example, closes his eyes and lies very still even if bitten gently. An attacking enemy thinks the opossum is dead, loses interest, and moves on. The hognose snake is another animal that "plays dead."

Other animals hide by blending into their background. This type of behavior is called camouflage. The green mantis, for example, has a long narrow green body. It looks like a blade of grass. In a similar way, the brown mantis blends in among twigs. The vine snake of the American tropics is another example. It looks like a hanging branch or jungle vine and moves gently as if it were a branch swaying in the wind. Among many animals, the arrangement of bands or stripes, the shape of parts of their bodies, and their overall coloring all serve to camouflage them.

Some animals are even able to change their color. Some fish are especially known for this ability. The Nassau grouper can change its color into eight different patterns in a few minutes. It changes color to match the tropical reef background of its habitat. Chameleons are well known for their ability to change color, although other lizards are even better at it. The little tree lizard *Anolis carolinensis* can change very rapidly from dark brown to green.

WARNING AN ENEMY

Other animals do not hide from an enemy but warn it of their own dangerous nature. In that way they often avoid attack. Here again, color is one method used. The North American cottonmouth, for example, assumes a threat posture and opens its mouth wide, revealing a white lining that contrasts with the snake's darkly patterned body. The South American frog *Dendrobates tinctorius* is very brightly colored and makes no effort to hide itself. Its bright coloring warns other animals of its very dangerous, highly venomous, nature.

Some animals use other methods to warn or threaten an enemy. Cobras, for example, often spread their head region into a "hood" shortly before striking. During the mating season, male birds often erect head feathers, call, and show other signs that attract females, but also warn other males to stay away.

Many animals use camouflage for protection. Top: a spotted blenny blending into its background. Bottom: a chameleon matching the green of the leaf.
Tom McHugh/PR

DEFENSE AND ATTACK

Some animals try to get away from danger as fast as possible. Some—the impala, the deer, the rabbit, and many others—can often outrun their attackers. Birds take to the air, flying away from a ground attacker. Some small animals flee into burrows, thickets, or small openings where larger predators cannot follow them.

In some cases, animals cannot hide from or avoid attack, and they themselves seek out battle. Each species has certain ways of fighting, and most are equipped with certain "weapons." Some snakes, for example, inject a poisonous venom that paralyzes or kills a victim. Many mammals and other animals have very sharp teeth, and some have sharp claws and sharp horns. Birds often use their beaks in fighting.

TRYING TO FIND A MATE

During the breeding season many species use different methods to attract a mate. As mentioned before, some animals, such as frogs, use vocal signals to call and attract a mate. Some, such as the silkworm moth and snakes, use odor to attract a mate.

In still other species, elaborate courtship rituals have been developed. In many birds, for example, the males are brightly colored. At the mating season, the males display their colorful feathers. Some, such as the peacock, open and erect special feathers to attract attention. Others like the grouse have colored sacs on each side of the head. They quickly inflate and deflate the sacs, making an audible noise to attract mates. Some male and female birds also engage in mutual pecking, nibbling, and dancing before mating.

Complex courtship rituals are not limited to birds. Some snakes perform complicated and very graceful dances before mating. In some species of spiders, males must perform a fixed courtship dance as they approach the females or else risk being eaten by the females. Some fish also perform certain rituals. The stickleback male, for example, dances around the female and then begins to nibble at her tail to nudge her into the nest he has built.

NEST BUILDING AND CARE OF THE YOUNG

In many kinds of animals the young must care for themselves shortly after they hatch from eggs or are born. In other species, however, one or both of the parents build a nest and care for and protect the young for a period of time.

Nest building is most elaborate among birds. The nests may be built on the ground or in a tree, or they may hang from a roof or branch. Some species construct the nest of twigs and leaves and carefully line them with feathers. Once the eggs are laid, one parent, usually the female, incubates the eggs. Sometimes, the male supplies the female with food while she incubates the eggs. As the young hatch the parents may give them food, often food that has been partially chewed.

The stickleback fish mentioned earlier also constructs an elaborate nest. The male builds the nest of plantlike material and glues it together with a threadlike substance that he secretes from his own kidney. After the female lays the eggs, he swims around them. Later he cares for the young.

Other animals have different ways of taking care of their young. Among ants and bees, certain members of each colony have the task of caring for the young. Kangaroos, opossum, and a few other mammals keep their newborn in a special pouch on the front of the mother's body.

ARE ANIMALS INTELLIGENT?

Intelligence is a very difficult word to define. It is sometimes explained as the ability to learn, or as the ability to respond to new circumstances, or as the ability to solve a problem. But these explanations do not really define the word. Learning, response to a situation, and problem solving can all exist without true intelligence. Intelligence, as we are using the word, is probably best defined as the ability to reason.

Many animals can learn certain ways of behaving. Like Pavlov's dog, they can be "taught" a particular way of acting. Many animals can also be presented with a problem and solve it through trial and error.

Scientists have developed several ways of testing an animal's ability to learn, to respond to circumstances, to adapt, to have an idea, and to reason. In most cases the tests reveal that animals develop conditioned reflexes or learn by repeated trial and error experiences and not through intelligence. A brief explanation of some of the tests will help explain the ability of some animals and also the nature of intelligence.

LEARNING THROUGH TRIAL AND ERROR

Many tests involve the use of a maze. A rat or other test animal is placed at the entrance of a maze. It must find the exit. As the animal moves along the maze, it finds a series of branches or forks. It must decide which fork to take. If it chooses the wrong fork and comes to a dead end, it must go back and try the other fork. It continues until it reaches the exit. After many errors the animal "learns" how to go through the maze without any wrong turns.

Some mazes are simple. Some are difficult, involving as many as 25 turns. Animals vary greatly in how quickly they learn how to go through a particular maze. In at least one test, white rats did better than college students. However, this proved only that rats are good at learning through

A cardinal feeding its young in a well-concealed cup-shaped nest.

Karl H. Maslowski/PR

Pygmy marmoset being given a three-object discrimination test. Tests such as this are being used to determine the intelligence and learning ability of many animals.

Harry F. Harlow, University of Wisconsin Primate Laboratory

repeated trial and error. It did not prove that they are intelligent.

PROBLEM SOLVING

In another type of test, animals must solve a particular problem. Some birds, squirrels, cats, and other animals have learned to open a door, open a lock, or solve other "problems" to get at what they want—usually food.

Still other tests determine whether an animal is just repeating a certain way of acting or whether it can remember a correct way. For example, can an animal once it has learned the answer to a problem, remember the answer and use it at a later date? Monkeys seemed to do well in these types of tests.

Such abilities are impressive. Even more impressive is adaptability. Some scientists think adaptability shows at least some intelligence. Monkeys seem to be able to adapt: they are able to change their normal patterns of behavior to adjust to new situations. When monkeys were presented with a series of problems, they solved the problems through trial and error. But each time they were given a new problem, they solved it faster and faster. In other words, the monkeys learned how to solve problems; they learned how to learn.

USING TOOLS

For many years the use of tools was considered a sign of intelligence. In fact, for many years scientists who trace the history of early peoples considered the use of tools as the beginning of modern mankind. Recent research, however, has cast some doubt on whether tool use can be considered a sign of high intelligence.

Primates are a group of animals that includes monkeys, chimpanzees, and large apes, as well as human beings. Some primates, beside humans, can use tools, and so can some other mammals and at least one bird.

Chimpanzees put in a room where food is placed on a high shelf, too high for them to reach, will pile boxes one on top of the other and then climb on the boxes to reach the food. The chimps, in other words, adapt to the situation and solve the problem by using the boxes as tools to get what they want.

If necessary, chimps will carry tool use another step. After piling up the boxes and climbing on them, they will use a stick to pull the wanted food closer to them.

Sea otters also use tools. They favor clams as food. A sea otter will hold the clams against its chest and use a rock to smash open the shell. In a very surprising finding, scientists at the University of Massachusetts have shown that a northern blue jay will use a feather or straw to rake food pellets closer to its cage where it can then reach them. Such findings hint that tool use is probably more widespread than has been thought.

ANIMAL BEHAVIOR

Family group of Gibraltar apes. Rank in social group is recognized through certain rituals, such as grooming seen here.

USE OF SYMBOLS AND LANGUAGE

The understanding of symbols shows an ability at abstract reasoning. Chimpanzees were given a series of tests involving the use of a vending machine and poker chips. The chimps learned that one grape came out of the machine each time they put a white chip into a slot. They next learned that two grapes came out when they used a blue chip. The chimps soon learned that blue chips were more valuable. They also learned that a red chip brought water, and they used it when they were thirsty. This experiment showed that the chimpanzees were able to understand symbols — in this case, chips. One color chip symbolized, or stood for, one thing, another for something else. The chimps understood this and remembered it at least for a time.

The understanding of symbols is essential for the use of language. In a series of experiments being conducted in several centers throughout the United States, chimpanzees are being taught language. Not a verbal language, but a sign language. Some of the chimps are being taught American Sign Language (ASL), the same language that deaf people use. Other chimps are being taught to "read" plastic chips that symbolize different words. And another chimp has been taught a special language called Yerkish, which she uses on a special computer typewriter.

COMMUNICATION "MODELS"

The matter of communication techniques could prove to be the central one in determining the degree to which animals might actually be aware of themselves. Some ethologists have proposed much more radical research along these lines. They suggest the development of "models" that animals could accept as communicating partners in experiments. Using such models, these scientists think that it might be possible to explore the mental worlds of different kinds of animals at their own level of experience. This is an exciting prospect for students of animal behavior. However, such experiments have only been proposed and much work remains before they can be put into practice.

INBORN ABILITY PLUS LEARNING

Adaptability, the use of tools, the understanding of symbols, and, lastly, language — all seem to be signs of intelligence. The field of exploring animal intelligence through observation in natural habitats and through laboratory studies is truly a fascinating experience.

ANIMAL MIGRATION

by John C. Pallister

When we speak of human migrations, we have in mind the movements of great bodies of men, women, and children from one place to another. These people may be driven by famine in their own land; they may be fleeing from a foe; they may be seeking freedom from persecution.

There have been numerous instances of such human migrations. The exodus of the Hebrews from Egypt in Biblical times is an ancient example. More recent ones include the migration of French Protestants to England in the seventeenth century to escape religious persecution and the flight of the American farmers in the 1930s from the dust bowl of the Great Plains westward to California.

When the world "migration" is applied to the journeys of animals, it generally has a more restricted meaning. In its narrowest sense, *animal migration* is the periodic traveling of a species between two locations.

MANY ANIMALS MIGRATE

The birds of the Northern Hemisphere offer perhaps the best-known examples of animal migration. Every fall most of them leave the localities where they build their nests, and fly south to spend the winter. Every spring they return north and build new nests in the neighborhood where they have nested before. They travel to the same winter quarters each year. They use the routes they have always followed. Such, at least, is their normal habit. Various accidents can break up the routine, however. Severe storms may drive them off course. A few members of the flock, worn out or injured, may spend a few days or the whole season in shelter far from their usual haunts. Changing climates may affect migration.

Birds are not the only animals that migrate. Caribou and wapiti, or American elk, change their feeding grounds according to the season. Seals travel long distances in order to breed; so do eels and salmon. Certain marine invertebrates and fishes move, according to the season, from deep water to shallow water and back again. Toads and some salamanders spend most of their time on land; but each spring they travel, often for some distance, to their favorite breeding waters. Marine reptiles such as sea snakes and turtles move from the water to the

Right: the tag on this least bat helps in the study of this animal's migration. Some bats migrate regularly on a north-to-south flight path. Far right: marking a king salmon by removing its dorsal fin. The salmon will return to its home stream to spawn.

shore to deposit their eggs. All these are regularly recurring two-way trips.

A few zoologists define migration so broadly as to include practically all group movements. They describe as *daily migrations* the flights of crows, gulls, starlings, and other birds from their roosts in search of food in the morning and the return flights at night. They also include certain one-way movements that most zoologists prefer to call irruptions, or *irruptive movements*. Among the best-known of these one-way trips is the mass movement of the Arctic rodents called lemmings. They migrate from the highlands, where they usually dwell, to the sea, where they meet their death by mass drowning.

In this article, we shall consider the round-trip migrations of certain representative birds, mammals, fishes, and other animals. We shall also deal with a few of the more interesting irruptive journeys.

STUDYING MIGRATION

The fact that certain creatures migrate has long been known. Comments on the comings and goings of birds, locusts, whales, seals, fishes, and turtles are found in early records. It was not until the eighteenth century, however, that investigators began to study animal migration methodically. The intensive study of the subject goes back only to the early years of the twentieth century.

To obtain detailed information about the migratory habits of any species, a great many observers are required. Their reports, sent in over a period of years, make up a massive accumulation of records. At the breeding locality of the species under observation, as many individuals as possible are captured, marked, and released. If and when they are recaptured, investigators are able to determine the distance the animals have traveled and the time they have taken.

There are various marking procedures. A leg of a bird or a bat is banded, or ringed, with an aluminum strip or a colored celluloid band. Butterfly wings are marked with a ticket punch or with a spot of paint. Tags are fastened to shrimp and some fishes. The ears of mammals are tattooed. Metal tubes are shot into the blubber of whales. Naturally the percentage that can be marked is small—and the percentage of marked animals recaptured is yet smaller. Another way of tracking animals is to attach small radio transmitters to them.

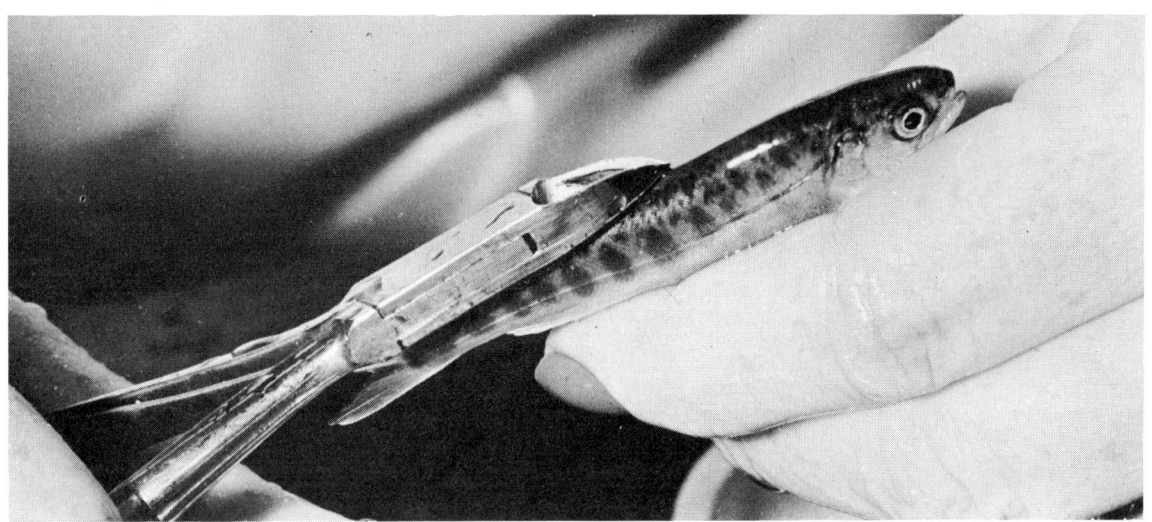
Cameron Thatcher from National Audubon Society

It is particularly difficult to find out why and how migration takes place. Why does an animal depart from the place of its birth and later return? What determines the time of departure, the direction of the journey, and its length? Seeking answers to these questions, scientists have not only made numberless observations, but have also engaged in a vast amount of physiological research and other experimental work.

It is evident that there is a relationship between migration and weather—particularly changes in temperature—and that food supply also influences migration. But this is not the whole story by any means. It is increasingly clear that the migration of a given species is determined particularly by what can be normally expected, on the basis of the accumulated experience of many generations. It is only when a change in the environment (such as abnormal weather or the destruction of breeding sites by human beings) persists over a period of years that a corresponding change in the migratory routine is likely to occur.

Researchers have tried to determine, without very much success, exactly how environmental factors influence the migratory trait. Most of the laboratory work in this field has been done with birds and fishes. Investigators have studied the effect of increasing or decreasing day length on a bird's hormones, which in turn influences molting and reproduction. They have noted variations in a fish's metabolism requiring a change in the salt content of the water in which it lives. They have examined periodic changes in the amount of secretion of the adrenal glands.

How does an animal find its way from station to station along its route? Trying to solve this problem, investigators have studied the effects of river and sea currents, polarized light, the earth's magnetic field, and the Coriolis force—the deflection caused by the rotation of the earth.

Some authorities question whether any one external stimulus, such as those just mentioned, could by itself account for the remarkable navigational ability of so many creatures, from warblers to whales.

On the other hand, for example, some authorities believe birds use the sun by day and the stars or constellations by night as their compass references in their migrations.

BIRD MIGRATIONS

Of all land creatures, birds are the best equipped for extensive travel because of their wings. The geography of the earth seems to provide the impetus for their migratory urge. Great land areas in the upper half of the Northern Hemisphere furnish abundant nesting sites, which rigorous winters make untenable. The long-distance

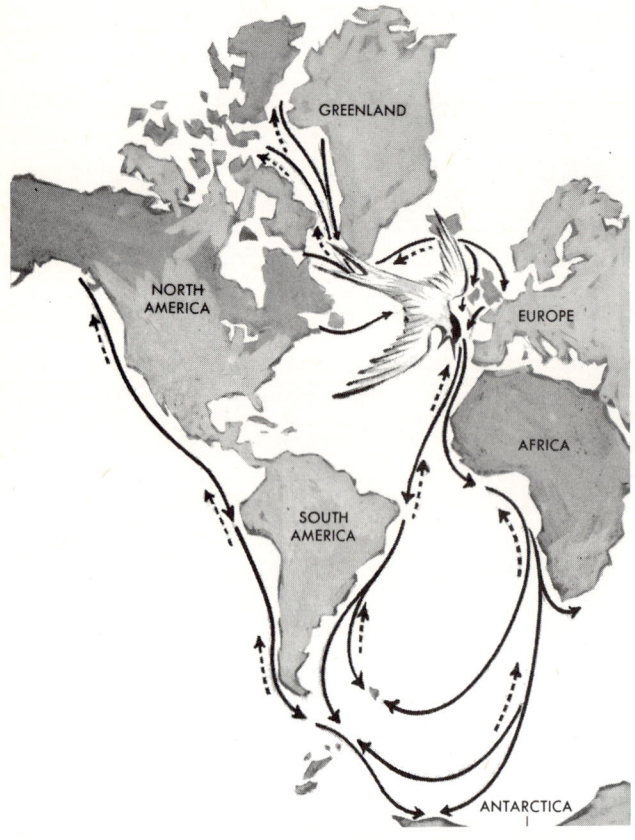

Map showing migration routes of the arctic terns of North America. These animals fly from pole to pole in their annual migration. Arctic terns hold the long-distance flight record of 35,000 kilometers.

migrators, therefore, are the birds that like cool weather, but not freezing weather. Birds that nest in the warmer areas do not migrate.

In the Southern Hemisphere, only the southern part of South America and the higher altitudes of the Andes have sufficient temperature changes to drive land birds toward warmer territories. There is little north-and-south bird migration in Australia or South Africa.

Some birds can stand very cold weather. The gyrfalcon and the snowy owl only occasionally leave their Arctic homes. The flightless emperor penguin lives the year round in Antarctica. Many birds can endure freezing temperatures, provided that the periods of low temperatures are not too protracted. They move only far enough away from their homes to escape the worst rigors of winter.

In general, birds that prey on other birds and on small mammals can live through a northern winter, and so can the larger seedeaters. Most of the small insect eaters and seedeaters migrate to territory that is not covered by snow.

The long-distance migration record of 35,000 kilometers is held by the arctic tern, *Sterna paradisaea*. This graceful bird builds its nest on rocky or sandy coasts, often around the Arctic Ocean and occasionally to within eight degrees of the pole. Arctic terns nesting along northern Greenland spend less than four months there. Then they fly eastward to northern Europe and southward along the coast to Africa. Some fly down the African continent to its very tip and then make their way to the Antarctic regions. Others cross over to Brazil from northern Africa. They fly down the South American coast and from there to their Antarctic winter range. Certain arctic terns of North America have taken to nesting farther south along both coasts.

Another Arctic Circle bird, the golden plover, *Pluvialis dominica,* builds its nest of reindeer moss in the wet tundras of Siberia, Canada, and Alaska. In early fall, the golden plovers start their flight south. Those from eastern North America fly down the Atlantic to the pampas of South America. In the spring, they follow the Central American coast to Louisiana and thence up the Mississippi Valley to the identical breeding grounds they had left. Alaskan golden plovers migrate to Hawaii. Those from Siberia winter in various islands of the Indian Ocean and also in Australia and New Zealand.

Among the birds that travel from northern Canada to southern Argentina are several species of the sandpiper family. Most species spend little time at their Arctic nesting places, arriving there in late May and departing south in July. The solitary sandpipers, *Tringa solitaria,* winter in various places. Some flocks go regularly to Bermuda, some to Cuba and other West Indian islands, and some to Argentina.

Many persons in North America have heard the loud honking of Canada geese, *Branta canadensis,* and have looked up to watch a flock in V formation flying overhead. These birds have a wide breeding range, from the Arctic Circle into the northern United States, and their winter quarters are equally widespread. Some that nest in Alaska winter in Japan. Those taking the Atlantic routes may winter anywhere along the coast of the United States and Mexico.

Storks may also migrate long distances. The white stork, *Ciconia ciconia,* spends the summer in northern Eurasia. It winters in Africa, Asia Minor, and India. The long journey takes its toll—thousands of storks die from starvation, storms and other bad weather, and, of course, the guns of hunters.

The small land birds are as capable of distant migration as are their larger relatives. One of the tiniest, the ruby-throated hummingbird, *Archilochus colubris,* nests in southern Canada and all through the eastern United States. Hummingbirds from the north winter in Florida and southern Louisiana; those from farther south spend their winters in Mexico and Central America. Some of the western hummingbirds occasionally migrate to the Gulf Coast and northern Florida, but generally they winter in southern Mexico.

The true songbirds make up the largest number of bird species. The majority of those that live in northern areas migrate. Some go only several hundred kilometers. Others fly incredibly long distances across the Caribbean into South America or from Scandinavia to Africa. One of these long-distance travelers, the scarlet tanager, *Piranga olivacea,* of North America, starts migrating in midsummer or in the latter part of the summer to Bolivia and Peru. Before leaving, the male changes its plumage to the olive green of its mate.

Nearly fifty species of the warbler family nest in the northern United States and Canada. Huge flocks containing several species of these small, charming birds pass regularly through the northeastern states. But since they travel at night they are rarely seen, and then usually because a storm or heavy fog has driven a flock against a city skyscraper. Many of these warblers fly across the Caribbean to Central and South America.

Migratory flight of bats. Much like birds, bats migrate south for the winter.

Flycatchers, swallows, and vireos also travel far from their nesting grounds. The varied members of the sparrow family, however, generally do not venture too far from home. They may even remain there. This family includes, besides the sparrows, the cardinals, grosbeaks, buntings, finches, goldfinches, and juncos. Some of them may nest rather high in the mountains, moving down to the protected valleys in order to escape the rigors of winter.

MAMMAL MIGRATIONS

Migration is not common among mammals, for most of them have found other ways of adapting to their environment. Some can endure severe seasonal changes. Others hibernate.

Bats. Curiously enough, the best example of migration by a present-day land mammal is furnished by one that can fly. In northern latitudes, as insects disappear in the fall, the bats that feed on them must either hibernate or move south. Some bats in the eastern United States travel a considerable distance to hibernate in large groups in limestone caves. Others migrate south and hibernate there.

Three species in eastern North America—the red bat, *Lasiurus borealis,* the larger hoary bat, *Lasiurus cinereus,* and the silver-haired bat, *Lasionycteris noctivagans* —make regular north and south migrations comparable to those of the birds. All three species are likely to migrate together in small or large groups, but the males and females travel separately. Their flight, as they migrate, is said to be steady, with a sailing or drifting motion quite different from their usual erratic flight patterns. They fly at a height of from 45 to 120 meters above ground. These bats travel day and night, usually along the coast, where their passage is regularly noted by lighthouse keepers. Occasionally migrating bats come aboard ships almost 100 kilometers from shore. Their winter quarters are from South Carolina to northern Florida.

Bison, caribou, and elk. Millions of bison, or American buffalo, *Bison bison,* once roamed throughout Canada, the central part of the United States, and northern

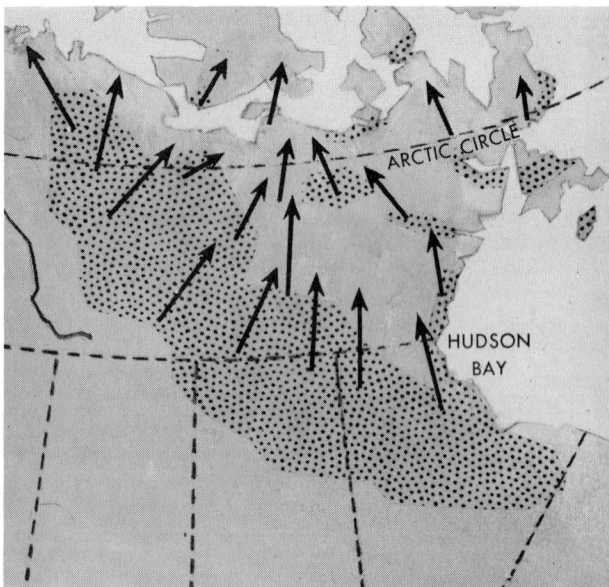

Adapted, with permission from the American Museum of Natural History, from Banfield

Above: map of migratory routes and grazing areas of the Barren Grounds caribou. Arrows indicate spring migration routes northward. Dots indicate where these animals graze in winter after they have migrated southward. Right: photo of the Barren Grounds caribou on the move in a July migration.

Mexico. The bison moved north and south with the seasons, over routes so well chosen that they were followed later by rail and automobile roadmakers. These magnificent animals were almost exterminated by the end of the nineteenth century. They were saved by conservation measures. Some 25,000 bison are now to be found in the United States and Canada, chiefly in government parks and reservations and in zoological gardens.

More fortunate than the bison, great herds of caribou, *Rangifer arcticus,* still exist in the Barren Grounds of Canada's Northwest Territories. Early in August, they collect by the thousands and start a somewhat circular counterclockwise migration. They move in a general southwesterly direction for several hundred kilometers, and return in a general northeasterly direction the following spring.

Farther south, the American elk, or wapiti, *Cervus canadensis,* finds room enough to maintain its normal habits in the

Charles J. Ott from National Audubon Society

great northwestern parks of the United States. About the middle of September, the wapiti begin to descend from the higher mountain pastures. The females and the immature males come down first, followed and then joined by the bulls. They spend the winter in sheltered valleys. As soon as the snow begins to retreat, the males start back toward the summer range, followed by the females and the newly born calves.

Lemmings. Among the most spectacular mass movements of land mammals are those of the true lemmings. As pointed out earlier, these movements are irruptions, rather than migrations. The true lemming, belonging to the genus *Lemmus,* is a small-eared, short-tailed, mouselike rodent about 12 centimeters long. It dwells in Lapland and the highlands of Norway and Sweden. When there is an excessive number of lemmings in their ordinary habitats, a mass movement toward lower-lying regions begins. As they advance, the animals disregard obstacles in their paths, swimming across rivers and even small lakes. Vast numbers are slaughtered by beasts and birds of prey—foxes, bears, weasels, cats, hawks, owls, and others. Many are exterminated by people, since they devour crops and other kinds of vegetation. Others fall victim to a disease called lemming fever. When the survivors reach the coast, they at once plunge into the sea and are drowned.

Seals. The fin-footed mammals spend most of their time in the ocean, but breed and bear their young on shore or on ice floes. Walruses do not migrate. Neither do most of the manatee group.

Species of the true, or earless, seals leave the water and spend a breeding season of several months on shore. One of these, the harp, or saddleback, seal, *Phoca groenlandica,* spends the summer in herds feeding far up along the coast of Greenland. In the fall, it migrates to Newfoundland, where the young are born the following spring, on land or on drifting ice.

The eared seals include the Alaskan

When population pressures reach a peak, lemmings migrate to the sea. The migrations are only one way, however: the animals run out of road when they reach the sea. They are forced in and they drown.

fur seals, *Callorhinus alascanus,* which make a long journey each year. The females and young males spend the winter in herds off the coast of California, while the adult males stay farther north along the shores of Canada and Alaska. In the spring, the males move north to the Pribilof Islands, where they are soon joined by the Californian group. Those females that are pregnant give birth to their young within a few days of arrival. Soon after weaning their calves, they are impregnated again and set out on their southerly migration.

Whales. Comparatively little was known about the migratory habits of most whales until the commercially valuable species were threatened with extinction. In 1925, the British organized a scientific expedition to the Antarctic in Captain Robert F. Scott's old ship, the *Discovery,* in order to study the habits of whales. Since that date, there has been continuous whale research in both the Atlantic and the Pacific.

In order to learn the migratory habits of these huge marine mammals, scientists shoot numbered metal tubes into the blubber of living whales. Whalers look for these tubes when they cut up whales, and send any tubes they recover, together with information about when and where they were found, to the appropriate agency. Whales have been taken 4,000 kilometers from the place where they were tagged, and as much as ten years after tagging.

Most whales are inhabitants of colder waters. All migrate seasonally toward warmer regions, some for considerable distances. The migrations of the humpback whale, *Megaptera novaeangliae,* are well known because it winters in coastal waters where it can be easily observed. Southern humpbacks migrate from Antarctic regions up the coasts of South Africa, Australia, and New Zealand, usually to the same location each season. Northern humpbacks move from Baffin Bay to the West Indies, from the Bering Sea to Mexico, and from Kamchatka to the Mariana Islands.

The largest whales, the blue, or sulfur-bottom, whales, *Sibbaldus musculus,* feed on plankton in the cold waters of the Arctic or Antarctic. When ice threatens, they move toward the tropics, where they eat little or nothing. There, the young are born in alternate years. Within a few months, the newborn whales are able to accompany their mothers back to the polar regions to feed.

FISH MIGRATIONS

Like birds, fishes are able to travel long distances with speed and ease. It is not so easy to follow their routes, however. Thus, we still know comparatively little about fish migration. People knew long ago, to be sure, that certain fishes arrived in large schools at the same place and at the same time, year after year. Why they came, whence they came, and where they went remained quite unknown until about the middle of the nineteenth century. At that time organized research on the habits of commercially important fish began under the auspices of fishing industries and governments.

The methods and techniques of this research have been enormously improved in the past decade or two. Radar can detect the presence of schools of fish at depths previously unexplorable. Machinery has been devised for fishing in deeper waters than ever before. As with birds and sea mammals, tagging has been the primary method used to determine migration courses. Investigators have also obtained information by examining fish scales, which tell something about the age and the migratory experiences of individuals.

When we look at the waters of ocean, river, or lake, we do not realize how varied they are. Freshness or saltiness, currents, depth—all combine to make the migratory habits of fish more complicated than those of land vertebrates. Like other animals, fishes need a suitable nursery for their young and sufficient food for themselves. If they cannot use their feeding areas as breeding grounds, they must migrate between the two. The different fish species follow a variety of routes as they migrate: some from the depths to the surface; some from offshore to inshore; some from salt to fresh water.

From ocean to river. Of the several fishes that leave the ocean and swim upstream to spawn in fresh water, the best known are the salmon. The main genera of salmon breed in fresh water and spend most of their lives in the sea.

In the spring, the adult salmon ascend the rivers where they were spawned. They leap over rapids and small falls until they

Alaskan fur seals at their breeding grounds in the Pribilof Islands off Alaska. Females and young males migrate north from the California coast as warm weather approaches. The older males stay north of the females and young males. When they reach the Pribilof Islands in spring, the females give birth. The older males and females mate once again. After the pup is weaned, the group heads south again for the winter.

Karl W. Kenyon from National Audubon Society

When salmon mature, they return from the sea to spawn in the freshwater streams where they were born. Their homing instinct is precise—they always find their home stream.

come to a place where shallow water rushes over a gravelly bed. Here the rapid flow aerates the water and the gravel purifies it to the high degree necessary for the embryo fish.

Sometimes the trip upstream may be several hundred kilometers, as was formerly the case along the Columbia River. Sometimes it may be much shorter, as along the rivers of Japan and eastern North America.

The salmon eat little or nothing once they leave the ocean. They are beaten and torn by the rocks and rapids. After spawning, therefore, most of them die. Not all succumb, however. A few lucky ones live to make the arduous trip back to sea. In fact, they may return a second, and very rarely a third, time. Scratches on the scales of a salmon show whether the fish has made the spawning trip and how many times.

The young salmon remain where they were spawned for a few years (the length of time depending upon the species) before following the stream to the sea. They spend several years in ocean waters, feeding voraciously and gaining size, strength, and fat for the return trip. When they finally reach the adult stage, they set out for the same stream and even for the same bed where they were spawned.

There are many man-made obstacles in the salmons' path. On certain rivers hydroelectric projects, with their dams and turbines, take a heavy toll of the migrating salmon. Attempts to build fishways to assist salmon over the dams have not been very successful.

Several other food fishes migrate in much the same way as the salmon. Among them is the American shad, *Alosa sapidissima,* a native of the Atlantic coast of North America but now introduced to the Pacific coast. It does not travel so far upstream as salmon do. It spawns in shallow water near the banks of rivers or lakes and then returns to the ocean. It may make the trip back to the same spawning area each spring for several years. The young shad remain in fresh water until fall before moving down to the sea. They reach maturity within a few years and then return to their birthplace to spawn.

The sea lamprey, *Petromyzon marinus,* an eellike creature with a round, sucking mouth, regularly migrates from the Atlantic coast of North America into the rivers to spawn, after which it dies. The young remain in the fresh water for three or four years before returning to the sea. Some sea lampreys remain in the St. Lawrence River or the Great Lakes and never migrate.

From fresh to salt water. Eels have a particularly interesting migration routine. The females of the Americal eel, *Anguilla rostrata,* grow up in the lakes, ponds, and rivers of eastern North and Central America and the West Indies. The males live in the brackish waters at the stream mouths.

After remaining for 7 to 12 years in fresh water, the females are mature and start in late summer on their migration to the sea. If they find the outlets dried up, they are able to travel for some distance overland, breathing through the skin. From the coast, males and females migrate to the Sargasso Sea, reaching it in midwinter. There they spawn and die.

By the second summer after their birth, the eel larvae, only five centimeters long, have reached the river mouths. Here the males remain, while the females begin their long journey upstream.

European eels follow a similar routine. But their migration takes much longer, since they also spawn in the Sargasso Sea and must travel farther to reach it from their former river abodes.

Within the ocean. The ocean fishes that supply the world with so much of its food travel in enormous schools between their feeding and spawning areas.

Cod feed on the rich plankton in the far north, coming south in the fall to spawn around Newfoundland and New England, or off the coast of Norway. In schools numbering many millions of fish, herring—both the Atlantic and Pacific species—travel inshore to spawn and back to the deeper waters to feed.

European sardines migrate with the seasons up and down the coast from North Africa to the British Isles. The California sardines make long, round-about trips between southern Alaska and Baja California (Lower California).

The tiny European anchovy leaves North Africa in early spring. Traveling close inshore, it goes all the way to Norway to deposit its oddly shaped eggs.

AMPHIBIAN MIGRATIONS

Most amphibians—toads, frogs, salamanders, newts, and their kin—lead a double life. They are born in water, where they remain until they have developed the ability to breathe on land. Then they climb ashore, later returning to the water to deposit the eggs of the next generation.

The American toad, *Bufo terrestris americanus*, spends the winter hibernating in a burrow that may be a long distance from the water. When spring comes, the males emerge from their burrows. They make their way by night to the body of water that will serve as the breeding place. As soon as they are established, they begin to call until females arrive.

Zollinger—Tierbilder Okapia

Eels also migrate over great distances to spawn. They are spawned in the ocean, but live until maturity in freshwater lakes and rivers. At maturity they return to the sea, spawn, and die.

ANIMAL MIGRATION 471

After the eggs have been deposited, male and female toads return to their feeding grounds. The tadpoles that hatch from the eggs require about two months to develop toad characteristics so that they can go ashore. They then remain on land until fully mature, before migrating back to the water, generally in the third summer of their lives.

Frogs have much the same breeding habits as toads. They do not travel as far, and they require deeper pools of water for their tadpoles.

The American spotted, or red, newt, *Triturus viridescens,* spends as much as three or four years on land as a red eft, while attaining maturity. In the early spring of its adulthood it loses its red coloring, turning a dull brownish green. Though retaining lungs, it develops a skin that aids in breathing. It migrates back to the water to breed and to spend the rest of its life. Three months after the larvae hatch, they have developed lungs and are ready to leave the water.

Some salamanders are completely aquatic. These include the giant salamander of Japan, *Megalobatrachus japonicus.* This creature, which reaches a length of 100 centimeters, lives in cool mountain streams. In midsummer the adult males leave their feeding areas, some 180 to 215 meters above sea level, and go upstream to a place as much as 600 meters above sea level, where the stream is only 1¼ or 1½ meters wide and 10 to 20 centimeters deep. They then dig deep burrows in the bank and wait for the females to arrive. After laying their eggs, the females return downstream to their feeding places. The males brood the eggs for the two or three weeks required to hatch them, before starting their return trip.

REPTILE MIGRATIONS

Comparatively few reptiles are known to migrate.

Turtles. Of the migrating species, all are marine except the huge and long-lived Galápagos turtles, *Geochelone elephantopus,* which dwell on the Galápagos Islands off the coast of Ecuador. These giants breed during the spring rainy season, depositing their eggs in the sandy soil of level valleys. Here they spend the summer.

When the dry season comes and the valley grasses become withered, the Galápagos turtles climb up the steep mountain slopes with their young to an altitude of 600 meters or more, where pastures are kept green by the moist trade winds. Next spring all the animals, young and old, descend to the flat again, using trails that appear to have been established for ages. Some of the turtles have to travel for days to reach the valleys from their preferred plateau.

All turtles deposit their eggs in burrows or nests on land. Therefore, female sea turtles have to migrate from the water every spring. The green turtles, *Chelonia*

Galápagos tortoises. These huge animals breed in valleys during the rainy season. When the dry season comes, they move to higher altitudes in order to feed on moist mountain vegetation.

mydas, with shells that look like huge hemlock cones, leave the ocean and travel up a stream for several kilometers to find a suitable sandy bank. The female of the North American freshwater yellow-bellied turtle, *Pseudemys scripta,* migrates in the summer some distance from the water to build a bottle-shaped nest for her eggs. The young hatch in the autumn and remain in the nest over winter. In the following spring, they dig out and go down to the water.

Snakes. Snakes that lead a marine life are apparently not so completely dependent as sea turtles on a land base for their nests. There are two subfamilies of sea snakes. The snakes of one of these—the Laticaudinae—are egg-layers and are found mainly around Australia and various adjacent islands. Some of these reptiles make seasonal trips to the shore, where the females deposit their eggs. They may spend some time inland before returning to the sea.

Snakes in the other subfamily, the Hydrophiinae, produce living young. They are more widely distributed. They are most abundant in Indonesian waters, where large schools of the small, brilliantly colored species are sometimes seen swimming in long lines, apparently to or from a breeding place. The females of a few of these species move on to small rocky islets to bring forth their young.

MIGRATION OF SNAILS AND CRABS

There are few examples of true migrations among invertebrates, or animals without backbones. Various marine snails migrate up rivers in the summer and retreat to the ocean in winter. Others come to the shallows off the coast in spring and seek deeper water when winter approaches. It is thought that such migrations may account, in part at least, for the evolution of freshwater species from marine forms and of land species from aquatic forms. These freshwater and land species may have developed from migrants that failed to return.

Most crabs are marine, but many can adapt themselves to fresh water. Some dwell upon the land. The land crabs are likely to return to the sea to deposit their eggs. In the spring, robber crabs, *Birgus latro,* living in various islands of the Pacific and Indian Oceans, can be seen moving steadily in a straight line, over instead of around obstructions, to the coast and into the water. After breeding, the crabs return to the land. The young larvae have to develop in the water through several stages before they are able to migrate inland.

The blue crab, *Callinectes sapidus,* spends the winter along the continental shelf off the east coast of North America. In the spring, it moves toward shore and often makes its way up streams through the brackish water in search of food. On the Pacific coast, tagging experiments on the coastal crab, *Cancer magister,* have shown that it may travel as far as 130 kilometers in six months.

INSECT MIGRATIONS

The monarch butterflies, *Danaus plexippus,* are true migrants. They are widely distributed throughout the United States and Canada and are now found in other countries as well. Their travels have attracted the attention of a great many people—particularly tourists in central California. Each October, hordes of the butterflies from unknown places in the north arrive at their usual wintering sites. The chosen sites are about a dozen small and isolated groups of trees along the central California coast. The insects settle on leaves and twigs about six meters from the ground. At Pacific Grove, California, it is estimated that 10,000 butterflies mass on five small branches, usually the same branches each year. They remain there throughout the winter. Cold spells kill many of them; on warm days, they fly around and occasionally sip from flowers. In April, a few individuals take off for the north. In a month's time, the trees are deserted.

As the monarch butterflies start north, the mass movement effect is lost. Apparently many individuals deposit eggs on their way north and die before they reach their place of origin. Butterflies developing from these eggs continue their course northward. Some of them also deposit eggs and die on the way. The butterflies that finally reach the northern breeding area include a few of the original migrants, some of the second generation, and most of the third.

It is sometimes difficult to tell whether a genuine migration takes place in the case of various invertebrates. Consider the seventeen-year locust, or cicada, for example. The females insert their eggs in the twigs of trees or shrubs. The young that hatch from the eggs are wingless creatures called nymphs. They drop to the ground and proceed to burrow into it. Thereafter they remain underground, traveling from one root to another and feeding on root tissues. After 17 years have elapsed, the nymphs make their way to the surface of the ground, climb up tree trunks, cling to the bark, and are transformed into winged creatures. What shall we call the trip from tree branches to underground roots and the return trip? Shall we say that a migration has taken place?

Everyone has seen the wingless generations of plant lice, or aphids (family Aphididae), on house or garden plants. Some of you may also have seen the swarming of a winged form. In the development of these different generations, aphids go through what is apparently a form of migration.

Many generations of aphids are produced in a summer, and any given one may differ considerably from those preceding and succeeding it. In one of the generations coming out in the spring, the females have wings. They fly away to a plant, usually an annual, belonging to a different species from that of the plant they left—usually a perennial. On the new host, several wingless generations will be born parthenogenetically (from unfertilized eggs), live, and die.

In the fall, when the annual plant is about to die, a generation of winged aphid males and females will be produced. These will mate and the females will fly back to the plant on which their ancestors lived. There a few generations will be born viviparously. That is, the female will produce living young from within her body. Then eggs will be laid; these will hatch the following spring. Winged females of one of the spring generations will seek out another annual.

ONE-WAY TRIPS

Entomologists generally give the name "migrations" to mass movements of insects, such as locusts, from their place of origin, even though flights of this kind are one-way trips, as far as we know.

The mass movements of locusts have been devastating the world for untold ages, but their actual pattern was not learned until the 1920s. At that time, B. P. Uvarov of the British Museum, working with a staff of trained observers, discovered that the migratory locust of the Old World was only a dark-colored form of the green, nonmigratory, solitary grasshopper. This is also true of the migratory locust of North America.

After several generations of normal green grasshoppers have increased the population of an area to a certain point, a large percentage of the succeeding generation will be dark-colored. These dark-colored insects will launch upon a few trial circulatory flights. They will then fly in a body toward new territory. In the United States, the locusts travel from the Rocky Mountains eastward. They fly high, coming down to feed and rest, and eventually reach the Missouri Valley, where they deposit their eggs.

It is not known whether or not the grasshoppers that hatch from these eggs survive, or, if they do survive, whether they or succeeding generations make their way back to the Rockies. In any event, enough normal green-colored grasshoppers have been left in the original breeding grounds to ensure that there will eventually be another migrating brood.

Some insect mass movements are still a complete mystery to entomologists. Some years ago, for example, the American naturalist Charles William Beebe and his staff watched spellbound as thousands of millions of all kinds of insects streamed south through the narrow Portachuelo Pass in north-central Venezuela. The migration continued night and day from May through September. It was apparently a one-way exodus, for no mass movement back to the north was ever noted.

FOSSILS

About 550,000,000 years ago: small crustaceans called trilobites crawled along the ocean floor.

175,000,000 years ago: large, often terrifying reptiles called dinosaurs roamed the land.

No human being has ever seen a living trilobite or dinosaur. How, then, can we say that these animals once lived on earth?

140,000,000 years ago: a pheasant-sized creature named *Archaeopteryx* climbed trees and glided from branch to branch. Scientists believe that this animal was the ancestor of all living birds.

50,000,000 years ago: flesh-eating mammals called miacids lived in the forest. Scientists say that miacids were the ancestors of today's carnivores.

But no human being has seen a living *Archaeopteryx* or miacid. How can scientists believe that modern animals had such ancestors?

As animals and plants die, they leave behind evidence of their existence. Some, such as jellyfish, may leave imprints in sand that slowly turns to rock. Others, such as trilobites and tree trunks, may actually turn into rock.

Such remains, impressions, and molds of animals and plants are called *fossils*. The great majority of fossils have been found in sedimentary rock—that is, rock formations derived from sediments laid down in streams and seas of past ages. Some fossils have been discovered in ice, frozen ground, and oil seeps.

We are able to reconstruct the history of life from the study of fossils. Fossils have given us a vast amount of information concerning ancient animal and plant life. They have also helped to determine the relative ages of the rock formations in which they have been found.

The study of fossils is called *paleontology*—the "science of ancient living things." This comes from the Greek words *palaios*, meaning "ancient," and *onta*, meaning "existing things."

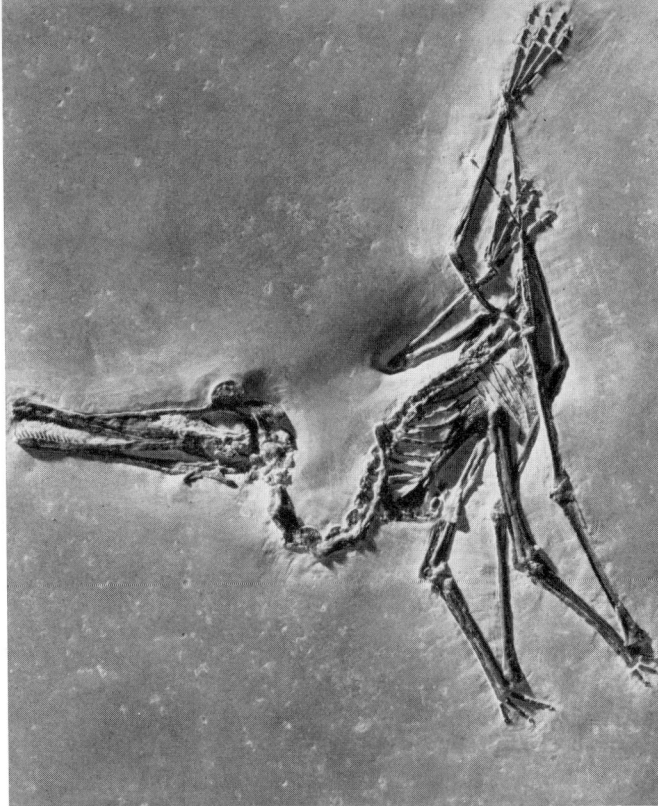

The American Museum of Natural History

The British Museum of Natural History

From rare finds of complete fossilized skeletons, paleontologists can make fairly accurate guesses concerning the animal's appearance. Top: a pterodactyl skeleton. Bottom: reconstruction.

KINDS OF FOSSILS

There are various kinds of fossils. Some represent the actual remains of animals or plants—usually the hard parts, such as teeth, bones, and shells, which have remained after the soft parts have decayed. It is true that entire bodies of woolly mammoths have been found preserved in ice and frozen mud in Siberia. But these particular remains are not old geologically.

The commonest animal fossils are shells. Many are shells of tiny, one-celled animals called foraminifers, so small that they must be studied under a microscope. These are often abundant in oil deposits and their study is particularly useful in the search for oil. Larger shells, such as those of snails and clams, are also common as fossils. So are the hard parts of corals, sea lilies, lamp shells, and relatives of the chambered nautilus. These shells are composed of calcareous (limy) material. They are often preserved in the rocks without chemical change.

Many times, however, drastic changes have taken place in the shells. Mineral-bearing water seeping through the rocks has replaced the lime with other sub-

Brachiopods were common forms of animal life from Cambrian time until the Devonian period.
Courtesy of the American Museum of Natural History

N.Y. State Museum and Science Service

Plant matter can be fossilized under certain conditions. The swirling designs are fossilized algae.

stances, especially silica (silicon dioxide). As a result, the fossil may retain the exact form of the original shell but contain none of the substance that composed the shell when the animal was alive.

Often the seeping ground water dissolves the buried shell fossil without replacing it. Then a cavity is left in the rock. This cavity preserves the shape of the fossil; it serves as a natural mold. By filling in the cavity in the rock, the paleontologist may obtain an accurate plaster cast of the original fossil.

Bones and teeth are also common animal fossils. Bones may be preserved without much change, except that the organic material that once filled cavities and microscopic spaces in the bone decays and disappears, leaving only the hard, mineral substance of the bone. Usually the spaces left empty by decay become filled with some other mineral, such as lime or silica. The bone becomes harder and heavier and is said to be *petrified*. Actually, however, it has not really turned to stone. The original hard, bony substance is still there and the only important change is that something has been added.

Teeth are usually preserved with even less change, because they are harder than

Courtesy of the American Museum of Natural History

Opalized wood. Wood may be petrified when water containing mineral salts seeps through the wood. The minerals remain in the woody tissues.

bone to begin with and contain fewer microscopic spaces. Only infrequently are original tooth and bone substances entirely replaced by other substances. Only infrequently, too, do they dissolve, leaving molds.

Fossil jellyfish and plants may be preserved as imprints. Impressions of leaves are especially common. These may be simply molds left after the surrounding rock has hardened and the substance of the leaf has decayed and disappeared. Often, however, some of the original leaf substance is left in the form of a brown or black film of carbon.

When a large mass of vegetation is buried, time, pressure, and heat in the crust of the earth may turn it into coal. The form of the original plants is seldom visible in a piece of coal. In certain cases, however, the structure of the plants is preserved and can be made visible again by special procedures.

A most interesting type of plant fossil is *petrified wood*. This is how it is formed: a stem, trunk, or sometimes a root is buried in the ground. Percolating water slowly fills porous spaces in the wood with mineral matter. Finally, the woody substance itself is replaced by minerals. The petrifying, or stone-making, substance is usually a form of silica, as in the well-known petrified forest in Arizona, in the United States.

There are various other types of fossils. Fossil footprints and trails are present in certain deposits. If an animal walked or crawled over soft mud or sand and this was covered over by another layer of sediment before the trail was wiped out, the track may still be preserved after the whole deposit has hardened into rock. Dinosaur tracks have been found in various places, and these have supplied a certain amount of information about dinosaurs whose bones are unknown.

Eggs are a rare type of fossil. Perhaps the most famous are the dinosaur eggs found in the Gobi Desert in Central Asia. Stomach stones, called gastroliths, may also be fossils, though this has been disputed by some. The gastroliths are highly polished pebbles. Supposedly, they were swallowed by dinosaurs and served to grind the food gulped down by the animals. They were polished in the course of the grinding process.

Molecules representing some of the original chemical constituents of once-living organisms have often been left in the rocks that at one time formed the environment where these organisms flourished. From such molecular remains, chemists have attempted to reconstruct the original

The Smithsonian Institution

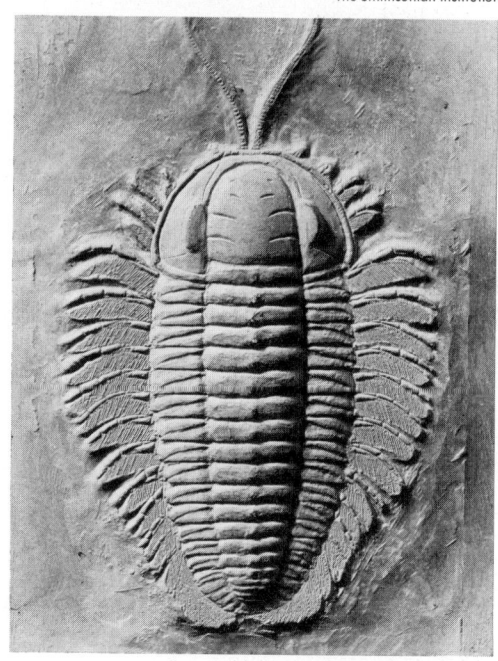

Courtesy of the American Museum of Natural History

Top: fossil of a species of giant fern. These plants, which grew 300 million years ago, are the basis of the coal industry. Bottom: a trilobite cast.

compounds and then compare them with those of modern organisms. In this way, some idea of the life processes of ancient plants and animals may be obtained.

GEOLOGICAL TIME SCALE

In our study of fossil life forms, we use the geological time scale, which has been worked out to show the events in the earth's history. The chief divisions in the scale are eras, periods, and epochs. Here we deal only with eras and periods.

The *eras*, the largest of the scale divisions, are separated from each other by great *revolutions*, which were times of extensive changes—mountain-building, continental uplifts, and retreats of the seas. The revolutions were not catastrophic events but, rather, a continuous series of changes that created new environmental conditions. These conditions caused the rise of new life forms and the extinction of many older forms that could not adapt. The process was repeated several times.

Within each of the eras, there were lesser events, known as *disturbances* (that is, disturbances in the earth's crust). They were less widespread and less far-reaching in their effects than revolutions, but they also marked definite breaks in the geological record. These disturbances separated the various *periods* of geological time.

PRE-CAMBRIAN TIME

As we shall see, the name Cambrian is given to the first period of the Paleozoic era. The tremendously long span of time that elapsed from the first rock records to the Cambrian period is generally called Pre-Cambrian time. It lasted until 600,000,000 years ago.

By the beginning of the Cambrian period, the invertebrates (animals without backbones) were already well along the evolutionary path. The question arises: "Where and when did these quite highly organized invertebrates originate?" Obviously there must have been ancestral types of all these forms in Pre-Cambrian time. Yet fossil remains of these ancestral forms are few and far between in Pre-Cambrian rock formations. Probably the chief reason is that these early animals were mostly soft-bodied and had few, if any, structures that would resist decay. By the beginning of the Cambrian period, the invertebrates began to possess shells and skeletal structures that could be preserved as fossils.

Most of the scanty remains of life in Pre-Cambrian time were of plants, particularly algae. It is generally believed that the formations known as *stromatolites* ("layer stones") represent the fossils of certain algae. According to one theory, these plants lived in colonies; limy substances settled upon the colonies and built up replicas of them. There is other evidence that algae existed in Pre-Cambrian time. Biscuit-shaped masses found in Pre-Cambrian rocks seem to consist of layers of shells that once covered the cells of algae. Remains of fungi have also been found.

There is less evidence of animal life in the Pre-Cambrian. Filled worm burrows indicate the existence of worms. Remains of sponges have been discovered in limestone formations about 1,000,000,000 years old. Other supposed Pre-Cambrian fossil remains are doubtful.

Pre-Cambrian time was ended by the Killarney Revolution, a time of worldwide continental uplift and erosion. From the end of Pre-Cambrian time to the beginning of the Paleozoic era, there is a tremendous break in the rock record.

THE PALEOZOIC ERA

The Paleozoic era lasted well over 300,000,000 years—from 600,000,000 years ago to 230,000,000 years ago. At its beginning, there were practically no land-dwelling animals or plants, and only a limited number of species were to be found in the sea. By the end of the era, both animals and plants had become established on land, and many different species flourished in the seas of the world. The Paleozoic era is divided into seven periods: the Cambrian, Ordovician, Silurian, Devonian, Mississippian, Pennsylvanian, and Permian.

The Cambrian period. The continents were invaded by the sea; by the end of the Cambrian, a good deal of what is now North America was under water. The climate was generally warm, though it is possible that glaciers appeared in certain areas. The period was ended by the Green Mountain Disturbance in northeastern North America and considerable mountain-making in the European continent.

Generally speaking, there is practically no direct evidence of land animals or plants in the Cambrian period, though primitive spore-bearing land plants seem to have occurred in several areas. Both animals and plants, however, were to be found in considerable abundance in the sea. The plants known as calcareous algae were quite widely distributed, and reefs were built up in various areas from their shells.

The only animals represented in the sea were invertebrates. The most highly developed forms were the trilobites. These were arthropods, probably distantly related to the modern horseshoe crab. The trilobites were quite flattened. They possessed a head, thorax, and tail. A pair of lengthwise grooves seemingly divided the animal into three distinct lobes, or rounded divisions; hence the name "trilobites." They generally ranged in length from two to ten centimeters. They lived on the bottom of shallow seas and probably fed on organisms found in the mud.

Another important Cambrian invertebrate group, the brachiopods, superficially resembled small clams, though they differed from them in the arrangement of their shells and in their internal structure. They dwelt in great profusion in various parts of the sea, anchoring themselves to the sea floor by means of stalks. More than 225 varieties of brachiopods still live in the seas today.

Spongelike animals called archaeocyathids ("old cups"), with outer walls and sometimes also inner walls of calcite (calcium carbonate), flourished during this period. Some looked like saucers, others like cylinders. They grew over one another and in the course of time built up massive reefs, which are found today in many parts of the world. One deposit in Australia is some 650 kilometers in length.

both photos, the Smithsonian Institution

Top: fossil of a cephalopod resembling a chambered nautilus shell. Bottom: insects were often caught in sticky residues like amber. When the amber hardened, the insects were preserved intact. Here, a fly has been caught.

Worms must have flourished in the Cambrian, since traces of their burrows are found in sandstone in widely separated areas. Other invertebrates play but a minor part in the Cambrian fossil record, not necessarily because they were rare but because they had not developed shells that could be preserved as fossils. As evidence of the wealth of such apparently minor forms, we find hosts of invertebrates preserved in the form of carbonized films in the shale beds of a quarry near Burgess Pass in British Columbia, Canada. These films show not only the outer shape but also the internal structure of glass sponges, worms, sea cucumbers, and other forms of animal life.

The Ordovician period. The continents, particularly North America, were widely submerged by shallow seas during this period. At the same time warping of the earth's crust lifted vast areas of the land above the ocean surface. As a result, the shorelines were constantly shifting. The period was ended by the Taconic Disturbance in eastern North America.

In the Ordovician period, life still flourished chiefly in the seas. There is still no conclusive evidence of either plant or animal life on land. As in the Cambrian period, the principal marine plants were the algae.

Trilobites and brachiopods were still among the most important sea-dwelling invertebrates. In fact, the trilobites attained their highest development at this time. A number of other invertebrates now became prominent. Among them were the graptolites, which formed branching structures. Floating graptolite colonies drifted across wide expanses of the ocean, and their fossil remains have been found in many different places. These mysterious creatures have been long extinct. True corals now appeared and formed reefs, which are preserved in limestone. Brachiopods evolved into a number of different species. Mollusks—gastropods, clams, and cephalopods—were also to be found in considerable numbers.

Echinoderms, the group to which modern starfishes belong, were represented by the crinoids, or sea lilies. These plantlike animals anchored themselves by means of stalks or stems to the bottom of the sea or to shells or other hard surfaces.

Primitive fishes first appeared in the Ordovician period. They were the earliest known vertebrates. These fishes, known as ostracoderms ("shell-skinned animals"), had no jaws. The mouth was simply a hole or slit in the head. Paired fins were absent. Thick plates covered the head. There were bony scales on the trunk and tail. The ostracoderms were related to the modern forms called lampreys and hagfishes.

The Silurian period. The sea alternately advanced and retreated during this period. A great deal of volcanic activity took place. The world climate was generally mild; some areas were arid. The end of the period was marked by the Caledonian Disturbance in Europe; it resulted in the rise of the Caledonian Mountains. Large-scale mountain building also took place in northern Africa and east central Asia.

In Silurian-period fossils we find the first signs of land-dwelling animals. These were scorpions and millipedes. The oldest definite land plants also date from the Silurian. Fragmentary remains of bits of stems and small leaves, going back to this period, have been found in England and Australia. Soft-tissued plants, such as certain algae, may also have appeared on land during this time, but they left no traces.

Marine invertebrates still dominated the seas, though primitive fishes were well represented. The trilobites and graptolites had declined considerably. Other invertebrate forms, however, showed a marked expansion. The corals developed many new species; the reefs formed by their skeletons were widely distributed. Sea lilies abounded. They were especially prolific in the central Mississippi valley in North America, where broken stems are everywhere to be found in limestone formations. The eurypterids, or sea scorpions, were also very common in certain areas. Most species were comparatively small, ranging up to 30 centimeters in length. A few species were far larger; *Pterygotus buffaloensis* attained a length of approximately two meters.

The Devonian period. The seas flooded large continental areas at one time or another during this period, but they subsided toward its end. The climate was generally mild; certain areas were very dry. There was marked volcanic activity. The Acadian Disturbance, an active period of mountain building, began about the middle of the period and brought it to a close.

Land plants, which had first definitely appeared in the Silurian period, became common in the Devonian. Some of the primitive Devonian trees have left petrified stumps more than 60 centimeters in diameter, which indicates that these plants must have been quite tall. Among the common trees were certain evergreens and scale trees. There were also primitive ferns, horsetails, and scouring rushes. The first forests now began to clothe land areas.

Courtesy of the American Museum of Natural History

Fossil remains of an alligator. This reptile lived in South Dakota during the Oligocene epoch, between 25,000,000 and 36,000,000 years ago.

As in earlier periods, animal life was particularly abundant in the sea. Marine invertebrates were still plentiful. Brachiopods reached the height of their development. Corals continued to flourish and to carry on their reef-building activities. In the Devonian, there were some unusually large corals; certain individuals were as much as 7½ centimeters in diameter and 60 centimeters high.

The ammonites, belonging to the cephalopod group, first appeared in this period. Their shells, which were divided into chambers, vaguely resemble rams' horns. (The Egyptian god Ammon is often represented with a ram's horns; hence the name of the animals.) Starfishes were increasingly well represented. Trilobites were on the decline.

Fishes teemed in the Devonian seas. The jawless ostracoderms still persisted. But now another type of fish, the placoderms ("plate skins"), came to the fore. They had heavy armor and well-developed jaws; some of them were sharklike in appearance. The earliest true sharks go back to the Devonian period. They were rather small, averaging only a meter or so in length.

Offshoots of the sharks, called arthrodires, attained large size, and were the biggest animals of the period. The best-known arthrodire was *Dinichthys,* which reached a length of about nine meters. With its large head and broad mouth, it looked somewhat like a catfish. Armor, made up of bony plates, covered its head; some of the plates covering the jaws took the place of teeth.

The true, or bony, fishes also appeared in the Devonian. They have become the dominant modern fishes. The interesting lungfishes were quite common at this time. In these bony fishes, the swim bladder was connected with the throat and formed a sort of primitive lung. If the ponds and streams in which they lived became stagnant and lacking in oxygen, the lungfishes would come to the surface and take a gulp of air, thus obtaining their oxygen supply.

Lungfishes belonged to the group called the Choanichthyes. The lobe-finned fishes, or Crossopterygii, also were members of this group. In these fishes, each of the paired fins had a lobed or rounded base of flesh, from which jointed bones extended.

The first land vertebrates appeared in the Devonian. These animals, which were descended from the lungfishes, were amphibians called stegocephalians, or "roof-headed animals," because their skulls were covered by bony armor. They crawled along on small weak legs, which extended outward from the body. One of the late-Devonian stegocephalians—*Ichthyostega*—attained a length of about 1½ meters.

The Mississippian and Pennsylvanian periods. These periods were characterized by the development of huge swamps. However, conditions varied widely from one area to another and at different times.

Plants related to modern ferns, club mosses, and scouring rushes thrived. There were also primitive seed plants, frequently referred to as seed ferns, for they had fernlike leaves but reproduced by seeds instead of spores.

In various swampy areas, the vegetation developed into dense, tropical forests. It is to those forests of the Mississippian and Pennsylvanian periods that we owe many of our coal deposits. For this reason, the name "Carboniferous (coal-bearing) period" is sometimes given to the two periods in question. We shall consider them together in these paragraphs.

In the dense forests, animal life was abundant. There were great numbers of insects, including dragonflies and cockroaches. They were remarkable for their large size. A dragonfly found in Belgium, for example, had a wing spread of about 75 centimeters. Most of the insects were primitive. The cockroaches were an exception, however. They were much like their modern descendants except that they were larger, attaining a length of ten centimeters or so.

Primitive amphibians, looking like gigantic salamanders, thrived in great profusion in the swamps. Toward the close of the Carboniferous period, there evolved from the amphibians a group of primitive land reptiles that were lizardlike in appearance. They became the ancestors of the great dinosaurs of the Mesozoic era and also of the mammals, which evolved later.

Life was as abundant as ever in the seas. The sea lilies (crinoids) flourished: their plates contributed to the thick limestone formations known as crinoidal limestones. The foraminifers emerged as important rock-makers. These tiny one-celled animals had shells, known as *tests,* with one or more openings in various sections. Pseudopods (extensions of soft living substance) protruded through the tests. Fishes were plentiful. The cestriacont ("shell-crushing") sharks reached a high stage of development. They were the ancestors of the modern Port Jackson shark.

The Permian period. The seas retreated during much of this period, and the continents rose. There were great extremes in climate; vast areas became arid, while in other regions there was heavy rainfall. At intervals, the climate became much colder and extensive regions became covered with sheets of ice.

The swamp-dwelling plants of previous periods were replaced more and more by hardier stocks that could withstand the cold. True conifers, or cone-bearing trees, became the most important group of woody plants. There were also great numbers of cycads, trees suggesting date palms.

Fossil of a eurypterid. Eurypterids are an extinct group of water scorpions.

N.Y. State Museum and Science Service

Insects decreased considerably in size and were more varied than before. The amphibious stegocephalians, which had first appeared in the Devonian, flourished, reaching lengths of about three meters.

Reptiles increased both in number and variety. They were all four-legged; they generally had long bodies and tails and short legs. Some were quick in their movements. Others were sluggish. The flesh-eating reptiles known as theriodonts are of particular interest because they were the ancestors of the mammals. Like mammals, they had teeth consisting of incisors, canines, and molars. Their skulls, too, were mammal-like.

The trilobites and certain other marine invertebrates that had once been abundant became extinct by the end of the Permian. The brachiopods, flourishing at the beginning of the period, dwindled greatly in importance by its end. Many coral varieties died out. Cephalopods were prominent; the forms known as ammonites showed steady development. Fishes were still abundant in the seas and freshwater forms became more important.

The Permian period and the Paleozoic era, of which the Permian formed part, came to an end with the Appalachian Revolution. There are indications that this revolution was a long drawn-out affair, lasting more than 1,000,000 years. It was a crucial period in the history of the earth and its living things. Great continental uplift began and large mountain ranges were formed. The continental seas were drained; deserts appeared in many regions. Many Paleozoic animal and plant species could not adapt to the new environments and became extinct.

THE MESOZOIC ERA

After the uplift of the continents in the Appalachian Revolution, at the close of the Paleozoic era, more settled conditions began to prevail. New groups of plants and animals now evolved. The primitive ferns and allied plants gave way to more modern types; the seed ferns to cone-bearing groups and cycads. The ancient amphibians disappeared and were replaced by reptiles. The latter came to dominate the earth in the course of the Mesozoic era. Hence this era is sometimes called the Age of Reptiles.

The Mesozoic lasted from 230,000,000 years ago to 63,000,000 years ago. It is divided into three periods: the Triassic, Jurassic, and Cretaceous.

Artist's conception of a group of Cladoselaches, which are believed to be the ancestors of modern sharks.

Courtesy of the American Museum of Natural History

The Triassic period. The climate ranged from mild to subtropical during much of this period. The forests were made up chiefly of conifers, much like the modern varieties, and of cycads. Ferns and scouring rushes abounded in the undergrowth. The flowering plants, or angiosperms, may have originated in this period.

Land vertebrates were evolving rapidly. The stegocephalians reached their highest development in the early Triassic, but they could not meet the competition of the reptiles, which could adapt far more effectively to different environments.

This period saw the rise of the reptiles known as dinosaurs ("terrible lizards"). The legs of other reptiles sprawled outward from the sides of the body. The legs of the dinosaurs were under the body, which was carried off the ground; hence they were able to move about rapidly. They could walk on all fours. But when speed was necessary, they could run on their powerful hind legs, with their thick tails providing balance. Dinosaurs did not reach their full development in the Triassic. Generally they were quite slender and did not exceed 4½ meters in length.

The reptiles also invaded the seas. Ichthyosaurs ("fish-lizards") became abundant in the seas of the late Triassic. They were long, sleek creatures, resembling dolphins. They had flipperlike limbs and were excellent swimmers. The ichthyosaurs fed on fish, squids, and other marine animals. They ultimately reached a length of nine meters and were the largest animals of the period. Another inhabitant of the Triassic seas was the plesiosaur ("almost-lizard"). It was a wide-bodied creature, which paddled with its flippers after the fashion of a turtle. It had a short tail but a very long neck, which it could dart out at its prey.

The ocean waters teemed with vast numbers of ammonites. They showed a great variety of forms and some of them attained considerable size. Clams and gastropods were common. Reef-building corals were active. Squidlike cephalopods made their appearance. The period saw the rise of modern varieties of starfishes and lobsters; however, they did not become abundant until later. Brachiopods were rapidly disappearing from the seas.

The Jurassic period. Shallow seas swept over much of the continental area. The climate was generally warm and humid. There were huge forests of pines and other conifers, ginkgoes, cycadlike trees, and tree ferns, with ferns and rushes in the undergrowth. The first definite evidence of flowering plants goes back to this period.

Reptiles now won complete domination over other land dwellers, they competed with the fishes in the sea, and they invaded the air. Some of the dinosaurs were very small. Others attained tremendous bulk and strength. Their brains, however, were negligible in size, weighing less than 500 grams, on the average.

Most dinosaurs were herbivorous (that is, plant-eating) and walked on four legs. The giant of the group was *Brachiosaurus,* which was 23 meters long, stood 13½ meters high, and weighed 85 tons or thereabouts. *Brontosaurus,* which was approximately 21 meters long and weighed some 35 tons, was also most impressive in size. It had a long neck and tail and walked on pillarlike legs. Related to *Brontosaurus* was *Diplodocus,* which reached a length of 26 meters but was comparatively slender, weighing only about 20 tons or so. Another herbivore, the sluggish *Stegosaurus,* developed armor. Two rows of bony vertical plates extended along the neck, back, and tail; the tail also had four upright spikes. However, the formidable armor protected only the back. The sides of the animal were covered with leathery skin and were quite vulnerable.

Other dinosaurs of the period were carnivores, or flesh-eaters. They walked on two legs, balancing themselves with their massive tails. Some of them were small; others were huge. Largest of all was the fierce *Allosaurus,* which was 10½ meters long. The lower jaw of this monster was hinged far back on the skull. As a result the animal could take huge bites and gulp down great chunks of food. It preyed on *Diplodocus* and other dinosaur vegetarians.

The reptiles known as pterodactyls ("winged fingers") took to the air. With their leathery wings and naked bodies, they were batlike. But they differed from bats in various important respects. For one thing, the bats are not reptiles, but warm-blooded mammals. Again, all the digits of the bat are extended to bear the wing membrane; only the fourth finger of the pterodactyl was thus extended, and the other digits were left free. Some pterodactyls had tails; others were tailless. In size, these reptiles ranged from small creatures no larger than sparrows to animals with a wing spread of more than a meter.

The first bird—*Archaeopteryx* ("ancient wing")—goes back to the Jurassic period. It was about as large as a crow. It had birdlike plumage but in other respects—its teeth, clawed wings, and long tail, among other things—it was reptilelike.

Primitive mammals now appeared on the scene. (They may go back to the late Triassic.) They were all small, hardly more than 30 centimeters or so in length. They occurred in many widely separated areas. All of them are now extinct. Among these mammalian pioneers, the pantotheres are especially noteworthy; they gave rise to the modern mammals called marsupials and placentals.

Ichthyosaurs and plesiosaurs, which had first appeared in the Triassic period, now dominated the seas. Among the other reptiles found in the sea or in streams were turtles and crocodiles. Of the marine invertebrates, the ammonites were at the peak of their development. Also plentiful during this time were the squidlike cephalopods known as belemnites. Certain sea-dwelling invertebrates had now practically attained their modern forms; among these were the corals, gastropods, true oysters, lobsters, sea urchins, and sponges.

The Cretaceous period. The sea spread widely over the continents in the earlier part of the period; in North America, it covered almost half the land. Then it retreated again. The climate was warm at first over most of the earth's land surface.

The Cretaceous period came to an end with the Laramide Revolution, marked by colossal worldwide mountain-building and accompanied by dropping temperatures.

In the course of the Cretaceous, flowering plants spread all over the world. Although conifers were still important, the landscape came to be dominated by deciduous trees (trees that lose their leaves after each growing season).

The dinosaurs still dominated the land until toward the end of the Cretaceous period. The largest flesh-eater of this period was the awe-inspiring *Tyrannosaurus rex*. It was some 14 meters long and 6 meters tall, and weighed about 10 tons. It preyed on herbivorous reptiles. Certain carnivorous dinosaurs were much smaller. *Struthiomimus* ("ostrich-imitator") was less than three meters tall when it stood erect. As its name indicates, it was quite ostrich-like in appearance, except for the fact that it had four legs and a long tail.

Other dinosaurs of the period were grotesque creatures. Some of them had birdlike feet; others, duck-type bills. *Pachycephalosaurus* ("thick-headed lizard"), one of the herbivores, was a particularly odd-looking creature. It had a dome-shaped skull adorned with knobs and spines—a formidable structure 63 centimeters long and 25 centimeters thick.

The snakes, which developed from lizard stock, first appeared in the Cretaceous. Their legs had degenerated (as had those of certain lizards) and in time were lost. The jaws were loosely attached to the skull and the mouth could be opened very wide. The snakes have become the most progressive of all modern reptiles.

The mammals were still small and unimpressive. This period marked the rise of two primitive mammal groups: the marsupials, or pouch-bearing mammals, and the insectivores, or insect-eaters. Probably all the chief modern orders of insects were represented, though fossil remains are rare.

Pterodactyls attained their greatest size in the Cretaceous period. The largest known form of flying reptile, a pterosaur, had a wing spread of more than 15 meters—as large as many airplanes. The true birds of the period generally had teeth. Among them were *Ichthyornis,* a shore bird, and *Hesperornis,* a large diving bird that had lost the power of flight.

The plesiosaurs were prominent in the Cretaceous seas. They attained their largest size during the period, reaching a length of 15 meters, of which half was taken up by the neck. Ichthyosaurs were not so important as in the preceding period. The fierce mosasaurs now became prominent. Superficially, they resembled the ichthyosaurs, but differed in various respects. The flippers were greatly reduced; the skin was scaly; and the mouth could be opened very wide because of extra joints in the lower jaw. The mosasaurs reached a length of more than ten meters.

Among the invertebrates, the ammonites and belemnites became less and less important. They became extinct by the end of the period.

The Laramide Revolution, which brought the Cretaceous period (and the Mesozoic era) to a close, marked a most drastic change in the annals of animal life. The dinosaurs and such marine reptiles as ichthyosaurs, plesiosaurs, and mosasaurs became extinct. Among the factors that contributed to their downfall was a drop in temperature, rising highlands, and the disappearance of many swampy tracts.

The Ornithischia were a group of bird-hipped dinosaurs. During Triassic times these reptiles were small, herbivorous, and walked on two legs. *Ornithosuchus* was a representative of the group.

British Museum (Natural History)

both photos, British Museum (Natural History)

The Ornithischian dinosaurs later adopted a quadripedal, or four-legged, lifestyle. The early armored dinosaur above was derived from a bipedal, or two-legged, ancestor similar to the one at right.

THE CENOZOIC ERA

The Cenozoic, the last of the great eras of geologic time, is also the shortest. It is commonly divided into two periods: the Tertiary, which began 63,000,000 years ago and makes up almost the entire era; and the Quaternary, only 1,000,000 years long, which extends to the present time.

The Tertiary period. The climate was fairly warm at first, but gradually became colder. Climatic belts developed. The sea intermittently covered parts of the continents, but on the whole the continents were assuming their present form. The period ended with widespread mountain-building known as the Cascadian Revolution. By the end of the Tertiary, the mountain ranges had attained substantially their present extent and shape.

Practically all the hardwood trees and conifers of today were represented by the beginning of the Tertiary period. Grasses had become common. There were vast prairies, where various animals grazed.

The Tertiary period marked the spectacular rise of the mammals. The early mammals of the Tertiary were small, short-legged, and flatfooted. Their jaws were rela-

FOSSILS 487

tively longer than at present and their brain capacity was limited.

In the course of mammalian evolution, there was a general increase in size in the various groups. The teeth became specialized and in most cases reduced in number, and the brain capacity was increased.

An example or two will illustrate the general course of development. The early horses were natives of North America. The oldest member of the horse tribe, *Eohippus* ("dawn horse"), was a tiny animal, scarcely more than 30 centimeters high. It was a slender creature, with three toes on its hind feet and four on its front feet. In the course of its evolution, the horse attained greater size and gradually lost toes. By the middle of the Tertiary, the animal had three toes on each foot. The toes on each side of the center one continued to shorten, until toward the end of the Tertiary there was but a single toe on each foot. By this time the horse had attained the size of a small pony. Sometime late in the Tertiary period or during the glacial period that followed, horses spread to Asia by means of a land bridge extending from Alaska to Siberia. Modern horses are apparently descendants of these ancient migrants.

Catlike and doglike forms had a common ancestor, but they were pretty well differentiated by the middle of the Tertiary period. They were small at first, but increased in size. By the end of the Tertiary, they were pretty close to the species of modern times. The doglike animals evolved into many different types—true dogs, wolves, raccoons, bears, and others. The catlike types developed into two diverging groups: the true cats (which came to include tigers, lions, pumas, leopards, and others) and the saber-toothed, or stabbing, cats. The upper canines of the latter were very long and protruded from the jaws and were used to stab and tear their prey.

The Tertiary period also marked the rise of primates. The earliest forms go back to about the close of the Cretaceous. They were small, tree-dwelling mammals, developing from primitive insect-eaters. They gave rise to animals that resembled today's lemurs and tarsiers. The latter had greatly enlarged brains, forward-looking eyes, and other monkeylike features; they were probably the ancestors of the true monkeys. The oldest known great ape was a form that most closely resembled the modern gibbon. By the end of the Tertiary, all the great apes—gorillas, chimpanzees, orangutans, and gibbons—were represented.

The Quaternary period. In this period there was a great Ice Age, divided into four epochs, in the course of which the glacial ice alternately advanced over a good deal of the earth's land areas and then retreated.

The last of the Ice-Age epochs came to an end only about 10,000 years ago, which is but a moment in geologic time. Life continued to flourish throughout the Ice Age, because the ice never covered more than about a quarter of the land area. Besides,

Dinosaurs laid eggs. Ancient eggs, however, are difficult to find. These eggs belonged to a *Protoceratops* that lived during the Cretaceous period.

Courtesy of the American Museum of Natural History

the interglacial periods (those between ice ages) were marked by warm climates. Various forms of life, however, became extinct as a result of the changing conditions.

Throughout most of the Quaternary period, great game animals roamed over much of North America and Europe. Various members of the elephant tribe were particularly impressive. Among these was the mastodon, which had reached its peak development in the Tertiary period. The big elephants called mammoths were majestic animals, some of which reached heights of more than four meters. The tusks of these huge beasts grew downward and then forward in a sweeping arch. Mammoths became extinct several thousand years ago.

Buffaloes, including a giant species, roamed the prairies. Wild pigs were to be found in many parts of North America, and modern types of carnivores were well represented. Some carnivores that were prominent during much of the Quaternary are now extinct. Among these were the saber-toothed cat and large species of wolf and lion.

Unquestionably, the most important development in the Quaternary was the rise of human beings. Their success has been due to several factors: a well-developed brain; an erect posture that leaves arms and hands free; remarkably versatile hands; and skillful use of tools.

According to present theories, modern man, *Homo sapiens sapiens*, appeared some 40,000 to 50,000 years ago. This form was preceded by *Homo sapiens neanderthalensis*, a heavy-boned, coarse-featured form. These early true humans were evolved from *Homo erectus*, fossil remains of which have been found in many parts of the world. One of the oldest finds was a skull found in Kenya and dated at about 1,000,000 years. Hand bones, very similar to those of modern man, have been found in Ethiopia and dated at 3,500,000 years.

The most primitive form believed to be a direct ancestor of modern man is *Ramapithecus*, jaw and teeth remains of which have been found in Pakistan and dated at about 10,000,000 years.

Courtesy of the American Museum of Natural History

British Museum (Natural History)

Artist's restorations based on fossil evidence. Top: *Dimetrodon,* a fin-backed reptile that lived during the Permian. Bottom: *Pteranodon,* a flying reptile.

selected readings

PLANT LIFE

Abbey, Edward, and editors of Time-Life Books. *Cactus Country.* New York: Time-Life, 1973; 184 pp., illus. — Emphasis on cacti, but the rest of desert life included as well; fine photography.

Branley, Franklyn M. *Roots Are Food Finders.* New York: Crowell, 1975; 33 pp., illus. — A basic introduction to the functions of plant roots, with instructions for interesting home experiments.

Davis, Bette J. *The World of Mosses.* New York: Lothrop, Lee & Shepard, 1975; 64 pp., illus. — Thorough coverage of mosses, from their natural habitat to terrarium use; for the younger reader.

Earle, Olive L., with Michael Kantor. *Nuts.* New York: Morrow, 1975; 63 pp., illus. — True nuts and others, accurately and clearly described for junior high school readers.

Grimm, William C., and Craig, Jean. *The Wondrous World of Seedless Plants.* New York: Bobbs-Merrill, 1973; 128 pp., illus. — Discussion of algae, fungi, liverworts, lichens, mosses, and ferns; for grades 7–12.

Heady, Eleanor B. *Plants on the Go: A Book About Seed Dispersal.* New York: Parents', 1975; 64 pp., illus. — Lively discussion of how seeds are spread; classification according to form of travel.

Holmes, Sandra. *Trees of the World.* New York: Bantam, 1975; 159 pp., illus. — Handy guide to the characteristics, distribution, and uses of trees; for nature lovers.

Kramer, Jack. *Indoor Trees.* New York: Hawthorn, 1975; 163 pp., illus. — Well-written guide to growing indoor plants, from selection to proper care.

Langer, Richard W. *Grow It Indoors: A Practical, Personal How-to and Why Guide to Growing Successful House plants.* New York: Saturday Review Press, 1975; 365 pp., illus. — Practical guide to growing all kinds of houseplants, with emphasis on proper selection.

Major, Alan. *Collecting and Studying Mushrooms, Toadstools and Fungi.* New York: Arco, 1975; 268 pp., illus. — The botany of fungi, including the edibility of various mushrooms; a list of habitats and their fungi; and glossary.

Milne, Lorus J. and Margery. *Because Of a Flower.* New York: Atheneum, 1975; 152 pp., illus. — Well-written and well-illustrated text exploring the role of flowering plants in nature.

Northern, Henry and Rebecca. *Ingenious Kingdom: The Remarkable World of Plants.* Englewood Cliffs, N.J.: Prentice-Hall, 1970; 274 pp., illus. — Fascinating volume on plants: biology, chemistry, evolution, ecology.

Pringle, Laurence. *Water Plants.* New York: Crowell, 1975; 32 pp., illus. — An introduction to the plants commonly found in freshwater ponds; for the younger reader.

Rahn, Joan Elma. *How Plants Are Pollinated.* New York: Atheneum, 1975; 135 pp., illus. — Clear and concise description of the mechanisms of pollination, for ages 10 to 15.

Silverstein, Alvin and Virginia. *Beans: All About Them.* Englewood Cliffs, N.J.: Prentice-Hall, 1975; 88 pp., illus. — Well-written book on beans as growing plants, with many possible experiments.

— — — — —*Oranges: All About Them.* Englewood Cliffs, N.J.: Prentice-Hall, 1975; 90 pp., illus. — Pleasant recounting of the history and the scientific cultivation of oranges; for junior high school.

Stangl, Martin. *Chilton's Encyclopedia of Gardening.* Radnor, Pa.: Chilton, 1975; 206 pp., illus. — Information-packed guide covering garden layouts, tools, kinds of plants, cultivation methods, pests and diseases.

ANIMAL LIFE

GENERAL

Batten, Mary. *The Tropical Forest: Ants, Ants, Animals, and Plants.* New York: Crowell, 1973; 130 pp., illus. — Very good discussion of the ecology of the tropical forest.

Bridges, William. *The New York Aquarium Book of the Water World.* New York: American Heritage, 1970; 287 pp., illus. — Introduction to the broad range of animal life that lives in, on, and over the water.

Cohen, Daniel. *Animal Territories.* New York: Hastings House, 1975; 95 pp., illus. — Interesting facts about and examples of the concept of territory among animals, from an evolutionary viewpoint.

Cosgrove, Margaret. *Messages and Voices: The Communication of Animals.* New York: Dodd, Mead, 1974; 144 pp., illus. — How animals use odors, sounds, movements, and other means to convey messages to one another; for high school students.

Farb, Peter. *Living Earth,* New York: Harper Torchbooks, 1969; 167 pp., illus. — Interesting and scientifically accurate discussion of life in the soil.

Ford, Barbara. *Can Invertebrates Learn?* New York: Messner, 1972; 96 pp., illus. — Good treatment of the way in which scientists go about their work, and of the nature of learning in animals; for junior high school readers.

Freedman, Russell. *Growing Up Wild: How Young Animals Survive.* New York: Holiday House, 1975; 63 pp., illus. — Good introduction to concepts of animal survival, from predator-prey relationships to regulation of body temperature; for the budding zoologist.

Hays, H. R. *Birds, Beasts and Men: A Humanist History of*

Zoology. New York: Putnam, 1972; 383 pp., illus. — Biographical history of animal study for the general reader.

Jarman, Cathy. *Atlas of Animal Migration.* New York: John Day, 1972; 124 pp., illus. — Striking presentation of information on kinds of animal migrations; for all ages.

Laycock, George. *Wild Travelers: The Story of Animal Migration.* New York: Four Winds, 1974; 110 pp., illus. — Review of a wide range of animals that migrate, from invertebrates to polar bears; good discussion of research on this subject; grade 7 and up.

Livaudais, Madeleine, and Dunne, Robert. *The Skeleton Book: An Inside Look at Animals.* New York: Walker, 1973; 31 pp., illus. — A guide to the different kinds of skeletons — from snakes' to humans' — using a handsome set of anatomical plates.

McGregor, Craig, and others. *The Great Barrier Reef.* New York: Time-Life, 1973; 184 pp., illus. — The history and geography of the Great Barrier Reef, and a description of its animal life; well illustrated.

Murie, Claus J. *Field Guide to Animal Tracks.* Boston: Houghton Mifflin, 2nd ed., 1975; 375 pp., illus — Tracks of all the North American mammals and some birds and reptiles, and all the things they can tell us; for all ages.

Scott, John Paul. *Animal Behavior.* Chicago: University of Chicago Press, 2nd ed., rev., 1972; 281 pp., illus. — Encyclopedic information on the subject, well illustrated and clearly written; for junior high school and up.

Simon, Seymour. *Life in the Dark: How Animals Survive at Night.* New York: Franklin Watts, 1974; 61 pp., illus. — Simple, clear, and accurate text for younger readers.

Street, Philip. *Animal Weapons.* New York: Taplinger, 1971; 176 pp., illus. — Description, by a zoologist, of how animals survive by means of bodily "weapons."

INVERTEBRATES

Anderson, Margaret J. *Exploring the Insect World.* New York: McGraw-Hill, 1974; 160 pp., illus. — How to learn about insects by simple experimentation and observation.

Aylesworth, Thomas G. *The World of Microbes.* New York: Franklin Watts, 1975; illus. — Includes bacteria, blue-green algae, slime molds, and some algae and protozoa.

Borror, Donald J., and Delong, Dwight M. *An Introduction to the Study of Insects.* New York: Holt, Rinehart & Winston, 3rd ed., 1971; 825 pp., illus. — Comprehensive text and reference on entomology.

Cloudsley-Thompson, J. L. *Spiders and Scorpions.* New York: McGraw-Hill, 1975; 48 pp., illus. — A simple and enthusiastic discussion of these arachnids for elementary school readers.

Cousteau, Jacques-Yves, and Diolé, Philippe. *Octopus and Squid: The Soft Intelligence.* Garden City, N. Y.: Doubleday, 1973; 304 pp., illus. — Developed from Cousteau's work aboard his *Calypso;* many handsome illustrations.

Dance, S. Peter, ed. *The Collector's Encyclopedia of Shells.* New York: McGraw-Hill, 1974; 288 pp., illus. — Comprehensive reference work as well as easy-to-use guide for the amateur collector.

Emmel, Thomas C. *Butterflies: Their World, Their Life Cycle, Their Behavior.* New York: Chanticleer/Knopf, 1975; 260 pp., illus. — The biology and life patterns of butterflies around the world, with beautiful color photographs.

Hess, Lilo. *A Snail's Pace.* New York: Scribner's, 1974; 48 pp., illus. — The behavior of snails, the formation of their shells, and how to raise snails, all treated in a well-illustrated text; for grades 5 – 9.

Hutchins, Ross E. *The Bug Clan.* New York: Dodd, Mead, 1973; 127 pp., illus. — Well-written account of the Hemiptera (true bugs) and the Homoptera (cicadas, aphids, and so forth); good photographs by the author.

Jenkins, Marie M. *The Curious Mollusks.* New York: Holiday House, 1972; 224 pp., illus. — The natural history of mollusks, for grades 8 – 12.

— — — — *Sea Star.* New York: Morrow, 1975; 48 pp., illus. — Careful, scientifically accurate description of the life of a common starfish and its relation to its environment; for the younger reader.

Nespojohn, Katherine V. *Worms.* New York: Franklin Watts, 1972; 84 pp., illus. — From earthworms to tapeworms, planarians to leeches; for the advanced elementary pupil.

Newell, Audrey. *Seashells in Action.* New York: Walker, 1973; 39 pp., illus. — The unusual aspects of 25 kinds of marine animals that live in shells; for grades 4 – 6.

Patent, Dorothy Hinshaw. *How Insects Communicate.* New York: Holiday House, 1975; 127 pp., illus. — A good introduction to instinct and behavior patterns, for the general reader.

Pitt, Valerie, and Cook, David. *A Closer Look at Ants.* New York: Franklin Watts, 1975; 30 pp., illus. — The story of ants as farmers, slave makers, hunters, and gardeners, in a well-written text with many color pictures.

Ross, Arnold, and Emerson, William K. *Wonders of Barnacles.* New York: Dodd, Mead, 1974; 78 pp., illus. — Thorough coverage of this abundant and significant but often overlooked crustacean; for grades 7 – 12.

Spoczynska, Joy O. I. *The World of the Wasp.* New York: Crane, Russak, 1975; 188 pp., illus. — Good, clearly written reference work on the natural history and behavior of all types of wasps; for grade 7 and up.

Taylor, Herb. *The Lobster: Its Life Cycle.* New York: Sterling, 1975; 80 pp., illus. — Also covers fishing and artifical breeding; for junior and senior high school students.

Villiard, Paul. *The Hidden World: The Story of Microscopic Life.* New York: Four Winds, 1975; 89 pp., illus. — Introduces the beginner to the microscopic world, with instructions in the techniques of microscopy.

Watson, Allan, and Whalley, Paul E. S. *The Dictionary of Butterflies and Moths in Color.* New York; McGraw-Hill, 1975; 296 pp., illus. — A reference work with excellent color photographs.

Zim, Herbert S., and Krantz, Lucretia. *Crabs.* New York:

Morrow, 1974, 64 pp., illus.—Simply written but with advanced information on the structure and behavior of crabs.

FISH, AMPHIBIANS, AND REPTILES

Blassingame, Wyatt. *Wonders of Frogs and Toads.* New York: Dodd, Mead, 1975; 80 pp., illus.—The many different kinds of frogs and toads around the world; illustrated with black-and-white pictures; for grade 5 and up.

Caras, Roger. *Sockeye: The Life of a Pacific Salmon.* New York: Dial, 1975; 135 pp.—The fascinating life cycle of a typical salmon provides a picture of species survival and adaptation in general.

Chace, G. Earl. *Wonders of Rattlesnakes.* New York: Dodd, Mead, 1975; 79 pp., illus.—Good introduction to rattlesnake ecology, for junior and senior high school students.

Guggisberg, C. A. W. *Crocodiles: Their Natural History, Folklore, and Conservation.* Harrisburg, Pa.: Stackpole, 1972; 203 pp., illus.—Crocodiles' behavior, role in nature, and endangerment by humans.

Halstead, Bruce W., and Landa, Bonnie L. *Tropical Fish: A Guide for Setting Up and Maintaining an Aquarium for Tropical Fish and Other Animals.* New York: Golden/Western, 1975; 160 pp., illus.—Clear and accurate source of information about aquariums, for junior high school and up.

Lauber, Patricia. *Who Needs Alligators?* Champaign, Ill.: Garrard, 1974; 63 pp., illus.—The world of the alligator, with many photos and maps; for all ages.

Lineaweaver, T., III, and Backus, R. *The Natural History of Sharks.* Philadelphia: Lippincott, 1970; 256 pp., illus.—Scientific and folkloristic study of sharks, with identification key; for the general reader.

Neugebauer, Wilbert. *Marine Aquarium Fish Identifier.* New York: Sterling, 1975; 256 pp., illus.—Description of the nature and care of fish that can be kept in saltwater aquariums, with some useful ideas for the beginner.

Scott, Jack Denton. *Loggerhead Turtle: Survivor from the Sea.* New York: Putnam, 1974; 64 pp., illus.—The loggerhead, its life and times; useful for the general reader.

Selsam, Millicent E., and Hunt, Joyce. *A First Look at Snakes, Lizards and Other Reptiles.* New York: Walker, 1975; 32 pp., illus.—An introduction to reptiles and the ways in which they differ from other animals; for the younger reader.

Simon, Hilda. *Frogs and Toads of the World.* Philadelphia: Lippincott, 1975; 128 pp., illus.—The many kinds of frogs and toads in the world and their very different forms of behavior, for grade 5 and up.

—————*Snakes: The Facts and the Folklore.* New York: Viking, 1973; 128 pp., illus.—Good source of general knowledge; snakes presented sympathetically.

Wheeler, Alwyne. *Fishes of the World: An Illustrated Dictionary.* New York: Macmillan, 1975; 366 pp., illus.—Useful, well-illustrated reference work.

White, William. *A Turtle Is Born.* New York: Sterling, 1973; 96 pp., illus.—The anatomy and different life histories of turtles and tortoises, for junior high school students.

Wise, William. *Monsters of the Deep.* New York: Putnam, 1975; 64 pp., illus.—Some amazing sea creatures, including the giant squid, the whale shark, poisonous fishes, the coelacanth, and others; for grades 5–9.

BIRDS

Anderson, John M. *The Changing World of Birds.* New York: Holt, Rinehart & Winston, 1973; 122 pp., illus.—Basically a book on bird ecology, with suggestions on observing birds as well; good reference for grades 6–10.

Callahan, Philip S. *The Magnificent Birds of Prey.* New York: Holiday House, 1974; 190 pp., illus.—Ecology, hunting behavior, form, and flight of birds of prey.

Dorst, Jean. *The Life of Birds,* Vols. 1 and 2. New York: Columbia University Press, 1974; 718 pp., illus.—Discusses the adaptations various species of birds have made to environments around the world; good diagrams, maps, charts, and bibliographies; for the well-informed general reader.

Ford, Barbara. *How Birds Learn to Sing.* New York: Messner, 1975; 96 pp., illus.—Some birds have to learn their songs very young or not at all, as this clearly written book explains.

Griffin, Donald R. *Bird Migration.* New York: Dover, 1974; 180 pp., illus.—Packed with information, for both the general reader and the specialist.

Hester, F. Eugene, and Dermid, Jack. *The World of the Wood Duck.* Philadelphia: Lippincott, 1973; 160 pp., illus.—Study of a beautiful North American wild duck, and a look at the work of a wildlife biologist.

Hunt, Bernice Kohn. *Pigeons.* Englewood Cliffs, N.J.: Prentice-Hall, 1973; 30 pp., illus.—A short but very interesting and accurate account of the domestic pigeon and its relatives.

Moseley, Katherine W. *Only Birds Have Feathers.* Irvington, N.Y.: Harvey House, 1973; 46 pp., illus.—The physiology and behavior of birds, their migrations, their usefulness to humans; good elementary introduction.

Pettingill, Olin Sewall, Jr. *Another Penguin Summer.* New York: Scribner's, 1975; 80 pp., illus.—Falkland Islands penguins and how they live; outstanding photographs.

Saunders, David. *Sea Birds.* New York: Grosset & Dunlap, 1973; 159 pp., illus.—Reference work on the four orders of seabirds, well illustrated and indexed.

Schick, Alice. *The Peregrine Falcons.* New York: Dial, 1975; 83 pp., illus.—The story of peregrines in the wild and bred in captivity, with an important environmental message derived from this endangered species.

Scott, Jack Denton, and Sweet, Ozzie. *That Wonderful Pelican.* New York: Putnam, 1975; 63 pp., illus.—The life history, behavior, and conservation of Florida's brown pelican, with many good photographs.

Turner, Ann Warren. *Vultures,* New York: McKay, 1973; 96 pp., illus.—Vultures presented as impressive and valuable birds, necessary scavengers in the world of nature.